Sustainable Management and Utilization of Sewage Sludge

Vishnu D. Rajput • Ajar Nath Yadav •
Hanuman Singh Jatav • Satish Kumar Singh •
Tatiana Minkina
Editors

Sustainable Management and Utilization of Sewage Sludge

Springer

Editors
Vishnu D. Rajput
Department of Soil Science, Academy of Biology and Biotechnology
Southern Federal University
Rostov-on-Don, Russia

Ajar Nath Yadav
Department of Biotechnology, Dr. Khem Singh Gill Akal College of Agriculture
Eternal University
Sirmour, India

Hanuman Singh Jatav
Department of Soil Science and Agriculture Chemistry
Sri Karan Narendra Agriculture University
Jobner, Jaipur, India

Satish Kumar Singh
Department Soil Science and Agricultural Chemistry
Banaras Hindu University
Varanasi, India

Tatiana Minkina
Department of Soil Science, Academy of Biology and Biotechnology
Southern Federal University
Rostov-on-Don, Russia

ISBN 978-3-030-85225-2 ISBN 978-3-030-85226-9 (eBook)
https://doi.org/10.1007/978-3-030-85226-9

This Springer imprint is published by the registered company Springer Nature Switzerland AG.
The registered company address is: Gewerbestrasse 11, 6330 Cham, Switzerland

Preface

The human population is increasing rapidly, and people are migrating from villages to urban areas for their livelihood. In third-world countries, around 2.1 billion of the populace will be residing in cities by 2030. The fact itself says that the population is burgeoning at an alarming rate, coupled with climate change and food security issues, which compels us to adopt an intensive farming system to supply the food requirements. The soil is a nonrenewable resource, and its formation takes thousands of years. It may be considered that the native inherent capacity of this versatile and worthy natural resource may get exhausted to fulfill the daily food demands of a rapidly increasing population.

In the coming era, new technological interventions or cropping systems have to adapt to meet increasing food demands, out of which intensive cultivation is one of them. The intensive cultivation will accelerate the nutrient depletion rate from the soil, leading to depletion of native fertility status, which will emerge in deficiency or lack of soil organic carbon. As in many countries, the shortage of macro- or micronutrients occurs in soil up to a threat level, more attention should be paid to managing soil fertility and recycling the soil wastes generated in urban areas to overcome this problem.

In today's era, sewage sludge may be a source of fertilizers that could improve soil fertility and productivity due to an array of nutrients and organic matter. Still, the presence of heavy metals in sewage sludge is a matter of concern. In agriculture, its proper utilization makes it suitable to fulfill the nutritional requirement for plants and the best option to manage the waste generated through various anthropogenic activities.

This book encloses the possible current knowledge and global scenario of sewage sludge for possible sustainable management. It compiles the different aspects of analytical methods, bioleaching approach, beneficial microbes for sustainable treatment of sewage sludge, biological and thermo-chemical treatment technologies, nutrient recovery technologies, biostabilization, health risk assessment, detoxification, socioeconomic aspects, sustainable use in restoring soil fertility, municipal waste management, and future possibilities for safe utilization. This new book could be handy with a bundle of scientific knowledge for faculty members, researchers, students, and policymakers associated with waste management.

Rostov-on-Don, Russia — Vishnu D. Rajput
Sirmour, India — Ajar Nath Yadav
Jobner, Jaipur, India — Hanuman Singh Jatav
Varanasi, India — Satish Kumar Singh
Rostov-on-Don, Russia — Tatiana Minkina

Contents

About the Editors

Vishnu D. Rajput is working as a Leading Researcher (Assoc. Prof.) at Sothern Federal University, Russia. His ongoing research is based on soil contaminations, i.e., potentially toxic elements, and metallic nanoparticles, and investigating the bioaccumulation, bio/geo-transformations, uptake, translocation, and toxic effects of metallic nanoparticles on plant physiology, morphology, anatomy, the ultrastructure of cellular and subcellular organelles, cytomorphometric modifications, and DNA damage. He has comprehensively detailed the state of research in environmental science in regard to "how nanoparticles/heavy metals interact with plants, soil, microbial community, and the larger environment." He has published (total of 229 scientific publications) 136 peer-reviewed highly rated full-length articles, 09 books, 40 chapters (Scopus indexed), and 29 conference articles. He is an internationally recognized reviewer (peer reviewed 137 manuscripts for internationally repute journals) and received an outstanding reviewing certificate by Elsevier and Springer. He is an editorial board member of various high impacted journals. He received "certificate for appreciation 2019" and "certificate of Honor 2020" by Southern Federal University, Russia, for outstanding contribution in academic, creative research, and publication activities. He has also received the prestigious "Highly Qualified Specialist" by SFedU and the Russian Ministry of Internal Affairs.

Ajar Nath Yadav is an assistant professor (senior scale) in the Department of Biotechnology, Dr. Khem Singh Gill Akal College of Agriculture, Eternal University, Baru Sahib, Himachal Pradesh, India. He has 6 years of teaching and 12 years of research experience in the field of microbial biotechnology, microbial diversity, and plant–microbe interactions. He obtained his doctorate degree in science (microbial biotechnology), jointly from the Indian Agricultural Research Institute, New Delhi, and Birla Institute of Technology, Mesra, Ranchi, India. He has made pioneering contributions in the area of microbial biotechnology, plant–microbe interaction, and environmental microbiology. Dr. Yadav has 250 publications with h-index 61, i10-index 173, and 10164 citations (Google Scholar). Dr. Yadav is editor of 21 books published by Springer, 4 books by Elsevier, and 5 books by CRC press, Taylor & Francis and 1 Wiley. Dr. Yadav has got 12 Best Paper Presentation Awards and 1 Young Scientist Award (NASI-Swarna Jayanti Purskar). Dr. Yadav received "Outstanding Teacher Award" in the 6th Annual Convocation 2018 by Eternal University, Baru Sahib, Himachal Pradesh. Dr. Yadav is currently handling two projects funded by DEST and HIMCOSTE. Presently, he is guiding 5 scholars for Ph.D. degree. Dr. Yadav is editor-in-chief for Journal of Applied Biology and Biotechnology. Dr. Yadav is Editor/Associate Editor for Annals of Microbiology [Springer-Nature], BMC Microbiology [Springer-Nature], Current Microbiology [Springer-Nature], Frontiers in Microbiology, International Journal of Microbiology, Plant Science Today and Microbial Biosystems. He has been serving as reviewer for 55 different national and international peer-reviewed journals. He has lifetime membership of the Association of Microbiologists of India and Indian Science Congress Council, India. Please visit https://sites.google.com/view/ajarnathyadav/ for more details.

Hanuman Singh Jatav is an Assistant Professor in the Department of Soil Science and Agricultural Chemistry, Sri Karan Narendra Agriculture University, Jobner Jaipur (Rajasthan), India. He has more than 4 years of teaching and 7 years of research experience in the field of sewage sludge, biochar, micronutrients, heavy metals, soil fertility management including M.Sc. (Ag.) and Ph.D. He has completed B.Sc. (Hons.) Agriculture from Maharana Pratap University of Agriculture and Technology, Udaipur-Rajasthan, India. Dr. Jatav has completed his M.Sc. (Ag.) and Ph.D. in Soil Science and Agricultural Chemistry from the Institute of Agricultural Sciences, Banaras Hindu University, Varanasi-U.P. India. At present, Dr. Jatav is an emerging person in the field of research transforming himself to an excellent scientific level by enhancing research capacity by promoting environmental sustainability. He has published more than 90 paper(s) including research papers, review papers, books, book chapters, and technical papers from various national and international reputed journals and publishers. He is serving as an Academic Editor and member of subject expertise in Springer, Taylor and Francis, Elsevier, Frontiers, CARJ, JEB, ICAR Journals, and Science Domain for the last five years. He is serving as an active member of more than 12 various national and international scientific societies. Dr. Jatav has been awarded a Certificate of Gratitude as an Ambassador by IUCN: International Union for Conservation of Nature-Gland, Switzerland (https://www.iucn.org/) in the World Conservation Congress during Global Youth Summit. He has been also awarded the best paper and poster presentation for his work. He has also attended more than 30 conferences; besides this, he has also completed more than 25 training pertaining to his research. Considering Dr. Jatav's work at an excellent level, he has been awarded as Outstanding Assistant Professor of the Sri Karan Narendra Agriculture University, Jobner Jaipur (Rajasthan). He may be contacted at E-mail-hanumaniasbhu@gmail.com; hsjatav.soils@sknau.ac.in.

Satish Kumar Singh is a former Head and currently working as Professor in Department of Soil Science and Agricultural Chemistry, Institute of Agricultural Sciences, Banaras Hindu University, Varanasi, India. He has more than 30 Years of experience in the field of agriculture in teaching /research /extension. His area of specialization is micronutrients nutrition and heavy metals pollution. He has completed his Ph.D. in the year 1992 from Banaras Hindu University. He is focusing his research work on sewage sludge management/ utilization in agriculture, particularly in rice-wheat system. He has guided 25 M.Sc. (Ag) and 7 Ph.D. students. He is Fellow of the Indian Society of Soil Science, New Delhi, and Fellow of the Indian Society of Agricultural Chemists, Allahabad. He has been involved in the process of designing the curriculum and is member of board of study of several universities. He is the editor of the *Journal of Indian Society of Soil Science* and *Agropedology*. He has been the Organizing Secretary of the conference and Course Director and Course Coordinator of the Refresher course of UGC and ICAR. He has 9 books, 10 book chapters, and 115 peer-reviewed publications to his credit in national and international journals of repute.

Tatiana Minkina is the Head of Soil science and Land Evaluation Department of Southern Federal University. She is the Head of International Master's Degree Educational Program "Management and estimation of land resources" (2015–2022, accreditation by ACQUIN). Her area of scientific interest is soil science, biogeochemistry of trace elements, environmental soil chemistry, soil monitoring, assessment, modeling, and remediation using physicochemical treatment methods. She was awarded in 2021 with Diploma of the Ministry of Education and Science of the Russian Federation for many years of long-term work for development and improvement of the educational process and significant contribution to the training of highly qualified specialists. Currently, she is handling projects funded by the Russian Scientific Foundation, Ministry of Education and Science of the Russian Federation, and Russian Foundation of Basic Research. She is Member of Expert Group of Russian Academy of Science, the International

Committee on Contamination Land, Eurasian Soil Science Societies, the International Committee on Protection of the Environment, and the International Scientific Committee of the International Conferences on Biogeochemistry of Trace Elements. Total scientific publications: 957 (total 957, 389 in English); Scopus—268, WoS—196, and citing index: Scopus—1976, h-index—21; WoS—1452, h-index—20. She is the invited editor of an Open Access Journal by MDPI *Water* (ISSN 2073-4441; impact factor: 3.229; SJR 0.718) and editorial board member of Geochemistry: Environment, Exploration, Analysis (ISSN: 1467-7873; impact factor: 1.545; SJR: 0.346) and *Eurasian Journal of Soil Science* (e-ISSN: 2147–4249; SJR 0.24).

Sewage Sludge Management for Environmental Sustainability: An Introduction

Jussara Borges Regitano, Mayra Maniero Rodrigues, Guilherme Lucio Martins, Júlio Flávio Osti, Douglas Gomes Viana, and Adijailton José de Souza

1 Introduction

Urban areas represent about 3% of the entire terrestrial surface but are home to more than 55% of the world population; and the tendency of greater people concentration in large urban centers will still increase in the coming decades (Edmondson et al. 2012; United Nations 2019). It should result in two-thirds of the world's population, i.e., 7 billion people, living in metropolitan areas by 2050 (Ritchie and Roser 2018). This social phenomenon intensifies the need to maintain efficient and sustainable water and sewage treatment systems to guarantee adequate water supply in large urban centers and to avoid contamination of the natural ecosystems (Buonocore et al. 2018).

However, the current diagnosis of the world population concerning access to sanitation services showed that only 45% have access to bathrooms with the collection, treatment, and adequate sewage transport (Safely managed); 29% make use of improved non-collective facilities that include a piped sewer system and the sewage is maintained in pits, septic tanks, and/or composting toilets (Basic); 8% make use of collective or shared sanitary facilities (Limited); 9% make use of latrines or buckets (Unimproved); and 9% still eliminate feces in the open, allowing contamination of water bodies and increase in the incidence of diseases (Open defecation) (Fig. 1) (WHO 2020).

Jussara Borges Regitano, Mayra Maniero Rodrigues and Guilherme Lucio Martins contributed equally with all other contributors.

J. B. Regitano (✉) · M. M. Rodrigues · G. L. Martins · J. F. Osti · D. G. Viana · A. J. de Souza
Department of Soil Science, "Luiz de Queiroz" College of Agriculture, University of São Paulo, Piracicaba, Brazil
e-mail: regitano@usp.br

V. D. Rajput et al. (eds.), *Sustainable Management and Utilization of Sewage Sludge*, https://doi.org/10.1007/978-3-030-85226-9_1

Fig. 1 Diagnosis of world sanitation access in the year of 2017. Adapted from WHO (2020)

Sewage sludge (SS) is the solid, semi-solid, or liquid residue generated during the treatment of domestic sewage in wastewater treatment plants (WWTPs), whose correct destination is one of the biggest challenges for sanitation services globally (Zhen et al. 2017; Raheem et al. 2018) since the worldwide 7.6 billion people excreta must be disposed of in some way. The generation of SS increases as public policies and investments are directed to enhance access to sanitation services, which also dictate sewage collection amounts as well as adequation of the adopted treatments and the final disposal strategy. It is evident that poor and populous regions as well as those having bad gross income distribution, such as Asia, Africa, and Latin America, among others, collect less sewage and thus produce proportionally much less SS than Europe, USA, and Canada (Table 1). Therefore, when these countries have sanitation systems, it does not necessarily reflect on adequate or even sufficient SS management. The scarcity or even the lack of updated data on SS generation in several countries makes it difficult to understand and identify the challenges associated with the management of this material.

Sewage sludge composition will depend mainly on the seasonal effects, adopted treatments' system, as well as its source (domestic, industrial, or a mix of both) (Tao et al. 2012; Nascimento et al. 2020). Generically, SS is composed by a complex mixture of organic components, such as proteins, lipids, carbohydrates, phenolic compounds, lignin, and cellulose (Zhang et al. 2018). After drying, on average, the SS presents 50–70% of organic matter (OM), 3–4% of nitrogen (N), and 0.5–2.5% of

Table 1 Generation of municipal sewage sludge (SS) in certain countries

Country/Region	Population (million)	Amounts of SS (thousands of dry metric tons)
European Union	446	8910
China	1440	6000[*]
United States	330	6510
Iran	83	650
Canada	36	550
Brazil	211	372

Adapted from: Mateo-Sagasta et al. (2015), [*]Source: Zhang et al. (2016)

phosphorus (P), as well as other micronutrients to plants, such as Zn and Cu (Rorat et al. 2019). SS has high microbial diversity, but *Proteobacteria, Bacteroidetes*, and *Firmicutes* are often the dominant phyla whereas *Clostridium, Treponema, Propionibacterium, Syntrophus*, and *Desulfobulbus* are often the dominant genera (Nascimento et al. 2018).

SS is essentially an organic residue, containing high microbial biomass as well as high organic matter, N, P, and Zn contents, among others. If properly treated and applied to land, it can improve soil's quality thus improving the productivity of agricultural crops and revegetation of disturbed ecosystems, such as mining tailing areas. Worldwide, SS is treated and applied to soils as either disposal or a recycling method. Although mineral fertilizers, based on fossil fuels, can properly supply nutrients to plants, tropical countries are highly dependent and not auto sufficient on these highly cost commodities, turning the use of SS for soil fertilization even more interesting since allows its reuse as well as the recycling of nutrients in the environment.

However, SS also has high amounts of potentially toxic elements in its composition, such as Cd, Cr, Cu, Co, Fe, Hg, Ni, Pb, organic pollutants, and emerging contaminants (Tyagi and Lo 2013; Cies□lik et al. 2015), besides a vast array of pathogens inherent to its origin, mostly human excreta. Infected people may excrete enteric pathogens, such as Coliforms, Enteric viruses, *Giardia*, and *Cryptosporidium* for months, but their peaks are under seasonal influence. Of particular interest, are the emerging pathogens such as the virus involved in the severe acute respiratory syndrome (SARS virus). The COVID-19, caused by SARS-CoV-2, has infected more than 130 million people since 2019 and caused >three million deaths (WHO 2021). The presence of viral particles in human excreta and wastewaters alerts about the possibility of new outbreaks or disease expansion due to virus mutation, thus urging adoption of appropriate strategies to properly collect, treat, and dispose wastewaters and SSs, mainly in socioeconomically vulnerable countries (Dhama et al. 2021; Donde et al. 2021). Their incorrect management can cause human and animal health problems (Sharma et al. 2017).

A broad overview of SS treatments, reuse, and disposal strategies is fundamental to guaranteed maintenance of ecosystem sustainability since it allows the adoption of public policies aiming at social well-being and environment preservation. Currently, SSs are mostly landfilled or amended to soils, but they are also incinerated or used in

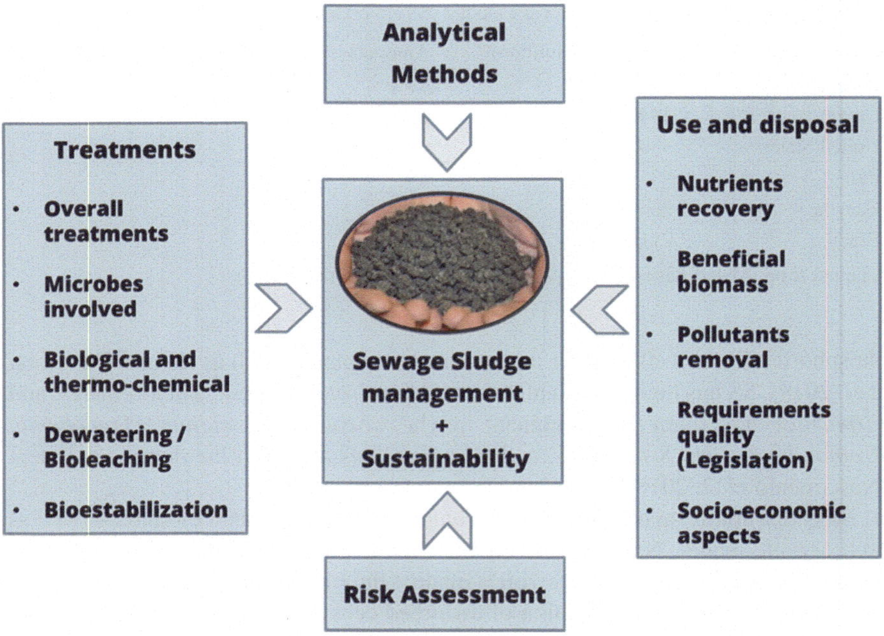

Fig. 2 Overview of major topics addressed in the book

construction (Kacprzak et al. 2017). All these disposal strategies have advantages and disadvantages. This book will touch on all these aspects but will emphasize newer treatment and disposal approaches, as summarized in Fig. 2. It will also address legislation, analytical methods, and risk assessment topics to properly reach its intent that is to review current science on SS management to assure environmental sustainability.

2 Brief history of Sanitation Services and Related Health Issues

The need to implement actions to treat wastes generated by society began nearly 10,000 years ago when humans left nomad life and began to settle in agricultural regions. Historically, the first primitive sanitation systems were settled in Ancient Mesopotamia (3500–2500 BC), using drainage channels to transfer domestic wastes from latrines to pits (Lofrano and Brown 2010). People of that region, which includes part of Iraq, Syria, Iran, and Turkey, were the first to face health problems related to the proliferation of parasitic diseases, such as *Schistosomiasis*, transmitted by water, soil, and food contaminated by not treating the sewage (McMahon 2015). Seven *Cholera* pandemics are also part of human history, responsible for thousands of deaths over 200 years (Deen et al. 2020), also directly associated with consumption of contaminated water and food, poor hygiene, and poor sanitation services (Taylor et al. 2015). Currently, it is estimated that 21,000–143,000 deaths are caused by cholera in the world, mainly in developing countries (Ali et al. 2015).

Several infectious diseases associated with poor sanitation persist in the twenty-first century, such as *Schistosomiasis* and *Ascaris lumbricoides*. Diarrhea is responsible for most deaths, ~1.4 million deaths annually of mainly children under 5 years old (Hughes and Koplan 2005; Freeman et al. 2017; Prüss-Ustün et al. 2019). The most alarming scenarios happen in countries like Chad and Madagascar, in which diarrhea kills >300 per 100,000 children, quite different from numbers seen in European countries (<1 per 100,000) (Troeger et al. 2018). Investments targeting access to adequate water supply and sanitation services can reduce diarrhea cases by more than a third (Bartram and Cairncross 2010). Other pathogens are found in the SS, such as *Salmonella* spp., *Shigella* spp., *Escherichia coli*, *Clostridium perfrigens*, *Campylobacter* spp., *Yersinia enterocolitica* (Arthurson 2008; Rorat et al. 2019). The major contamination routes are related to accidental ingestion of soil applied SS-residues and consumption of poorly sanitized fresh foods (Van Frankenhuyzen et al. 2013). Land workers may also be contaminated during SS-application, but its risk lessen over time (Brooks et al. 2012). Complaints regarding health problems related to SS exposure are common (Keil et al. 2011; Viau et al. 2011).

The number of diseases triggered by multi-resistant bacteria has grown considerably in recent years (Nicolas et al. 2019), resulting in ~700,000 deaths annually (de Oliveira et al. 2020). The Discovery of new antibiotics is quite rare in the last decades, thus increasing infection risks associated with bacteria resistant to multiple antibiotics (Tacconelli et al. 2018). Antibiotic resistance is promoted by antimicrobial resistance genes (ARGs) that encode several defense mechanisms against toxic effects of the antimicrobials (Sui et al. 2016). SS is an important source of ARGs' dissemination through its application to soils (Bondarczuk et al. 2016). About 40–90% of the administered antibiotics are not metabolized in the body, thus being excreted in the feces, and ending up in the sewer network; from which ~70% is sorbed to the SS (Sun et al. 2019).

SS has high N and P contents, mostly in the organic forms. However, organic-N is readily mineralized to ammonium (NH_4^+) and then to nitrate (NO_3^-). Nitrate is very soluble in water and its excess amounts are leached to groundwaters and may even reach aquifers and cause methemoglobinemia, also known as blue baby, in young infants (Knobeloch et al. 2000; Fan et al. 2014). In addition, SS is applied based on the needs of N by the crops, which is much higher than that of P. Therefore, excess of P can accumulate in the soils after years of SS application leading to eutrophication of surface water resources. P is transported mainly via runoff, attached to fine soil particles (Hua and Zhu 2020).

Depending on its source and adopted treatments, SS may have high hazardous trace element contents, mostly known as heavy metals (Chanaka Udayanga et al. 2018). The most concerning ones are Zn, Cu, Cd, Ni, Pb, Hg, Mo, and As. Their contents in the SS will depend on the amounts of industrial wastes imputed into the municipal sewage system. Therefore, SS application without criteria can accumulate these elements in the soils since they do not degrade, and then be up taken by crops or transported to water resources offering risks to human and animal health (Duan et al. 2017). However, only 4 out of 19 sludges from WWTPs from São Paulo, the most populated state from Brazil, presented concerning contents of either Zn or Ni, but these elements are nutrients to plants (Nascimento et al. 2020). When performed at appropriate loading rates, farmland application of the SS often results in far less pollution than its landfilling. Currently, there is no large-scale technology used for the removal of such elements from the SS (Geng et al. 2020), but there are technologies to reduce metal inputs into the sewage (Pepper et al. 2006). In addition, metal contamination can be diluted when SS is used as the organic matrix in the manufacture of organomineral fertilizers (Kominko et al. 2017). SS having high Zn, Cu, and Ni should be regarded for this purpose since these elements are micronutrients to plants (Nascimento et al. 2020).

Organic toxic substances, such as polycyclic aromatic hydrocarbons (PAHs), pesticides, dioxins, flame retardants, plasticizers, and surfactants are often found in the SS (Poulsen and Bester 2010; Ozcan et al. 2013) and concerns are gaining grounds in recent years (Regkouzas and Diamadopoulos 2019) since many of them can cause mutagenic effects, endocrine and reproductive system dysfunctions, immunological impairment, and developmental defects (Venegas et al. 2021). Not all of them are easily degraded, many are persistent and mobile in the environment (Čadková et al. 2020). Biological stabilization treatments fail to effectively degrade all types of xenobiotic molecules present in the SS, many times showing absent or only partial degradation (Poulsen and Bester 2010; Gonzalez-Gil et al. 2016).

The past allows us to understand the importance of maintaining and expanding basic sanitation services around the world, even more, when facing a highly globalized and interconnected society such as ours. Despite technological and scientific advances, ~2.4 billion people still do not have access to basic sanitation services, such as sewage collection and treatment, mainly in economically vulnerable countries such as those in Sub-Saharan Africa, Asia, and Latin America (Freeman et al. 2017; Morgan et al. 2017).

3 Sewage Sludge Treatments and Associations with Microbial Population

SS presents a great microbial diversity (Aida et al. 2015). The high organic load stimulates the development of a saprobic microbiota that is important in its treatment (Rorat et al. 2019). However, the presence of human and animal pathogens, as well as heavy metals and organic pollutants, raises concerns about its health safety, thus requesting further treatments especially when the land application is intended (Rorat et al. 2019). SS-land application increased as an important disposal strategy since ocean damping and landfilling started to be restricted, mainly in developed countries. The SS is called biosolid when it is treated to meet land-application standards; and

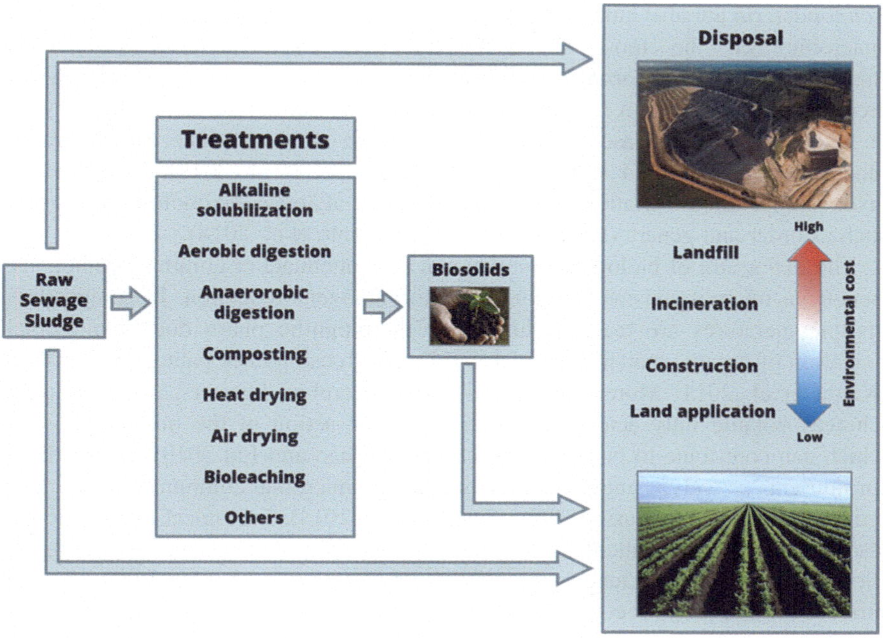

Fig. 3 Sewage sludge treatments and disposal techniques as well as its environmental costs

there are two classes of biosolids: (i) Class A: treated to reduce pathogens below detection levels and can be used without site restrictions, and (ii) Class B: treated to reduce pathogens, but still have them at detectable levels and, therefore, its use is site restricted (Part 503 rule, EPA 1993). Therefore, the row SS must endure further treatments before its reuse or disposal (Fig. 3).

Alkaline stabilization (liming) refers to the addition of lime to rise sludge-pH to >12 for at least 2 h, thus eliminating odors and inhibiting pathogenic bacteria and virus, but not parasites. Heat treatment involves temperatures up to 260 °C for 30 min, under pressure, thus dewatering the sludge and killing pathogens and parasites (Pepper et al. 2006). Biotic treatments, such as aerobic and anaerobic digestions and composting, involve the participation of microorganisms to reduce pathogens and toxic substances in the SS (Kelessidis and Stasinakis 2012; Nascimento et al. 2018). Treatment systems naturally select the best-adapted microorganisms to the conditions (Lloret et al. 2016; Goberna et al. 2018). In other words, factors such as pH, presence of organic pollutants and heavy metals, and type of biological treatment dictate sludge microbial community, either in terms of diversity or structure, or both (Nascimento et al. 2018). Moreover, the presence of toxic compounds in industrial sludges tends to decrease microbial community diversity compared with those of domestic origin (Ibarbalz et al. 2013). For aerobic SS, the most abundant bacterial phyla are *Proteobacteria* and *Bacteroidetes*, but *Acidobacteria, Firmicutes, Chloroflexi, Planctomycetes, Verrucomicrobia*, and

Actinobacteria are also quite frequent (Xia et al. 2018; Nascimento et al. 2018). For anaerobic SS, the most abundant phyla are *Proteobacteria, Firmicutes, Bacteroidetes*, and *Chloroflexi*, but methanogenesis is promoted exclusively by archaea from the phylum *Euryarchaeota* (Luo et al. 2009; Guo et al. 2015; Fykse et al. 2016). In this case, biogas production is attractive since contributes to biodigesters' operational and maintenance costs (Li et al. 2017). Nevertheless, taxon composition variations are usually greater at lower taxonomic categories, such as order and genera (Hu et al. 2012; Nascimento et al. 2018).

The major aim of biological stabilization is to attenuate or eliminate pathogens. Beneficial microorganisms can contribute to pathogens reduction. In composting, high temperatures are reached during the thermophilic phase due to microbial oxidation of organic matter (Liu et al. 2018), which can reduce pathogenic bacteria (Khalil et al. 2011; Moretti et al. 2015). In anaerobic digesters, pH is reduced whereas volatile fatty acids are produced by the action of the microorganisms, which can contribute to pathogens inactivation (Zhao and Liu 2019). Competition for nutrients as well as antagonism among sludge microbial communities may also reduce pathogens (Arthurson 2008; Scaglia et al. 2014). Biological treatments are also capable to either attenuate or eliminate organic pollutants present in the SS (Semblante et al. 2015; Dubey et al. 2021), such as pharmaceutical and personal care products (Pérez-Lemus et al. 2019), since these molecules can be degraded by specific microbial groups (Lü et al. 2021) that are capable to use them as C source and energy for growth and reproduction (Margot et al. 2015).

Microorganisms also open opportunities for eliminating SS-hazardous trace elements, through bioleaching. Specific microorganisms, such as *A. ferrooxidans* and *A. thiooxidans*, are used to oxidize reduced sulfur (S) compounds to sulfuric acid. SS acidification enhances solubilization of several metals, such as Zn, Cu, Cr, Cd, Pb, Mn, and Ni (Camargo et al. 2016), which can then be extracted by leaching (Gu et al. 2017; Gu and Bai 2018). This technology is not routinely used but ratifies the role of beneficial microorganisms on sewage sludge management (Zhou et al. 2013).

3.1 Impact of the SS on Soil Microbiota

Soil application of the SS alters its microbial community in the short-term by mixing microbes from the sludge and the soil (Wolters et al. 2018). However, the structure of the soil microbial communities tends to return to their initial state over the months after its application, even if changes are still noticeable (Cytryn et al. 2011; Wolters et al. 2018). In the long run, successive application of sludge can alter the soil microbiota through changes in soil attributes, such as pH reduction that favors Gram-positive bacteria (Börjesson et al. 2014). Although the soil type dictates its microbial structure, the accumulation of heavy metals, such as Zn and Cu, can also alter soil microbial structure (Macdonald et al. 2011) and reduce its microbial biomass (Charlton et al. 2016) in long term. Moreover, distinct SS stabilization processes, such as composting and anaerobic digestion, can produce substrates with different impacts on soil microbiota (Mattana et al. 2014; Lloret et al. 2016).

SS amendment to soils tends to stimulate soil microbial activity and enhances microbial biomass, which may contribute to pathogens suppression by competition between beneficial and pathogenic microorganisms (Heck et al. 2019). Additionally, volatile fatty acids are released during microbial decomposition of the SS, which can be toxic to pathogens (Pinto et al. 2013). However, depending on the pathogen, the SS amendment may either suppress or even stimulate the infection (Bettiol and Ghini 2011; Ghini et al. 2016). Thus, more research is needed to show the microbial factors associated with the SS that promote pathogen suppression (De Corato 2020).

4 Methodological Aspects

The use of SS in agriculture can cause a series of changes in soil behavior, whether due to the presence of organic matter, hazardous trace elements, organic pollutants, or exogenous microorganisms. A wide variety of methodologies can be used to evaluate the chemical and biological attributes of the SS and to help predict its impacts on the environment. However, different methodologies may provide different results. Therefore, each country or region often has its guidelines and methodologies, and they also set their own threshold values for all concerning contaminants that can adversely impact public health. It is important because establishes quality indicators for the generated SS since it may impact soil microbiota and its functions, nutrients recycling, and environmental contamination. SS characterization will indicate further treatment needs, better disposal strategy as well as management practices, and most relevantly will support local legislation (Fig. 4).

For example, the United States Environmental Protection Agency (US-EPA) guides specific methodologies that must be used for determining concentrations of macronutrients (P, K, Ca, Mg, S) and micronutrients (Fe, Zn, Cu, Mn, Ni, B, Mo), as well as potentially toxic trace elements (Al, Ba, Cr, Pb, As, Se and Cd) and strongly recommends studies for organic contaminants (pesticides, pharmaceuticals, surfactants, hormones, polyaromatic hydrocarbons, polychlorinated biphenyl, solvents, plasticizers, and volatile compounds) (Hu et al. 2020; Moško et al. 2021). The inorganic elements are extracted by acid digestion and the macro and micronutrients are usually analyzed either by spectrophotometry (P) or atomic absorption spectrophotometry (K, Ca, Mg, Fe, Zn, Cu), or others, while hazardous trace elements are analyzed either by assisted microwave extraction or atomic emission spectrometry with inductively coupled plasma (ICP-AES). The choice is made according to the analyzed element and the available technology (US-EPA 2000, 2007; Schütte et al. 2015; Guedes et al. 2015). The use of optical techniques, such as X-ray diffraction (XRD) and X-ray fluorescence (XRF), is less common but allows to evaluate the SS attributes as well as element speciation that will dictate its toxicity and risks to the environment (Uysal and Kuru 2013; Wilfert et al. 2018).

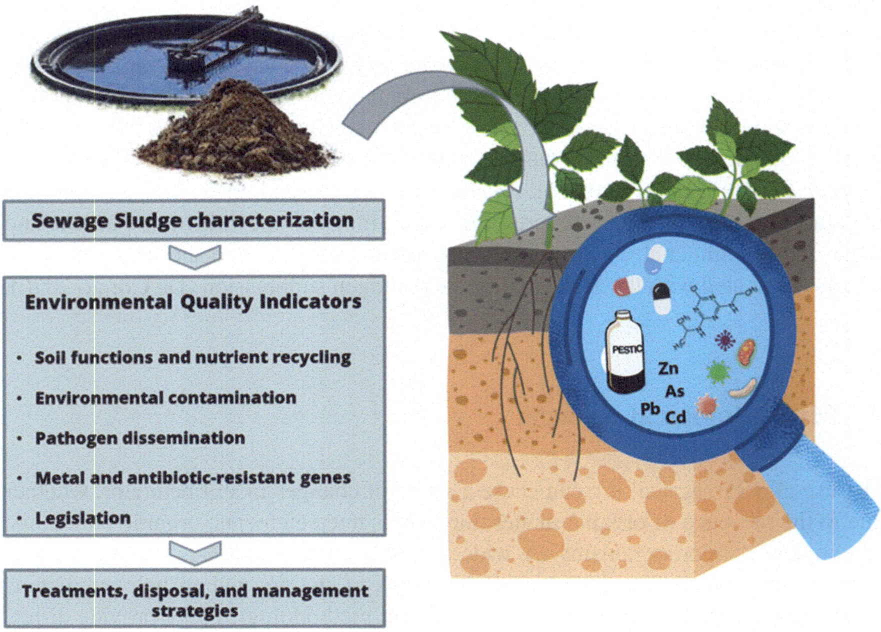

Fig. 4 Sewage sludge characterization methods and their implications accessing environmental quality

The methods to evaluate SS biological attributes have been gaining more relevance in the last years due to their spread and subsequent cost reductions. The microbiota composition seems to work as an indicator of SS quality providing answers to environmental changes that occur after its amendment (Zhang et al. 2010; Wolinska and Stepniewsk 2012). For example, basal soil respiration and carbon from microbial biomass provide information on the population and activity of soil microbial population (Vance et al. 1987; Menyailo et al. 2003). These methods are based on C and other elements biogeochemical cycles, regulated in large part by the activity of microbial enzymes (Siebielec et al. 2018; Melo et al. 2018). Therefore, enzymatic activities play a key role in assessing soil quality by showing its activity and functionality, as well as the cycling of nutrients (Pavan Fernandes et al. 2005; Silva et al. 2019). In soil-plant system, the enzymatic analyses show that SS amendments may enhance energy supply to plants, organic matter degradation, and nutrients recycling (mainly N, P, and S), thus resulting in more resilient environments having greater endurance against biotic and abiotic stresses (Wyrwicka and Urbaniak 2016; Siebielec et al. 2018; Ma et al. 2019) (Table 2).

The functionality of soil can be assessed by the microorganism's activity resulted from its gene expression, i.e., the transcription of its DNA into RNA. The use of molecular techniques allows for an evaluation of predominant microbial groups and

Table 2 Sewage Sludge (SS) amendments and their effects on soil-plant systems evaluated through the activity of certain specific enzymes

Enzyme	Effects on soil-plant system	References
Amylase	Higher polysaccharides and xenobiotics degradation, Increased soil activity.	Pavan Fernandes et al. (2005), Melo et al. (2018)
Cellulase	Higher cellulose, hemicellulose, and lignin degradation at low SS application.	Melo et al. (2018), Ma et al. (2019)
Invertase	Increased complex sugars hydrolysis to assimilable forms by plants and soil microorganisms.	Hu et al. (2011), Melo et al. (2018)
Catalase	Greater protection against oxidative stress, Higher soil microbial activity.	Xue and Huang (2013), Wyrwicka and Urbaniak (2016)
Dehydrogenase	Higher microbial activity due to oxidizing organic compounds	Siebielec et al. (2018), Hamdi et al. (2019)
Protease	Enhanced biological mineralization of N-organic compounds	Xue and Huang (2013), Hamdi et al. (2019)
Urease	Higher N-mineralization	Pavan Fernandes et al. (2005), Arif et al. (2018), Ma et al. (2019)
Phosphatase	Higher P-mineralization	Arif et al. (2018), Siebielec et al. (2018)
Arylsulfatase	Transformation of S-organic to plant and microbes assimilable forms	García-Sánchez et al. (2016), Ma et al. (2019)

their soil functions at highly accurate levels, with large information on DNAs, RNAs, and proteins (Delgado-Baquerizo et al. 2016; Newcomb et al. 2017; Saleem et al. 2019), as summarized in Table 3. In other words, it is important to assess soil biodiversity to understand the predominant organisms, what they do and how they behave. The genomic methods provide a broad but specific view on how microbes respond to changes in soil management while the proteomic and transcriptomic methods allow identifying functions performed by bacteria and fungi by evaluating their proteins and RNA molecules (Biswas and Sarkar 2018). Proteins mediate the most important functions that occur in the community, so proteomics is the best technique for assessing the impacts of the SS application (Xu and Geelen 2018; Rutgersson et al. 2020). Finally, antibiotic and metal resistance genes (ARGs and MRGs, respectively) may disseminate in soils after SS amendments, but it may be better evaluated by real-time PCR (qPCR) than genomics (Rutgersson et al. 2020). For qPCR, one or more genes can be used for each evaluated function (Stalder et al. 2014; Pal et al. 2015).

However, classic approaches based on culture medium are still important to complement DNA analyzes in SS-amended soils (Van Frankenhuyzen et al. 2013; Xie et al. 2016). For example, they can be used to assess the persistence of ARGs and MRGs groups (Li and Zhang 2010; Wang et al. 2015; Zhang et al. 2018).

Table 3 Main molecular techniques, their specific targets, and evaluated effects

Methods	Targets	Evaluated effects	References
Fingerprint and amplicon sequencing	16S rDNA (Archaea Bacteria); ITS (fungi); 18S rDNA (protists)	Changes in microbial structure and diversity.	Biswas and Sarkar (2018), Paul et al. (2018), Guo et al. (2020)
Fingerprint qPCR	16 s rDNA; ITS; functional genes (*amoA*; *mcrA*; *tetG* etc.)	Abundance of bacteria and fungi. Specific functions, Presence of pathogens, etc.	Van Frankenhuyzen et al. (2013), Xie et al. (2016)
Metagenomics	Study of the collective genome of the total microbiota of a given habitat.	Structural: study structure of uncultivated microbial population, Functional: aims to identify genes that encode a function of interest.	Franzosa et al. (2015), Alves et al. (2018)
Metabolomics	Microbial metabolism (functions performed by microbial processes)	Presence of microbial metabolites and activity level (lignin degradation, methanogenesis, sulfate reduction, etc.).	Beale et al. (2016), Rutgersson et al. (2020)
Proteomics	Proteins	Proteins involved in specific functions, such as solubilization of organic P.	Bastida et al. (2019)

5 Legislation

Effective means are needed to regulate the use and reuse of SSs. Legislation that establishes acceptable limits for toxic pollutants is fundamental to warranty SS safe use in agriculture (Tables 4 and 5) or even prevent SS recycle in soils, directing it to other purposes (Fig. 4). The presence of hazardous trace elements and certain toxic organic compounds depends on industrial discharges handled by WWTPs and can be at least to a certain extent controlled, but pharmaceuticals and personal care products are directly eliminated in the domestic sewer requesting proper SS management.

Although certain hazardous trace elements are considered micronutrients to plants, they are persistent in the environment and at high contents can cause damage to ecosystems, thus turning their monitoring fundamental (Collivignarelli et al. 2019b). Even after its recent update, Brazil has one of the most restrictive and strict legislation in the world regarding allowed threshold values for these elements and their application conditions (Table 4), ignoring the country's vast diversity of soils and climates and making the use of SS unfeasible in many cases (Bittencourt 2018). Brazil is a huge farming country offering great opportunities for SS-land application in agriculture and forestry, or to recover degraded areas, but it is still minimally used.

Concerns with organic pollutants in the SS is growing, whose origin is diverse and comes from plastics and derivatives, solvents, preservatives, medical drugs, and personal care products, among others. It also involves dioxins, such as

Table 4 Threshold values (mg kg^{-1}) for hazardous trace elements in sewage sludges aiming land application according to European (EEC 278/1986), Brazilian (CONAMA 498/2020), and North American (US EPA 40 CFR Part 503) legislations

Trace elements	Legislations' threshold values (mg kg^{-1}) Europe EEC 278/1986	Brazil CONAMA 498/2020	USA US EPA 40 CFR Part 503
As	NR[a]	41	75
Cd	20 to 40	39	85
Cr	NR[a]	1000	NR[a]
Cr^{6+}	NR[a]	NR[a]	NR[a]
Cu	1000 to 1750	1500	4300
Fe	NR[a]	NR*	NR[a]
Hg	16 to 25	17	57
Mo	NR[a]	50	75
Ni	300 to 400	420	420
Pb	750 to 1200	300	840
Se	NR[a]	36	100
Zn	2500 to 4000	2800	7500

[a]NR: Not regulated

Table 5 Threshold values (mg kg^{-1}) for certain toxic organic compounds in the sewage sludges aiming land application in certain European countries according to Directive 86/278/EEC

Country	Organic compounds PCB	AOX	LAS	NPE/PAH	PCDD/F
Germany	0.1[a]	400	–	–	100
France	0.8[b]	–	–	2-5[c]	–
Italy	0.8	–	–	6	25
Austria	0.2–1	500	–	6	50–100
Sewden	0.4[a]	–	–	3 [d]	–
Portugal	0.8	–	5000	6	100
Denmark	0.2[a]	–	1300	3[d]	–
Belgium	0.6–0.8[a]	–	–	3-20	20

[a]For each congener; [b]Sum of seven congeners: PCB 28, 52, 101, 118, 138, 153, and 180; [c]Different values for different compounds (fluoranthene-5, benzo(b)fluoranthene-2.5, benzo(a)-pyrene-2); [d]Different values for the capacity of wastewater treatment plants (WWTPs) (expressed in population equivalent); *PCB* polychlorinated biphenyls, *AOX* absorbable organic halogens, *LAS* linear alkylbenzene sulfonates, *NPE* nonylphenol ethoxylates, *PAH* polycyclic aromatic hydrocarbons, *PCDD/F* polychlorinated dibenzo-p-dioxins and furan. Adapted from: Collivignarelli et al. (2019a)

polychlorinated biphenyls (PCBs), and polycyclic aromatic hydrocarbons (PAHs) (Rigby et al. 2021), which tend to be persistent in the environment (not biodegraded) and can cause serious ecological harms even at low concentrations (Zuloaga et al.

2012). They are regulated just in certain richer and more criterion's countries, mostly in Europe (Table 5).

6 Sewage Sludge Disposal Techniques

SS disposal strategy varies considerably among the most populous nations in the world. For example, ~60% of the SS generated in the USA is land applied, mostly as Class B biosolid; the remaining are either sent to landfills or incinerated. In African countries, except for South Africa, SS is disposed of mainly in landfills or directly dumped into the environment without prior treatment. In Japan, ~70% of the SS is incinerated whereas the rest is sent to landfills. In China, land application is also the most frequent destination (Shaddel et al. 2019). European countries are directing efforts to reduce landfilling to 50%, sending the remaining material to the land application (Collivignarelli et al. 2019a). SS disposal varies according to socioeconomic and environmental criteria that are specific to each country or region, and all of them have distinct environmental costs (Fig. 3) as well as advantages and disadvantages (Table 6).

Table 6 Main disposal strategies for sewage sludges and their advantages and disadvantages

Disposal strategies	Advantages	Disadvantages
Landfilling	It may be cheap and easy.	Despite considered "easy and cheap", modern landfills are expensive because of regulations and location, Risk of water and soil contamination by the leachate rich in hazardous trace elements and organic pollutants, High greenhouse gas emissions, Requires large areas, Preclude nutrient recycling.
Land application	Improves physical, chemical, and biological attributes of the soils, Allows nutrients recycling, mainly N, P, and micronutrients, such as B, Cu, Mn, and Zn, Inputs organic matter, Reduces pressure for new landfills.	Although not likely, may cause soil and water contamination with hazardous trace elements and other organic pollutants, High transport and application costs.
Incineration	Reduces SS volumes, Generates heat and energy, Reduces pressure for new landfills, Eliminates pathogens.	High cost, Requires use of fossil fuels, Emission of polluting gases, Ashes must be reused or sent to landfills.
Construction	Reuses SS that is unsuitable for agriculture, as bricks, tiles, or other building materials, Either raw SS or its ashes can be used.	Raw SS generates fewer resistant materials, SS must be dried before incineration and ash production.

Sources: LeBlanc et al. (2009), Świerczek et al. (2018), Chung et al. (2020)

Landfilling is discouraged in several countries but is still the main SS destination worldwide (Urban and Isaac 2018) and should be replaced by more sustainable disposal alternatives. However, ~40% of the world SS are still dumped in the open or landfills (Kaza and Yao 2018). Incinerated-SS ashes must be disposed of in landfills due to their high contents of hazardous trace elements resulted from their industrial origin; therefore, their use in agriculture is forbidden. Land application aims to align safe disposal with nutrients recycle, thus improving soil fertility. In Brazil, biosolid amendments to soil improved several crop yields, such as cotton (Samaras et al. 2008), coffee (Tezotto et al. 2012; Martins et al. 2015), sugar cane (Franco et al. 2008; Nogueira et al. 2013), sunflower (Figueiredo Lobo and Grassi Filho 2009; Nascimento et al. 2013), castor bean (Chiaradia et al. 2009), corn (Yada et al. 2020), soybeans (dos Santos Ferreira et al. 2018), wood production (Abreu-Junior et al. 2020), among others. To mitigate adverse effects on agroecosystems, SSs are often composted and, more recently, may also be used as the organic matrix in the manufacture of organomineral fertilizers. Composting reduces organic loads, unpleasant odors (Maulini-Duran et al. 2013), pathogens (Kulikowska 2016), and organic pollutants (including ARGs and emergent pollutants) as well as heavy metal contents (Wang et al. 2017; Giagnoni et al. 2020; Guo et al. 2020). The manufacture of SS-based organomineral fertilizers seems to be a viable and safe alternative for SS disposal in agriculture because they are applied at much smaller rates than the SS *in natura*, thus decreasing the spread of hazardous substances in the environment. It turns feasible to transport over long distances; eases field application (Magela et al. 2019); attenuates seasonal effects on sludge composition, thus allowing to adjust application rates to crop needs and soil fertility (Deeks et al. 2013); slows nutrients release to plants (Kominko et al. 2017); and decreases leaching and runoff of contaminants and nutrients to water resources (Savci 2012).

SS incineration generates high amounts of ashes that are composed of Si, Al, Ca, Fe, P, and, to a lesser extent, Hg, Cd, Sb, As, and Pb (Donatello and Cheeseman 2013), often sent to landfills or reused in construction materials. This thermal method is adopted mainly in Germany, Slovenia, Austria, Belgium, and Netherlands (Stunda-Zujeva et al. 2018), but the world's largest incineration plant is located in Hong Kong, China (Swann et al. 2017). SS ashes have been increasingly explored in construction materials, such as ecological cement, bricks, tiles, and ceramic materials. This way, hazardous persistent trace elements are immobilized within the manufactured material, making their use sustainable and safe (Martínez-García et al. 2012; Chang et al. 2020). China allocates ~16% of its SS to manufacture construction materials (Wei et al. 2020).

7 SS-Nutrient Recovery Technologies and Their Biostimulant Action

Adoption of technologies to recover nutrients from the SS has been attracting more and more attention due to the availability of new technologies, landfill restrictions, and growing demands of nutrients to crops, especially P. The world is highly dependent on mineral fertilizers, mainly the P-fertilizers, as nutrient sources for food production, however, it is estimated that the largest P-mine in the USA will be depleted in just 20 years and that world reserves are limited to about 60 to 250 years. Its higher demand and limited supply caused an abrupt increase in the cost of rock phosphate and may even affect geopolitical balance when nations start competing for the remaining reserves, such as happen for petroleum. In the short term, it seems that the only way out is to recycle as much phosphate as possible and this invariably will involve creating plants for processing human and livestock wastes, such as SSs due to their high nutrient contents, mainly N and P. Tropical countries are highly dependent on imported mineral fertilizers but produces high amounts of municipal organic wastes that should be better intended, either reused or recycled. Modern agriculture is highly dependent on nutrients and mineral fertilizers, turning the production system more expensive and unsustainable (Tyagi and Lo 2013; Raheem et al. 2018). Therefore, the adoption of technologies to recover nutrients is becoming mandatory in modern and environmentally friendly agriculture.

Currently, there are several types of technologies adopted to recover P from the SS based either on its direct use or of its ashes (Cordell et al. 2011). Despite their benefits, often associated with water quality and food as well as environmental security, P-recover technologies still have a lot of space to improve since little P is recovered, especially when facing their high implementing costs (Table 7) (Mayer et al. 2016). Another major limitation is that recovered P can be in unavailable forms, i.e., in forms having low solubility or not assimilable by plants, such as hydroxyapatite and struvite (Table 7). Anyway, SS is likely the organic residue with the greatest appeal for P recycle in a sustainable manner due to its global production scale (Havukainen et al. 2016; Cieślik and Konieczka 2017). The available technologies for P removal from SS as well as their benefits and limitations are summarized in Table 7.

SS can be used to produce biostimulants that can improve metabolic and enzymatic systems of the plants, mainly in the initial growth stages, thus increasing crop yields (Xu and Geelen 2018; Fels et al. 2019). SS-biostimulants are rich in humic and fulvic acids, hydrolyzed proteins, and inorganic elements, as well as beneficial bacteria and fungi (du Jardin 2015). Humic substances can also be extracted from composted CC and favor root growth and crops productivity due to the increases in rhizosphere microbial activity and soil organic fraction (Pascual et al. 2010); whereas hydrolyzed proteins precipitate the heavy metals within the SS, releasing peptides and amino acids that have great action as plant stimulants (Colla et al. 2015; Tejada et al. 2016).

Table 7 Available technologies for P recovery as well as product composition, advantages, and limitations

Methods of P-recovery	Product composition	P contents	Advantages	Limitations	References
Direct SS use	Raw sewage sludge	0.5–0.7%	Low costs.	Expansive transport, Heavy metals accumulation, Spread of pathogens and other pollutants.	Tyagi and Lo (2013), Alvarenga et al. (2015)
Acid leachates	Hydroxy-apatite	2.6–4.0%	Low Cd content, High P content.	High investments costs, Pathogen contamination, High chemical consumption.	Shi et al. (2014), Zheng et al. (2020)
Alkaline leachates	Struvite	11–26%	High P content, Low metal contents.	Low water solubility, High chemical consumption.	Kataki et al. (2016), Munir et al. (2017)
Incineration	Sewage sludge ashes	9–15%	Low mass, High P content, Less pollutants.	Low P availability, High operational cost, High metals concentration.	Krüger and Adam (2014), Kirchmann et al. (2017)
Composting	Sewage sludge compost	2.8–3.6%	Low-cost method, High P residual, OM benefits.	Low P availability.	Alvarenga et al. (2015)
Aerobic/ Anaerobic digestion	Biogas/dry sludge	1.0–1.5%	High nutrient, Pathogen inactivation.	High operational costs.	Borowski and Szopa (2007), Tomei et al. (2011)
Enhanced biological P-removal	Polyphosphate sludge	5.6–7.3%	Rapid P recovery, High P content.	High operational costs.	Angela et al. (2011), Roldán et al. (2020)

8 Future Perspectives

The increasing generation of SS has risen the interest of modern society since it is often related to the outbreak of transmissible diseases, especially after its association to the most recent pandemic experienced in human history caused by the SARS-CoV2 virus. Therefore, the SS needs sustainable management before its disposal in the environment. Several nations are experiencing great technological advances in SS sustainable management, but access to quality water and basic sanitation services are far from becoming reality for many more economically vulnerable countries, such as those in Africa, Latin America, and Asia. However, public policies that guarantee access to these services tend to be intensified in the coming years. For example, a new Sanitation Legal Framework was recently settled in Brazil to warranty the universalization of the sanitation services. The expectation is that more accessible technologies and social pressure for decent water supply and effluent treatment will bring new advances to these vulnerable countries.

The improvement of consolidated techniques, such as composting, as well as the adoption of simple practices, such as its use in the manufacture of organic mineral fertilizers, have the potential to reduce the dependence of many countries on mineral fertilizers, thus turning their agriculture a viable activity. The expansion of new eco-friendly technologies dedicated to the recovery of nutrients (mainly P) and the production of biostimulants in association with new biological treatment approaches are also expected. However, SS management often rises public concerns about risks of environmental contamination by hazardous trace elements and toxic organic compounds, whose analytical protocols and threshold values are well established for land application in most countries. Several studies in the last decades have been changing the perception of the advantages of using high-quality SS in agriculture based on its low contamination risks, especially when compared to landfilling. Based on that, some countries are even revising their tables with higher acceptable values for potentially toxic elements, but SS must be previously treated to avoid the spread of less regulated emerging contaminants, such as antibiotics and their resistance genes in the environment. Finally, the use of other SS potentialities, such as for energy generation or the manufacture of construction materials, besides land application, will be essential to sustain and conserve the environment.

9 Conclusions and Recommendations

When handled correctly, the SS has several uses. Its use in civil construction, electric and thermal energy generation plants, and agriculture as soil fertilizers will be intensified in the coming decades to avoid landfilling and contamination of water resources. Land application of high-quality SS is still the most promising disposal strategy since it reduces mineral fertilizer needs and improves soil quality by supplying organic matter and stimulating its microbiota, thus expanding basic sanitation services through efficient public policies targeted to the reality of each region. Therefore, the reuse of the SS must be aligned with new technologies and associated with sustainable management practices, avoiding landfilling as a major disposal option. In other words, SS has to start being seen as a high-value product instead of just a waste.

References

Abreu-Junior CH, de Oliveira MG, Cardoso PHS et al (2020) Sewage sludge application in eucalyptus urograndis plantation: availability of phosphorus in soil and wood production. Front Environ Sci 8:1–14. https://doi.org/10.3389/fenvs.2020.00116

Aida AA, Kuroda K, Yamamoto M et al (2015) Diversity profile of microbes associated with anaerobic sulfur oxidation in an upflow anaerobic sludge blanket reactor treating municipal sewage. Microbes Environ 30:157–163. https://doi.org/10.1264/jsme2.ME14105

Ali M, Nelson AR, Lopez AL, Sack DA (2015) Updated global burden of cholera in endemic countries. PLoS Negl Trop Dis 9:e0003832

Alvarenga P, Mourinha C, Farto M et al (2015) Sewage sludge, compost and other representative organic wastes as agricultural soil amendments: benefits versus limiting factors. Waste Manag 40:44–52. https://doi.org/10.1016/j.wasman.2015.01.027

Alves LDF, Westmann CA, Lovate GL, et al (2018) Metagenomic approaches for understanding new concepts in microbial science. Int J Genomics 2018. https://doi.org/10.1155/2018/2312987

Angela M, Béatrice B, Mathieu S (2011) Biologically induced phosphorus precipitation in aerobic granular sludge process. Water Res 45:3776–3786. https://doi.org/10.1016/j.watres.2011.04.031

Arif MS, Riaz M, Shahzad SM et al (2018) Fresh and composted industrial sludge restore soil functions in surface soil of degraded agricultural land. Sci Total Environ 619–620:517–527. https://doi.org/10.1016/j.scitotenv.2017.11.143

Arthurson V (2008) Proper sanitization of sewage sludge: a critical issue for a sustainable society. Appl Environ Microbiol 74:5267–5275. https://doi.org/10.1128/AEM.00438-08

Bartram J, Cairncross S (2010) Hygiene, sanitation, and water: forgotten foundations of health. PLoS Med 7:e1000367

Bastida F, Jehmlich N, Martínez-Navarro J et al (2019) The effects of struvite and sewage sludge on plant yield and the microbial community of a semiarid Mediterranean soil. Geoderma 337: 1051–1057. https://doi.org/10.1016/j.geoderma.2018.10.046

Beale DJ, Karpe AV, McLeod JD et al (2016) An "omics" approach towards the characterisation of laboratory scale anaerobic digesters treating municipal sewage sludge. Water Res 88:346–357. https://doi.org/10.1016/j.watres.2015.10.029

Bettiol W, Ghini R (2011) Impacts of sewage sludge in tropical soil: a case study in Brazil. Appl Environ Soil Sci 2011:1–11. https://doi.org/10.1155/2011/212807

Biswas R, Sarkar A (2018) 'Omics' tools in soil microbiology: the state of the art. Springer, Singapore, pp 35–64

Bittencourt S (2018) Agricultural use of sewage sludge in Paraná State, Brazil: a decade of national regulation. Recycling 3:50–53. https://doi.org/10.3390/recycling3040053

Bondarczuk K, Markowicz A, Piotrowska-Seget Z (2016) The urgent need for risk assessment on the antibiotic resistance spread via sewage sludge land application. Environ Int 87:49–55. https://doi.org/10.1016/j.envint.2015.11.011

Börjesson G, Kirchmann H, Kätterer T (2014) Four Swedish long-term field experiments with sewage sludge reveal a limited effect on soil microbes and on metal uptake by crops. J Soils Sediments 14:164–177. https://doi.org/10.1007/s11368-013-0800-5

Borowski S, Szopa JS (2007) Experiences with the dual digestion of municipal sewage sludge. Bioresour Technol 98:1199–1207. https://doi.org/10.1016/j.biortech.2006.05.017

Brooks JP, McLaughlin MR, Gerba CP, Pepper IL (2012) Land application of manure and class B biosolids: an occupational and public quantitative microbial risk assessment. J Environ Qual 41: 2009–2023. https://doi.org/10.2134/jeq2011.0430

Buonocore E, Mellino S, De Angelis G et al (2018) Life cycle assessment indicators of urban wastewater and sewage sludge treatment. Ecol Indic 94:13–23. https://doi.org/10.1016/j.ecolind.2016.04.047

Čadková Z, Száková J, Mukhtorova D, et al (2020) The response of soil nematode Caenorhabditis elegans on the sewage sludge-derived micropollutants. J Hazard Mater 384. https://doi.org/10.1016/j.jhazmat.2019.121468

Camargo FP, Sérgio Tonello P, dos Santos ACA, Duarte ICS (2016) Removal of toxic metals from sewage sludge through chemical, physical, and biological treatments—a review. Water Air Soil Pollut 227. https://doi.org/10.1007/s11270-016-3141-3

Chanaka Udayanga WD, Veksha A, Giannis A et al (2018) Fate and distribution of heavy metals during thermal processing of sewage sludge. Fuel 226:721–744. https://doi.org/10.1016/j.fuel.2018.04.045

Chang Z, Long G, Zhou JL, Ma C (2020) Valorization of sewage sludge in the fabrication of construction and building materials: a review. Resour Conserv Recycl 154:104606. https://doi.org/10.1016/j.resconrec.2019.104606

Charlton A, Sakrabani R, Tyrrel S et al (2016) Long-term impact of sewage sludge application on soil microbial biomass: an evaluation using meta-analysis. Environ Pollut 219:1021–1035. https://doi.org/10.1016/j.envpol.2016.07.050

Chiaradia JJ, Chiba MK, de Andrade CA et al (2009) Yield and plant nutrition of castor bean in crop rotation with sugarcane in a sewage sludge-treated soil. Rev Bras Ciência do Solo 33:701–709

Chung C, Choi K, Kim C et al (2020) Overview of the policies for phasing out ocean dumping of sewage sludge in the Republic of Korea. Sustainability 12:4553. https://doi.org/10.3390/su12114553

Cieślik B, Konieczka P (2017) A review of phosphorus recovery methods at various steps of wastewater treatment and sewage sludge management. The concept of "no solid waste generation" and analytical methods. J Clean Prod 142:1728–1740

Cies⍰lik BM, Namies⍰nik J, Konieczka P (2015) Review of sewage sludge management: standards, regulations and analytical methods. J Clean Prod 90:1–15. https://doi.org/10.1016/j.jclepro.2014.11.031

Colla G, Nardi S, Cardarelli M et al (2015) Protein hydrolysates as biostimulants in horticulture. Sci Hortic (Amsterdam) 196:28–38

Collivignarelli MC, Abbà A, Frattarola A et al (2019a) Legislation for the reuse of biosolids on agricultural land in Europe: overview. Sustainability 11:6015

Collivignarelli MC, Canato M, Abbà A, Carnevale Miino M (2019b) Biosolids: what are the different types of reuse? J Clean Prod 238

CONAMA, (Environment National Council) (2020) Define criteria and procedures for the production and application of biosolids in soils, and other measures. (In portuguese). http://www2.mma.gov.br/port/conama/legiabre.cfm?codlegi=749. Accessed 19 Set 2020

Cordell D, Rosemarin A, Schröder JJ, Smit AL (2011) Towards global phosphorus security: a systems framework for phosphorus recovery and reuse options. Chemosphere 84:747–758. https://doi.org/10.1016/j.chemosphere.2011.02.032

Cytryn E, Kautsky L, Ofek M et al (2011) Short-term structure and functional changes in bacterial community composition following amendment with biosolids compost. Appl Soil Ecol 48:160–167. https://doi.org/10.1016/j.apsoil.2011.03.010

De Corato U (2020) Disease-suppressive compost enhances natural soil suppressiveness against soil-borne plant pathogens: a critical review. Rhizosphere 13:100192. https://doi.org/10.1016/j.rhisph.2020.100192

de Oliveira MS, Oshiro-Junior JA, Sato MR et al (2020) Polymeric nanoparticle associated with ceftriaxone and extract of schinopsis brasiliensis engler against Multiresistant enterobacteria. Pharmaceutics 12:1–18. https://doi.org/10.3390/pharmaceutics12080695

Deeks LK, Chaney K, Murray C et al (2013) A new sludge-derived organo-mineral fertilizer gives similar crop yields as conventional fertilizers. Agron Sustain Dev 33:539–549. https://doi.org/10.1007/s13593-013-0135-z

Deen J, Mengel MA, Clemens JD (2020) Epidemiology of cholera. Vaccine 38:A31–A40. https://doi.org/10.1016/j.vaccine.2019.07.078

Delgado-Baquerizo M, Maestre FT, Reich PB et al (2016) Carbon content and climate variability drive global soil bacterial diversity patterns. Ecol Monogr 86:373–380. https://doi.org/10.1002/ecm.1216/suppinfo
Dhama K, Patel SK, Yatoo MI, et al (2021) SARS-CoV-2 existence in sewage and wastewater: a global public health concern? J Environ Manag 280. https://doi.org/10.1016/j.jenvman.2020.111825
Donatello S, Cheeseman CR (2013) Recycling and recovery routes for incinerated sewage sludge ash (ISSA): a review. Waste Manag 33:2328–2340. https://doi.org/10.1016/j.wasman.2013.05.024
Donde OO, Atoni E, Muia AW, Yillia PT (2021) COVID-19 pandemic: water, sanitation and hygiene (WASH) as a critical control measure remains a major challenge in low-income countries. Water Res 191:116793. https://doi.org/10.1016/j.watres.2020.116793
dos Santos FJ, Volschan I, Cammarota MC (2018) Co-digestion of sewage sludge with crude or pretreated glycerol to increase biogas production. Environ Sci Pollut Res 25:21811–21821. https://doi.org/10.1007/s11356-018-2260-3
du Jardin P (2015) Plant biostimulants: definition, concept, main categories and regulation. Sci Hortic (Amsterdam) 196:3–14
Duan B, Zhang W, Zheng H et al (2017) Comparison of health risk assessments of heavy metals and as in sewage sludge from wastewater treatment plants (WWTPs) for adults and children in the urban district of Taiyuan, China Int J Environ Res Public Health 14. https://doi.org/10.3390/ijerph14101194
Dubey M, Mohapatra S, Tyagi VK et al (2021) Occurrence, fate, and persistence of emerging micropollutants in sewage sludge treatment. Environ Pollut 273:116515. https://doi.org/10.1016/j.envpol.2021.116515
Edmondson JL, Davies ZG, McHugh N et al (2012) Organic carbon hidden in urban ecosystems. Sci Rep 2:1–7. https://doi.org/10.1038/srep00963
EEC (European Economic Community) (1986) On the protection of the environment, and in particular of the soil, when sewage sludge is used in agriculture. Council Directive 86/278. https://op.europa.eu/en/publication-detail/-/publication/f76faa39-2b27-42f2-be1e-9332f795e324/language-en. Accessed 19 Set 2020
El Fels L, El Hayany B, Aguelmous A et al (2019) The use of microorganisms for the biodegradation of sewage sludge and the production of biocompost for sustainable agriculture. Springer, Cham, pp 301–316
Fan W, Huo G, Li X et al (2014) Impact of diet in shaping gut microbiota revealed by a comparative study in infants during the first six months of life. J Microbiol Biotechnol 24:133–143. https://doi.org/10.4014/jmb.1309.09029
Figueiredo Lobo T, Grassi Filho H (2009) Sewage sludge levels on the development and nutrition of sunflower plants. Rev la Cienc del suelo y Nutr Veg 9:245–255
Franco A, Marques MO, de Melo WJ (2008) Sugarcane grown in an Oxisol amended with sewage sludge and vinasse: nitrogen contents in soil and plant. Sci Agric 65:408–414
Franzosa EA, Hsu T, Sirota-Madi A et al (2015) Sequencing and beyond: integrating molecular "omics" for microbial community profiling. Nat Rev Microbiol 13:360–372. https://doi.org/10.1038/nrmicro3451
Freeman MC, Garn JV, Sclar GD et al (2017) The impact of sanitation on infectious disease and nutritional status: a systematic review and meta-analysis. Int J Hyg Environ Health 220:928–949. https://doi.org/10.1016/j.ijheh.2017.05.007
Fykse EM, Aarskaug T, Madslien EH, Dybwad M (2016) Microbial community structure in a full-scale anaerobic treatment plant during start-up and first year of operation revealed by high-throughput 16S rRNA gene amplicon sequencing. Bioresour Technol 222:380–387. https://doi.org/10.1016/j.biortech.2016.09.118
García-Sánchez M, Klouza M, Holečková Z et al (2016) Organic and inorganic amendment application on mercury-polluted soils: effects on soil chemical and biochemical properties. Environ Sci Pollut Res 23:14254–14268. https://doi.org/10.1007/s11356-016-6591-7

Geng H, Xu Y, Zheng L et al (2020) An overview of removing heavy metals from sewage sludge: achievements and perspectives. Environ Pollut 266:115375. https://doi.org/10.1016/j.envpol.2020.115375
Ghini R, Fortes N, Navas Cortés J et al (2016) Combined effects of soil biotic and abiotic factors, influenced by sewage sludge incorporation, on the incidence of corn stalk rot. PLoS One 11: e0155536. https://doi.org/10.1371/journal.pone.0155536
Giagnoni L, dos Anjos Borges LG, Giongo A et al (2020) Dolomite and compost amendments enhance cu phytostabilization and increase microbiota of the leachates from a cu-contaminated soil. Agronomy 10:1–14. https://doi.org/10.3390/agronomy10050719
Goberna M, Simón P, Hernández MT, García C (2018) Prokaryotic communities and potential pathogens in sewage sludge: response to wastewaster origin, loading rate and treatment technology. Sci Total Environ 615:360–368. https://doi.org/10.1016/j.scitotenv.2017.09.240
Gonzalez-Gil L, Papa M, Feretti D et al (2016) Is anaerobic digestion effective for the removal of organic micropollutants and biological activities from sewage sludge? Water Res 102:211–220. https://doi.org/10.1016/j.watres.2016.06.025
Gu C, Bai Y (2018) Heavy metal leaching and plant uptake in mudflat soils amended with sewage sludge. Environ Sci Pollut Res 25:31031–31039. https://doi.org/10.1007/s11356-018-3089-5
Gu XY, Wong JWC, Tyagi RD (2017) Bioleaching of heavy metals from sewage sludge for land application. Curr Dev Biotechnol Bioeng Solid Waste Manag 241–265. https://doi.org/10.1016/B978-0-444-63664-5.00011-3
Guedes P, Magro C, Couto N et al (2015) Potential of the electrodialytic process for emerging organic contaminants remediation and phosphorus separation from sewage sludge. Electrochim Acta 181:109–117. https://doi.org/10.1016/j.electacta.2015.03.167
Guo J, Peng Y, Ni BJ et al (2015) Dissecting microbial community structure and methane-producing pathways of a full-scale anaerobic reactor digesting activated sludge from wastewater treatment by metagenomic sequencing. Microb Cell Factories 14:1–11. https://doi.org/10.1186/s12934-015-0218-4
Guo Z, Wan S, Hua K et al (2020) Fertilization regime has a greater effect on soil microbial community structure than crop rotation and growth stage in an agroecosystem. Appl Soil Ecol 149. https://doi.org/10.1016/j.apsoil.2020.103510
Hamdi H, Hechmi S, Khelil MN et al (2019) Repetitive land application of urban sewage sludge: effect of amendment rates and soil texture on fertility and degradation parameters. Catena 172: 11–20. https://doi.org/10.1016/j.catena.2018.08.015
Havukainen J, Nguyen MT, Hermann L et al (2016) Potential of phosphorus recovery from sewage sludge and manure ash by thermochemical treatment. Waste Manag 49:221–229. https://doi.org/10.1016/j.wasman.2016.01.020
Heck DW, Ghini R, Bettiol W (2019) Deciphering the suppressiveness of banana Fusarium wilt with organic residues. Appl Soil Ecol 138:47–60. https://doi.org/10.1016/j.apsoil.2019.02.021
Hu J, Lin X, Wang J et al (2011) Microbial functional diversity, metabolic quotient, and invertase activity of a sandy loam soil as affected by long-term application of organic amendment and mineral fertilizer. J Soils Sediments 11:271–280. https://doi.org/10.1007/s11368-010-0308-1
Hu M, Wang X, Wen X, Xia Y (2012) Microbial community structures in different wastewater treatment plants as revealed by 454-pyrosequencing analysis. Bioresour Technol 117:72–79. https://doi.org/10.1016/j.biortech.2012.04.061
Hu Y, Xia Y, Maio F Di, et al (2020) Investigation of polycyclic aromatic hydrocarbons (PAHs) formed in three-phase products from the pyrolysis of various wastewater sewage sludge PAHs content distribution and TEQ values in pyrolysis liquid products of sludge from different sources. https://doi.org/10.1016/j.jhazmat.2020.122045
Hua K, Zhu B (2020) Phosphorus loss through surface runoff and leaching in response to the long-term application of different organic amendments on sloping croplands. J Soils Sediments 20: 3459–3471. https://doi.org/10.1007/s11368-020-02675-3
Hughes JM, Koplan JP (2005) Saving lives through global safe water. Emerg Infect Dis 11:1636

Ibarbalz FM, Figuerola ELM, Erijman L (2013) Industrial activated sludge exhibit unique bacterial community composition at high taxonomic ranks. Water Res 47:3854–3864. https://doi.org/10.1016/j.watres.2013.04.010

Kacprzak M, Neczaj E, Fijalkowski K et al (2017) Sewage sludge disposal strategies for sustainable development. Environ Res 156:39–46. https://doi.org/10.1016/j.envres.2017.03.010

Kataki S, West H, Clarke M, Baruah DC (2016) Phosphorus recovery as struvite: recent concerns for use of seed, alternative Mg source, nitrogen conservation and fertilizer potential. Resour Conserv Recycl 107:142–156

Kaza S, Yao L (2018) At a glance: a global picture of solid waste management

Keil A, Wing S, Lowman A (2011) Suitability of public records for evaluating health effects of treated sewage sludge in North Carolina. N C Med J 72:98–104. https://doi.org/10.18043/ncm.72.2.98

Kelessidis A, Stasinakis AS (2012) Comparative study of the methods used for treatment and final disposal of sewage sludge in European countries. Waste Manag 32:1186–1195. https://doi.org/10.1016/j.wasman.2012.01.012

Khalil AI, Hassouna MS, El-Ashqar HMA, Fawzi M (2011) Changes in physical, chemical and microbial parameters during the composting of municipal sewage sludge. World J Microbiol Biotechnol 27:2359–2369. https://doi.org/10.1007/s11274-011-0704-8

Kirchmann H, Börjesson G, Kätterer T, Cohen Y (2017) From agricultural use of sewage sludge to nutrient extraction: a soil science outlook. Ambio 46:143–154. https://doi.org/10.1007/s13280-016-0816-3

Knobeloch L, Salna B, Hogan A, et al (2000) Blue babies and nitrate-contaminated well water. Gd Rounds Environ Med 675–678

Kominko H, Gorazda K, Wzorek Z (2017) The possibility of organo-mineral fertilizer production from sewage sludge. Waste Biomass Valorization 8:1781–1791. https://doi.org/10.1007/s12649-016-9805-9

Krüger O, Adam C (2014) Recovery potential of German sewage sludge ash. Waste Manag 45: 400–406. https://doi.org/10.1016/j.wasman.2015.01.025

Kulikowska D (2016) Kinetics of organic matter removal and humification progress during sewage sludge composting. Waste Manag 49:196–203. https://doi.org/10.1016/j.wasman.2016.01.005

LeBlanc RJ, Matthews P, Richard RP (2009) Global atlas of excreta, wastewater sludge, and biosolids management: moving forward the sustainable and welcome uses of a global resource. Un-habitat

Li B, Zhang T (2010) Biodegradation and adsorption of antibiotics in the activated sludge process. Environ Sci Technol 44:3468–3473. https://doi.org/10.1021/es903490h

Li C, Wang X, Zhang G et al (2017) Hydrothermal and alkaline hydrothermal pretreatments plus anaerobic digestion of sewage sludge for dewatering and biogas production: bench-scale research and pilot-scale verification. Water Res 117:49–57. https://doi.org/10.1016/j.watres.2017.03.047

Liu H, Pu C, Yu X et al (2018) Removal of tetracyclines, sulfonamides, and quinolones by industrial-scale composting and anaerobic digestion processes. Environ Sci Pollut Res 25: 35835–35844. https://doi.org/10.1007/s11356-018-1487-3

Lloret E, Pascual JA, Brodie EL et al (2016) Sewage sludge addition modifies soil microbial communities and plant performance depending on the sludge stabilization process. Appl Soil Ecol 101:37–46. https://doi.org/10.1016/j.apsoil.2016.01.002

Lofrano G, Brown J (2010) Wastewater management through the ages: a history of mankind. Sci Total Environ 408:5254–5264. https://doi.org/10.1016/j.scitotenv.2010.07.062

Lü H, Chen XH, Mo CH et al (2021) Occurrence and dissipation mechanism of organic pollutants during the composting of sewage sludge: a critical review. Bioresour Technol 328:124847. https://doi.org/10.1016/j.biortech.2021.124847

Luo H, Sun Z, Arndt W et al (2009) Gene order phylogeny and the evolution of methanogens. PLoS One 4:2–6. https://doi.org/10.1371/journal.pone.0006069

Ma C, Hu B, Wei MB et al (2019) Influence of matured compost inoculation on sewage sludge composting: enzyme activity, bacterial and fungal community succession. Bioresour Technol 294:122165. https://doi.org/10.1016/j.biortech.2019.122165

Macdonald CA, Clark IM, Zhao FJ et al (2011) Long-term impacts of zinc and copper enriched sewage sludge additions on bacterial, archaeal and fungal communities in arable and grassland soils. Soil Biol Biochem 43:932–941. https://doi.org/10.1016/j.soilbio.2011.01.004

Magela MLM, de Camargo R, Lana RMQ et al (2019) Efficacy of organomineral fertilizers derived from biosolid or filter cake on early maize development. Aust J Crop Sci 13:662–670. https://doi.org/10.21475/ajcs.19.13.05.p1132

Margot J, Rossi L, Barry DA, Holliger C (2015) A review of the fate of micropollutants in wastewater treatment plants. Wiley Interdiscip Rev Water 2:457–487. https://doi.org/10.1002/wat2.1090

Martínez-García C, Eliche-Quesada D, Pérez-Villarejo L et al (2012) Sludge valorization from wastewater treatment plant to its application on the ceramic industry. J Environ Manag 95: S343–S348. https://doi.org/10.1016/j.jenvman.2011.06.016

Martins DR, de Camargo OA, Melo LCA et al (2015) Nutritional status of commercial coffee crop following the application of sewage sludge as a soil conditioner. Rev Ciências Agrárias/ Amazonian J Agric Environ Sci 58:248–256

Mateo-Sagasta J, Raschid-Sally L, Thebo A (2015) Global wastewater and sludge production, Treatment and Use Wastewater Econ Asset an Urban World 15–38. https://doi.org/10.1007/978-94-017-9545-6_2

Mattana S, Petrovičová B, Landi L et al (2014) Sewage sludge processing determines its impact on soil microbial community structure and function. Appl Soil Ecol 75:150–161. https://doi.org/10.1016/j.apsoil.2013.11.007

Maulini-Duran C, Artola A, Font X, Sánchez A (2013) A systematic study of the gaseous emissions from biosolids composting: raw sludge versus anaerobically digested sludge. Bioresour Technol 147:43–51. https://doi.org/10.1016/j.biortech.2013.07.118

Mayer BK, Baker LA, Boyer TH, et al (2016) Total value of phosphorus recovery. https://doi.org/10.1021/acs.est.6b01239

McMahon A (2015) Waste management in early urban southern Mesopotamia. Sanit Latrines Intest Parasites Past Popul Farnham 19–40

Melo W, Delarica D, Guedes A et al (2018) Ten years of application of sewage sludge on tropical soil. A balance sheet on agricultural crops and environmental quality. Sci Total Environ 643: 1493–1501. https://doi.org/10.1016/j.scitotenv.2018.06.254

Menyailo OV, Lehmann J, Da Silva CM, Zech W (2003) Soil microbial activities in tree-based cropping systems and natural forests of the Central Amazon, Brazil. Biol Fertil Soils 38:1–9. https://doi.org/10.1007/s00374-003-0631-4

Moretti SML, Bertoncini EI, Abreu-Junior CH (2015) Composting sewage sludge with green waste from tree pruning. Sci Agric 72:432–439. https://doi.org/10.1590/0103-9016-2014-0341

Morgan C, Bowling M, Bartram J, Lyn Kayser G (2017) Water, sanitation, and hygiene in schools: status and implications of low coverage in Ethiopia, Kenya, Mozambique, Rwanda, Uganda, and Zambia. Int J Hyg Environ Health 220:950–959. https://doi.org/10.1016/j.ijheh.2017.03.015

Moško J, Pohořelý M, Cajthaml T, et al (2021) Effect of pyrolysis temperature on removal of organic pollutants present in anaerobically stabilized sewage sludge. Chemosphere 265. https://doi.org/10.1016/j.chemosphere.2020.129082

Munir MT, Li B, Boiarkina I et al (2017) Phosphate recovery from hydrothermally treated sewage sludge using struvite precipitation. Bioresour Technol 239:171–179. https://doi.org/10.1016/j.biortech.2017.04.129

Nascimento AL, Sampaio RA, Fernandes LA et al (2013) Yield and nutrition of sunflower fertilized with sewage sludge stabilized by different processes. Rev Ceres 60:683–689

Nascimento AL, Souza AJ, Andrade PAM et al (2018) Sewage sludge microbial structures and relations to their sources, treatments, and chemical attributes. Front Microbiol 9:1–11. https://doi.org/10.3389/fmicb.2018.01462
Nascimento AL, de Souza AJ, Oliveira FC, et al (2020) Chemical attributes of sewage sludges: relationships to sources and treatments, and implications for sludge usage in agriculture. J Clean Prod 258. https://doi.org/10.1016/j.jclepro.2020.120746
Newcomb CJ, Qafoku NP, Grate JW, et al (2017) Developing a molecular picture of soil organic matter-mineral interactions by quantifying organo-mineral binding. Nat Commun 8. https://doi.org/10.1038/s41467-017-00407-9
Nicolas I, Bordeau V, Bondon A et al (2019) Novel antibiotics effective against gram-positive and -negative multi-resistant bacteria with limited resistance. PLoS Biol 17:1–23. https://doi.org/10.1371/journal.pbio.3000337
Nogueira TAR, Franco A, He Z et al (2013) Short-term usage of sewage sludge as organic fertilizer to sugarcane in a tropical soil bears little threat of heavy metal contamination. J Environ Manag 114:168–177. https://doi.org/10.1016/j.jenvman.2012.09.012
Ozcan S, Tor A, Aydin ME (2013) Investigation on the levels of heavy metals, polycyclic aromatic hydrocarbons, and polychlorinated biphenyls in sewage sludge samples and ecotoxicological testing. Clean (Weinh) 41:411–418. https://doi.org/10.1002/clen.201100187
Pal C, Bengtsson-Palme J, Kristiansson E, Larsson DGJ (2015) Co-occurrence of resistance genes to antibiotics, biocides and metals reveals novel insights into their co-selection potential. BMC Genomics 16:964. https://doi.org/10.1186/s12864-015-2153-5
Pascual I, Azcona I, Aguirreolea J et al (2010) Growth, yield, and fruit quality of pepper plants amended with two sanitized sewage sludges. J Agric Food Chem 58:6951–6959. https://doi.org/10.1021/jf100282f
Paul D, Kumar S, Mishra M et al (2018) Molecular genomic techniques for identification of soil microbial community structure and dynamics. Springer, Singapore, pp 9–33
Pavan Fernandes SA, Bettiol W, Cerri CC (2005) Effect of sewage sludge on microbial biomass, basal respiration, metabolic quotient and soil enzymatic activity. Appl Soil Ecol 30:65–77. https://doi.org/10.1016/j.apsoil.2004.03.008
Pepper IL, Brooks JP, Gerba CP (2006) Pathogens in biosolids. Adv Agron 90:1–41. https://doi.org/10.1016/S0065-2113(06)90001-7
Pérez-Lemus N, López-Serna R, Pérez-Elvira SI, Barrado E (2019) Analytical methodologies for the determination of pharmaceuticals and personal care products (PPCPs) in sewage sludge: a critical review. Anal Chim Acta 1083:19–40
Pinto Z, Morandi M, Bettiol W (2013) Induction of suppressiveness to Fusarium wilt of chrysanthemum with composted sewage sludge. Trop Plant Pathol 38:414–422. https://doi.org/10.1590/S1982-56762013005000026
Poulsen TG, Bester K (2010) Organic micropollutant degradation in sewage sludge during composting under thermophilic conditions. Environ Sci Technol 44:5086–5091. https://doi.org/10.1021/es9038243
Prüss-Ustün A, Wolf J, Bartram J et al (2019) Burden of disease from inadequate water, sanitation and hygiene for selected adverse health outcomes: an updated analysis with a focus on low- and middle-income countries. Int J Hyg Environ Health 222:765–777. https://doi.org/10.1016/j.ijheh.2019.05.004
Raheem A, Sikarwar VS, He J et al (2018) Opportunities and challenges in sustainable treatment and resource reuse of sewage sludge: a review. Chem Eng J 337:616–641
Regkouzas P, Diamadopoulos E (2019) Adsorption of selected organic micro-pollutants on sewage sludge biochar. Chemosphere 224:840–851. https://doi.org/10.1016/j.chemosphere.2019.02.165
Rigby H, Dowding A, Fernandes A, et al (2021) Concentrations of organic contaminants in industrial and municipal bioresources recycled in agriculture in the UK. Sci Total Environ 765. https://doi.org/10.1016/j.scitotenv.2020.142787
Ritchie H, Roser M (2018) Urbanization. Our world data

Roldán M, Bouzas A, Seco A, et al (2020) An integral approach to sludge handling in a WWTP operated for EBPR aiming phosphorus recovery: Simulation of alternatives, LCA and LCC analyses. Water Res 175. https://doi.org/10.1016/j.watres.2020.115647
Rorat A, Courtois P, Vandenbulcke F, Lemiere S (2019) Sanitary and environmental aspects of sewage sludge management. In: Industrial and Municipal Sludge. Elsevier, pp. 155–180
Rutgersson C, Ebmeyer S, Lassen SB et al (2020) Long-term application of Swedish sewage sludge on farmland does not cause clear changes in the soil bacterial resistome. Environ Int 137. https://doi.org/10.1016/j.envint.2019.105339
Saleem M, Hu J, Jousset A (2019) More than the sum of its parts: microbiome biodiversity as a driver of plant growth and soil health. Annu Rev Ecol Evol Syst 50:145–168
Samaras V, Tsadilas CD, Stamatiadis S (2008) Effects of repeated application of municipal sewage sludge on soil fertility, cotton yield, and nitrate leaching. Agron J 100:477–483
Savci S (2012) An agricultural pollutant: chemical fertilizer. Int J Environ Sci Dev 3:73
Scaglia B, D'Imporzano G, Garuti G et al (2014) Sanitation ability of anaerobic digestion performed at different temperature on sewage sludge. Sci Total Environ 466–467:888–897. https://doi.org/10.1016/j.scitotenv.2013.07.114
Schütte T, Niewersch C, Wintgens T, Yüce S (2015) Phosphorus recovery from sewage sludge by nanofiltration in diafiltration mode. J Memb Sci 480:74–82. https://doi.org/10.1016/j.memsci.2015.01.013
Semblante GU, Hai FI, Huang X et al (2015) Trace organic contaminants in biosolids: impact of conventional wastewater and sludge processing technologies and emerging alternatives. J Hazard Mater 300:1–17. https://doi.org/10.1016/j.jhazmat.2015.06.037
Shaddel S, Bakhtiary-Davijany H, Kabbe C, et al (2019) Sustainable sewage sludge management: from current practices to emerging nutrient recovery technologies. Sustainability 11. https://doi.org/10.3390/su11123435
Sharma B, Sarkar A, Singh P, Singh RP (2017) Agricultural utilization of biosolids: a review on potential effects on soil and plant grown. Waste Manag 64:117–132
Shi W, Feng C, Huang W et al (2014) Study on interaction between phosphorus and cadmium in sewage sludge during hydrothermal treatment by adding hydroxyapatite. Bioresour Technol 159:176–181. https://doi.org/10.1016/j.biortech.2014.02.108
Siebielec G, Siebielec S, Lipski D (2018) Long-term impact of sewage sludge, digestate and mineral fertilizers on plant yield and soil biological activity. J Clean Prod 187:372–379. https://doi.org/10.1016/j.jclepro.2018.03.245
Silva E, Melo M, Sombra K, et al (2019) Organic nitrogen in agricultural systems, pp 1–22
Stalder T, Barraud O, Jové T et al (2014) Quantitative and qualitative impact of hospital effluent on dissemination of the integron pool. ISME J 8:768–777. https://doi.org/10.1038/ismej.2013.189
Stunda-Zujeva A, Kreicbergs I, Medne O (2018) Sustainable utilization of sewage sludge: review of technologies. In: Key Engineering Materials. Trans Tech Publ, pp. 121–125
Sui Q, Zhang J, Chen M et al (2016) Distribution of antibiotic resistance genes (ARGs) in anaerobic digestion and land application of swine wastewater. Environ Pollut 213:751–759. https://doi.org/10.1016/j.envpol.2016.03.038
Sun Y, Qiu T, Gao M et al (2019) Inorganic and organic fertilizers application enhanced antibiotic resistome in greenhouse soils growing vegetables. Ecotoxicol Environ Saf 179:24–30. https://doi.org/10.1016/j.ecoenv.2019.04.039
Swann L, Downs D, Waye M (2017) Waste to energy solution–The sludge treatment facility in Tuen Mun, Hong Kong. Energy Procedia 143:500–505
Świerczek L, Cieślik BM, Konieczka P (2018) The potential of raw sewage sludge in construction industry – a review. J Clean Prod 200:342–356. https://doi.org/10.1016/j.jclepro.2018.07.188
Tacconelli E, Carrara E, Savoldi A et al (2018) Discovery, research, and development of new antibiotics: the WHO priority list of antibiotic-resistant bacteria and tuberculosis. Lancet Infect Dis 18:318–327. https://doi.org/10.1016/S1473-3099(17)30753-3
Tao J, Wu S, Sun L et al (2012) Composition of waste sludge from municipal wastewater treatment plant. Procedia Environ Sci 12:964–971. https://doi.org/10.1016/j.proenv.2012.01.372

Taylor DL, Kahawita TM, Cairncross S, Ensink JHJ (2015) The impact of water, sanitation and hygiene interventions to control cholera: a systematic review. PLoS One 10:e0135676
Tejada M, Rodríguez-Morgado B, Gómez I et al (2016) Use of biofertilizers obtained from sewage sludges on maize yield. Eur J Agron 78:13–19. https://doi.org/10.1016/j.eja.2016.04.014
Tezotto T, Favarin JL, Azevedo RA et al (2012) Coffee is highly tolerant to cadmium, nickel and zinc: plant and soil nutritional status, metal distribution and bean yield. F Crop Res 125:25–34
Tomei MC, Rita S, Mininni G (2011) Performance of sequential anaerobic/aerobic digestion applied to municipal sewage sludge. J Environ Manag 92:1867–1873. https://doi.org/10.1016/j.jenvman.2011.03.016
Troeger C, Blacker BF, Khalil IA et al (2018) Estimates of the global, regional, and national morbidity, mortality, and aetiologies of diarrhoea in 195 countries: a systematic analysis for the Global Burden of Disease Study 2016. Lancet Infect Dis 18:1211–1228
Tyagi VK, Lo S-L (2013) Sludge: a waste or renewable source for energy and resources recovery? https://doi.org/10.1016/j.rser.2013.05.029
United Nations U (2019) World population prospects 2019
Urban RC, Isaac R d L (2018) WTP and WWTP sludge management: a case study in the metropolitan area of Campinas, southeastern Brazil. Environ Monit Assess 190:584. https://doi.org/10.1007/s10661-018-6972-0
US-EPA (1993) Standards for the use or disposal of sewage sludge (40 Code of Federal Regulations Part 503) US Environmental Protection Agency, Washington, DC
US-EPA (2000) Inductively coupled plasma-atomic emission spectrometry 1.0 1:220
US-EPA (2007) Microwave assisted acid digestion of sediments, sludges, soils, and oils SW-846
Uysal A, Kuru B (2013) Magnesium ammonium phosphate production from wastewater through box-behnken design and its effect on nutrient element uptake in plants. Clean (Weinh) 41:447–454. https://doi.org/10.1002/clen.201200231
Van Frankenhuyzen JK, Trevors JT, Flemming CA et al (2013) Optimization, validation, and application of a real-time PCR protocol for quantification of viable bacterial cells in municipal sewage sludge and biosolids using reporter genes and *Escherichia coli*. J Ind Microbiol Biotechnol 40:1251–1261. https://doi.org/10.1007/s10295-013-1319-x
Vance ED, Brookes PC, Jenkinson DS (1987) Microbial biomass measurements in forest soils: The use of the chloroform fumigation-incubation method in strongly acid soils. Soil Biol Biochem 19:697–702. https://doi.org/10.1016/0038-0717(87)90051-4
Venegas M, Leiva AM, Reyes-Contreras C, et al (2021) Presence and fate of micropollutants during anaerobic digestion of sewage and their implications for the circular economy: a short review. J Environ Chem Eng 9. https://doi.org/10.1016/j.jece.2020.104931
Viau E, Bibby K, Paez-Rubio T, Peccia J (2011) Toward a consensus view on the infectious risks associated with land application of sewage sludge. Environ Sci Technol 45:5459–5469. https://doi.org/10.1021/es200566f
Wang X, Cui H, Shi J et al (2015) Relationship between bacterial diversity and environmental parameters during composting of different raw materials. Bioresour Technol 198:395–402. https://doi.org/10.1016/j.biortech.2015.09.041
Wang Y, Li C, Tu C et al (2017) Long-term no-tillage and organic input management enhanced the diversity and stability of soil microbial community. Sci Total Environ 609:341–347. https://doi.org/10.1016/j.scitotenv.2017.07.053
Wei L, Zhu F, Li Q et al (2020) Development, current state and future trends of sludge management in China: based on exploratory data and CO_2-equivaient emissions analysis. Environ Int 144:106093. https://doi.org/10.1016/j.envint.2020.106093
WHO (2020) State of the world's sanitation: an urgent call to transform sanitation for better health, environments, economies and societies
Wilfert P, Dugulan AI, Goubitz K et al (2018) Vivianite as the main phosphate mineral in digested sewage sludge and its role for phosphate recovery. Water Res 144:312–321. https://doi.org/10.1016/j.watres.2018.07.020

Wolinska A, Stepniewsk Z (2012) Dehydrogenase activity in the soil environment. In: Dehydrogenases. InTech
Wolters B, Fornefeld E, Jechalke S et al (2018) Soil amendment with sewage sludge affects soil prokaryotic community composition, mobilome and resistome. FEMS Microbiol Ecol 95:1–14. https://doi.org/10.1093/femsec/fiy193
Wyrwicka A, Urbaniak M (2016) The different physiological and antioxidative responses of zucchini and cucumber to sewage sludge application. PLoS One 11. https://doi.org/10.1371/journal.pone.0157782
Xia Y, Wen X, Zhang B, Yang Y (2018) Diversity and assembly patterns of activated sludge microbial communities: a review. Biotechnol Adv 36:1038–1047. https://doi.org/10.1016/j.biotechadv.2018.03.005
Xie WY, McGrath SP, Su JQ et al (2016) Long-term impact of field applications of sewage sludge on soil antibiotic resistome. Environ Sci Technol 50:12602–12611. https://doi.org/10.1021/acs.est.6b02138
Xu L, Geelen D (2018) Developing biostimulants from agro-food and industrial by-products. Front Plant Sci 871
Xue D, Huang X (2013) The impact of sewage sludge compost on tree peony growth and soil microbiological, and biochemical properties. Chemosphere 93:583–589. https://doi.org/10.1016/j.chemosphere.2013.05.065
Yada MM, De Melo WJ, de Melo VP (2020) Trace elements in soil, plant and grain of corn plants cultivated in Latosols after sixteen years with application of sewage sludge. Eng Sanit e Ambient 25:371–379. https://doi.org/10.1590/s1413-41522020150124
Zhang J, Gao D, Bin CT et al (2010) Simulation of substrate degradation in composting of sewage sludge. Waste Manag 30:1931–1938. https://doi.org/10.1016/j.wasman.2010.04.004
Zhang QH, Yang WN, Ngo HH et al (2016) Current status of urban wastewater treatment plants in China. Environ Int 92–93:11–22. https://doi.org/10.1016/j.envint.2016.03.024
Zhang W, Mo Y, Yang J et al (2018) Genetic diversity pattern of microeukaryotic communities and its relationship with the environment based on PCR-DGGE and T-RFLP techniques in Dongshan Bay, southeast China. Cont Shelf Res 164:1–9. https://doi.org/10.1016/j.csr.2018.05.006
Zhao Q, Liu Y (2019) Is anaerobic digestion a reliable barrier for deactivation of pathogens in biosludge? Sci Total Environ 668:893–902. https://doi.org/10.1016/j.scitotenv.2019.03.063
Zhen G, Lu X, Kato H et al (2017) Overview of pretreatment strategies for enhancing sewage sludge disintegration and subsequent anaerobic digestion: current advances, full-scale application and future perspectives. Renew Sust Energ Rev 69:559–577. https://doi.org/10.1016/j.rser.2016.11.187
Zheng X, Ye Y, Jiang Z et al (2020) Enhanced transformation of phosphorus (P) in sewage sludge to hydroxyapatite via hydrothermal carbonization and calcium-based additive. Sci Total Environ 738:139786. https://doi.org/10.1016/j.scitotenv.2020.139786
Zhou J, Zheng G, Wong JWC, Zhou L (2013) Degradation of inhibitory substances in sludge by *Galactomyces* sp. Z3 and the role of its extracellular polymeric substances in improving bioleaching. Bioresour Technol 132:217–223. https://doi.org/10.1016/j.biortech.2012.12.179
Zuloaga O, Navarro P, Bizkarguenaga E et al (2012) Overview of extraction, clean-up and detection techniques for the determination of organic pollutants in sewage sludge: a review. Anal Chim Acta 736:7–29. https://doi.org/10.1016/j.aca.2012.05.016

Ecological and Health Risk Assessment in Sewage Irrigated Heavy Metal Contaminated Soils

Anandkumar Naorem, Boris Huirem, and Shiva Kumar Udayana

1 Introduction

1.1 Heavy Metal Contamination Due to Sewage Irrigation

The soil system is structured by several factors including both natural and anthropogenic activities. The present agricultural era has been facing several soil-related constraints such as low soil fertility. In addition to supplementing soil nutrients, it is also essential to avoid the loading of toxic elements in the soil. It is ascribed due to the transfer of these toxic elements to our human body through the food chain. A large growing body of literature has been continuously studying soil contamination as one of the most crucial environmental problems on a global scale (Rostami et al. 2020). These contaminants when transmitted through the food chain can jeopardize human health through direct and/or indirect pathways (Mohammadi et al. 2019). Heavy metals (hereafter HM) are recognized as one of these contaminants through industrial effluents, use of pesticides and fertilizers, sewage irrigation etc. (Nagajyoti et al. 2010). Owing to its high resistance towards decomposition, HMs are classified under 'persistent environmental pollutants'. One of the main sources of HM entry in the soil is long-term sewage irrigation, which is commonly followed in several parts

A. Naorem (✉)
ICAR-Central Arid Zone Research Institute, RRS-Bhuj, India
e-mail: Anandkumar.naorem@icar.gov.in

B. Huirem
Department of Food Engineering, Bidhan Chandra Krishi Viswavidyalaya, Mohanpur, West Bengal, India

S. K. Udayana
MS Swaminathan School of Agriculture, Centurion University of Technology and Management, Paralakhemundi, Gajapati, Odisha, India

V. D. Rajput et al. (eds.), *Sustainable Management and Utilization of Sewage Sludge*, https://doi.org/10.1007/978-3-030-85226-9_2

of the country. Sewage irrigation is commonly practiced due to some obvious reasons:

(i) non-availability of good quality irrigation water;
(ii) shortage of irrigation water;
(iii) nutrients contained in sewage water.

Along with the plant nutrients, a large amount of HMs is delivered into the soil system through sewage irrigation. The soil system interacts with HMs depending on the chemical properties of both soil and HMs. The plant available HMs are taken up by the plants and get translocated in different parts of the plant. The question here often arises on where it is stored maximum in the edible part of the plant? Even though the HMs are accumulated in the edible part of the plant, it is not less harmful to the environment. *For example*, even if no or less concentration of HM is accumulated in rice grains, feeding rice straw to animals or using it further for mushroom cultivation indirectly transfer the HM load to human beings. Therefore, the priority must be to restrict the entry of the HMs inside the plant cells. Through the food chain, the HMs get bioaccumulated and disturbs the body functioning and negatively affects several human organs. Understanding the contamination levels of these HMs in soil and plant is the first step in remediation of HM contaminated sites. Calculating the ecological indices from these levels depict how much the environment is contaminated and whether remediation is needed for safe cultivation of crops. One of the main goals in human nutrition is to stop consumption of HM rich foods. Therefore, it is equally essential to calculate some indices that reflect the safe levels of pollution and whether the pollutant level is carcinogenic or non-carcinogenic risk to human health.

1.2 Quality Analysis of Sewage Water for Agricultural Use

Although sewage water contains ample amount of essential nutrients, it also delivers certain contaminants. Proper analysis of the sewage water quality is a must to understand the characteristics of sewage water. Generally, sewage water contains high amount of water with small concentrations of dissolved and/or suspended solids. The sewage quality used for irrigation can be interpreted through various levels of quality variables shown in Fig. 1. The most common parameter used in sewage quality analysis is TDS (total dissolved solids). 450-2000 mg/l refers to slight-moderate degree of restriction on use as irrigant, below which is safe and greater than this range refers to severe restriction. In addition to it, for sewage water to be classified under safe irrigant, it must contain safe levels of trace elements, otherwise could negatively affect crop production (Fig. 2).

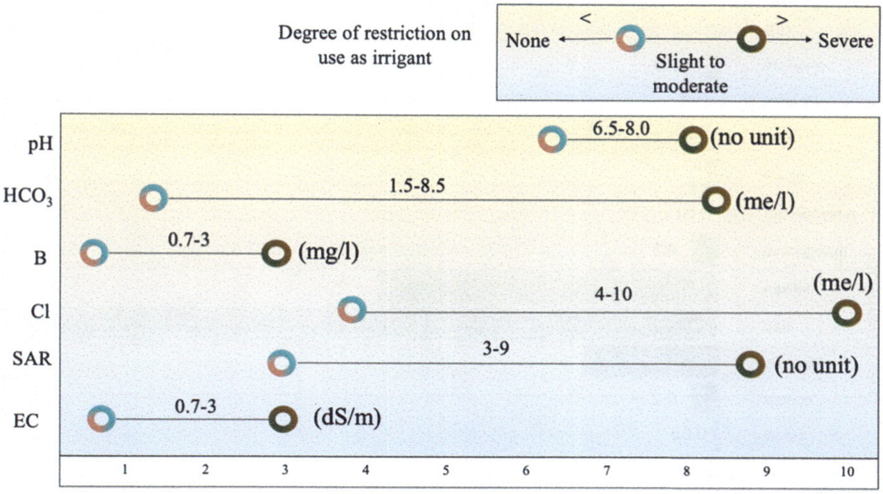

Fig. 1 Interpretations of values of sewage quality parameters according to FAO guidelines; The value depicts the degree of restriction of use as irrigant. The lollipop range refers to slight to moderate degree of restriction, above which refers to severe degree while below this range indicates no restriction; *B* Boron concentration, *HCO₃* Bicarbonate, *Cl* chloride, *SAR* Sodium adsorption ratio, *EC* electrical conductivity

2 Ecological and Human Health Risk Assessment

Several health implications are still reported due to exposure to HMs such as cardiovascular diseases, bone deformation, infertility, neurotoxicity and blindness (Rafiee et al. 2020). Excessive accumulation of cadmium (Cd) in the kidney affects the urinary tract. Some HMs, *for example*, cadmium, mercury (Hg) and metalloids such as arsenic (As) are dangerous to humans even in a very meager amount (Gupta et al. 2018). Having mentioned its toxicity, some of the HMs are essential for the growth and functioning of living organisms (Chabukdhara et al. 2017). *For example*, copper (Cu) is an important micronutrient essential for normal plant growth and development. It is also a cofactor of several enzymes. Cu also takes a crucial role in photosynthesis, development of reproductive organs and respiration. However, it is essential in trace amounts for plants. Otherwise, a higher load of Cu in plant destroys the structure and function of the plant cell membrane. Toxic amount of Cu also leads to damage to the plant's antioxidative system and chloroplast. Similarly, zinc (Zn) is another essential plant micronutrient involved in plant physiological improvements. On the other hand, excessive Zn in plant damages the plant root system and limits plant growth. A high concentration of HMs also causes negative impacts on plant growth and development. Lead (Pb) stress in plants results in to change in cell membrane permeability, disturbs enzyme activities, negatively affects mitosis and DNA damage.

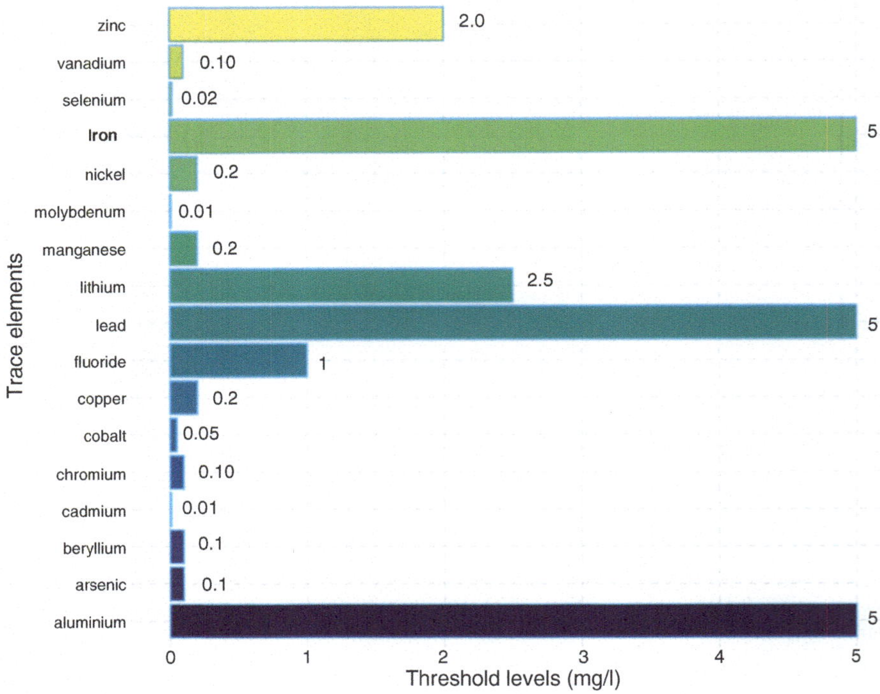

Fig. 2 Threshold levels of trace elements (mg/l) found in sewage water above which is regarded as contaminant in sewage water

Therefore, to control and prevent HMs entry into the human body and our food, it is the prime step to assess the concentrations of these HMs in soil, plant and the human body. Spatial distribution and contamination levels of each HMs are identified to promote public health (Hu et al. 2013). Risk assessments (hereafter RA) are carried out to understand the magnitude of injury from high HM concentration in a system. In other words, RA is a scientific tool that can be applied in environmental legislation. RA is a process that estimates the possible probability of event occurrence (here soil contamination) and its related magnitude of adverse health impacts due to exposure to the contaminated soil over a specified period. RA involves estimation of contamination level of HM in air, soil, sediments, water, plants and human bodies and determination of the possible negative effects on living organisms. First, the number of contaminants must be identified followed by their respective concentration. Secondly, the contamination levels of any areas are calculated through several indices such as the Ecological Risk Index (RI). These indices aid in assessing any region's HM contamination status (Kamani et al. 2017). Zhou et al. (2014) identified the sources of HMs found in cultivated soils and estimated the Ecological Risk of the study region. Only the ecological risk of Cd and Hg were found to be significant (Zhou et al. 2014). A human health risk assessment conducted in Pakistan showed only Cd at a harmful level for human health (Khan et al. 2013).

The chapter aims to provide a systematic compilation of ecological indices and risk assessment indices related to HM contamination in soil and plants.

3 Ecological Monitoring Indices

There are two types of HM contamination related ecological indices: single element and multi-element. Single element pollution indices depict how much an element is concentrated in the study site as compared to a background value. Examples of single element pollution indices are contamination factor, Index of geo-accumulation and enrichment factor. Due to few limitations of single element pollution indices, multi-element pollution indices were introduced to assess HM contamination in soil and sediments. Some examples under this group comprise contamination degree, pollution index, modified pollution index.

3.1 Heavy Metal Pollution Index

Hakanson (1980) introduced pollution index (PI) to determine the toxicity potential of each of the HM pollution. PI was developed by adding an arbitrary weightage value for each HM. The weightage range between 0–1 and its choice relies on the importance of the HM parameter in estimating the contamination level. PI is calculated by Eq. (1) in Table 5 (Mohan et al. 1996). The weights are inversely proportional to the recommended values for each variable. The sub-index is calculated using Equation no. (2) in Table 6. HM pollution level can be interpreted through PI (Fig. 1).

3.2 Geo-accumulation Index

Muller proposed geo-accumulation index (Igeo) to calculate the contamination levels of HMs in soil samples relative to the concentration of the particular HM during the pre-industrial era (Fig. 1). It reflects the geochemical index for HM contamination. A coefficient of 1.5 is added to amplify the effect of any possible change in Bn, owing to soil lithology and ground factor effects. Rostami et al. (2020) investigated the contamination levels of HMs through pollution indices, Igeo as a reference to estimate the extent of HM contamination. Except for Cd and As showed negative Igeo. Wei and Yang (2010) found higher Igeo (>1) in the case of Hg and Cd in agricultural soils. Igeo is calculated using the Equation no. (3) in Table 5.

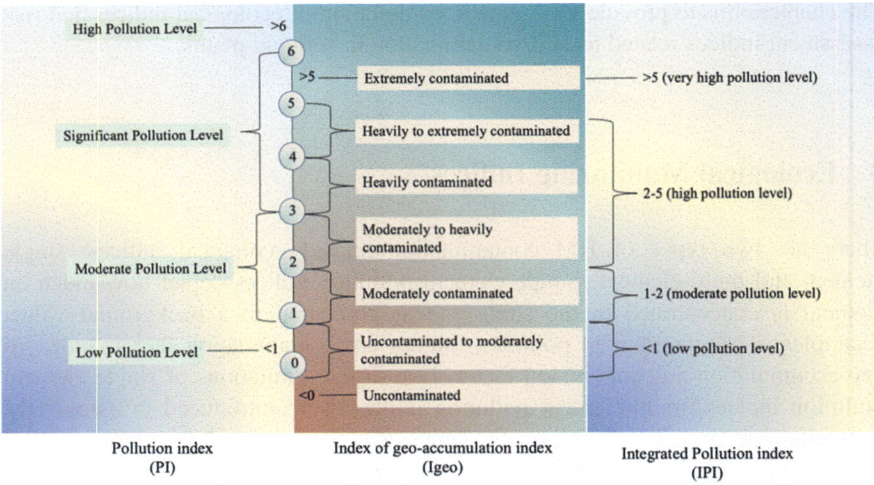

Fig. 3 Interpretations of values in different indices used to measure ecological risk of HM pollution (*left*: pollution index, *middle*: geo-accumulation index, *right*: integrated pollution index)

3.3 Integrated Pollution Index

All the PIs of all HMs included in the study can be used to calculate its mean value which is termed as Integrated Pollution Index (IPI). An IPI value of <1 depicts a low HM pollution level, 1–2 represents moderate HM pollution. An IPI value of 2–5 and >5 indicates high and extreme high HM pollution respectively (Fig. 3) (Chen et al. 2005). The IPI of collective HMs in a study site is calculated using Equation no. (4) in Table 5.

3.4 Potential Ecological Risk Index

Hakanson (1980) proposed the Potential Ecological Risk Index (PERI) to estimate biological toxicity. PERI comprehensively provides an estimate of environmental risk to HM pollution. PERI was introduced to assess the ecological hazard index, calculated by the ratio of HM content in soil or sediment to the maximum background value of the respective HM before industrialization. PERI indicates the sensitivity of the living biota to toxic levels of HM and depicts the ill effects due to contamination of several HMs. Based on PERI values, four levels of HM risk are identified: high (>600), considerable (300–600), moderate (150–300) and low (<150) (Fig. 4) (Maanan et al. 2015). The PERI is calculated using Equation no. (5) in Table 5.

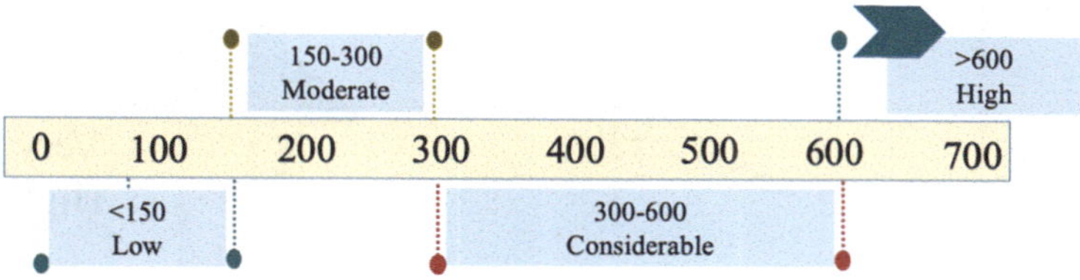

Fig. 4 PERI levels indicating the biological toxicity in response to HM contamination

Table 1 Classes of HM pollution based on HEI values

Class	HEI
Low	<10
Medium	10–20
High	>20

3.5 *Heavy Metal Evaluation Index*

Heavy metal evaluation index (HEI) represents an overall quality of soil or water samples concerning HM contamination (Edet and Offiong 2002) and is calculated using Equation no. (6) in Table 5. Three classes of HM contamination based on HEI values is shown in Table 1.

3.6 *Ecological Risk Factor or Single Metal Ecological Risk*

Ecological risk factor (E^i_r) indicates the risk of an ecology due to HM contamination. It addresses only one element at a time and therefore it is also referred to "single metal ecological risk factor". There are three classes of ecological risk based on E^i_r values (Fig. 5). E^i_r is calculated using Equation no. (7) in Table 5.

Ecological risk factor includes the HM concentration in the sample along with the toxic response factor (T_r^i) of each element (Table 2). T_r^i is the toxicity level of each HM in the environment.

3.7 *Enrichment Factor*

The enrichment factor (EF) of a HM in a soil sample is estimated to understand the possible source of contamination. It is calculated by comparing the HM concentration in the sample to a predetermined concentration of a control sample element such as Al, Fe and Mn. EF greater than 1 indicates the significant contribution of anthropogenic activities in its contamination in soil (Fig. 6). EF is calculated using Equation no. (8) in Table 5.

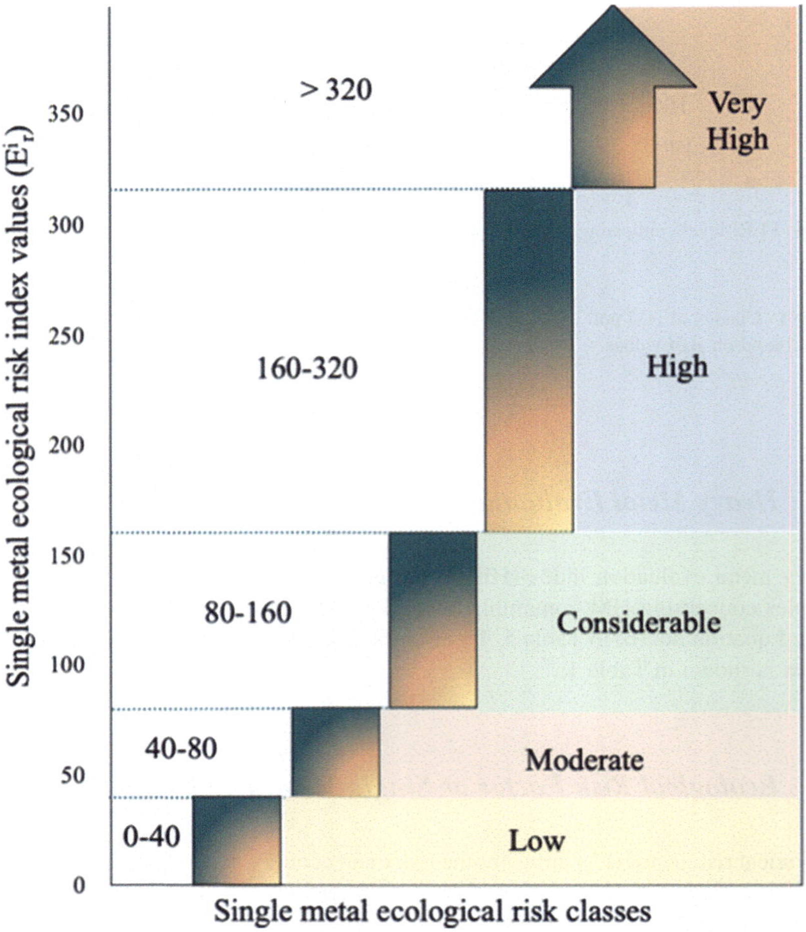

Fig. 5 The bars indicate the minimum to maximum values of each class of E^i_r. Left side of the bar numerically depicts the limits of values. Right side of the bars interprets the class of the study site based on E^i_r values

Table 2 Toxic response factor of HMs and metalloids

Toxic response factor (T_r^i)	
Element	T_r^i Value
Mercury	40
Cadmium	30
Arsenic	10
Lead	5
Copper	5
Nickel	5
Chromium	2
Zinc	1

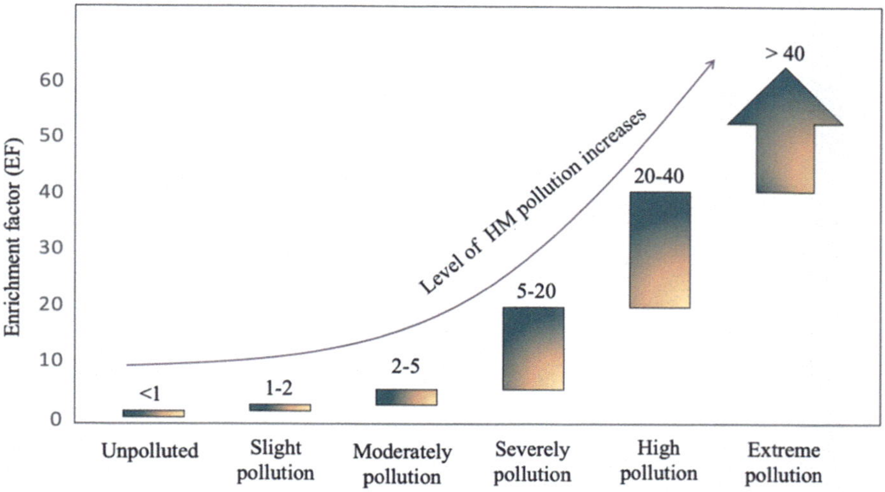

Fig. 6 Levels of HM pollution based on EF values. With the increase of EF values, HM pollution increases

3.8 Contamination Factor and Degree

Contamination factor (C_f) and contamination degree (C_d) represents general contamination of an environment with a respective pollutant (Fig. 7). Cf is calculated by the ratio of an HM concentration in a soil sample to the background value for the respective HM. C_d is the overall summation of all C_fs of respective HMs. It is to be noted that C_d might underestimate the levels of HM contamination and therefore other indices must be calculated along with to confirm the levels of HM contamination. Contamination factor (C_f) and contamination degree (C_d) are calculated using Equation no. (9) and Equation no. (10) respectively in Table 5.

3.9 Contamination Index

The contamination index (C_{di}) calculates the extent of overall HM contamination of soil samples by summing up the effects of several soil quality parameters (Prasanna et al. 2012). There are three classes of C_{di} based on its value, reflecting the level of HM contamination (Table 3). It is calculated using Equation no. (11) in Table 5.

3.10 Heavy Metal Index

Multivariate statistical tools have been increasingly used in HM contamination studies. It includes Principal component analysis (PCA) before classifying the

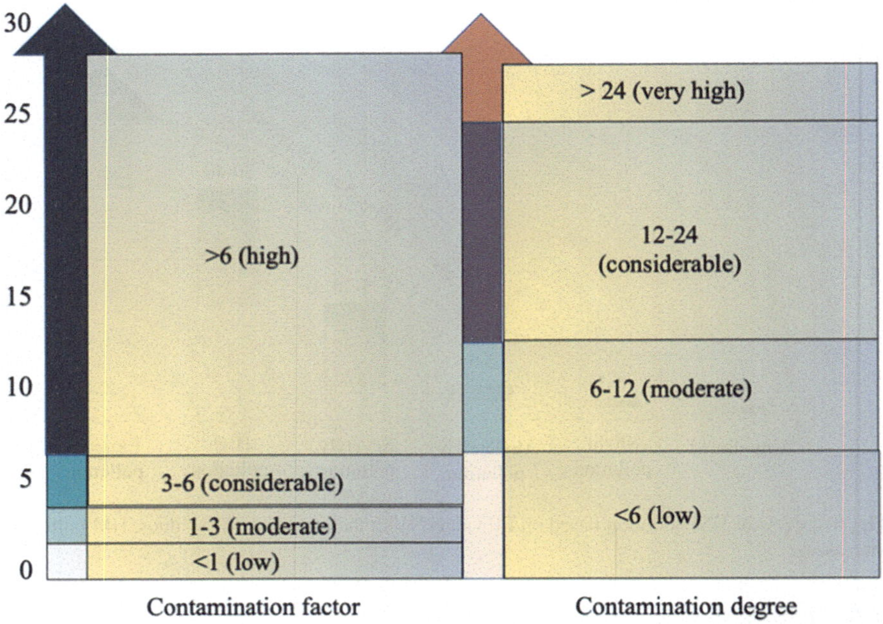

Fig. 7 Interpretations of C_f (left) and C_d (right) values into HM contamination groups

Table 3 Three classes of C_{di} (from lowest to highest)

Classes	Contamination index values
Low	<1
Medium	≥1–<3
High	>3

contamination sites based on hierarchical cluster analysis (HCA). Through these tools, it is easy to identify the contaminated sites based on the same possible pollution sources. PCA applies weights on the water quality data whereas HCA classifies the monitoring sites into clusters of similar characteristics. PCA aids in data reduction. Heavy metal index (HMI) is computed using the weight assigned to each water quality parameters. HMI is calculated using Equation no. (12) in Table 5.

3.11 *Nemerow Pollution Index*

Nemerow Pollution Index (NI) indicates the pollution levels of several HM in a study site. This index includes the mean and maximum values of single factor pollution index and identifies the HM with high contamination. NI has three five classes of HM pollution (Table 4). NI is calculated using Eq. (13) in Table 5.

Table 4 HM pollution levels based on NI values

Class	Pollution degree	NI
0	No pollution	≤ 0.5
1	Clean	0.5–0.7
2	Warm	0.7–1.0
3	Polluted	1.0–2.0
4	Medium pollution	2.0–3.0
5	Severe pollution	>3.0

4 Heavy Metal Transfer Within Soil-Plant Continuum

4.1 Accumulation Factor or Bio-accumulation Factor

Accumulation factor (AF), also known as a bioaccumulation factor (BAF) is defined as the ability of any plant grown in a HM contaminated site to take up the HM its tissues. AF is calculated as the ratio of HM concentration in plant roots to the HM concentration in the soil. The AF value of Cr in rice was 0.04 (Han et al. 2020). The AF of Cr is generally affected by soil chemical properties. High soil pH affects speciation of Cr. In normal soil pH, Cr is found as Cr (VI) form. Low soil pH favours the accumulation of stable Cr (III) with low mobility and toxicity in soils. However, a higher soil pH, especially alkaline soils, leads to the formation of Cr (IV). So, higher soil pH in Cr-contaminated site can increase the AF by enhancing the mobility of the element in the soil and easy uptake through plant roots. However, plants can suffer from injury or root damage if more total Cr content is more than 75–100 ppm. The AF value of Cd in rice was 3.158. AF of Cd is affected by soil pH as Cd absorption by plants is negatively correlated with soil pH value. Cd mostly exists in water-soluble form. Cd competes with H^+ ion in the plant root surface. In higher soil pH, plant roots release positive ion-binding sites that facilitate Cd binding and absorption (Zhang et al. 2007). The redox value (Eh) is also the main contributor to Cd availability to plant roots. Eh is positively correlated with the water-soluble Cd content in the soil and Cd absorbed by rice roots. The AF values of Pb were also lower than AF of other heavy metals in rice, owing to lower mobility of Pb in soil and lesser exchangeable forms of Pb in the soil. Pb is absorbed by plant roots through heteroplastic and symplastic pathways. The heteroplastic pathway involves absorption of dissolved Pb from the soil solution by the plant roots while the symplastic method encompasses Pb absorption by plant roots through low-affinity cation transporters, calcium channels, calmodulin etc. Cu seems to have higher AF values than that of other HMs. AF is calculated using Equation no. (14) in Table 5.

4.2 Translocation Factor

Unlike AF, the translocation factor (TF) determines the plant ability to translocate the HM within the plants. To calculate TF, HMs must enter the plant roots. TF is

Table 5 Ecological monitoring indices to evaluate HM contamination in soil

Sl. No	Ecological indices	Equations	Depictions	References
1	Heavy metal pollution index (HPI)	$HPI = \frac{\sum_{i=1}^{n} W_i Q_i}{\sum_{i=1}^{n} W_i}$	Q_i = sub-index of the ith parameter W_i = unit weight of the ith parameter n = number of parameters considered	Mohan et al. (1996)
2	Sub-index (Q_i)	$Q_i = \sum_{i=1}^{n} \frac{(M_i - I_i)}{(S_i - I_i)} \times 100$	M_i = HM concentration in soil I_i = ideal HM concentration S_i = standard values of the ith parameter	Rahman et al. (2020)
3	Geo-accumulation index (I_{geo})	$I_{geo} = Log_2(C_n/1.5B_n)$	C_n = soil HM concentration B_n = background concentration of respective HM	Rahman et al. (2012)
4	Integrated pollution index (IPI)	$IPI_i = (IP_1 + IP_2 + IP_3 + \ldots + IP_n)/n$	PI = pollution indices for n^{th} types of HMs	Wei et al. (2009)
5	Potential ecological Risk Index (PERI)	$PERI = \sum_i^m E_r = \sum_i^m T_r \times C_f = \sum_i^m T_r \times C_s/C_{ref}$	C_s = soil HM concentration C_f = contamination factor C_{ref} = background concentration of HM E_r = environmental risk factor T_r = metal toxic factor	Brady et al. (2015)
6	Heavy metal evaluation index (HEI)	$HEI = \sum_{i=1}^{n} \frac{H_c}{H_{mac}}$	H_c = monitored value H_{mac} = maximum permissible level (MAC) of the ith parameter	Edet and Offiong (2002)
7	Ecological risk factor (Er_i)	$Eri = Tr_i \times Cf_i$	Tr_i = toxic-response factor for HM Cf_i = contamination factor	Tesleem et al. (2018)
8	Enrichment factor (Ef)	$Ef = (C_i/C_{ref})/(B_i/B_{ref})$	C_i/C_{ref} = ratio of concentration between HM and a reference element in the soil B_i/B_{ref} = ratio of concentration between HM and a reference element in the background sample	Qingjie et al. (2008)
9	Contamination Factor (C_f)	$C_f = C_s/C_{ref}$	C_s = soil HM concentration C_{ref} = background concentration	Brady et al. (2015)

10	Contamination degree (C_d)	$C_d = \sum_{i=1}^{n} Cf_i$	C_f = contamination factor	Hakanson (1980)
11	Contamination index (C_{di})	$C_{di} = \sum_{i=1}^{n} C_{fi}$ $C_{fi} = \frac{C_{Ai}}{C_{Ni}} - 1$	C_{fi} = contamination factor, C_{Ai} = analytical value C_{Ni} = upper permissible concentration of the ith component N = 'normative value' C_{Ni} = MAC	Backman et al. (1997)
12	Heavy metal index (HMI)	$HMI = \sum_{i=1}^{n} \left[P_i \times \frac{M_i}{S_i}\right] \times 100$	P_i = weight given to each ith parameter M_i = actual values of HM S_i = permissible limit values	Dash et al. (2019)
13	Nemerow Pollution Index (NI)	$NI = \left(\frac{[1/n \sum (C_i/S_i)]^2 + [\max (C_i/S_i)]^2}{2}\right)^{1/2}$	n = number of indices C_i = measured content of HM S_i = standard value	Li et al. (2001)
14	Accumulation Factor (AF)	$AF = \frac{Metal_{root}}{Metal_{soil}}$	$Metal_{root}$ = root HM concentration (dry mass) $Metal_{soil}$ = soil HM concentration (dry soil)	Moradi et al. (2016)
15	Translocation factor (TF)	$TF = \frac{Metal_{shoots}}{Metal_{root}}$	$Metal_{shoots}$ = shoot HM concentration (dry mass) $Metal_{root}$ = root HM concentration (dry mass)	Moradi et al. (2016)
16	Pollution index (PI)	$PI = C_s/B_n$	C_S = soil HM concentration B_n = background value	Hakanson (1980)
17	Modified pollution index (MPI)	$MPI = \sqrt{\frac{(Ef_{avg})^2 + (Ef_{max})^2}{2}}$	Ef_{avg} = average of enrichment factors Ef_{max} = maximum enrichment factor	Brady et al. (2015)
18	Single-Factor Pollution Index (C_f^i)	$C_f^i = \frac{C_D^i}{C_R^i}$	C_D^i = measured soil HM content C_R^i = background content of a HM	Wang et al. (2011)
19	Compound pollution index (CPI)	$CPI = \sum_{i=1}^{m} \frac{C_f^i}{m}$	C_f^i = single-factor index of a HM m = number of HM types	Zhong et al. (2015)

calculated as the ratio of HM concentration in plant shoot to HM concentration in plant root (Singh et al. 2010). Han et al. (2020) determined Cr concentration in rice plant, being accumulated highest in root > leaves> grain> stem> husk. The TF values of Cd and Cr in rice were 0.1247 and 0.2940 respectively (Han et al. 2020). In the case of Cd, Han et al. (2020) showed the highest accumulation in root followed by leaves and stem. The highest Pb accumulation in rice was found to be in root followed by stem and leaves. After the absorption of Pb through rice roots, it is transported to different parts of the plants through two main pathways: transfer from xylem parenchyma to vessels and movement in vessels. Most of the total Pb (>90%) absorbed by plant roots stays in the root tissues and a meagre amount of this total absorbed Pb gets transferred to above-ground plant parts through symplastic methods. TF values on the type of plant species or type grown in HM contaminated sites. For example, Chen et al. (2016) showed higher translocation of HMs in the soil-rice system than soil-wheat. The plant type also differs in its choice of HM uptake. Wheat could transfer more Zn, Cu and Cd from roots to grain while canola limited the Cu and Cd uptake. TF is calculated using Equation no. (15) in Table 5.

Few other ecological indices such as pollution index, modified pollution index, single factor pollution index and compound index can be used for further understanding of HM pollution in soil (Equation nos. 16-19 respectively in Table 5).

5 Human Health Risk Assessment Due to Heavy Metal Soil Contamination

5.1 Average Daily Dose

The risk assessment methodology from US EPA indicates that as there are three ways of human exposure to pollutants: ingestion, inhalation and dermal contact, three types of ADD must be calculated. Siriwong (2006) calculated the magnitude, duration and frequency of human exposure to HM contamination by using average daily dose (ADD). ADD accounts for both the non-carcinogenic as well as carcinogenic risks of human exposure to HM pollution. Based on it, the toxicity responses were identified for each HMs (USEPA IRIS 2011; Wongsasuluk et al. 2014). ADD is calculated using Equation nos. (1–3) given in Table 6.

5.2 Estimated Daily Intake of HM

Although HMs are found in trace amount in edible parts of plants frequent intake of the harvested portion or by-products may increase the HM concentration in the human body. Estimated daily intake (EDI) of HMs, therefore, depends on the concentration of HM in plant parts and amount of consumption. The unit of EDI is

Table 6 Risk assessment indices to calculate possible effects of HM contamination on human health

Sl. No	Risk assessment indices	Equations	Depictions	References
1	Average daily dose (ADD) through ingestion	$ADD_{ing} = \frac{(C \times R_{ing} \times CF \times ED)}{(BW \times AT)}$	C = concentration of an element in soil (mg/kg) R_{ing} = ingestion rate of soil (mg/day) EF = exposure frequency (days/year) ED = exposure duration (years) BW = average body weight (kg) AT = average time (days) CF = conversion factor (kg/mg) R_{inh} = inhalation rate (m^3/day) PEF = particle emission factor (m^3/kg) SA = the surface area of the skin that contacts the dust (cm^2) SL = skin adherence factor for the dust mg/cm^2 ABS = dermal absorption factor (chemical specific)	Siriwong (2006)
2	Average daily dose (ADD) through inhalation	$ADD_{inh} = \frac{(C \times R_{inh} \times EF \times ED)}{(PEF \times BW \times AT)}$		
3	Average daily dose (ADD) through dermal contact	$ADD_{derm} = \frac{(C \times SA \times CF \times SL \times ABS \times ED)}{(BW \times AT)}$		
4	Estimated daily intake of heavy metals (EDI)	$EDI = \frac{C_{heavy\ metal} \times W_{food}}{BW}$	$C_{heavy\ metal}$ (μ g g^{-1} dry weight) = concentration of heavy metals in edible portion of plants; W_{food} = daily average consumption of crops in this region, and BW = average adult body weight (60 kg)	Zhuang et al. (2009)
5	Chronic daily intake (CDI)	CDI = (C × DI)/(BW)	C = concentration of heavy metal in drinking water (mg L^{-1}) DI = average daily intake rate (2.0 L/day/person) and BW = body weight (72 kg)	USEPA (1992); Chrowtoski (1994)

(continued)

Table 6 (continued)

Sl. No	Risk assessment indices	Equations	Depictions	References
6	Hazard quotient (HQ)	$HQ = CDI/RfD$	CDI = chronic daily intake RfD = oral reference dose (mg kg^{-1}/ day) for each HM	USEPA (1999)
7	Target hazard quotient (THQ)	$THQ = \frac{EF \times ED \times FI \times MC}{RfD_o \times BW \times AT} \times 10^{-3}$	EF = exposure frequency (365 days year⁻1); ED = exposure duration (70 years for adults); FI = food ingestion rate (g $person^{-1}$ day^{-1}); MC is heavy metal concentration in food (μg g^{-1}); RfD_o = oral reference dose (mg kg^{-1} day^{-1}); BW = average body weight (60 kg for adult); AT = average time for non carcinogens (365 days $year^{-1}$× number of exposure years)	Chien et al. (2002)
8	Hazard index (HI)	$HI = \sum HQ$	HQ = hazard quotients	USEPA IRIS (2011)
9	Lifetime average daily dose (LADD) through inhalation	$LAAD_{inhala} = \frac{C \times EF}{AT \times PEF} \times \left(\frac{Ing_{child} \times ED_{child}}{BW_{child}} \times \frac{Ing_{adult} \times ED_{adult}}{BW_{adult}}\right)$	C = concentration of heavy metal in soil (mg/kg); Ing = ingestion rate of soil (mg/d); CF = conversion factor (m^3/kg); EF = exposure frequency (d/a); ED = exposure duration (a); AT = time period over which the dose is averaged (d), and it is derived by using pathway-specific period of exposure for non-carcinogenic effects (ED × 365 days/ year) and 74.8-year life-time for carcinogenic effects (74.8 years × 365 days/year)	Roy et al. (2019)
10	Lifetime average daily dose (LADD) through dermal contact	$LAAD_{dermal} = \frac{C \times CF \times EF \times SL \times ABS}{AT} \times \left(\frac{SA_{child} \times ED_{child}}{BW_{child}} \times \frac{SA_{adult} \times ED_{adult}}{BW_{adult}}\right)$		
11	Lifetime average daily dose (LADD) through ingestion	$LAAD_{ingest} = \frac{C \times CF \times EF}{AT} \times \left(\frac{Ing_{child} \times ED_{child}}{BW_{child}} \times \frac{Ing_{adult} \times ED_{adult}}{BW_{adult}}\right)$		

			averaging time; BW = average body weight (kg); Inh = inhalation rate (m^3/d); PEF = particle emission factor (m^3/kg); SA = skin surface area (cm^2); SL = soil adsorption coefficient to skin (mg/(cm^2·d)); and ABS = dermal absorption factor.	
12	Carcinogenic risk index (CR)	$CR = ADD \times SF$	ADD = average daily intake (mg/(kg-day)); SF (dimension-less) = carcinogenic slope factor (per mg/kg-day)	Cui et al. (2018)

presented as ug kg^{-1} body weight day^{-1} (Zhuang et al. 2009). EDI is calculated using Equation no. (4) in Table 6.

5.3 *Chronic Daily Intake*

De Miguel et al. (2007) calculated CDI and found the highest CDI values in Fe followed by Zn, Pb, Mn and Cu. The general threshold level of CDI is 1×10^{-6} (USEPA IRIS 2011). Higher CDI values indicate an increase in HM concentration in the sample. A drinking water sample with higher CDI may indicate contamination due to run-off from agricultural fields that affects water quality. CDI is calculated using Equation no. (5) in Table 6.

5.4 *Non-Carcinogenic Risk Assessment*

5.4.1 Hazard Quotient

After ADD is calculated from three possible pathways, a hazard quotient (HQ) is calculated that highlights the non-carcinogenic effects of human exposure to HM pollution and is calculated by using ADD and oral reference dose (RfD). The RfD relative to ADD value will give an idea of adverse effects on human health. If ADD is lower than RfD, there would not be any such adverse effects of HM contamination on human health and vice-versa (US EPA 1993). If HQ <1, no adverse effects on human health is predicted whereas HQ > 1 shows high risk on human health (US EPA 1986) (Fig. 8). HQ is calculated using Equation no. (6) in Table 6.

5.4.2 Target Hazard Quotient

Chien et al. (2002) calculated target hazard quotient (THQ). If THQ <1, it is interpreted as no risk to human health and if THQ > 1, some degree of human health risk exists (Fig. 8). THQ is calculated using Equation no. (7) in Table 6.

5.4.3 Hazard Index

All the individual HQs of every heavy metal are summed up to calculate the hazard index (HI). The ratio of HI to HQ can interpret the risk levels of HM exposure on human health (Fig. 8). If the ratio > 1, it refers to an unacceptable risk of non-carcinogenic effects whereas ratio < 1 depicts an acceptable risk level (US EPA 2001). HI is calculated using Equation no. (8) in Table 6.

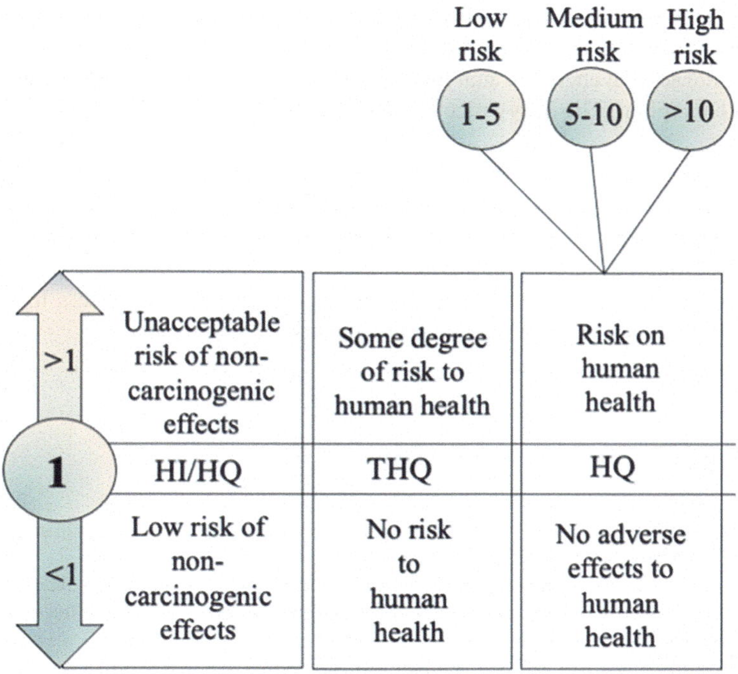

Fig. 8 Risk assessment indices with 1 as the central point of interpretation

5.5 Carcinogenic Risk Assessment

The carcinogenic risk assessment involves the lifetime average daily dose (LADD) to evaluate the carcinogenic risk of HM contamination. Like ADD, LADD is calculated through three pathways: ingestion, inhalation and dermal contact (Equation nos. 9–11 in Table 6). The carcinogenic risk (CR) is calculated using Equation no. 12 in Table 6 (Cui et al. 2018). The threshold value of CR is 1×10^{-6}. The safe point lies between 10^{-4} to 10^{-6} and exceeding 10^{-4} shows high carcinogenic risks and needs immediate intervention.

6 Conclusions

The chapter discusses all the heavy metal-related assessment indices used for ecological monitoring and human health risk evaluation. It can be summarised as given in following points:

- Heavy metal contamination in soil system and plant tissues due to continuous sewage irrigation has been an old practice being followed in several countries.

- Through food chain, these heavy metals get accumulated in human body, leading to severe diseases and organ failure.
- There are several chemical analytical techniques to assess the heavy metal content in soil and plant. The heavy metal concentration data are used to calculate several indices.
- Such indices can express the heavy metal contamination level of any sites, thus letting us to compare the pollution levels within a site or between sites.
- Risk assessment addresses the risk levels of heavy metal pollution to human health through different pathways.
- Understanding of all the indices entices readers and policy makers to have an overall idea on soil pollution and its associated human health risks.

References

Backman B, Bodis D, Lahermo P et al (1997) Application of a groundwater contamination index in Finland and Slovakia. Environ Geol 36:55–64

Brady JP, Ayoko GA, Martens WN et al (2015) Development of a hybrid pollution index for heavy metals in marine and estuarine sediments. Environ Monit Assess 187(5):306. https://doi.org/10.1007/s10661-015-4563-x

Chabukdhara M, Gupta SK, Kotecha Y et al (2017) Groundwater quality in Ghaziabad district, Uttar Pradesh, India: multivariate and health risk assessment. Chemosphere 179:167–178. https://doi.org/10.1016/j.chemosphere.2017.03.086

Chen TB, Zheng YM, Lei M et al (2005) Assessment of heavy metal pollution in surface soils of urban parks in Beijing, China. Chemosphere 60:542–551

Chen H, Yuan X, Li T et al (2016) Characteristics of heavy metal transfer and their influencing factors in different soil-crop systems of the industrialization region, China. Ecotoxicol Environ Saf 126:193–201

Chien LC, Hung TC, Choang KY et al (2002) Daily intake of TBT, Cu, Zn, Cd and As for fishermen in Taiwan. Sci Total Environ 285:177–185

Chrowtoski PC (1994) Exposure assessment principles. In: Patrick DR (ed) Toxic air pollution handbook. Van Nostrand Reinhold, New York, p 154

Cui ZW, Wang Y, Zhao N et al (2018) Spatial distribution and risk assessment of heavy metals in paddy soils of Yongshuyu irrigation area from Songhua river basin, Northeast China. Chin Geogr Sci 28(5):797–809. https://doi.org/10.1007/s11769-018-0991-1

Dash S, Borah SS, Kalamdhad A (2019) A modified indexing approach for assessment of heavy metal contamination in Deepor Beel, India. Ecol Indic 106:105444

De Miguel E, Iribarren I, Chacón E et al (2007) Risk-based evaluation of the exposure of children to trace elements in playgrounds in Madrid (Spain). Chemosphere 66:505–513

Edet AE, Offiong OE (2002) Evaluation of water quality pollution indices for heavy metal contamination monitoring. A study case from Akpabuyo-Odukpani area, Lower Cross River Basin (southeastern Nigeria). GeoJournal 57:295–304. https://doi.org/10.1023/B:GEJO.0000007250.92458.de

Gupta SK, Ansari FA, Nasr M et al (2018) Multivariate analysis and health risk assessment of heavy metal contents in foodstuffs of Durban, South Africa. Environ Monit Assess 190(3):151. https://doi.org/10.1007/s10661-018-6546-1

Hakanson L (1980) An ecological risk index for aquatic pollution control. A sedimentological approach. Water Res 14(8):975–1001. https://doi.org/10.1016/0043-1354(80)90143-8

Han C, Xie W, Chen C et al (2020) Health risk assessment of heavy metals in soils before rice sowing and at harvesting in southern Jiangsu Province, China. J Chem 2020:7391934, 1–13. https://doi.org/10.1155/2020/7391934

Hu Y, Liu X, Bai J et al (2013) Assessing heavy metal pollution in the surface soils of a region that had undergone three decades of intense industrialization and urbanization. Environ Sci Pollut Res 20(9):6150–6159. https://doi.org/10.1007/s11356-013-1668-z

Kamani H, Mahvi AH, Seyedsalehi M et al (2017) Contamination and ecological risk assessment of heavy metals in street dust of Tehran, Iran. Int J Environ Sci Technol 14(12):2675–2682. https://doi.org/10.1007/s13762-017-1327-x

Khan K, Lu Y, Khan H et al (2013) Heavy metals in agricultural soils and crops and their health risks in Swat District, northern Pakistan. Food Chem Toxicol 58:449–458. https://doi.org/10.1016/j.fct.2013.05.014

Li JJ, Feng MH, Yu L (2001) Assessment on the situation of water quality in Liaodong Bay shallow waters. Mar Environ Sci 3:42–45

Maanan M, Saddik M, Maanan M et al (2015) Environmental and ecological risk assessment of heavy metals in sediments of Nador lagoon Morocco. Ecol Indic 48:616–626

Mohammadi F, Samaei MR, Azhdarpoor A et al (2019) Modelling and optimizing pyrene removal from the soil by phytoremediation using response surface methodology, artificial neural networks, and genetic algorithm. Chemosphere 237:124486. https://doi.org/10.1016/j.chemosphere.2019.124486

Mohan SV, Nithila P, Reddy SJ (1996) Estimation of heavy metals in drinking water and development of heavy metal pollution index. J Environ Sci Health 31(2):283–289. https://doi.org/10.1080/10934529609376357

Moradi A, Honarjoo N, Najafi P et al (2016) A human health risk assessment of soil and crops contaminated by heavy metals in industrial regions, central Iran. Hum Ecol Risk Assess 22(1):153–167. https://doi.org/10.1080/10807039.2015.1056293

Nagajyoti PC, Lee KD, Sreekanth TVM (2010) Heavy metals, occurrence and toxicity for plants: a review. Environ Chem Lett 8(3):199–216. https://doi.org/10.1007/s10311-010-0297-8

Prasanna MV, Praveena SM, Chidambaram S et al (2012) Evaluation of water quality pollution indices for heavy metal contamination monitoring: a case study from Curtin Lake, Miri City, East Malaysia. Environ Earth Sci 67:1987–2001. https://doi.org/10.1007/s12665-012-1639-6

Qingjie G, Jun D, Yunchuan X et al (2008) Calculating pollution indices by heavy metals in ecological geochemistry assessment and a case study in parks of Beijing. J China Univ Geosci 19(3):230–241. https://doi.org/10.1016/S1002-0705(08)60042-4

Rafiee A, Delgado-Saborit JM, Sly PD et al (2020) Environmental chronic exposure to metals and effects on attention and executive function in the general population. Sci Total Environ 705:135911. https://doi.org/10.1016/j.scitotenv.2019.135911

Rahman SH, Khanam D, Adyel M et al (2012) Assessment of heavy metal contamination of agricultural soil around Dhaka Export Processing Zone (DEPZ), Bangladesh: implication of seasonal variation and indices. Appl Sci 2:584–601. https://doi.org/10.3390/app2030584

Rahman MATMT, Paul M, Bhoumik N et al (2020) Heavy metal pollution assessment in the groundwater of the Meghna Ghat industrial area, Bangladesh, by using water pollution indices approach. Appl Water Sci 10:186. https://doi.org/10.1007/s13201-020-01266-4

Rostami S, Kamani H, Shahsavani S et al (2020) Environmental monitoring and ecological risk assessment of heavy metals in farmland soils. Hum Ecol Risk Assess 1–14. https://doi.org/10.1080/10807039.2020.1719030

Roy S, Gupta SK, Prakash J et al (2019) Ecological and human health risk assessment of heavy metal contamination in road dust in the National Capital Territory (NCT) of Delhi, India Environ Sci Pollut Res 1–14. https://doi.org/10.1007/s11356-019-06216-5

Singh R, Singh DP, Kumar N et al (2010) Accumulation and translocation of heavy metals in soil and plants from fly ash contaminated area. J Environ Biol 31(4):421–430

Siriwong W (2006) Organophosphate pesticide residues in aquatic ecosystem and health risk assessment of local agriculture community. Dissertation, Chulalongkorn University, Thailand

Tesleem OK, Akinade SO, Mustapha TJ et al (2018) Heavy metal contamination and ecological risk assessment in soils and sediments of an industrial area in southwestern Nigeria. J Health Pollut 8 (19):180906

USEPA (1986) Superfund public health evaluation manual, 540, Washington, DC, p 1–86

USEPA (1992) Guidelines for exposure assessment, EPA/600/Z-92/001. Risk Assessment Forum, Washington, DC

USEPA (1999) Guidance for performing aggregate exposure and risk assessment office of Pesticide Programs, Washington, DC

USEPA (2001) Baseline human health risk assessment. Vasquez Boulevard and I-70 superfund site Denver, Denver (Co)

USEPA (United States Environmental Protection Agency) (1993) Reference dose (RfD): description and use in health risk assessments, Washington, DC, USA

USEPA IRIS (2011) US Environmental Protection Agency's Integrated Risk Information System. Environmental Protection Agency Region I, Washington DC. http://www.epa.gov/iris/

Wang Y, Yang Z, Shen Z et al (2011) Assessment of heavy metals in sediments from a typical catchment of the Yangtze River, China. Environ Monit Assess 172:407–417

Wei B, Yang L (2010) A review of heavy metal contaminations in urban soils, urban road dusts and agricultural soils from China. Microchem J 94(2):99–107. https://doi.org/10.1016/j.microc.2009.09.014

Wei B, Jiang F, Li X et al (2009) Spatial distribution and contamination assessment of heavy metals in urban road dusts from Urumqi, NW China. Microchem J 93:147–152

Wongsasuluk P, Chotpantarat S, Siriwong W et al (2014) Heavy metal contamination and human health risk assessment in drinking water from shallow groundwater wells in an agricultural area in Ubon Ratchathani province, Thailand. Environ Geochem Health 36:169–182

Zhang CH, Yan-Ling HU, Ting JU et al (2007) Effect of solution pH on Cd absorption of rice seedlings. Guangxi Agric Sci 38(4):391–394

Zhong S, Geng H, Zhang F et al (2015) Risk assessment and prediction of heavy metal pollution in groundwater and river sediment: a case study of a typical agricultural irrigation area in Northeast China. Int J Anal Chem 2015(921539):1–11. https://doi.org/10.1155/2015/921539

Zhou L, Yang B, Xue N et al (2014) Ecological risks and potential sources of heavy metals in agricultural soils from Huanghuai Plain, China. Environ Sci Pollut Res 21(2):1360–1369. https://doi.org/10.1007/s11356-013-2023-0

Zhuang P, Zou B, Li NY et al (2009) Heavy metal contamination in soils and food crops around Dabaoshan mine in Guangdong, China: implication for human health. Environ Geochem Health 31:707–715

Bioleaching Approach for Enhancing Sewage Sludge Dewaterability

Subhashisa Praharaj, Sagar Maitra, Akbar Hossain, Lalichetti Sagar, Ajar Nath Yadav, Usha Das, Tanmoy Shankar, Biswajit Pramanick, and Dinkar Gaikwad

1 Introduction

A huge volume of sewage and sludge are produced as a result of the biological wastewater treatment process. Sewage sludge is the final solid byproduct of the wastewater treatment course. Rapid urbanization, rising population, systematic wastewater disposal system has led to production of a large amount of sewage and sludge (Ghavidel et al. 2017; Pathak et al. 2009; Wu et al. 2020). The amount of sewage sludge production is also likely to upsurge with the developments of high performance biological and chemical wastewater development processes (Kwarciak-Kozłowska 2019). As a huge amount of sewage and sludges are produced around the globe (Table 1), hence its sustainable and ecologically safe management is very crucial. The two commonly followed disposal strategies for municipal sewage sludge management include reuse and final disposal (Grobelak et al. 2019). The inconsistency in composition, contamination by pathogens and micropollutants, and

S. Praharaj · S. Maitra (✉) · L. Sagar · T. Shankar · D. Gaikwad
M.S. Swaminathan School of Agriculture, Centurion University of Technology and Management, R.Sitapur, Odisha, India
e-mail: sagar.maitra@cutm.ac.in

A. Hossain
Bangladesh Wheat and Maize Research Institute, Dinajpur, Bangladesh

A. N. Yadav
Department of Biotechnology, Dr. Khem Singh Gill Akal College of Agriculture, Eternal University, Baru Sahib, Sirmour, Himachal Pradesh, India

U. Das
G.B. Pant University of Agriculture and Technology, Pantnagar, Uttarakhand, India

B. Pramanick
Dr. Rajendra Prasad Central Agricultural University, Pusa, Samastipur, Bihar, India

V. D. Rajput et al. (eds.), *Sustainable Management and Utilization of Sewage Sludge*, https://doi.org/10.1007/978-3-030-85226-9_3

Table 1 Sewage sludge product in different countries

Country	Sewage sludge (dry) (Thousand metric ton)
USA	6514
China	2966
Germany	2000
Japan	2000
Netherlands	1500
UK	1500
Italy	1000

Source: LeBlanc et al. (2008)

presence of high organics and water make it very difficult to manage this waste (Westerhoff et al. 2015; Wiśniowska 2019).

The land application is well thought-out as a suitable option of sludge disposal as it adds organic matter to soil and helps in nutrient recycling (Gu and Wong 2004; Henry and Frasad 2006; Dabrowska and Rosinska 2012). Moreover, it can also be used for soil improvement as it can affect physicochemical and biological soil properties (Ghavidel et al. 2017). Sewage sludge may be beneficially used as a fertilizer after proper processing and decontamination. Further, it creates an opening for the advantageous recycling of organic matter as well as nutrients, principally, nitrogen and phosphorus. The sewage sludge also contains lots of micronutrient like iron, zinc, manganese, copper etc. which are essential for growth and productivity of plants. Their presence in appropriate amount will not be harmful to plants (Ignatowicz 2017). However, contamination by poisonous metalsinclusive of Pb, Cu, Cd, Ni, Cr and Zn is a restraining factor for use of these biosolids to the land (Ban et al. 2018; Islam et al. 2015; Saha et al. 2017; Xu et al. 2017; Zou et al. 2013). Sewage sludges are usually contaminated with heavy metals; thus, making it unfit for direct application of arable field. Application of polluted sewage sludge with a higher concentration of heavy metals may lead to heavy metal uptake by crop plant and its subsequent entry into the human and animal food chain (Gu and Wong 2004). When plant products are consumed which are contaminated with heavy metals it causes many chronic diseases and metabolic disorder in human being (Ozoreshampton et al. 2005). Considering this agricultural application of sewage sludge should be carefully monitored to avoid any buildup of heavy metals and pathogen in the agricultural field. Strict regulations regarding the agricultural application of sludge only if they meet some quality standard can be very useful in this regard. It will ensure both beneficial applications of sewage sludge applications while ensuring no negative impact of the same.

Sewage from different sources like industrial wastewater, household and urban runoffs add heavy metals to the sewage system. There is a noteworthy requirement for state-of-the-art sludge treatment methodsfor removal ofpoisonous metals and thus improvement in sludge digestibility to exploit the advantageous reuse potential and minimize expensesincurred for transportation and disposal (Du 2015). The solid phase of the sludge may contain tightly bound heavy metals. To make the sewage sludge meet the standard of agricultural application, heavy metal removal is very important. Leaching of heavy metals have been studied employing chemical and biological methods. The chemical methods targets leaching of heavy metals. In the process of leaching, inorganic and organic acids are used (Jenkins et al. 1981; Marchioretto et al. 2002; Veeken and Hamelers 1999). Though it could help in dissolving some of the heavy metals, its efficacy varies based on the metals to be dissolved and characteristics of the sludge (Ghavidel et al. 2017). Bioleaching of sewage sludge has been found to be an environmentally safe and efficient method. Bioleaching helps in transferring heavy metals from their solid state into liquid. Metals can be recovered from the liquid phase and treated sludge can be used for farming purpose (Marchenko et al. 2018).

Bioleaching causes improvement of the sludge dewaterability (Ban et al. 2018; Liu et al. 2016; Song and Zhou 2008) However, the mechanism behind this is still completely not completely understood. As sludge contains high amount of water, hence dewatering is very vital to decrease the sludge bulk, easier transport and ease in their further treatment or disposal (Wong and Gu 2004; Wu et al. 2020). As bioleaching serves as an ecologically safe alternative to other processes that involves use of harmful chemicals, hence its use can be promoted in the wastewater disposal systems. The bio-oxidation of energy substance leads to bio-acidification which helps in the sludge-borne metal elimination in the bioleaching process. It also helps improving the dewaterability of sludge at a suitable pH level (Liu et al. 2012b).

Improper handling of sewage sludge or any other biowaste has serious negative environmental implications. In addition to environmental pollution, it may also affect human, plant and animal health (Ban et al. 2018). Many countries across the globe have developed their guidelines and regulation regarding handling, managing and disposing of the bio-waste (Bastian 1997). Though many technologies have been developed for handling huge volume of sewage sludge still; it requires continuous refinement and improvement. Finding a technology which is economically cheap and viable, causes minimum or no damage to the environment and allows the best possible reuse or recycling of sewage sludge will be very useful. The technology should also meet the prescribed regulation of the respective country regarding sewage sludge disposal. Many technologies have been found suitable in lab scale. Their performance in industrial-scale must be tested for handling huge amounts of sewage sludge.

Multiple factors have affected the bio-solid disposal decision making. Increase in the volume of biosolids generated, availability of advanced and more efficient equipment, as well as management practice, industrial pretreatment improving the quality of biosolids, advances in biosolid research and development, rise in the cost of land, labor and energy in urban areas, changing regulatory policy and growing concern on ecosystem health among general public, are some of the key factors that have influenced the processing and disposal strategy of biosolids in the past and are also expected to influence the decision making in future (Bastian 1997).

As understanding the sewage sludge disposal and management improves through advances in research and development; their large-scale adaptation can help to address the negative environmental aspects of sewage sludge management and disposal. Even in the intermediary stages of sewage sludge management such as

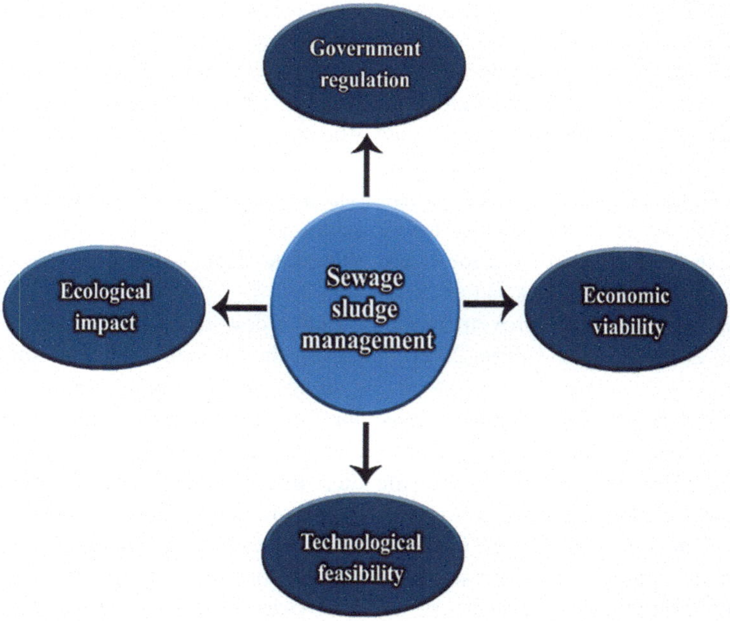

Fig. 1 Factors affecting sewage sludge management decisions

sludge conditioning and dewatering, more eco-friendly techniques can be followed to reduce the overall negative impact on the environment and improve efficiency. Sludge bioleaching, a technique that has been highly successful in removing heavy metals (Pathak et al. 2009; Camargo et al. 2016; Ghavidel et al. 2017; Mehrotra and Sreekrishnan 2017; Sreekrishnan and Tyagi 1994), has been also found a way for improvement of the dewaterability of sewage sludge (Pathak et al. 2009; Liu et al. 2012a). Moreover, the process is ecologically viable and economically cheaper, but, the mechanism of bioleaching for improvement sludge dewaterability is still unclear. In the chapter, the process of bioleaching, its process and mechanism and role in improving the dewaterability of sludge have been discussed, based on available literature and researches. Few areas have also been outlined in which further research can be carried out on the aspect (Fig. 1).

2 Characteristics of Sewage Sludge

Sewage sludge is the by-product of the wastewater treatment process. Sludge consists of inorganic and organic compounds. Sewage sludge is containing different heavy metals such as (Bonfiglioli et al. 2014) and the heavy metal content varies from 0.5 to 2.0% on the dry weight basis with a higher limit in exceptional cases may even reach up to 6% (Pathak et al. 2009). Sludge contains heavy metals and makes it

Table 2 Physico-chemical properties of sewage sludge

Parameters	Value
pH	5–8
Moisture (%)	Up to 95
Total Solids (TS) (%)	0.83–12
Volatile solid (% of TS)	30–88
Nitrogen (% of TS)	1.5–6
Phosphorus (% of TS)	0.8–11
Potassium (% of TS)	0.4–3
Protein (% of TS)	15–41
Silica (% of TS)	10–20

Source: Pathak et al. (2009)

Table 3 Significance of different sludge parameters

Sludge parameter Groups	Significance
Physical parameters	Processability and Handleability can be known
Chemical parameters	Nutrient/metal concentration etc. Suitability for different use like agricultural application
Biological parameters	Information on microbial activities, Organic matter content etc. Presence of any specific pathogens Safety evaluation for agricultural application

unsuitable for the direct applications to the agricultural field. Sewage sludge with a high concentration of heavy metal, when used in the agricultural field, also has the risk of entering into the food chain of human and animal. Some physical and chemical properties of sewage sludge have been given in Tables 2 and 3.

The physicochemical properties of sewage sludge are very important for their treatment. Solids concentration is a vital variable that can dictate the amount of sludge to be treated. Organic content (volatile solids), nutrients content, pathogens (may cause soil-borne disease or soil sickness), metals (may enter the food chain), and toxic organics are few important sludge characteristics that decide the suitability of sludge for beneficial use, then availability of nutrient elements such as N, P, and K is an important parameter especially when the sludge is used as fertilizer or conditioner in croplands. Trace elements in sludge refer to elements which in small concentration can either be beneficial or detrimental for crops and animals (Al-Malack et al. 2008). Wastewater sludge or biosolid characterization is very essential as wide variation exists in their quality basedon the origin of the solids, type of processing or treatments it has gone through and extent of aging. Sludge can be conventionally characterized based on physical, chemical and biological parameters (Yan et al. 2009) (Table 3). It should be kept in mind that all parameters interact with each other and also affect them to a different extent. However; these conventional sludge characteristics can give some useful information especially about their suitability for agricultural application and thus hold great significance (Table 4).

Table 4 Concentration of heavy metals in sludge (mg/kg dry sludge solids)

Country	Zn	Cu	Ni	Cr	Cd	Pb	Reference
India	870–1510	280–543	192–293	102–8110	41–54	91–129	Pathak et al. (2009)
China	783.4–3096.3	131.2–394.5	49.3–95.5	45.8–78.4	5.9–13	57.5–109.3	Dai et al. (2007)

3 Dewatering of Sewage Sludge

Sewage sludge, an inadvertently producedderivative of wastewater treatment process andit may contain numerous harmful materials, heavy metals, organic pollutants and pathogens (Pathak et al. 2009; Lema and Suarez 2017; Wu et al. 2020). Municipal wastewater treatment may lead to the production of a large amount of sewage sludge material. Rising urbanization and organized waste disposal system from such urban areas has increased the sewage sludge production multifold. In fact, high-income countries having good infrastructure and treatment technologies produce a large mass of wastewater sludge per person. In middle-income countries having comparatively less developed disposal infrastructure and where wastewater is collected and treatedin less volume because ofless produtionof per capita wastewater sludge at the national level (LeBlanc et al. 2008). Disposal of such huge amount of sewage sludge in an environmentally sustainable as well as economically viable way is a big task that requires immediate attention. Improper sewage sludge disposal leads to pollution risks.

As sewage sludges are very bulky hence, dewatering provides multiple advantages such as minimizing sludge volume, facilitating transport, enhancing calorific value and minimizing the creation of leachate at landfill sites (Mahmoud et al. 2011; Wu et al. 2020). However, the hydrated colloidal structure of microbial aggregates makes it difficult to dewater the sludge. The interaction between water and sludge solid at varying strength depending on the way water is attached to sludge affects dewaterability to a great extent. The organic phase in sludge keeps water surroundedin a biopolymeric network and behaves in a different manner from the wholesale water (Heukelekian and Weisburg 1956).

In the activated sludge process, the secondary settling tank produces a huge quantity of waste activated sludge whose dewatering becomes very difficult (Neyens and Baeyens 2003). Currently the technique of dewatering that includes the addition of polymers followed by mechanical dewatering comprises of enough moisture that makes the sludge bulky and difficult to transport. Moreover, low water content of the sludge will make it easier for subsequent reutilization/recycling or disposal of sludge cake (Liu et al. 2012a). Sludge cake with higher dry solid content is desirable as it increases the energy proficiency of burning, reduces the obligation of a supplemental bulking agent during conversion into compost and diminishes the quantity of leachate in landfill site (Lo et al. 2001). For minimizing the cost of disposal, ease of transportation and better utilization, producing dewatered sludge with high dry solid percentage is of utmost importance. To achieve high dry solid percentage, sludge conditioning is required. Sludge conditioning helps to alter sludge structure and physical states of water in sludge; thus, allowing better dewatering of the sludge so that high dry solid percentage can be achieved.

Sewage sludge conditioning is very essential for improvement of the dewaterability of sludge before mechanical dewatering. The conditioning of sludge is traditionally done by using organics or inorganic chemicals (Liu et al. 2012a). Bioleaching is an efficient and economically viable alternative that uses the microbial method for removing sludge borne heavy metals and improving sludge dewaterability. Sludge conditioning alters the structure of sludge and physical conditions of water in the sludge. It changes the bound water in sludge to free water, thus improves the dewaterability of sludge (Liu et al. 2012a).

Many conditioning techniques were developed in the past to better the sludge dewaterability. Physical conditioning refers to technology involving non-chemical conditioning reagents or energy inputs (freeze-thawing, heating, sonication or electric field and so on) while, chemical conditioning refers to altering the physicochemical properties of sludge using chemicals (Wu et al. 2020) Physical and chemical approaches were commonly used such as ultrasonic, microwave, hydrothermal and chemical conditioning (Liu et al. 2012a). Use of microwave radiation disrupts the sludge microbial cells subsequently releasing the bound water in sludge. Microwave treatment also reduced the specific resistance to filtration (SRF) of conditioned sewage sludge (Wojciechowska 2005). Contrasting results were with respect to the effect of ultrasonic treatment on sludge dewatering. Na et al. (2007) reported improved dewaterability of ultrasonic treated waste activated sludge while Wang et al. (2006) showed reduced sludge dewaterability due to ultrasonic disintegration (Liu et al. 2012a). Hydrothermal conditioning of sludge reduces the water content, thus improving the dewaterability of sludge (Xun et al. 2009). In order to enhance sludge dewaterability, Chemical conditioning can be done which uses the addition of chemicals such as ferric chloride and calcium oxide to sludge (Chen and Wu 2009; Krishnamurthy and Viraraghavan 2005; Liu et al. 2012a).

The physical and chemical methods followed for conditioning usually alter the organic matter content and thermal value. Incorporation of a large number of inorganic substances severely lessens the organic matter or thermal value of dry sludge (Liu et al. 2012a). Considering this, anappropriate, cost-effective, environmentally sustainable methodfor improvement of dewaterability of sewage sludge is very essential. In this direction, bioleaching technology has been found as a suitable technology. Bioleaching technique involves microbes such as *A. thiooxidans* and *A. ferrooxidans* which successfully removes heavy metals and improves sludge dewaterability (Liu and Zhou 2009).

As mentioned above, it is imperative to state that the conditioning of sludge is very decisive for improving the dewaterability of sludge. Conditioning advances sludge dewaterability. But conditioning concurrently alters extracellular proteins or polysaccharides, releases nutrients like nitrogen and phosphorus and influences the chemical speciation of heavy metals. Hence, the study on the impact of sludge conditioning is important on consequent utilization and disposal processes (Wu et al. 2020).

4 Dewaterability of Sewage Sludge

Multiple factors such as rheological property, particle size distribution, micromorphology and porosity, surface charge, and EPS affects the dewaterability of sludges (Wu et al. 2020; Yan et al. 2009). The dewaterability is commonly indicated using indices like capillary suction time (CST) and Specific resistance of filtration (SRF). As, relative portions of various fractions of water and binding strength of solid-water are very crucial factors for dewaterability (Kopp and Dichtl 2001; Wu et al. 2020); hence, an index integrating both was proposed by Wu et al. (2020) to evaluate the sludge dewaterability.

Dewaterability can be subject to the different fractions of water in sewage sludge, which are characterized by their strength with which they are physicallly bonded with the solids (Kopp and Dichtl 2001). The water present in a sewage sludge suspension can be grouped into different categories, such as interstitial water, free water, surface water and intracellular water. Free water is not bound to particles, whereas, water is bound by the capillary forces in the sludge flocs in case of interstitial water. Surface water, on the other hand, is bound by the adhesion. Intracellular water contains the water in cells and water of hydration. The free water represents the largest fraction of sewage sludges. As the free water is not adsorbed by sludge particle, are free to move and not influenced by capillary forces; hence, this form of water can be separated by mechanical dewatering process (Kopp and Dichtl 2001).

Particle size distribution is an important factor deciding sludge water dewaterability. The sludge particles can be categorized into four: true colloidal solid (0.001–1.0 μm), supracolloidal solid (1–10 μm), fragile settable solid (10 to 100 μm) and rigid settable solid (>100 μm). In the supracolloidal solid and true colloidal solid, the particle fractions are very small size and have been found to increase the filtration resistance and thus negatively affect the dewatering of sludge. Although the true colloidal solid and supracolloidal solid account for a very small fraction of the total solid concentration of sludge. The enhancement of supracolloidal solid by 50% may lead to an upsurge in SRF by 100%. However, in the fragile and rigid settable solids, the particle fraction having comparatively larger size, have less effect on SRF of sludge (Shao et al. 2009; Wu et al. 2020). This suggests that particle size distribution is vital that decides dewaterability of sludge.

The surface charge is vital in deciding sludge dewaterability. Surface charge affects the sludge particle interactions which in turn influences thedistribution of particle size and ease of solid-liquid separation (Christensen et al. 2009). Further, pH also affects the dewaterability of sludge in multiple ways. Lowered pH damages the integrity of microbial cells and reduces SRF; thus, influencing dewaterability (Jin et al. 2004; Wu et al. 2020).

The metabolic products accumulating on the bacterial cell surface are known as Extracellular polymeric substances (EPS). These act as a layer of protection or deference for cells against any unfavourable and harsh environment. They also act as carbon and energy sources during the time of starvation (Liu and Fang 2002) EPS refers to "some compounds of high molecular weight distributed both outside of cells and in the interior of microbial aggregates, accounting for 50 to 80 % wt of the total organic fractions in sludge" (Wu et al. 2020). Biochemicals produced by microorganisms, cellular materials or products are generated from cell lysis or organic matter in the medium form EPS matrix (Liu and Fang 2002; More et al. 2014; Sheng et al. 2010). EPS is mainly composed of carbohydrate and protein.

However small amounts of humic substance, uronic acid and DNA is also present (Liu and Fang 2002; Wu et al. 2020). Extracellular proteins component in sludge plays a more important role in dewaterability of sludge than humic substances and polysaccharides (Houghton and Stephenson 2002; Shao et al. 2009; Wu et al. 2020; Yuan et al. 2011; Zheng et al. 2016).

The content ratio of polysaccharides and protein also significantly affects the hydrophilicity/hydrophobicity with surface charge of sludge flocs. However; the humic substances and nucleic acid do not show any significant impact on sludge dewaterability (Wu et al. 2020). Considering these facts, it is clear that EPS is one very important parameter that decides the dewaterability of sludge.

5 Bioleaching

Bioleaching refers to the "solubilization of metals from solid substrates either directly by the metabolism of leaching bacteria or indirectly by the products of metabolism" (Rulkens et al. 1995). Bioleaching is an effective and environmentally responsive procedure for treating of polluted sewage sludges. In addition to improved dewaterability, bioleaching could remove 85% and 40% of copper and chromium from sludge respectively (Liu et al. 2012b). Bioleaching has also been found to improve sludge dewaterability as specified by the fact that SRF of bioleaching sludge reduces significantly when compared to fresh sludge (Liu et al. 2012a). The dewatering of sludge may be improved by 4–ten-fold using bioleaching, a microbial conditioning method (Gao et al. 2017). After completion of bioleaching process, the moisture percentage of sludge cake reduces to around 60 during diaphragm filter press. In fact, bioleaching is considered as an eco-friendly and highly efficient dewatering technique due to little or no addition of chemical flocculants to the sludge (Liu et al. 2016).

The conventional sewage sludge treatment involves the use of organic or inorganic flocculants such as PAM and $FeCl_3$ and mechanical dewatering, which produces dewatered sewage of 80% moisture content or higher (Liu et al. 2012a, b, Shi et al. 2015). The dewatered sludge of high moisture content are too bulky to transport, requires high energy during drying and incineration, and enhancesthe cost of bulking agent for preparation of compost (Shi et al. 2015). Bioleaching is an excellent alternative to physical and chemical treatment of sludge as it reduces cost, ecologically safer compared to other available options and generates very low waste derivatives and waste solution (Chen and Lin 2004). Cost reduction by almost 80% can be achieved by bioleaching as compared to the chemical. The bioleaching could also effectively destroy pathogenic bacteria and reduce the odours (Pathak et al. 2009; Shi et al. 2015).

5.1 Microorganisms Involved in Bioleaching

Various kinds of microorganisms perform in the process of bioleaching. However, most experiments on bioleaching of sewage and sludge have used pure cultures of precise microorganisms (*A. ferrooxidans* and *A. thiooxidans*). But somenative microorganisms of sludge are also used for providing sufficient source of energy (Pathak et al. 2009). Though multiple bacterial species has the capacity of oxidizing reduced sulphur compounds their use in bioleaching process has been limited due to multiple factors. For example, bacterial species from family Chlorobiaceae, Ectothiorhodospiraceae, Chromatiaceae, and Rhodospirillacea can oxidise hydrogen sulphide. However; they can't efficiently oxidize elemental sulphur. Similarly, many bacterial species which have capacity to oxidize thiosulphate can't oxidize elemental sulphur due to lack of suitable enzymatic system. Though many bacterial species have been found to be capable of oxidizing both thiosulphate and elemental sulphur; their metabolic activity has not been found to be at par with *Thiobacillus* species (Blais et al. 1993).

Based on temperature requirement, bacteria involved in bioleaching can be grouped into mesophiles and thermophiles. The most commonly used mesophiles used include sulphur oxidizing bacteria (*A. thiooxidans*) and iron-oxidizing bacteria (*A. ferrooxidans*). These bacteria are chemolithotrophic and obtain their energy by oxidation of reduced sulfur compounds and ferrous iron. At higher temperature bacteria like *Sulfobacillus thermosulfidoxidans* and related species which are moderate thermophiles, utilize the higher temperature for quicker bioleaching rate. Some bacteria mainly of the sulfurous genus are extreme thermophiles which can grow at the temperature as high as 70 °C and use either sulfur or thiosulfate as the source of energy (Pathak et al. 2009). Tolerance to high acidity and ability of the oxidation of insoluble iron and sulfur compounds makes *A. thiooxidans* and *A. ferrooxidans*, the most widely used microorganism for metal bioleaching.

A. thiooxidans utilizes the reduced inorganic S instead of Fe^{2+} for energy and it prefers a pH 0.5 to 5.5 with the optimum being pH 2 to 3.5, and can drastically reduce the leaching medium pH to as low as 1.5 to 1.0 or even lower. *A. ferrooxidans* varies from *A. thiooxidans* originate energy from the oxidation process of Fe^{2+} as an electron donor with reduction sulphur compounds. In case of devoid of oxygen, *A. ferrooxidans* can multiply on reduced inorganic sulphur compounds using Fe^{3+} as an alternative electron acceptor (Liu et al. 2018). The *A. thiooxidans* and *A. ferrooxidans* get energy from the oxidation of substances causing acidification as well as heavy metals solubilization. After completion of the bioleaching, heavy metals are separated, harmful pathogenic bacteria are killed and odour is eliminated (Chen and Lin 2004; Wang et al. 2010).

5.2 Mechanism of Bioleaching

During the bioleaching process, the microorganisms like *Acidithiobacillu sthiooxidans* and *Acidothiobacillus ferrooxidans*oxidise the Sulphur and iron compounds. In the bioleaching process for heavy metals from sewage sludge, different sources of energy such as $FeSO_4$, FeS_2 and S^0 have been provided to the bacteria (Pathak et al. 2009; Wong and Gu 2004). Under acidic medium, the heavy metals get solubilized and thus can be subsequently recovered. The ability of *A.thiooxidans* and *A. ferrooxidans* to survive under extreme acidic condition and to oxidise insoluble compounds of iron and sulphur makes them the most widely used microorganism for metal bioleaching. Metal dissolution in the bioleaching can be achieved in two ways *i.e.*, direct bioleaching and indirect bioleaching (Ghavidel et al. 2017) in a sulphur based bioleaching process or iron-based bioleaching process.

5.2.1 Sulphur Based Bioleaching Process

In this process, metal sulfide dissolution takes place by direct or indirect mechanism. In direct bacterial leaching, bacteria directly oxidizes the metal sulfide into metal sulfate. In this process, bacteria come in close contact with the metal sulfides in sludge and help in the oxidation process. Metal sulphides such as CuS, and ZnS are solubilized in this mechanism. In the indirect mechanism, the elemental sulphur and reduced sulfur are oxidized into H_2SO_4 by sulfur-oxidizing bacteria ensuing in reduced pH of the sludge medium. The low pH environment favours the solubilization of metal. In an indirect method, Acidithiobacillus takes an active part in the oxidation of reduced sulphur or elemental sulphur, while solubilization of metals takes place chemically without the bacterial involvement (Pathak et al. 2009).

5.2.2 Iron-Based Bioleaching Process

Like sulphur based bioleaching, iron-based bioleaching also involves direct and indirect mechanisms, in which the oxidation of reduced iron and sulphur takes place. In direct mechanism, non-ferrous metallic sulphides are oxidized into soluble metal sulfate by *A. ferrooxidans*. The indirect mechanism involves the oxidation of Fe^{2+} to Fe^{3+} by bacteria in the liquid phase and leaching of Fe^{3+} through a chemical reaction. Direct contact between bacteria and mineral surface is not needed in this process (Pathak et al. 2009). The first step in the indirect mechanism where oxidation of Fe^{2+} to Fe^{3+} takes place, involves to active participation of *A. ferrooxidans* while the subsequent step occurs chemically without the involvement of bacteria (Fig. 2).

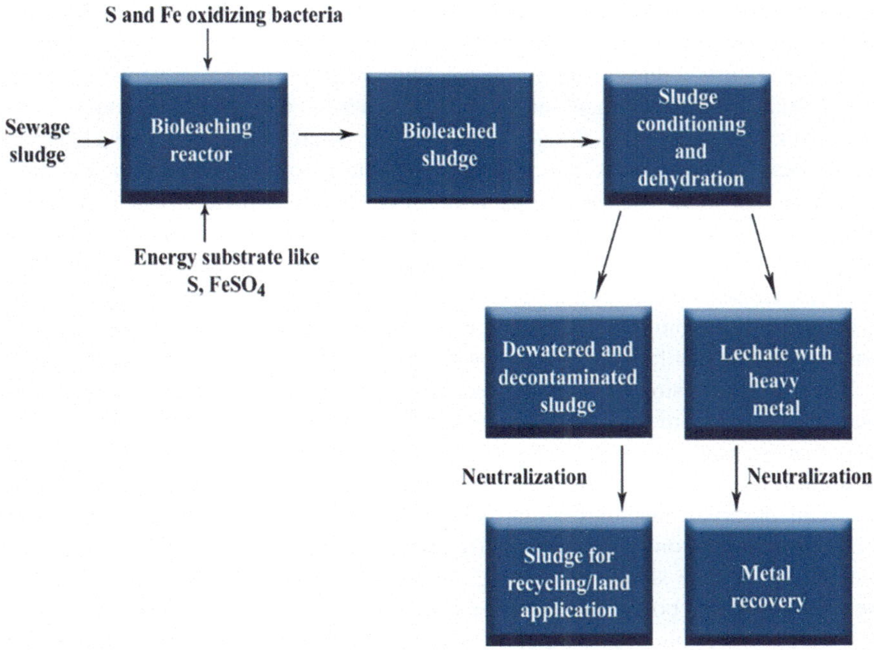

Fig. 2 Bioleaching for sludge dewatering and heavy metal decontamination

5.3 *Effect of Bioleaching on Dewaterability*

Though bioleaching has been found very effective in improving dewaterability still, the mechanism of it is not very clear. Though improved dewaterability has been attributed to lower EPS, an explanation for such low EPS was not clear. Work done by Liu et al. (2016) provides some explanation regarding improved dewaterability of bioleaching sewage sludge. Liu et al. (2016) noted that this low EPS of bioleached sludge was observed to the shift of microbes in completion of the process of bioleaching. As the process of bioleaching proceeded, a shift in the microbial community was observed. The bio-substitution resulted in gradual increase in the population of *A. ferrooxidans* with bioleaching process. The EPS content of the *A. ferrooxidans* was lower compared to the raw sludge and as the bioleached sludge was dominated by *A. ferrooxidans*, hence, it resulted in low EPS. This low EPS content improved the dewaterability of bioleached sludge.

Zeta potential can be measured to know the surface charge of sludge that affects settling property of sludge. Liu et al. (2016) observed a nearly constant zeta potential of −40 mV for raw sludge throughout the culture period while the zeta potential of bioleached sludge increased significantly over time. This might be due to the increase in H^+ in the acidic environment of bioleached sludge. After bioleaching, the sludge also resulted in a significant decrease in SRF after 72 h of treatment. A lower SRF show increased dewaterability of bioleached sludge. The improvement in dewaterability by bioleaching can also be accredited to the flocculation effect of Fe^{3+}, transformed from the Fe^{2+} added (Liu et al. 2016).

A. thiooxidans oxidizes sulphur to sulphuric acid producing an acidic environment. With the development of the acidic environment, neutralization of zeta potential takes place in the sludge. Moreover, the acidic environment developed in the process also enhances flocculation and settling of sludge and in this way improving the sludge dewaterability (Gao et al. 2017). Particle size and structure affect the dewaterability of sludge to a large extent. Analysis of sludge structure using an optical microscope reveals that sludge samples subjected to bioleaching treatment show a change in the structure of sludge from flocculent to granular (Shi et al. 2015). The bioleaching treatments also resulted in larger particle size and denser structure that favours sludge dewaterability (Shi et al. 2015; Wu et al. 2020).

From the discussion so far, it is clear that bioleaching may significantly ensure qualitative improvement in the dewaterability of sewage sludge. The acidic conditions that occuring the bioleaching process bring the sludge floc surface potential closer to zero as compared to the untreated sludge, thus improving dewaterability. It has been observed that bioleaching shows better performance in comparison to chemical acidification, as it significantly affects extracellular polymeric substances that impede sewage sludge dewatering (Marchenko et al. 2018). Excessive EPS in the form of loosely bound EPS declines sludge dewaterability and results in meagre separation of biosolids and water (Liu et al. 2016). Increased dewaterability at low EPS content in sludge was also reported by Houghton and Stephenson (2002). High EPS concentration results in an increase in the viscosity of sludge; thus, reducing its filterability. Bioleaching improves dewaterability, however, excessive bio-acidification that leads to a fall in pH as well as an enhancement in bioleaching time unexpectedly reduces the dewaterability of sludge. Hence, to achieve optimum efficiency of sludge dewatering and removal of metals from sludge, pH ~ 2.4 was suggested as the optimum endpoint for dewatering (Liu et al. 2012a, b). Bioleaching can be used as a useful tool in sewage sludge management as it provides the dual advantage of heavy metal decontamination and improved dewaterability. The water content of bioleached sludge can be easily lowered down to 60% by subsequent dewatering technique (Fig. 3).

6 Future Scope

The mechanism by which dewatering is improved due to bioleaching is still not completely understood (Liu et al. 2016). A majority of the bioleaching studies were carried out in batch mode under controlled condition. For a wider scale of field conditions, a more detailed study on bioleaching with continuous mode of operation as well as sludge digestion needs to be done. It can help in the treatment of a bulk

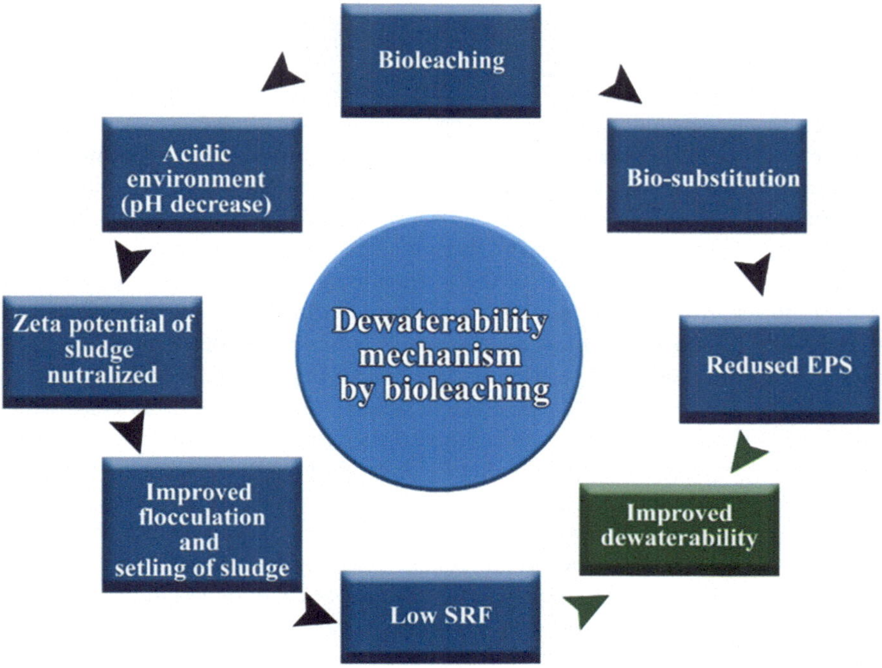

Fig. 3 Mechanisms of improved dewaterability by bioleaching

amount of sludge in a single operation. The process of dewatering can be developed commercially into a more viable by employing an efficient yet cost-effective and recoverable source of sulphur such as sulphur tablets or rods. Different indigenous bacteria strain should be tested for their efficiency in bioleaching process.

The scope of genetically improved strains of microorganisms can also be studied for further improvement of the speed and efficiency of bioleaching. The efficiency of a mixed culture of bacteria against single bacterial culture in bioleaching can be studied with respect to heavy metal decontamination and dewaterability. The potential loss of nutrients such as N and P, from sludge during the bioleaching process has been reported (Blais et al. 2004; Shanableh and Ginige 1999; Wong and Gu 2004). The strategy should be developed to avoid excessive loss of beneficial nutrients during bioleaching process. Though several physicochemical properties have been found to have a correlation with sludge dewaterability, a combinedapproach for improving dewaterability of sludge is still not available (Wu et al. 2020). Different approaches such as physical, chemical or bioleaching approaches for improving sludge dewaterability have been developed. However, selecting a suitable conditioning strategy for sludge of specific physicochemical character is still unclear (Wu et al. 2020).

7 Conclusion

Rising population, increasing urbanization, improved sanitation measures has directed to the generation of a bulk quantity of sewage sludge across the globe. The sewage sludge production is also expected to grow bigger in coming years especially in low and middle-income countries where sanitation is gaining more importance. Government regulations in many countries direct strict measuresfor sustainable wastewater management. There is also a growing public concern regarding ecosystem health. This makes us think of a strategy that will cause minimal or no damage to the environment and ensure proper sewage sludge disposal and/reuse. Considering a high heavy metals content, the presence of pathogens and strong odour direct application of sewage sludge to field application is limited. Sludge bioleaching has been proven as an effective strategy for reducing metal contamination. However, the sludge bioleaching has also been found as an effective method in improving the dewaterability of sludge, thus providing a win-win situation. Bioleaching is also an ecofriendly approach that involves the use of no or minimum flocculant. To add to it bioleaching is also a cost-effective approach as compared to chemical methods. Hence; the overall benefits make it a sustainable alternative to other methods. However, more understanding of the effect of bioleaching in improving sludge water dewaterability is required. As most of the works have been conducted on a lab-scale, industrial-scale experimentation and practice will give a better picture regarding its efficiency and applicability on a large scale.

References

Al-Malack MH, Abuzaid NS, Bukhari AA (2008) Physico-chemical characteristics of municipal sludge produced at three major cities of the eastern province of Saudi Arabia. J King Saud Univ Eng Sci 20(1):15–26

Ban Y, Li L, Liu C, Yan Y, Gao J, Zhang J, Gao J (2018) Enhancing sludge dewatering and heavy metal removal by bioleaching with $Na_2S_2O_3$ as substrates. Water Sci Technol 78(7):1545–1555

Bastian RK (1997) The biosolids (sludge) treatment, beneficial use, and disposal situation in the USA. Eur Water Pollut Contr 2(7):62–79

Blais JF, Tyagi RD, Auclair JC (1993) Bioleaching of metals from sewage sludge: microorganisms and growth kinetics. Water Res 27(1):101–110

Blais JF, Meunier N, Mercier G, Drogui P, Tyagi JF (2004) Pilot plant study of simultaneous sewage sludge digestion and metal leaching. J Environ Eng 130(5):516–525

Bonfiglioli L, Bianchini A, Pellegrini M, Saccani C (2014) Sewage sludge: characteristics and recovery options. DIN, Facoltà di Ingegneria, Università di Bologna, pp 1–21

Camargo FP, Tonello PS, Santos ACA, Duarte ICS (2016) Removal of toxic metals from sewage sludge through chemical, physical and biological treatments-a review. Water Air Soil Pollut 227(12):1–11

Chen SY, Lin JG (2004) Bioleaching of heavy metals from livestock sludge by indigenous sulfur-oxidizing bacteria: effects of sludge solids concentration. Chemosphere 54(3):283–289

Chen JY, Wu XW (2009) Research of wastewater sludge organic conditioning concentration and inorganic conditioning dewatering technology. Chinese J Environ Eng 3(3):529–532

Christensen ML, Hjorth M, Keiding K (2009) Characterization of pigslurry with reference to flocculation and separation. Water Res 43:773–783

Dabrowska L, Rosinska A (2012) Change of PCBs and forms of heavy metals in sewage sludge during thermophilic anaerobic digestion. Chemosphere 88(2):168–173

Du F (2015) Novel sludge treatment technology enhances toxic metal removal, pathogen reduction, sludge digestibility and dewaterability, Doctoral thesis University of Queensland, School of Chemical Engineering, Advanced Water Management Centre, Australia, p. 99

Gao J, Shen Y, Li L, Gao J, Li Y, Liu C, Chen J (2017) Enhancing dewaterability of sewage sludge by the application of tween-20 during bioleaching: performance evaluation and mechanistic study. Dry Technol 36(7):780–789

Ghavidel A, Rad SN, Alikhani HA, Sharari M, Ghanbari A (2017) Bioleaching of heavy metals from sewage sludge, direct action of *Acidithiobacillus ferrooxidans* or only the impact of pH. J Mater Cycles Waste Manag 20(2):1179–1187

Grobelak A, Czerwińska K, Murtaś A (2019) General considerations on sludge disposal, industrial and municipal sludge. In: Prasad MNV, de Campos Favas PJ, Vithanage M, Mohan SV (eds) Industrial and municipal sludge. Butterworth-Heinemann, pp. 135–153. https://doi.org/10.1016/B978-0-12-815907-1.00007-6

Gu X, Wong JW (2004) Identification of inhibitory substances affecting bioleaching of heavy metals from anaerobically digested sewage sludge. Environ Sci Technol 38(10):2934–2939

Henry JG, Frasad D (2006) Biosolids from two-stage bioleaching could produce compost for unrestricted use. Environ Technol 27(6):665–672

Heukelekian AJM, Weisburg E (1956) Bound water and sludge bulking. Sew. Ind Waste 28:558–574

Houghton JI, Stephenson T (2002) Effect of influent organic content on digested sludge extracellular polymer content and dewaterability. Water Res 36(14):3620–3628

Ignatowicz K (2017) The impact of sewage sludge treatment on the content of selected heavy metals and their fractions. Environ Res 156:19–22

Islam MS, Ahmed MK, Raknuzzaman M, Habibullah-Al-Mamun M, Islam MK (2015) Heavy metal pollution in surface water and sediment: a preliminary assessment of an urban river in a developing country. Ecol Indic 48:282–291

Jenkins RL, Scheybeler BJ, Smith ML (1981) Metals removal and recovery from municipal sludge. J Water Pollut Control Fed 53:25–32

Jin B, Wilén B, Lant P (2004) Impacts of morphological, physicaland chemical properties of sludge flocs on dewaterability of activatedsludge. Chem Eng J 98(1–2):115–126

Kopp J, Dichtl N (2001) Influence of the free water content on the dewaterability of sewage sludges. Water Sci Technol 44(10):177–183

Krishnamurthy S, Viraraghavan T (2005) Chemical conditioning for dewatering municipal wastewater sludges. Energy Sources 27(1–2):113–122

Kwarciak-Kozłowska A (2019) Co-composting of sewage sludge and wetland plant material from a constructed wetland treating domestic wastewater. In: Prasad MNV, de Campos Favas PJ, Vithanage M, Mohan SV (eds) Industrial and municipal sludge. Butterworth-Heinemann, pp. 337–360. https://doi.org/10.1016/B978-0-12-815907-1.00015-5

LeBlanc RJ, Matthews P, Richard RP (2008) Global atlas of excreta, wastewater sludge and biosolids management: moving forward the sustainable and welcome uses of a global resource. United Nations Humans Settlements Programme (UN-HABITATAT), Nairobi, Kenya

Lema J, Suarez S (2017) Innovative wastewater treatment & resource recovery technologies: impacts on energy, economy and environment, 1st edn. IWA Publishing London

Liu H, Fang HH (2002) Extraction of extracellular polymeric substances (EPS) of sludges. J Biotechnol 95(3):249–256

Liu FW, Zhou LX (2009) Enhancing dewaterability of municipal sewage sludge through the combined approaches of bioleaching and Fenton reaction. Acta Sci Circumst 29(5):974–979

Liu F, Zhou J, Wang D, Zhou L (2012a) Enhancing sewage sludge dewaterability by bioleaching approach with comparison to other physical and chemical conditioning methods. J Environ Sci 24(8):1403–1410
Liu F, Zhou L, Zhou J, Song X, Wang D (2012b) Improvement of sludge dewaterability and removal of sludge-borne metals by bioleaching at optimum pH. J Hazardous Mater 221:170–177
Liu H, Yang S, Shi J, Xu X, Liu H, Fu B (2016) Towards understanding the dewatering mechanism of sewage sludge improved by bioleaching processing. Sep Purif Technol 165:53–59
Liu C, Ma Q, Zhou X, Lai H, Li L (2018) Bioleaching of heavy metals from sludge by mixed strains. In IOP conference series: earth and environmental science 208(1):12077
Lo LMC, Lai KCK, Chen GH (2001) Salinity effect on mechanical dewatering of sludge with and without chemical conditioning. Environ Sci Technol 35(23):4691–4696
Mahmoud A, Olivier J, Vaxelaire J, Hoadley AF (2011) Electro-dewatering of wastewater sludge: influence of the operating conditions and their interactions effects. Water Res 45(9):2795–2810
Marchenko O, Demchenko V, Pshinko G (2018) Bioleaching of heavy metals from sewage sludge with recirculation of the liquid phase: a mass balance model. Chem Eng J 350:429–435
Marchioretto MM, Bruning H, Loan NTP, Rulkens WH (2002) Heavy metals extraction from anaerobically digested sludge. Water Sci Technol 46:1–8
Mehrotra A, Sreekrishnan TR (2017) Heavy metal bioleaching and sludge stabilization in a single-stage reactor using indigenous acidophilic heterotrophs. Environ Technol 38(21):2709–2724
More TT, Yadav JSS, Yan S, Tyagi RD, Surampalli RY (2014) Extracellular polymeric substances of bacteria and their potential environmental applications. J Environ Manag 144:1–25
Na S, Kim YU, Khim J (2007) Physiochemical properties of digested sewage sludge with ultrasonic treatment. Ultrasonics Sonochem 14(3):281–285
Neyens E, Baeyens J (2003) A review of thermal sludge pre-treatment processes to improve dewaterability. J Hazar Mater 98(1–3):51–67
Ozoreshampton M, Stansly PA, Obreza TA (2005) Heavy metal accumulation in a sandy soil and in pepper fruit following long-term application of organic amendments. Compost Sci Utilizati 13(1):60–64
Pathak A, Dastidar MG, Sreekrishnan TR (2009) Bioleaching of heavy metals from anaerobically digested sewage sludge. J Environ Sci Health Part A 43(4):402–411
Rulkens WH, Grotenhuis JTC, Tichý R (1995) Methods for cleaning contaminated soils and sediments. In: Förstner U, Salomons W, Mader P (eds) Heavy metals: problems and solutions. Springer, Berlin, pp 165–191. https://doi.org/10.1007/978-3-642-79316-5_11
Saha S, Saha BN, Pati S, Pal B, Hazra GC (2017) Agricultural use of sewage sludge in India: benefits and potential risk of heavy metals contamination and possible remediation options–a review. Int J Environ Technol Manag 20(3–4):183–199
Shanableh A, Ginige P (1999) Impact of metals bioleaching on the nutrient removal biosolids. Water Sci Technol 39(6):175–181
Shao L, He P, Yu G, He P (2009) Effect of proteins, polysaccharides, and particle sizes on sludge dewaterability. J Environ Sci 21(1):83–88
Sheng GP, Yu HQ, Li XY (2010) Extracellular polymeric substances (EPS) of microbial aggregates in biological wastewater treatment systems: a review. Biotechnol Adv 28(6):882–894
Shi C, Zhu N, Shang R, Kang N, Wu P (2015) Simultaneous heavy metals removal and municipal sewage sludge dewaterability improvement in bioleaching processes by various inoculums. World J Microbiol Biotech 31(11):1719–1728
Song XW, Zhou LX (2008) The influence of bioleaching on dewaterability of municipal sewage sludge. Acta Sci Circumst 28(10):2012–2017
Sreekrishnan TR, Tyagi RD (1994) Heavy metal leaching from sewage sludges: a techno-economic evaluation of the process options. Environ Technol 15:531–543
Veeken AHM, Hamelers HVM (1999) Removal of heavy metals from sewage sludge by extraction with organic acids. Water Sci Technol 40(1):129–136

Wang F, Ji M, Lu S (2006) Influence of ultrasonic disintegration on the dewaterability of waste activated sludge. Environ Prog 25(3):257–260

Wang S, Zheng G, Zhou L (2010) Heterotrophic microorganism *Rhodotorula mucilaginosa* R30 improves tannery sludge bioleaching through elevating dissolved CO_2 and extracellular polymeric substances levels in bioleach solution as well as scavenging toxic DOM to *Acidithiobacillus* species. Water Res 44(18):5423–5431

Westerhoff P, Lee S, Yang Y, Gordon GW, Hristovski K, Halden RU, Herckes P (2015) Characterization, recovery opportunities, and valuation of metals in municipal sludges from US wastewater treatment plants nationwide. Environ Sci Technol 49(16):9479–9488

Wiśniowska E (2019) Sludge activation, conditioning, and engineering. In: Prasad MNV, de Campos Favas PJ, Vithanage M, Mohan SV (eds) Industrial and municipal sludge. Butterworth-Heinemann, pp. 181–199. https://doi.org/10.1016/B978-0-12-815907-1.00009-X

Wojciechowska E (2005) Application of microwaves for sewage sludge conditioning. Water Res 39(19):4749–4754

Wong JW, Gu XY (2004) Enhanced heavy metal bioleaching efficiencies from anaerobically digested sewage sludge with coinoculation of *Acidithiobacillus ferrooxidans* ANYL-1 and *Blastoschizomyces capitatus* Y5. Water Sci Technol 50(9):83–89

Wu B, Dai X, Chai X (2020) Critical review on dewatering of sewage sludge: Influential mechanism, conditioning technologies and implications to sludge re-utilizations. Water Res 115912

Xu Y, Zhang C, Zhao M, Rong H, Zhang K, Chen Q (2017) Comparison of bioleaching and electrokinetic remediation processes for removal of heavy metals from wastewater treatment sludge. Chemosphere 168:1152–1157

Xun R, Wang W, Qiao W (2009) Water distribution and dewatering performance of the hydrothermal conditioned sludge. Environ Sci 30(3):851–856

Yan S, Bala-Subramanian S, Tyagi RD, Surampalli RY (2009) Wastewater sludge characteristics. sustainable sludge management: production of value added products. Am Soc Civil Eng Resto Virgi USA 6–36

Yuan HP, Cheng XB, Chen SP, Zhu NW, Zhou ZY (2011) New sludge pretreatment method to improve dewaterability of waste activated sludge. Bioresour Technol 102(10):5659–5664

Zheng G, Huo M, Zhou L (2016) Extracellular polymeric substances level determines the sludge dewaterability in bioleaching process. J Environ Eng 142(2):4015060

Zou D, Chi Y, Dong J, Fu C, Wang F, Ni M (2013) Supercritical water oxidation of tannery sludge: stabilization of chromium and destruction of organics. Chemosphere 93(7):1413–1418

Potential Role of Beneficial Microbes for Sustainable Treatment of Sewage Sludge and Wastewater

Tabinda Athar, Anamika Pandey, Mohd. Kamran Khan, Zulfiqar Ahmad Saqib, Mah Jabeen, Shumila Shahid, Mehmet Hamurcu, Sait Gezgin, Vishnu D. Rajput, and Maria A. Elinson

1 Introduction

Intensification in the industrial activities and accelerated development in the urban and semi-urban areas is generating high levels of organic, and inorganic contaminants that are specifically discharged to the wastewater and sewage networks (Atashgahi et al. 2015). Sewage sludge not only contains different levels of contaminants but also different bacterial communities. This microbial community may vary depending on treatment conditions, industrial activities, and sewage origin but these microbial communities have significant potential to treat the sewage sludge for safe use for arable land, agricultural production, horticultural uses, and industrial uses by extracting valuable products. But sustainability of use is greatly dependent

T. Athar · Z. A. Saqib
Institute of Soil and Environmental Sciences, University of Agriculture Faisalabad, Faisalabad, Pakistan

A. Pandey (✉) · M. K. Khan (✉) · M. Hamurcu · S. Gezgin
Faculty of Agriculture, Department of Soil Science and Plant Nutrition, Selcuk University, Konya, Turkey
e-mail: anamika@selcuk.edu.tr; mkkhan@selcuk.edu.tr

M. Jabeen
Punjab Agriculture Extension Department, Lahore, Pakistan

S. Shahid
Department of Soil Science, The Islamia University of Bahawalpur, Bahawalpur, Pakistan

V. D. Rajput
Academy of Biology and Biotechnology, Southern Federal University, Rostov-on-Don, Russia

M. A. Elinson
Bashkir State University, Ufa, Russia

V. D. Rajput et al. (eds.), *Sustainable Management and Utilization of Sewage Sludge*, https://doi.org/10.1007/978-3-030-85226-9_4

on system efficiencies, implementation of policies, and cost associated with technological processes (Shchegolkova et al. 2018).

In the twenty-first century, the sewage sludges generation and their disposal are considered as greatest challenges for environmental protection, changing climate, human safety, ecosystem functioning, biodiversity, and national and international economies. Water is a must for the existence of life, from maintenance and operation of households to extensive agricultural and industrial use. It is our ethical, ecological and political responsibility to critically think about management of natural resources. Due to shortage now it is vital to safeguard the good quality water as there is a significant loss of its share below the surface and in the landscaping, water channels. Industrial, and domestic effluent is an environmental concern even if it is going through a sewerage system and treated in the urban wastewater treatment plants and eventually released openly to environment.

A sizeable quantity of refused water is generated by domestic community companies, urban local builds, and industries, that water without suitable handling discharged into neighbouring water bodies, lakes, and rivers, cause water pollution. This mess-up handling of wastewater causes many challenges such as lack of energy for the treatment of discarded water as well as deficiency of freshwater. The overall challenge is to plan such schemes which are not only important treat wastewater but also helpful in the generation of many energy compelled products. Therefore, this chapter is focused to highlight the significance of microorganisms and microbial techniques for the sustainable, efficient, and cost-effective management of wastewater and sewage sludge.

2 Role of Microorganisms for Sustainable Management of Sewage Sludge and Wastewater

Biological treatments aim to decompose toxic organic compounds (pharmaceutical compounds, xenobiotics, and petroleum derivatives) and decrease pathogens' population, lessen the effects on environment and human beings. Among biological processes in wastewater treatment plants (WWTPs), activated sludge (AS) processes are widely used across the world and for removal of pollutants and they are being used for more than a century, due to their biomass retention capabilities, high toxin degradation, and nutrient removal efficiencies (Xia et al. 2018).

The AS procedures have been studied extensively and the recent results have revealed that full-scale system of AS offers a core microbiome and its activities offers good decontamination of wastewater. By-product of sewage sludge usually have greater microbial diversities, and it may vary depending on origin of sewage, industrial activity, treatment conditions (e.g., redox conditions, and liming). The processes of activated sludge depend on the capability of microorganisms for the utilization of organic material as basis of carbon and/or source of energy and other essential minerals for growing processes, and plays crucial roles in the biodegradation of organic materials, removal of specific nutrients such as ammonia, sulfate, phosphate, and nitrate and conversion of hazardous compounds to less toxic products (Rahimi et al. 2020).

Biological processes and treatments are dependent on nematodes, bacteria, and other soil organisms that causes decomposition of organic wastes by utilizing normal and specific cellular processes. Typically, wastewater and sewage sludge have high amount of organic matters, like partially digested foods, garbage, and waste. Moreover, it also contains numerous pathogenic organisms, toxins, and heavy metals. The major purpose of biological wastewater treatment is creation of a system that can easily collect the end results of decomposition for further disposal activities.

These microorganisms cause decomposition of organic pollutants to get food and energy. They may stick together during this whole process and causes the creation of flocculation effects that in turn allows the settlement of organic matter and organic residues in the solution. In this way it is essentially helpful for the safe and easy management of sewage sludge because it can be dewatered with great ease and can be disposed of as solid waste. Therefore, the role of microorganisms for the treatment of sewage sludge and wastewater is greatly helpful to achieve food security and environmental sustainability. Microorganisms may not cause the complete mineralization of all toxic compounds but causes a significant conversion of more toxic products to the less toxic ones and thus protects environment, plants, animals, and humans from the risks associated with contamination.

2.1 Composition and Structure of Bacterial Communities and Their Control

Relative abundance, occurrence and activities of several microbial communities in sewage sludge greatly affects the stable practices and operations of biological based wastewater treatment plants (WWTPs).

Although there is always a significant variation in the community composition of microbes, but this variation is associated with alterations in functional capabilities and structural dimensions of microbial communities. Functional stabilities in the microbial communities have been recognized as major factors that affects the efficiency of wastewater treatment (Wang et al. 2014). Microbial diversity and changes in the community structures greatly affects the functional stability and performance of WWTPs. While, there are many factors that causes modulation of community structures of microbes in WWTPs. Usually this variation is based on the presence of different types of bacterial niches such as autotrophs, heterotrophs, and chemotrophs, and the sources of effluents.

Since the past decade, there have been intensive studies to check the functional stabilities and other properties in the activated sludge in WWTPs. These studies are especially focused on dealing with sewage sludge and wastewater in the WWTPs. Communities that are directly involved for biological treatment of dangerous substances such as from industrial waste harbours different populations of microbes which are specifically adapted to numerous stresses in these systems. Effluents from textile industries also contains higher levels of dyeing additives, dyes, and varying degree of other chemicals. Some of these contaminants are non-biodegradable, mutagenic, toxic, and carcinogenic and therefore can pose major threats to environment and health.

Generally, textile wastewaters have low ratio of biological oxygen demand/chemical oxygen demand (BOD/COD) that is around 20%. Moreover, there is a varied range of pH (4–12) and therefore it may include numerous inhibitor composites that can exert hampering effects on biological treatment of wastewater, adsorbable organic halogens, active substances (e.g., chlorine compounds) (AOX) and higher concentrations of salts. This entirely makes it difficult to treat textile wastewater and thus greater care and management is required to achieve sustainability. Furthermore, microbial communities in the wastewater are different than the communities present in the industrial wastewater and requires different kinds of handling techniques and management approaches.

Studies have shown that in the municipal (domestic sewage), predominnat species was Proteobacteria phylum (21–65%) that belongs to Betaproteobacteria that represents a specific class of microorganisms responsible for degradation of organic matter and cycling of nutrients. This sewage also contained other less dominant taxa such as Bacteroidetes, Chloroflexi and Acidobacteria. Whereas, proteobacteria were found to be abundant in the sewage generated by the industrial activities. This sludge was reported to contain a higher level of obstinate compounds coming from petroleum refineries, pharmaceutical industries, factories for animal feeds, textiles, and others.

Conditions of biological treatments are other modulating factors and studies have reported that microorganisms were abundantly found in the anaerobic-aerobic and anaerobic systems as compared to the aerobic systems. While the abundance of proteobacteria was reported to be more in the aerobic environment. Whereas the abundance of Bacteroidetes was reported to be more in the bioreactors provided with anaerobic conditions. Furthermore, there are some chemical attributes such as concentrations of micronutrients, pH presence of different types of toxic compounds such as heavy metals, and other inorganic, and organic pollutants and oxidation and reduction conditions in the biological treatments can directly affect structure of bacterial communities in the sludge.

Like in Brazil, the sulfur oxidoreductive bacterial community was composed of 22 different families, and could have been clustered by the chemical characteristics, such as S, Zn, K, N, Mn, and P and sewage sources (Meyer et al. 2016). Studies have also reported that temperature also impacts the diversity of microbial communities and their structures in WWT plants. Temperature is most important among the physical factors as it is key players for the determination of survival rates of microbial communities. Moreover, it greatly controls the composition of hydrocarbons and therefore due attention should be given to this factor for effective and efficient management.

Biological enzymes always have good participation for the degradation pathways but require optimal temperature for their functioning. Every little or major change in the temperature will directly affect the metabolic turnover and thus management situation may fluctuate. Moreover, temperature specification is also important for the breakdown processes of different compounds. Increased temperature is always associated with the increased rate of microbial activities and maximum activities can be sustained at the optimal temperatures. These activities are declined with further decrease or increase in the temperature and are eventually stopped after reaching a maximum limit. Scientific studies have also reported that cold temperature produces effects on the growth of microbes by decreasing the availability of water, changing the energetics, reducing the molecular motion, and increasing the concentration of solutes due to reduced water availability. It also has been reported that adaptation of different communities of microbes to lower temperatures is a problematic scenario for WWT systems.

Likewise, pH of specific compound that is either basic, acidic or alkaline in nature of the compound, exerts its own effects on metabolic activities of microbes and may also affect the efficiency of removal process. Low or high pH values also causes negative results on microbial communities and their metabolic processes because these creatures are greatly intolerant to even little fluctuation in pH. Similarly, microbial communities and their activities are also dependent on the concentration of oxygen because some species requires oxygen and some do not requires oxygen for their survival and bio-degradation processes. While, bio-degradation can be carried in both anaerobic and aerobic conditions because oxygen is an essential requirement for various living creatures and some that does not requires oxygen may have developed slight tolerance. Studies have reported that the metabolism of hydrocarbons is greatly improved due to presence of oxygen.

The balance for essential nutrients is also important for growth, survival, and reproduction related activities of microbes. An optimal balance is not only important for these processes but is also required to accelerate the efficiency and rates of biodegradation. Nutrient balancing is especially important for P and N as they can improve the efficiencies of biodegradation by optimization of C:N:P ratios in the sewage sludge, wastewater, and soil systems. Microorganisms also needs different nutrients such as P, N, and C for their growth, development, survival, and functioning. Addition of appropriate quantity of these nutrients is important strategy to improve their metabolic activity and functioning and thus the process of biodegradation can be greatly accelerated in the colder regions. The process of biodegradation is especially limited in the aquatic environments due to limitation of nutrients. The microbes that feed on oil also needs nutrients for their growth, and development. Usually, these nutrients are available in their surroundings and in the natural sources but their concentration is low so they must be augmented by some external sources for better functioning and activities of these microbes.

Biotic factors also exert direct influences on the degradation of different organic compounds due to the competition between numerous species of microorganisms for limited sources of carbon and predation of microbes by bacteriophages and protozoa, or due to antagonistic interactions between numerous microorganisms. The degradation rates of contaminants is also dependent on the levels, concentrations, and types of contaminants, and the amount of catalyst for the specific degradation reactions. In this specific context amount of catalyst present indicates the specific number of organisms that can metabolize different contaminants. Moreover, the production of enzymes by the cells is also important factor that affects the overall degradation and stabilization of contaminants. The specific expression of enzymes by the reduced or improved rate of contaminants degradation also have significant importance to predict and measure the enzymatic activities and degradation of pollutants. The major biological factors in this context are size of bacterial population, community composition, gene transfer, interaction of different microbial and other communities, enzymatic activities and mutation.

Despite of the significant progress for the effective management of sewage sludge and waster water by microbial processes there are some kinds of associated shortcomings. Scientific data is also indicating that issues related to structures of microbial communities can be managed in the WWTPs but it only involves and manages smaller populations and samples. Other than this majority of designs and scientific knowledge is only being applied to pilot systems and bioreactors in the laboratories (Saia et al. 2016). Controlled operational conditions such as flow of effluents, aeration, and temperature can easily affect the diversity of microbial communities (Muszyński and Załęska-Radziwiłł 2015; Muszyński et al. 2013). Whereas most of the scientific studies are based on using conventional techniques and therefore only 60–90% of populations of microbes have been cultured. Emerging molecular, biotechnological, and bioinformatics techniques should be implied for better understanding about community structures and their functioning. Therefore, the tolerance, survival, and working capabilities of these microbes can also be improved by using latest technological solutions. Furthermore, ecological role of microbial communities can also be significantly improved, and thus environmental protection can be attained on sustainable basis.

2.2 *Microbial Activities*

In the different spatial and temporal conditions, effluent treatment plants (ETPs) may contain different level of microbial communities as dynamic associations. The various co-existing populations of microbes in wastewaters vary with the operational conditions of reactor. Their involvement for overall degradation of pollutants may cause unprecedented controls for bioremediation of contaminants and effluents (Manefield et al. 2005).

Microbes perform sewage sludge and wastewater treatment either through aerobic digestion or anaerobic digestion.

1. Aerobic, in these microorganisms need oxygen for decomposition of organic matter to microbial biomass and carbon dioxide.
2. Anaerobic, in these microbes do not require oxygen for decomposition of organic matter, and often produces excess biomass, carbon dioxide, and methane.

Anaerobic digestion is an auspicious biotechnology for highly polluted wastewater with organic contaminants and hence contains higher amount of substances that can be degraded biologically. The substrate digestion in anaerobic reactors causes significant lessening in the total contents of volatile solids and additionally the weight and volume of the substrate. The process of anaerobic digestion is complex and consists of several biochemical based processes and are systemically mediate by the interconnected communities of microbes from Archaea and bacterial domains and some smaller percentages of viruses and eukaryotes. These biochemical based transformations offer significant degradation of complex organic compounds to the reduced and oxidized forms of carbon such as methane, and carbon dioxide (Batstone and Virdis 2014).

Anaerobic digestion process comprises of four steps such as hydrolysis, acidogenesis, acetogenesis, and methanogenesis. Hydrolysis causes the cleavage of the complex biopolymers by the action of extracellular enzymes of specific fermentation causing bacteria to smaller monomers like proteins, lipids, soluble organic matter, and polysaccharides, these are all degraded and the final products are further treated through acidogenesis to produce volatile fatty acids (VFAs). During acidogenesis, monomers are converted by fermentative bacteria mainly into volatile fatty acids like alcohols, propionic acids, butyric acids, acetic acids and lactic acid, and also molecular hydrogen.

Mostly, these bacteria belong to the category of obligate anaerobes but sometimes facultative anaerobes may also be there. During the process of acidogenesis, volatile fatty acids are converted into molecular hydrogen, CO_2 and acetate. Acetogenic bacteria anaerobically oxidized the hydrolysis products other than acetate such as H_2 and CO_2, i.e., propionate, alcohols, aromatic acids, longer-chain fatty acids and aromatic acids into acetate. During the process of methanogenesis, methane gas is produced by three main ways, methylotrophic, acetolactic, and hydrogenotrophic by methanogenic archaea.

Aerobic respiration is though most effective way to decompose organic matter and waste material but may not cause complete breakdown of effluents. Porous solid materials are present in the tanks, where biofilms can be easily developed, thus enhancing the numbers of microorganisms and thus the efficacy of decomposition process. The produced material is solid in nature and is known as activated sludge is formed during this process, containing a mixture of undigested materials and microbes. Since all of vital microbes are present in it to break down incoming waste, some of it is also added to new batches of sewage. Mostly the addition of aerobic bacteria is for the new treatment plants in the aerated environment. Free oxygen is used by this bacterium within the water for degradation of contaminants in the wastewater and then convert it into energy to be used for growth, development, and reproduction.

3 Strategies to Using Microbial Techniques for Wastewater Treatment

Global reports of subsurface heavy metal pollution of water have become a major health problem and, in this regard, proper knowledge of the source of wastewater and its biological, physical, and chemical aspects are important. It is very important to identify the appropriate strategies for treatment. Microorganisms are partly the key to reducing pollution and maintaining the stability of biological systems. The implementation of biological wastewater treatment technology has many advantages as opposed to other treatment measures, relatively inexpensive costs, minimal emissions, and less detrimental effects on the environment. Also, biological wastewater treatment technology is reaping economic benefits against both chemical as well as physical therapy technologies, in terms of rehabilitation costs and investment of capital (Mittal et al. 2011).

3.1 Biological Treatment Techniques

In Aerobic treatment, the pond contains bacteria and algae that can survive in the aerobic state. Cao and Li (2011) proposed electrolysis involving the biological oxidation procedure for wastewater which contains alkyl-benzene sulfonate. Souza et al. (2011) used bio-activated carbon to treat refinery wastewater for reuse. These bio-compounds can bear a wide range of pH and temperature. Furthermore, they are more suitable in environmental and Petrochemical implementation than synthetically made surfactants because of naturally producing macromolecules like fatty acids, lipoproteins, and glycolipids.

In the Anaerobic technique, the pond is involved in the fermentation procedure which is especially effective in eliminating organic compounds from the solution and eliminating high concentrations of BOD and COD. The anaerobic treatment system has been working in wastewater of industry treatment for several years. In this process, biogas produced that containing methane as well as carbon dioxide. This process can take place in places where organic matter is available but redox potential is lower. Both aerobic and anaerobic processes can be employed in the treatment of dirty water (Table 1).

Table 1 Comparison among aerobic and anaerobic systems

Factors	Aerobic	Anaerobic
Temperature	Low	High
Nutrient necessity	High	Low
Effluent quality	High	Medium
Odor	Low potential	High potential
Energy demand	High	Moderate

3.2 Microalgae Role in Treatment of Wastewater

Microalgae contain photosynthetic ability that change solar energy into biomass and have the efficacy to absorb important nutrients like N, P in a short time. The microalgae involve in the treatment of sewage systems in a wide range and are called tertiary treatment procedures that can extract organic ions. Extract of organic ion get by biologically or chemically (Abdel-Raouf et al. 2012).

3.3 Microbial Electro Remediation Technique

Major portion of metal wastewater produced by humans and industries. The metal-laden polluted water causes critical environmental and health issues and should be properly managed to avoid negative consequences. A significant volume of metal-loaded wastewater is produced due to industrial and human activities, for removal of metal ions strict instructions have been clasped to avoid contamination. An ideal approach is not only to remove metals but collecting and retrieving them during the treatment procedure. Although there are numerous ways to treat or remove arsenic from water and wastewater, recent research has led to the development of technological approaches that seem economically realistic, low cost, and friendly for the environment.

Therefore, there is an urgent need for serious research and development in this direction so that modern techniques can be further advanced, and its scope of application can be extended to the real situation in the direction of pollution prevention. Traditional techniques such as precipitation, coagulation, and the removal of metals are needed to find a solution, which is generally considered less effective. It is costly to treat methods of activated carbon-based absorption, ion exchange, and membrane technologies that involve large amounts of industrial pollutants and wastewater that contain large amounts of heavy metal ions that cannot be operated on a large scale.

3.4 *Microbial Treatment of Sewage Sludge Concerning Specific Organic and Inorganic Contaminants*

Sewage sludge is a semi-solid material and remaining of municipal and industrial wastewater. The word "septage" is also known as the simple treatment of sewage, but it is involved in a clean system from the site, like as a septic-tank. The treatment of sewage sludge describes various methods that are used to dispose and manage that sewage sludge produced during the treatment of sewage. That sewage is usually with more amount of water with lower quantity of solid materials. In Primary sludge, settleable solids are eliminated during primary treatment. While in secondary sludge the secondary clarifiers include in treated sewage sludge from bioreactors of secondary treatment.

3.5 *Treatment Processes for Organic Contaminants*

The sewage sludges are treated by using various types of techniques, its objective is to lessen the organic matter amount and various harmful microorganisms which cause diseases inside solids. Most techniques include aerobic and anaerobic systems. Sludge technique about 50% and also provide biogas which is a good source of energy (Cao and Li 2011).

3.6 *Anaerobic Technique*

Anaerobic is a bacterial procedure when oxygen is not present. This procedure includes a thermophilic technique where the sludge is fermented at 55 °C temperature inside the tank or 36 °C in a mesophilic system. MAD (Mesophilic anaerobic digestion) is also an easy method for the treatment of sewage sludge. In this method, the sludge is fed into big tanks and retain for a minimum of 12 days which allow digestion procedure to digest sludge. These are including acidogenesis, hydrolysis, methanogenesis, and acetogenesis.

In that procedure, the complex sugars and proteins splits into various compounds like methane, carbon dioxide, and water (Souza et al. 2011). Anaerobic produces biogas with a major amount of methane which is used to run engines and provide heat to the tank. Methane production is the best advantage in this process. While in liquid sewage sludge the denitrifying bacteria convert nitrate to dinitrogen which removes that nitrate from sludge. The solid sewage in primary treatment separated is anaerobically fermented by bacteria.

3.7 Aerobic Technique

This process occurs when oxygen is present that directly involved in the continuation of the procedure of activated sludge. In this technique, the bacteria digest organic matter and release carbon dioxide. In absence of organic matter, the bacteria starting die and other bacteria used them as food. This process stage is called *endogenous respiration*. Then reduction of Solids occurs in this stage. Because aerobic technique happens faster as compared to anaerobic and the aerobic capital cost is also lower. Aerobic technique can also be obtained by using jet aerators that oxidize sludge. Excellent bubble spread is usually a more cost-effective method of dispersal but plugging is usually a problem due to sedimentation in small air holes. Coarse bubble spread is commonly used in tanks of activated sludge or the flocculation.

3.8 Treatment Processes for Inorganic Contaminants

Inorganic contaminants from wastewater are removed by Bio-absorption and Bioaccumulation. Bio-absorption is a fast and reversible passive adsorption mechanism. The inorganic contaminants like metals are retained by physiochemical interactions like (adsorption, ion exchange, precipitation, crystallization, and complexation) between the metals and functional groups of cell surface (Fosso-Kankeu and Mulaba-Bafubiandi 2014). Several factors affects bio-absorption of contaminants like pH, biomass concentration, ionic strength, particle size, temperature, and other ions present in solution.

While bioaccumulation includes both extracellular and intracellular processes. In general, bio-absorption is inexpensive as biomass can be produced from industrial waste and offers significant benefit of regeneration. On the other hand, bioaccumulation is expensive because the processes occur in living cells whose reuse is limited. Bacteria also causes elimination of heavy metals from wastewater through functional groups present in their cell wall-like aldehydes, ketones, and carboxyl groups and thus produce less chemical sludge. Brown and red algae are also being used as bio-absorbents, and the use of yeasts and fungi has also been reported for absorption.

3.9 Microbial Ecology of Sewage Sludge and Wastewater Treatment

Biological treatment of wastewater and sewage sludge is most important biotechnology implementations as the driver of the critical systems microorganisms are key to its success. So, the study of dirty water microorganisms is of clear importance. However, the significance of treated wastewater reactors is overlooked as a model system for the environment of microbes. No doubt, the microbial environment of bioprocesses is of great importance for the performing bioprocesses, especially in WWT (wastewater treatment).

Microorganisms have their characteristics during the treatment of wastewater, and they focus on the procedure that is used. There are several types of therapies, including biases, anaerobic therapies, and aerobic procedures that involve protozoa and bacteria, but their fate is unnecessary. The condition of the fungus has a endurance rate such as low pH and low nitrogen which makes the fungus well-thought-out wastewater treatment. Thus, the fungus has the potential to impair the ability to settle sludge due to its fibrous structure, which can affect this process. The rotifer presence at the beginning of treatment of wastewater is the best sign as it can absorb dispersed organic matter and bacteria (Pagnanelli et al. 2009).

3.10 Ponds Stabilization

Waste consolidation ponds are an unconventional system for treating wastewater. This stabilization of wastewater, known as biological treatment, that can work well when equipment maintenance is limited, and directly promotes better thickening of sludge. The proper architecture will help in the cultivation of algae and bacteria which will effectively and completely remove the organic waste in the water thus reducing the problem during the treatment and wastewater disposal (Vaajasaari and Joutti 2006).

3.11 Structural Units of Bacteria

Heterotrophic bacteria have a significant role in organic matter removal from wastewater treatment. That bacteria work in the treatment of wastewater in clusters such as biofilm or granule and floc.

3.12 Flocs

Floc is sludge that forms a bacterial colony by attaching to cells and pollutes wastewater through physiological chemical processes. Flocs contain bacteria and EPS. The content of microorganisms and factors mediates flux stability because environmental stress causes the floc to disintegrate.

4 Wastewater as an Exceptional Resource of Renewable Energy

Wastewater which is produced from different sources is enriched with many different nutrients, minerals, organic matter different metabolites which are used for the progress of many microorganisms, algae, and various plants that are used to generate renewable energy products. Methane gas is released when organic stuff was decomposed in an oxygen-free atmosphere (Koch et al. 2015). When the solid slush is treated via thermal hydrolysis, a large quantity of methane gas can make. Then waste is entered into an anaerobic digester, starts breakdown, and obtains the final product in the form of methane gas which utilize as natural gas (Maragkaki et al. 2017). A distinctive wastewater has a 0.5 kg/m^3 COD value and tentatively can produce 1.47 to 107 J/kg which oxidized to CO_2 and water while energy density of wastewater is 0.74 to 107 J/m^3.

5 Strategies for Energy Adoption from Wastewater

Processing to their capability as energy basis procedure streams must be distinguished by succeeding input, intermediate and output streams that acceptable to technical possibilities for recovering energy from wastewater.

5.1 Inputs

Organic content of carbonaceous dissolved and suspended that was in wastewater ways its energy potential as a chemical nature. The absorbance of carbon dioxide through sunlight energy and usage of wastewater for growth media is due to their inorganic components (Bhatia et al. 2019).

5.2 Intermediates

Many intermediate compounds were prepared by green plants, algae, and microorganisms which are used as a storehouse of biochemical energy. For their activities, these compounds are not only used by microorganisms but also by animals to meet the necessity of energy. These compounds can also be used to produce various energy products for example gaseous methane or hydrogen, biodiesel, or liquid ethanol, or solid dry biomass by the use of specific microorganisms (Evcan and Tari 2015).

5.3 Outputs

Particularly fuel provided by methane can be utilized to generate electricity, heat, and even in propulsion automobiles. The presently provided system is the least effective and it can transform 25–35% of thermal energy to electrical energy and it causes energy losses (Kassongo and Togo 2011; Naina Mohamed et al. 2020). To make it better and efficient joined heat and power solicitation is suggested also skills that transformation of inputs towards intermediates with the assistance of carbon-bound energy into biodiesel, biogas and finally convert into outputs with the help of gasification. Further, it changes the inputs directly into outputs by heat and microbial fuel cells for generation of electricity (Ungureanu et al. 2020).

6 Beneficial Energy Products Generation from Wastewater

If wastewater treatment is controlled, it can produce many valuable stuffs. To obtain valuable material from wastewater biological wastewater system is commonly used. Varieties of products that can be specifically utilized in the form of biofuels are produced from wastewater skills, which are described in Table 2.

Table 2 Energy products recycled from various sources of wastewater, their operational feature and characteristics

Bioenergy produced	Source of wastewater	Operational conditions	References
Biogas	Sewage Sludge Waste from municipal source	Use of anode and neutral red graphite and modified bacteria Dynamic Membrane Filter By using anode reactor pH range: 6.8–7.3	Rahimnejad et al. (2015) Quek et al. (2017) Xu et al. (2018)
Biodiesel	Diary Wastewater Sludge Textile Wastewater Domestic Wastewater	Sludge dewatering and drying Bioremediation of Microalgal followed by lipid and biodiesel production Cultivation of Microalgae Nostoc sp., Chlorella sp.	Leandro et al. (2019) Fazal et al. (2018) Mostafa et al. (2012)
Microbial Fuel Cell	Urban Wastewater Sewage Sludge Starch Processing Wastewater	Presence of salt bridge, two-chamber of graphite electrode Microbial fuel cell: brush electrode, graphite fiber brush electrodes Carbon paper anode	Slate et al. (2019) Liang et al. (2011) Lv et al. (2014), Malaeb et al. (2013)

7 Mechanisms Involved for Recovering the Renewable Energy Products

To produce a specific type of energy stuffs many values added energy objects from wastewater are digested with the help of many microorganisms. The components which are left after digestion are further treated with the help of many physical as well as chemical methods to get more energy. A huge variety of energy products also be taken from wastewater sewerages (Khalid et al. 2011). Methods that are used to produce energy from wastewater are described below:

7.1 *Production of Biogas*

Anaerobic handling of wastewater treatment offers the capacity to speedily remove organic contents of waste while decreasing the energy usage of dealing method and the manufacture of sludge and microbial biomass (Cavinato et al. 2011). It is a difficult procedure that includes various reactions in the absence of oxygen like methanogenesis, acetogenesis, and hydrolysis (Bhatia et al. 2020a, b). On a vast variety of waste discharges anaerobic digestion is very useful effluents like sewage sludge, industrial wastewater, domestic wastewater, it is also beneficial for the alteration of useful products into different forms such as biohydrogen and methane (Parihar and Upadhyay 2016).

Formation of slush in wastewater generate by-product which takes more dealing. The decrease in sludge and energy utilization are the two points that make it economically striking for industrial and municipal waste streams to reflect direct anaerobic pre-treatment of wastewater. This digestion is exaggerated by many reasons like temperature (25–350C), pH (~7), C/N ratio, carbon sources, moisture, and nitrogen. Just because of fewer disposals AD of manure sludge is treating plants and is eco-friendly also. Significant degradable organic components are also produced by direct anaerobic treatment (Manyuchi et al. 2018). Effluents that are produced by anaerobic treatment are not directly throwing into receiving water and they require aerobic polishing. The average ambient temperature of the wastewater influences anaerobic dealing design quality. Effective anaerobic treatment of wastewater as low as 150C is achievable but the use of anaerobic digestion is not reserved in contemplation below 120C (Bhatia et al. 2017).

7.2 *Microbial Fuel Cells*

Bacterial oxidation process involved in microbial use cell in which bacteria oxidized organic matter to treat the wastewater and play a role in cation exchange process of cathode and anode in which electron transfer through electricity production due to difference in potential coupled with flow of electrons (Rahimnejad et al. 2015). Microbial fuel cells are the advanced and emerging technology that has been successfully operated in pure as well as in mixed cultures and enriched by activated sludge from wastewater treatment plants (Forss et al. 2017). This technology is eco-friendly due to the already presence of bacteria in wastewater to produce electricity as a catalyst. Despite this advantage, its advancement is hindered due to low power and high cost and valuable products.

Waste activated sludge (WAS) present a major ongoing disposal challenge and by-product of activated sludge-based water treatment for water management authorities worldwide. Conventional waste-activated sludge has been used in agricultural practices such as preparation of land, soil health, offensive odors, and disease risks from toxic chemicals and pathogens. These restricted chemicals hinder the acceptance in public for this adaptation (Egan 2013). Sustainable waste-activated sludge consists of the recovery and reuse of value-added products and also has the potential to minimize environmental as well as human harmful impact. This implementation is usually have been applied in the agriculture sector due to high nutrient and organic matter. It is also a rich source of making of methane gas by anaerobic digestion when mixed in primary sludge.

The land application and implementation of urban wastewater is necessary as compared to rural wastewater is necessary because of the accumulation of industrial effluents which results in too much contamination in the environment as well as in human health which is harmful to the food chain and animal health (Campbell 2000). The production of electricity by thermal energy is also a popular and sustainable way for WAS management (Rulkens 2007). However, it is a cost-effective procedure as additional fuel is required to maintain additional facilities and requirements due to higher energy utilization due to high moisture contents and lower heating values of biosolids (Wang et al. 2008). There is a fundamental challenge for specific biorefinery approaches due to presence of all compound of WAS system in the heterogenous and single complex mixtures.

7.3 Amino Acids and Proteins

Waste activated sludge can be collected as a source of protein and amino acid which consist of organic compounds in the type of protein, lipids, and polysaccharides. It contains almost 70–80% protein fraction in which 50% dry weight of bacterial cell is present (Raunkjaer et al. 1994). Consequently, protein derived from waste activated sludge present an impactful and potential source which is the main source for production of feed of animals compared with traditional source of protein. However, detoxification of sludge is carried out for removal of heavy metals, sterilization process (Adebayo et al. 2004).

The solubilization of intracellular material is an effective way for recovering protein from waste-activated sludge. Thermal digestion is one of the easiest methods which are supported by the centrifugal separation. This process increases the wastewater sludge decomposing ability and break down of decomposing sludge and lysing of the microbial cell. Thermal digestion has a notable profit such as low-cost treatment, no use of additional waste, no use of reagents for waste degradation, use of effective heat exchange. The efficiency of chemical and mechanical treatment on protein extraction is described in two activated sludge and check the compatibility with quantification method (Ras et al. 2008).

The efficacy can be improved by applying mechanical and chemical treatments and various extraction protocols and similar approaches. Triton treatment is used to extract the protein and show the significant hydrophobic interactions linking protein with extracellular polymer matrix. The waste activated sludge amino acid is friendly for the environment due to associated amphiphilic molecular structure which contains carboxyl groups and amino acids. The use of amino acids as powerful inhibitors to regulate destructive responses in a few unique metals in acidic media has been confirmed by a progression of examinations (Khaled 2010).

7.4 Bio-Pesticides

Nitrogen, Carbon and phosphorus is the enriched nutrient source from waste activated sludge which is the potential and feasible medium of growth for microbial accumulation to produce valuable metabolic products. *B. thuringiensis* (Bt) can produce the proteinaceous Para poral crystal inclusion during spore formation which is called endotoxin which is the most famous bio-pesticide globally (Bravo et al. 2001). The production of these bacteria depends on the growth medium of nutrients sources like nitrogen, carbon, protein, and yeast sources (Lisansky et al. 1993). Reuse of waste-activated sludge as a medium for Bt production depends upon its utilization in agriculture for pest control and economical as well as compatible exercises. Three possible strategies are important for Bt production process as Fermentation, recovery, and formulation of products. Several factors like pH, dissolved oxygen concentration, C/N ratio, foaming, and inoculum sludge which have an impact on the production of bio-pesticide.

7.5 Bio-Flocculants and Bio-Surfactants

Bio-surfactants and bio-flocculants are the significant metabolic products during microbial transformation. Microorganisms secreted polysaccharides, cellulose derivatives, and lipids which are consumed in mineral and chemical industries such as food and wastewater treatments (Flores et al. 1997; Jegou et al. 2001). Bio-flocculating activities are non-toxic and degradable for humans and the environment as compared with synthetic flocculants (Yokoi et al. 1996). Due to the high cost of these by-products, the use of this treatment is limited due to its association with the supplies of organic sucrose and glucose. Waste activated sludge is one of the best reservoir for separation of bio-flocculant producing microorganisms during aerobic process occur naturally. Moreover, a variety of different bacterial strains of bio-flocculants have been isolated from waste activated sludge named as *Bacillus cereus*, *Achromobacter* sp., *Agrobacterium* sp., *Enterobacter* sp., *Pichia membranifaciens*, *Exiguobacterium acetylicum*, *Rhodococcus erythropolis*, *Solibacillus silverstris*, *Saccharomycete* spp. etc. (Wang et al. 2014).

8 Microbial Fuel Cells for Improved Bioremediation

Microbial fuel cells (MFCs) can be used as power resource and as a tool for bioremediation. By creating an electrical connection between the anaerobic sediments and aerobic water column, an MFC can increase the metabolism rates of bacteria in the sediment, allowing the bacteria to break down complex molecules they would not be able to consume otherwise. When bacteria break down organic matter, it produces CO_2, protons, and electrons. The bacteria required more energy to breakdown the organic matter by using electron acceptors with high electric potentials, in the form of oxygen. Ideally, the bacteria donate electrons to oxygen molecules, which can combine with hydrogen to produce water.

However, when bacteria live in sediments, which are typically anaerobic environments, they are not able to easily access oxygen. Some bacteria can access electrons from the oxygen in the water column by using natural shuttles such as iron oxide materials (Li and Yu 2015). However, these electron transfers are weak due to low concentrations of electron mediating substances in the sediment (Li and Yu 2015). Similarly, some cable bacteria can form chains to reach the surface of the sediments (Schauer et al. 2014). However, in most cases, bacteria must transfer their electrons to less energetically favourable reactions, such as sulphate, which lowers the amount of energy they receive and inhibits their ability to break down difficult to degrade organic matter.

An MFC functions like common batteries. However, the chemistry in the MFC is catalysed by the metabolism of bacteria. An MFC is created by placing the anode into the sediment, an anaerobic environment, while the cathode is placed in the water column, an anaerobic environment. The following reaction, catalysed by the consumption of organic matter (CH_2O) by bacteria, occurs at the anode:

$$CH_2O + H_2O- > CO_2 + 4H^{++}\ 4e^-$$

The anode is the electron acceptor, which transfers the microbes' electrons from the sediments to the water column through a wire where the electrons are donated to oxygen. The following reaction occurs at the cathode:

$$4H^{++}\ 4e^- + 2O_2 - > H_2O$$

By living on the MFC anode, microbes can utilize the reaction with the greatest electric potential and break down organic matter that cannot be broken down under anaerobic conditions or is decomposed slowly. Earlier research on MFCs focused on creating batteries that can produce a current to power another instrument necessary to monitor the site while cleaning up biotoxins (Santoro et al. 2017). Additionally, MFCs can be utilized by wastewater treatment plants to decompose organic matter, particularly sulfides, and produce electricity (Du et al. 2007). However, another use for MFCs is to use a similar design to increase rates of bioremediation.

9 Algae and Cyanobacteria for Wastewater and Sewage Treatment

Micro-algae such as eukaryotic algae and cyanobacteria are sustainable and energy-intensive for biological treatment process which is environmentally friendly and are used worldwide (Singh et al. 2015). Micro-algae use in wastewater treatment is cost-valuable and renewable source of biomass for the biological fixation of carbon dioxide (Almomani et al. 2019). Microalgae are historically seen as difficult and cost-effective to remove and cause problems that lead to create dangerous disinfection by-products. Algae have the potential to improve wastewater and wastewater treatment plant effluent and generate biomass for biofuels (Arbib et al. 2014). During wastewater treatment, algae integrated into the secondary treatment process as well as in tertiary treatment. During secondary treatment, algae need low aeration due to solar irradiation which is difficult due to turbid conditions (Humenik and Hanna 1970).

In the tertiary treatment process, it is cost-effective, and the generation of biofuels or other useful products may be offset. During tertiary treatment, the process has direct access to sunlight which improves to removal of nutrients that remained in secondary treatments in which algae is settled by other biosolids while in tertiary, it required additional harvesting (Van Den Hende et al. 2011). For drinking water safety, production of biofuel, or wastewater treatment, algae are used as a biomass in which chlorophyll is used to check the concentration of algae, and its growth. Chlorophyll can be quantified by using autofluorescence or absorption methods on the instruments (Held 2011).

10 Current Challenges

Along with advantages, several challenges are still hinder the implementation of wastewater and sewage sludge treatment related to industrial application as well as consumption of energy in the cultivation process. Pumping and Aeration systems are most conventional approaches for wastewater treatment are frequently used to culture micro-algae to create turbulent flow to enhance the gaseous exchange and environmental performance. Nowadays sewage sludge treatment is becoming a difficult strategy due to its cost-effective nature in terms of its constituents (Muhammad and Rohani 2011). Production of biorefinery from waste activated sludge offers advantages for sludge management in the future by adopting a treatment pathway for a sustainable production system and looking forward to future for value-added by-products. Acceptable challenges and issues which are associated with bio-refinery production from waste activated sludge included:

1. Enzyme and protein production is highly cost-effective in terms of heavy metal toxicity and pathogenic (Kalogo and Monteith 2008).
2. Selection of heavy metal tolerant microbial strains is needed to progress and recovery of metabolic products such as bio-pesticides, bioplastics, bioflocculants, and bio-surfactants need further optimization of operational parameters.
3. Feasible growth environment and wastewater matrix for treating waste-activated sludge with specific bioproducts should also be refined as it exerts direct influences on growth, functioning, and survival of microbial communities.
4. Purification and efficient work needed more development during the treatment of wastewater and would improve in overall biorefinery approaches and sludge management practices.
5. In an anaerobic digester, low temperature is also a crucial challenge because microorganisms need optimum temperature15–35 °C for their growth and multiplication and if there is any change from this range, then kinetics of the overall mechanism disturb (Malaeb et al. 2013; Bhatia et al. 2017).

Stabilization of sewage sludge to be used for arable land is a major challenge in both developed and underdeveloped world as it generally contains a good level of organic and inorganic pollutants and pathogenic microorganisms that can cause serious consequences for human beings, animals, and surroundings. After stabilization, it must be properly analysed for risk assessment to determine its safety profile. Bacterial communities may differ due to temporal and spatial factors so the determination of the functional potential of these communities according to prevailing climatic conditions, and soil type is also a serious challenge that requires specific attention of regional research institutes and organizations.

11 Future Prospects

The latest advancement in science and technology has greatly revolutionized the sewage sludge management practices over the past decades and thus there has been greater contribution for environmental protection, maximum positive and safe use of biosolids and residues for agricultural purposes, and human safety. However, the accelerated costs associated with the microbial techniques and other biological processes have always been a major concern and needs significant attention from scientific communities, policymakers, and institutional organizations.

Improved production of biogas, advanced dewatering techniques, controlled thermal, and landfilling processes are greatly being applied practically. Sometimes these costs may exceed 50% of the total amount as for the treatment of wastewater. Problems of high cost can be resolved by using reuse and recovery practices to hit the mark of sustainability. The sustainable management of sewage sludge revolves around six major practices such as improving the value of sewage sludge by various techniques (especially biological), beneficially using the compounds of organic carbon, and other inorganic compounds, decrease of total volume of sludge, recovering phosphates, and other essential nutrients from sewage sludge, changing the scenario of microbial treatment by using different strains, and combinations of different biological, physical, and chemical practices for sustainable sludge management (Picture 1).

Sustainable management of sewage sludge has become a serious issue around the globe and there should be direct and target-oriented studies to manage sewage sludge without causing any serious implications for human beings, animals, and environmental protection. Biological processes and the use of microbial techniques are greatly helpful for the biological conversion of chemical energy of the sewage sludge to good quality and methane-rich biogas. Sewage sludge contains nutrients in the form of proteinaceous materials as can be used as an exceptional plant fertilizer for direct application onto the soil. Therefore, a good revenue can be generated by using microbial and biological treated sewage sludge.

Sewage sludge is a precious source of essential nutrients and carbon contents and can be utilized as an amendment for improvement of soil health and overall fertility. But right integration of desired amount and providing safe and well-managed sludge

Picture 1 Future prospects for management of sewage sludge and wastewater by microbial techniques

to the growers and farmers is important to eliminate the hazards of toxicity. Even though microbial technologies and biological processes are significantly helpful to produce safe and high-quality products and end-products, but the safety profile of these products must be ensured to get maximum benefits. Moreover, the resulting products must be properly tested by following the regulatory measures and standards.

The sewage sludge contains different types and levels of pollutants such as endocrine-disrupting chemicals, pharmaceutical contaminants, nanoparticles, contaminants in personal care products, pesticides, fertilizers, and micropollutants. There is always a variable number of effluents in sewage sludge and the use of microbial techniques should be properly optimized according to the level of contamination and hazards. Moreover, the microbial niches also need proper optimization according to prevailing climatic conditions, and treatment conditions. There should be proper and well-documented efforts to interlink, and interconnect the energy, food, and water in the nexus system.

Management strategies should be optimized in such a way that there is no negative effect on the climate, environment, and ecosystem functioning. The protection of the ecosystem and climate is not only dependent on the quality of water but on productivity also. Therefore, there should be optimized and well-planned proposals for the nexus of microbial niches for the sustainable management of sewage sludge. Nexus of microbial niches should be significantly capable of removing both unknown and known pollutants.

The provision of quality and easily available food sources to different microbial species is a key target to ensure maximum removal efficiency by microorganisms. Provision of food source is essential for a diverse range of microorganisms such as polyphosphate accumulating organisms, denitrifies, nitrifiers, and heterotrophs. The microbial communities should be cultivated through an optimized series of aerobic, anoxic, and anaerobic reactors to increase the removal efficiencies. Currently, microbial techniques for the removal of different contaminants are not focused on the targeted removal of multiple contaminants so scientific communities should be more focused on the synergistic removal of different contaminants for safe and effective handling of sewage sludge.

There should be a significant focus on bioinformatics and novel microbial techniques for uncultured microbial functioning. Niches related to the novel functioning of microbes may have a greater level of variation than the conventional and cultured microbes so proper investigation about microbial interactions, diversity, and metabolic kinetics should be properly evaluated and studied. Further implementation of technological solutions as per community standards of microbes can help to produce significant beneficial results. The functioning of microbes for sewage sludge management and wastewater treatment processes can be greatly improved by the clarification of biological mechanisms. So, therefore there should be a good collaboration between scientific communities, and international organizations for data sharing and improved understanding and development of working standards.

A combination of long-term operating systems and diverse microbial communities can significantly help to discover unknown functions of microbes and can also help for the development and optimization of different strategies. Natural environments like intertidal zones can also provide alternate and are valuable sources for microbial functions, and metabolisms. Identification of microbial metabolism and functioning is important for combining proteomics, RNA, and DNA-based techniques for regulatory strategies and exploring the ecological functioning of microbes. There should be proper consideration for the regulation of amino acids, vitamins, and micronutrients along with elucidation of metabolic pathways and macronutrient cycles.

Furthermore, proper attention should be given to operational control processes and system designing. The designing of new functional systems must be capable of following scientific and technological rules to provide more time, space, and substrate ingredients for diverse functioning. Also, there should be a controlled and optimized focus to control environmental conditions to achieve efficiency and performance for different microbes. Interdisciplinary cooperation can also play a key role to achieve the purpose of sewage sludge and wastewater management.

12 Conclusion

Sewage sludges always contain a good level of bacterial diversity and the identification of bacterial community structures and chemical attributes is significantly important to target the desired efficiency of treatment and production of end products. There has been a good potential for the treatment of sewage sludge and wastewater by using microbial techniques. But this potential has not been fully explored due to the diversity of microbial species, their chemical attributes, and different functioning under different climatic and geographical conditions. A collaboration between researchers, scientific communities, and international students will be essentially helpful to achieve the goal of sewage sludge management on a sustainable basis.

References

Abdel-Raouf N, Al-Homaidan AA, Ibraheem IB (2012) Microalgae and wastewater treatment. Saudi J Biol Sci 19:257–275

Adebayo O, Fagbenro O, Jegede T (2004) Evaluation of Cassia fistula meal as a replacement for soybean meal in practical diets of *Oreochromis niloticus* fingerlings. Aquac Nutr 10:99–104

Almomani F, Judd S, Bhosale RR, Shurair M, Aljaml K, Khraisheh M (2019) Integrated wastewater treatment and carbon bio-fixation from flue gases using *Spirulina platensis* and mixed algal culture. Process Saf Environ Prot 124:240–250

Arbib Z, Ruiz J, Alvarez-Diaz P, Garrido-Perez C, Perales J (2014) Capability of different microalgae species for phytoremediation processes: wastewater tertiary treatment, CO_2 bio-fixation and low-cost biofuels production. Water Res 49:465–474

Atashgahi S, Aydin R, Dimitrov MR, Sipkema D, Hamonts K, Lahti L, Maphosa F, Kruse T, Saccenti E, Springael D (2015) Impact of a wastewater treatment plant on microbial community composition and function in a hyporheic zone of a eutrophic river. Sci Rep 5:1–13
Batstone DJ, Virdis B (2014) The role of anaerobic digestion in the emerging energy economy. Curr Opin Biotechnol 27:142–149
Bhatia SK, Kim SH, Yoon JJ, Yang YH (2017) Current status and strategies for second generation biofuel production using microbial systems. Energy Conserv Manag 148:142–1156
Bhatia SK, Bhatia RK, Jeon JM, Kumar G, Yang YH (2019) Carbon dioxide capture and bioenergy production using biological system – a review. Renew Sust Energ Rev 110:143–158
Bhatia RK, Ramadoss G, Jain AK, Dhiman RK, Bhatia SK, Bhatt AK (2020a) Conversion of waste biomass into gaseous fuel: present status and challenges in India. Bioenergy Res 13:1046–1068
Bhatia RK, Sakhuja D, Mundhe S, Walia A (2020b) Renewable energy products through bioremediation of wastewater. Sustainability 12:7501
Bravo A, Likitvivatanavong S, Gill SS, Soberon M (2001) *Bacillus thuringiensis*: a story of a successful bioinsecticide. Insect Biochem Mol Biol 41:423–431
Campbell H (2000) Sludge management–future issues and trends. Water Sci Technol 41:1–8
Cao XZ, Li YM (2011) Treatment of linear alkylbenzene sulfonate (LAS) wastewater by internal electrolysis-biological contact oxidation process. Water Sci Technol J Int Assoc Water Pollut Res 64:147–154
Cavinato C, Bolzonella D, Fatone F, Cecchi F, Pavan P (2011) Optimization of two-phase thermophilic anaerobic digestion of biowaste for hydrogen and methane production through reject water recirculation. Bioresour Technol 102:8605–8611
Du Z, Li H, Gu T (2007) A state-of-the-art review on microbial fuel cells: a promising technology for wastewater treatment and bioenergy. Biotechnol Adv 25:464–482
Egan M (2013) Biosolds management strategies: an evaluation of energy production as an alterntive to land application. Environ Sci Pollut Res 20:4299–4310
Evcan E, Tari C (2015) Production of bioethanol from apple pomace by using cocultures: conversion of agro-industrial waste to value added product. Energy 88:775–782
Fazal T, Mushtaq A, Rehman F, Ullah Khan A, Rashid N, Farooq W, Rehman MSU, Xu J (2018) Bioremediation of textile wastewater and successive biodiesel production using microalgae. Renew Sust Energy Rev 82:3107–3126
Flores ER, Perez F, De la Torre M (1997) Scale-up of *Bacillus thuringiensis* fermentation based on oxygen transfer. J Ferment Bioeng 83:561–564
Forss J, Lindh MV, Pinhassi J, Welander U (2017) Microbial biotreatment of actual textile wastewater in a continuous sequential rice husk biofilter and the microbial community involved. PLoS One 12:16
Fosso-Kankeu E, Mulaba-Bafubiandi AF (2014) Implication of plants and microbial metalloproteins in the bioremediation of polluted waters: a review. Phys Chem Earth Parts A/B/C 67:242–252
Held P (2011) Monitoring of algal growth using their intrinsic properties. Biofuel Res Appl Note 1–5
Humenik FJ, Hanna GP (1970) Respiratory relationships of a symbiotic algal-bacterial culture for wastewater nutrient removal. Biotechnol Bioeng 12:541–560
Jegou S, Douliez JP, Molle D, Boivin P, Marion D (2001) Evidence of the glycation and denaturation of LTP1 during the malting and brewing process. J Agric Food Chem 49:4942–4949
Kalogo Y, Monteith H (2008) State of science report: energy and resource recovery from sludge. Alexandria, VA, Water Environment Research Foundation
Kassongo J, Togo CA (2011) Evaluation of full-strength paper mill effluent for electricity generation in mediator-less microbial fuel cells. Afr J Biotechnol 10:15564–15570
Khaled K (2010) Corrosion control of copper in nitric acid solutions using some amino acids A combined experimental and theoretical study. Corros Sci 52:3225–3234

Khalid A, Arshad M, Anjum M, Mahmood T, Dawson L (2011) The anaerobic digestion of solid organic waste. Waste Manag 31:1737–1744

Koch K, Helmreich B, Drewes JE (2015) Co-digestion of food waste in municipal wastewater treatment plants: effect of different mixtures on methane yield and hydrolysis rate constant. Appl Energy 137:250–255

Leandro MJ, Marques S, Ribeiro B, Santos H, Fonseca C (2019) Integrated process for bioenergy production and water recycling in the dairy industry: selection of Kluyveromyces strains for direct conversion of concentrated lactose-rich streams into bioethanol. Microorganisms 7:545

Li WW, Yu HQ (2015) Stimulating sediment bioremediation with benthic microbial fuel cells. Biotechnol Adv 33:1–12

Liang M, Tao H, Li SF, Li W, Zhang LJ, Ni JR (2011) Treatment of Cu^{2+-} containing wastewater by microbial fuel cell with excess sludge as anodic substrate. Environ Sci Technol 32:179–185

Lisansky S, Quinlan R, Tassoni G (1993) *Bacillus Thuringiensis* production handbook: laboratory methods, manufacturing, formulation, quality control, registration. CPL Scientific Ltd., Berkshire

Lv Z, Xie D, Li F, Hu Y, Wei C, Fen C (2014) Microbial fuel cell as a bio-capacitor by using pseudo-capacitive anode materials. J Power Sources 246:642–649

Malaeb L, Katuri KP, Logan BE, Maab H, Nunes SP, Saikaly PE (2013) A hybrid microbial fuel cell membrane bioreactor with a conductive ultrafiltration membrane for wastewater treatment. Environ Sci Technol 47:11821–11828

Manefield M, Griffiths RI, Leigh MB, Fisher R, Whiteley AS (2005) Functional and compositional comparison of two activated sludge communities remediating coking effluent. Environ Microbiol 7:715–722

Manyuchi MM, Chiutsi P, Mbohwa C, Muzenda E, Mutusva T (2018) Bio ethanol from sewage sludge: a biofuel alternative. S Afr J Chem Eng 25:123–127

Maragkaki AE, Fountoulakis M, Gypaki A, Kyriakou A, Lasaridi K, Manios T (2017) Pilot-scale anaerobic co-digestion of sewage sludge with agro-industrial by-products for increased biogas production of existing digesters at wastewater treatment plants. Waste Manag 59:362–370

Meyer DD, de Andrade PAM, Durrer A, Andreote FD, Corção G, Brandelli A (2016) Bacterial communities involved in sulfur transformations in wastewater treatment plants. Appl Microbiol Biotechnol 100:10125–10135

Mittal P, Jana S, Mohanty K (2011) Synthesis of low-cost hydrophilic ceramic–polymeric composite membrane for treatment of oily wastewater. Desalination 282:54–62

Mostafa SSM, Shalaby EA, Mahmoud GI (2012) Cultivating microalgae in domestic wastewater for biodiesel production. Not Sci Biol 4:56–65

Muhammad NS, Rohani S (2011) Lipid extraction and biodiesel production from municipal sewage sludges: a review. Renew Sust Energy Rev 15:1067–1072

Muszyński A, Łebkowska M, Tabernacka A, Miłobędzka A (2013) From macro to lab-scale: Changes in bacterial community led to deterioration of EBPR in lab reactor. Open Life Sci 8: 130–142

Muszyński A, Załęska-Radziwiłł M (2015) Polyphosphate accumulating organisms in treatment plants with different wastewater composition. Archt Civil Eng Environ 4:99–106

Naina Mohamed S, Ajit Hiraman P, Muthukumar K, Jayabalan T (2020) Bioelectricity production from kitchen wastewater using microbial fuel cell with photosynthetic algal cathode. Bioresour Technol 295:122226

Pagnanelli F, Mainelli S, Bornoroni L, Dionisi D, Toro L (2009) Mechanisms of heavy-metal removal by activated sludge. Chemosphere 75:1028–1034

Parihar RK, Upadhyay K (2016) Production of biohydrogen gas from dairy industry wastewater by anaerobic fermentation process. Int J Appl Res 2:512–515

Quek PJ, Yeap TS, Ng HY (2017) Applicability of upflow anaerobic sludge blanket and dynamic membrane-coupled process for the treatment of municipal wastewater. Appl Microbiol Biotechnol 101:6531–6540

Rahimi S, Modin O, Mijakovic I (2020) Technologies for biological removal and recovery of nitrogen from wastewater. Biotechnol Adv 43:107570

Rahimnejad M, Adhami A, Darvari S, Zirepour A, Oh SE (2015) Microbial fuel cell as new technology for bioelectricity generation: a review. Alex Eng J 54:745–756

Ras M, Girbal-Neuhauser E, Paul E, Sperandio M, Lefebvre D (2008) Protein extraction from activated sludge: an analytical approach. Water Res 42:1867–1878

Raunkjaer K, Hvitvedjacobsen T, Nielsen PH (1994) Measurement of pools of protein, carbohydrate and lipid in domestic wastewater. Water Res 28:251–262

Rulkens W (2007) Sewage sludge as a biomass resource for the production of energy: overview and assessment of the various options. Energy Fuel 22:9–15

Saia FT, Souza TS, Duarte RTD, Pozzi E, Fonseca D, Foresti E (2016) Microbial community in a pilot-scale bioreactor promoting anaerobic digestion and sulfur-driven denitrification for domestic sewage treatment. Bioprocess Biosyst Eng 39:341–352

Santoro C, Arbizzani C, Erable B, Ieropoulos I (2017) Microbial fuel cells: from fundamentals to applications. A review. J Power Sour 356:225–244

Schauer R, Risgaard-Petersen N, Kjeldsen KU, Bjerg JJT, Jorgensen BB, Schramm A, Nielse LP (2014) Succession of cable bacteria and electric currents in marine sediment. ISME J 8:1314–1322

Shchegolkova N, Shurshin K, Pogosyan S, Voronova E, Matorin D, Karyakin D (2018) Microalgae cultivation for wastewater treatment and biogas production at Moscow wastewater treatment plant. Water Sci Technol 78:69–80

Singh B, Bauddh K, Bux F (2015) Algae and environmental sustainability, 1st edn. Springer, New Delhi

Slate AJ, Whitehead KA, Brownson DA, Banks CE (2019) Microbial fuel cells: An overview of current technology. Renew Sust Energ Rev 101:60–81

Souza BM, Cerqueira AC, Sant'Anna GL Jr, Dezotti M (2011) Oil-refinery wastewater treatment aiming reuse by advanced oxidation processes (AOPs) combined with biological activated carbon (BAC). Ozone Sci Eng 33:403–409

Ungureanu N, Vladut V, Biris SS (2020) Capitalization of wastewater-grown algae in bioethanol production. In proceedings of the 19th International scientific conference. Engineering for rural development, Latvia university of life sciences and technologies, Jelgava, Latvia, 20–22 March 2020; pp. 1859–1864

Van Den Hende S, Vervaeren H, Desmet S, Boon N (2011) Bioflocculation of microalgae and bacteria combined with flue gas to improve sewage treatment. New Biotechnol 29:23–31

Vaajasaari K, Joutti A (2006) Field-scale assessment of phytotreatment of soil contaminated with weathered hydrocarbons and heavy metals (9 pp). J Soils Sediments 6:128–136

Wang H, Brown SL, Magesan GN, Slade AH, Quintern M, Clinton PW (2008) Technological options for the management of biosolids. Environ Sci Pollut Res 15:308–317

Wang L, Lee DJ, Ma F, Wang A, Ren N (2014) Bioflocculants from isolated strain or mixed culture: role of phosphate salts and Ca^{2+} ions. J Taiwan Inst Chem Eng 45:527–532

Xia Y, Wen X, Zhang B, Yang Y (2018) Diversity and assembly patterns of activated sludge microbial communities: a review. Biotechnol Adv 36:1038–1047

Xu H, Yang B, Liu Y, Li F, Shen C, Ma C, Tian Q, Song X, Sand W (2018) Recent advances in anaerobic biological processes for textile printing and dyeing wastewater treatment: a mini review. World J Microbiol Biotechnol 34:165

Yokoi H, Shiraki M, Hirose J, Hayashi S, Takasaki Y (1996) Flocculation properties of xanthan produced by *Xanthomonas campestris*. Biotechnol Tech 10:789–792

Biological and Thermo-chemical Treatment Technologies for Sustainable Sludge Management

Pooja Bhadrecha, Shilpy Singh, Indu Kumari, and Deepika Bhatia

Abbreviations

AGS	aerobic granular sludge
BP	biological predation
CAV-OX	cavitation oxidation process
COD	chemical oxygen demand
HTL	hydrothermal liquefaction
MBR	membrane bioreactor
MLSS	mixed liquor suspended solids
SCWG	supercritical water gasification
SCWO	supercritical water oxidation
SONIWO	sono-chemical degradation followed by wet air oxidation
SRT	sludge retention time
TMP	trans-membrane pressure
WWTP	waste water treatment plant

P. Bhadrecha (✉)
University Institute of Biotechnology, Chandigarh University, Mohali, Punjab, India

S. Singh
Department of Biotechnology, Mangalmay Institute of Management and Technology, Greater Noida, Uttar Pradesh, India

I. Kumari
Department of Environmental Sciences, School of Earth and Environmental Sciences, Central University of Himachal Pradesh, Kangra, Himachal Pradesh, India

National Institute of Pathology, New Delhi, India

D. Bhatia
Department of Biotechnology and Medical Sciences, Baba Farid Group of Institutions, Deon, Bathinda, India

V. D. Rajput et al. (eds.), *Sustainable Management and Utilization of Sewage Sludge*, https://doi.org/10.1007/978-3-030-85226-9_5

1 Introduction

Nature has been programmed to gift all the living organisms with enormous wonderful sources. The air, soil, water, plants all are blessed with tremendous amounts of resources. But rapid urbanization and increasing human activities have been deceiving them all by directly or indirectly polluting them. Pollutants generated via all the human, industrial, agricultural and other such activities are dumped into waterbodies, directly or indirectly. The primary and secondary stage treatments of wastes generated from municipal, paper and pulp mills' wastewater is referred as 'sewage sludge'. The increased rate of sewage sludge production, reduced landfill space, increased environmental hazards and rules and regulations for the disposal of sewage sludge for better environment has become a major problem worldwide (Rulkens 2008). The catchment area and infrastructure of treatment facility defines the properties and quantities of the generated sewage sludge. Non-toxic organic compounds, organic fraction which contain carbon source (utilized for biogas production), nutrients content, mainly nitrogen (ammonium) and phosphorus are the major constituents of sewage sludge. In addition to these, sewage sludge is contaminated with some very harmful heavy metals (concentrations range 1000 ppm to 1 ppm), polycyclic aromatic hydrocarbons, polychlorinated biphenyls, dioxins, linear-alkyl-sulfonates, nonyl-phenols, pesticides and pathogenic and microbial pollutants, trace organic chemicals like industrial chemicals, consumer/cosmetics products and pharmaceuticals in the aquatic environments which should be removed from recycling process (Spinosa 2004). The inorganic components of sewage sludge are water and other compounds which may contain Mg, Ca, aluminates and silicates, and their concentrations vary from trace values to even 95%. Therefore, sludge generated from sewage water can be a source of energy and fertilizer if its components are segregated with efficient methods. Therefore, extraction of useful materials or energy from sludge can be improved by its preliminary reduction in the initial phase only. There are conventional, alternative and hybrid methods used for the treatment of sludge. Conventional methods include land application, landfilling, incineration and anaerobic digestion. The alternative methods are thermochemical conversions which include pyrolysis, gasification, hydrothermal liquefaction and other supercritical methods. The hybrid methods for sludge treatment are wet oxidation and nanofiltration, Ultrasound/H_2O_2 or ozone, Ultraviolet/ H_2O_2 or ozone, Ozone/H_2O_2, Photo-Fenton processes, sono-photochemical oxidation, catalytic advanced oxidation processes, use of advanced oxidation processes in conjunction with biological oxidation, SONIWO (sono-chemical degradation followed by wet air oxidation), and CAV-OX (Cavitation Oxidation Process) (Gogate and Pandit 2004). Land application and landfilling has been the prominent methods for the disposal of sludge water. Advantage of landfilling is that the harmful pathogens present in the sludge remain covered which avoid spread of diseases by vectors (Syed-Hassan et al. 2017). A significant need of the hour is to minimize energy uptake for the employed methods to dewater sludge (Bień and Bień 2015). Incineration method involve the treatment of sludge at high temperature which reduces it volume up to 70% and destroys toxic compounds and pathogens as well. Anaerobic digestion consists of multiple phages like hydrolysis, acidogenesis, acetogenesis and methanogenesis which converts organic part of sludge into biogas and nutrient part into digestate (Zhang et al. 2014). The biological treatment of anaerobic digestion is a combination of O_3 treatment, use of microbes and annelids (protozoa, metazoan, earthworm), aerobic and anaerobic composting, advanced dewatering processes and complexing agents or microbial leaching. Thermochemical conversion method is an alternative to conventional ones which include pyrolysis, gasification, HTL (hydro-thermal liquefaction), SCWG (supercritical water gasification) and SCWO (super-critical water oxidation) (Bora et al. 2020). Pyrolysis and HTL can convert organic component of sludge into bio-oil which itself is an efficient energy source. Sludge decomposition without oxidizing agents is increasingly used in pyrolysis method. Fossil, electric and radiations can be a source of input energy for pyrolysis under different conditions (Callegari et al. 2020). Gasification is operated in devoid of oxygen to obtain maximum gas output (Werle and Sobek 2019). Thermochemical conversion has dual advantage as energy and valuable organic nutrients are also recovered in this. Further, the integrated pretreatment of chemical and thermal of sewage sludge has given more effective anaerobic digestion by increased methaneproduction and sludge stabilization. Sewage sludge has significant amount of energy and nutrients embedded which can be extracted and utilized in the agriculture and industrial sector.

Solid particles suspended in wastewater, organic materials (biodegradable), pathogens, nutritious compounds, nondegradable organic compounds, heavy metals and dissolved inorganic compounds constitute the major contaminants of wastewaters. The suspended solids are removed by sedimentation, filtration, flotation, coagulation and land treatment systems. Specific biological processes are required to get rid of biodegradable organic materials, and exclusive disinfection processes are employed to get rid of pathogens. Numerous methods to remove biological organic nutrients and methods of physical and chemical approaches are applied to control the contents of nutrients. Exclusive chemical treatments are a must to restrain heavy metals form the wastewaters. Ion exchange and reverse osmosis help to extinguish dissolved inorganic compounds. The optimized approach to treat wastewater comprises four basic stages: preliminary treatment, primary treatment, secondary treatment and tertiary treatment. The main intent of preliminary treatment is to get rid of big objects, non-biodegradable stuff and grits, thereby protecting the equipment from damages and blockages. Primary treatment aims at sedimentation of the suspended and floating materials, in primary clarifiers. Secondary treatments target at removal of suspended, colloidal and dissolved organic and inorganic materials by various biological, physical and chemical processes. Tertiary treatments refine the effluents from secondary treatment in such ways that they can be reused and/or discharged safely, via processes like absorption, oxidation and disinfection. Table 1 enlists various inorganic, organic, biological and radioactive contaminants present in wastewaters, with the adverse effects caused them (Sharma and Bhattacharya 2017). Fluorides, arsenic, mercury, copper, chromium, lead, antimony, nitrate, asbestos, selenium, barium, beryllium and cyanide are among the hazardous

Table 1 Types of contaminants present in wastewaters (Source: Sharma and Bhattacharya 2017)

Contaminants		Sources	Hazards
Inorganic Contaminants	Fluoride	Pharmaceutical products	Alzheimer's disease, dementia, retarded growth in children
	Arsenic	Natural deposits, agricultural and industrial wastes	Arsenicosis, partial paralysis, blindness
	Mercury	Seepage from industries and run-off from agricultural lands	Neurological disorders, retarded growth in children, abortions, issues in endocrine system
	Copper	Rock, soil and household corrosions	Permanent kidney and liver damage
	Chromium	Outdated mining sites and inappropriate waste disposal	Liver and kidney damage, respiratory issues, dermatitis
	Lead	Corrosion in municipal water system	Delayed development in children; high blood pressure and kidney issues in adults
	Antimony	Flame retardant industries	Affects blood cholesterol and glucose levels
	Nitrate	Fertilizers and sewage	Shortness of breath and blue skin
	Asbestos	Minute fibers of asbestos in environment	Risk of certain cancers
	Selenium	Through food and soil	Loss of sense and control of arms and legs
	Barium	Discharged through naturally occurring minerals in grounds	Harmful for heart and cardiovascular system
	Beryllium	Run-off from mining, processing plant's discharge, improper waste disposal	Damages bones and lungs, cancer threats
	Cyanide	Inappropriate waste disposal	Harmful for spine, brain, liver
Organic Contaminants	Pesticides	Agriculture and public hygiene sources	Damages liver and disturbs the nervous system
	Volatile Organic Chemicals	Industrialization and human activities	Cancer, liver and kidney damage, birth defects, productive disorders
	Dyes	Industrialization	Eutrophication, Several cancers
	Emerging Organic Pollutants	Pharmaceuticals, industries, personal care products, plastics	Cancers, endocrine disruptions
Biological Contaminants	Algae	Increased phosphorous in water bodies enhance their growth and division	Stale taste and odor of water; Congested filters; Liberate toxins harmful for liver, skin and nervous system
	Bacteria (Pathogenic)	Contaminated, untreated wastewaters	Typhoid, dysentery, cholera, gastroenteritis

(continued)

Table 1 (continued)

Contaminants		Sources	Hazards
	Protozoan	Sewage water, animal feces in water	Diarrhea, nausea, fatigue, dehydration, headaches
	Viruses	Untreated disposal from contaminated areas	Hepatitis, polio
Radiological Contaminants	Radioactive Elements	Run-off water from industries, soil, rocks	Cancer

inorganic contaminants which lead to some very severe health issues. Organic contaminants like pesticides, volatile chemicals and dyes are very well known for their harm to environment and human health. Microorganisms are among biological contaminants which cause severe infections to humans, whereas, other contaminants like radioactive elements are carcinogenic.

2 Sewage Sludge

Sewage refers to wastewater that comprises of wastewaters from humans, industries, animals and agriculture. Wastewater constitutes of 99.9% water and 0.1% solids that are either dissolved or suspended. These solids incorporate excretes, food wastes, household chemical products, plastics, metals, sand, domestic wastes (Gray 2005; Lin 2007). Sewage sludge can be defined as residues generated by the treatment of wastewater. Major categories of sludge are the primary and secondary sludge. Tanks used for settling the suspended particles used at wastewater treatment plants (WWTP) produce primary sludge. When this primary sludge is treated biologically, it becomes the secondary sludge, which is also referred as 'biological sludge' (Ren 2004; Sanin et al. 2011). Several chemicals and chemical approaches are employed to treat sludge, and they result in generation of 'chemical sludge' (Turovskiy and Mathai 2006).

The processing and disposal of sewage sludge are the chief criteria for designing and functioning of a wastewater treatment plant. Diminution of the sludge volume and stabilizing the organic stuff in the by-products and final products are two major goals of a WWTP. The stabilized sludge should not have unpleasant odor and shouldn't be a health hazard. The final costs in pumping and storage can be effectively slashed if the sludge volumes are small. Utilization in agriculture, incineration and landfills are among the most common methods to dispose the sewage sludge at primary level. However, there are issues related to each technique, in terms of both health hazards and environmental effects (Fytili and Zabaniotou 2008). Management of waste water sludge can be done by some conventional method as well as advanced thermodynamic conversions, as per the requirement of the treatment and availability of the resources. Conventional methods include land application, landfilling, incineration and anaerobic digestion; alternative

Waste Water Sludge Management

Conventional Methods

- Land Application (unrestrained nutrient recycling)
- Landfilling (negligible nutrient or energy recovery)
- Incineration (partial energy recovery)
- Anaerobic Digestion (partial resource recovery via biogas)

Thermodynamic Conversions

- Pyrolysis (resource recovery by bio-oil, bio-char gas)
- Gasification (extensive energy recovery through syngas)
- Hydrothermal Liquefaction (resource recovery through bio-oil, liquids, hydro-char)
- Other methods (energy recovery through products)

Fig. 1 Management of wastewater sludge though conventional and advanced thermodynamic methods (Source: Yuan et al. 2011)

thermodynamic conversion methods are pyrolysis, gasification, thermodynamic liquefaction and various such methods (Yuan et al. 2011). Figure 1 enlists various conventional and advanced methods with their recovery efficiencies. Whatever the source of waste water is, it requires to be treated well enough before further disposals and/or reuses. A technically proficient wastewater treatment method should aim at the following criteria: components of the wastewater should be transformed into harmless materials which won't be risky to dispose; the disposed end products should not be threat to the environment and public health; the worthy components should be fruitfully harvest and/or recycled; the whole process should be economically feasible (Samer 2015). Biochar from sewage sludge is popular because of some amazing efficiencies like it improves soil qualities, it minimizes uptake of heavy metals and is hence beneficial for agricultural perspective. The only retracting concern is high cost of bio-char disposal. Sewage sludge has wide range of applications in surface assimilation of harmful compounds like antibiotics, heavy metals, textile dyes and phenolic compounds. There use minimizes agricultural pollution, hence playing major role in agriculture and climate change (Singh et al. 2020).

Sludge reduction by biological methods results in noticeable decline in production of secondary pollutants. Interestingly, some other unconventional methods are also being employed to treat wastewaters, and one such documented case study used the aquatic worm *L. variegatus*, reporting 33% reduction in suspended solids (Basim et al. 2016). Spinosa et al. (2011) have proposed a brilliant system for sustainable management of sludge, as demonstrated in Fig. 2. A standard management system for sewage sludge includes digesting the sludge via anaerobic approach, that produces energy that may be utilized at different levels of the process. Energy is required for the processes of mechanical dewatering and thermal drying, used for removing excess moisture content, which has to be managed separately as by-product. Pyrolysis and gasification are the thermo-chemical processes which yield syngas, that can be utilized. Energy is also evolved during these steps. Ceramsite and adsorbents can be produced via the method of thermal processing,

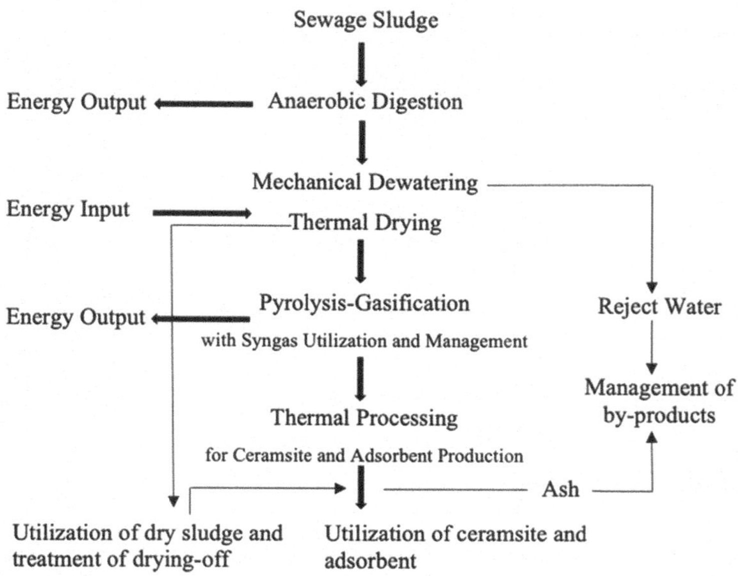

Fig. 2 Sustainable Sewage Management (Source: Spinosa et al. 2011)

and can be used further. Through-out the process, ash and water are resealed as by-products, whose management is also a considerable task which demands specific handling.

3 Biological Methods for Treating and Stabilizing Sewage Sludge

Treatment of sludge starts with mixing the primary, secondary and tertiary sludge, which may have 1–4% total solids (suspended and dissolved) and is termed as 'raw sewage sludge'. Presence of pathogens, decomposable and unsteady components tends to make raw sludge as a hazard to the human health and environment. But, various treatments are available and used to stabilize the sludge, which results in reduced pathogenicity and increased solid content. Some of the processes most commonly used to stabilize and reduce pathogen levels in sewage sludge are displayed in Fig. 3, which represents the scheme employed for sludge treatment processes, incorporating mechanical, biological and thermo-chemical techniques.

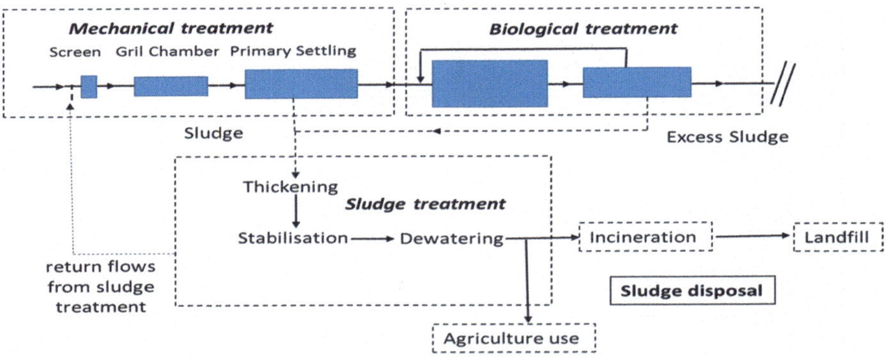

Fig. 3 Simplified scheme of sludge treatment process

3.1 Sludge Thickening

Thickening of sludge is the first step in sludge processing andrefers to the process of dividing the solid and liquid phases, decreasing the sludge volume (Bień et al. 2009). Thickening of sludge results in up to 92–94% water diminution. This is the water which was earlier present in free-state (primary water) or became available after the conditioning process. A proper treatment of thickening helps reduce the investments and the plant's functioning costs, which thereby applies on he whole treatment process, decreasing the operational costs. As per the requirement for treating the sludge, different categories of sludge thickening are available, viz., gravitational thickening of sludge, mechanical thickening of sludge and flotational thickening of sludge, as described below.

Gravitational thickening, as the name suggest, utilizes the gravitational force and hence thickening is achieved by sedimentation and compressing the sludge. Both primary as well as secondary tanks are used for settling and the unit is referred as 'gravitational thickeners'. The process of gravitational thickening can be accomplished either periodically or in a continuous manner, which demands for regular supply and removal of sludge. This whole process can be escalated employing 'slow-speed rod mixers' which can divide as well as reorganize sludge to fill the 'in-between sludge particles' spaces with water (Podedworna and Umiejewska 2008). Sludge should not be stored for long in the gravitational thickeners, else it may cause decomposition of organic compounds which releases gas bubbles. This produces floating sludge might emit unpleasant odors (Bień and Wystalska 2011). So, when a dry matter of almost 4–6% is desired from the sludge, the process of gravitational thickening is opted. But, in case of sludge which can hardly be thickened just by the gravitational force and also in case of time-saving, mechanical thickening is suggested. Mechanical thickening is achieved by the utilization of additional forces, like centrifugal force, which speeds up the process of thickening. Various equipment like belt thickener, drum thickener and thickening centrifuges are available for the same (Wójtowicz et al. 2013). Just like in the case of dewatering centrifuges, the thickening centrifuges also operate on the basis of centrifugal forces. If additional flocculants is not available, mechanical thickening process can yield dry matter of 5–7%, whereas a dry-matter content of 6–8% can be produced with the addition of flocculants (Podedworna and Umiejewska 2008). In case of flotational thickening, as the name suggests, the sludge floats upwards like a layer of floating stuff, and it is then removed with the help of special sweeps (Bień et al. 2004). Depending on the mechanism of floating, the process can be biological flotational thickening, chemical flotational thickening or air flotational thickening (Podedworna and Umiejewska 2008; Bień and Wystalska 2011).

3.2 Sludge Digestion

Once all the solids present in the sludge (dissolved and suspended) are accumulated, digestion of the sludge can be initiated. Sludge digestion refers to the biological process of transforming the organic solid content present in the sludge into decomposed stable forms (Appels et al. 2011; Nasir et al. 2012). The process of aerobic digestion requires oxygen for microorganisms to ingest organic contents present in the sludge, followed by converting them into carbon dioxide, water and biomass. This whole process required very précised selection of microorganisms as per he impurities to be treated, in a well-designed set-up. These microorganisms may vary in terms of oxygen requirement, nutrient requirement and their functioning conditions so that they can completely mineralize the organic content to methane, bicarbonate or carbon dioxide (Gujer and Zehnder, 1983). Different steps involved for digesting the sludge in an anaerobic manner are hydrolysis, acidogenesis, acetogenesis and methanogenesis (production of acetate and hydrogen methogenic substrate), methanogenesis (Meulepas et al. 2005). The benefits of employing the process of aerobic digestion are: (i) it is feasible with sludge containing higher moisture levels, (ii) the biogas produced has higher energy content, (iii) negligible carbon emission (iv) residues can be utilized as fertilizers, (v) costs for transportation and disposal are waived-off, (vi) numerous methods are available for pre- and post-treatments, (vii) low sludge production and (viii) lower nutrient and chemical requirements (Oladejo et al. 2019; Ahammad et al. 2013). But there are some drawbacks related to this process, some of which are: (i) over-all reaction time is longer, the process results in formation of various organic pollutants, (ii) lesser conversion efficiency, (iii) the treatment premises are left with a polluted odor, (iv) public health and environment are left at risk, (v) higher capital and maintenance costs (Oladejo et al. 2019). Tarpani et al. (2020) have reported the agricultural applications of sludge digested in an anaerobic manner to be the lowest negative effects on environment. The technology can be employed on various organic waste streams like bio-wastes, organic fractions of mixed wastes, dewatered sewage sludge, non-recyclable papers and other such market wastes (Feodorov 2016). Figure 4 enlists the various methods used in aerobic and anaerobic process, for biological treatment of waste water (Ahammad et al. 2013).

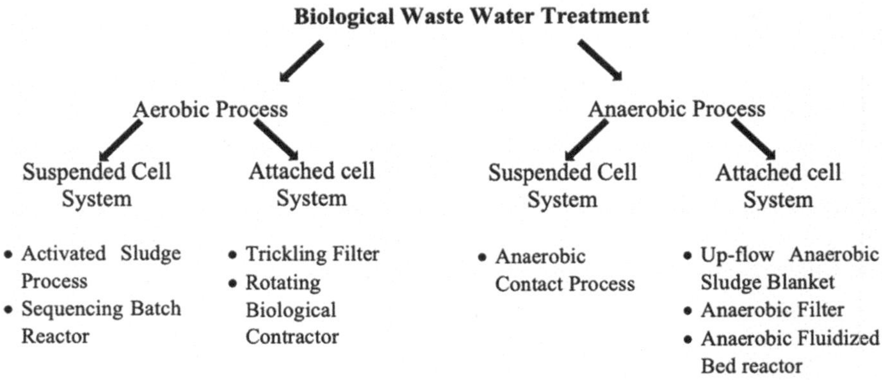

Fig. 4 Techniques used for biological treatment of waste water (Source: Ahammad et al. 2013)

The process of aerobic digestion is doesn't need external heating and has a high degradation rate as compared to the physical and chemical methods of pre-treatments, and, is therefore opted as a better alternative (Jang et al. 2014). Waste sludge generated through aeration-implanted systems, filtered sludge and/or mixture of activated sludge can be treated by aerobic digestion. Provision of uncontaminated oxygen, conventional set-up and thermophilic microorganisms are a must for aerobic digestion. A favorable process is bio-augmentation, which employs proteolytic bacteria (e.g. *G. stearothermophilus*) with activated sludge and it stands out with great overall stabilization results (Dumas et al. 2010). The process is also chosen in case of sludge with materials which can-not be degraded via anaerobic pathways (Carrère et al. 2010). Power and standards required for the operation and maintenance of aerobic treatment set-up costs more than the anaerobic systems because they are generally provided with additional aeration systems and contact stabilization units. In terms of conversion efficiency, both aerobic and anaerobic (conventional) treatments result in transformation of almost 50% of the organic content into liquid and gaseous products. But, if the anaerobic digestion system is employed after thermal hydrolysis, 60–70% of the solid contents can be transformed into liquid and gaseous states (Pagilla et al. 1996). Figure 5 summarizes the steps involved in aerobic and anaerobic processes for digesting sludge.

Dumas et al. (2010) used thermophilic microorganisms to analyze the consequences on mesophilic aerobic system and discovered lowered content of suspended solid particles (39–83%) by employing aerobic-anaerobic process, but without any improvements in methane production. Also, solid reduction in pre-treatment stage was noted to be influenced by aerobic oxidation of organic contents. Another experimentation of addition of thermophiles in an anaerobic digester was carried out by Miah et al. (2005) where they recorded 21–112% higher methane yield and 4–44% higher contents of volatile solids. In a similar way, post-treatment via aerobic pathways have been documented to enhance the reduction of volatile solid (Novak et al. 2011; Tomei et al. 2011). Other studies have reported that highest results were recovered on application of low sludge with lower extents of biodegradability (Miah

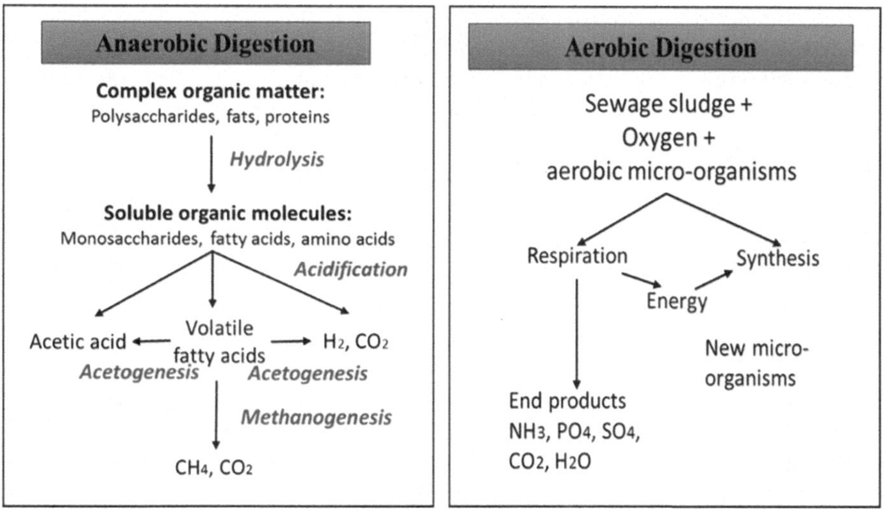

Fig. 5 Aerobic and anaerobic treatment procedure showing potential steps of sludge processing

et al. 2005). Opting for anaerobic process has been reported to give highest recoveries of methane (Jang et al. 2014; Dumas et al. 2010; Miah et al. 2005; Pagilla et al. 1996). Study conducted by Jang et al. (2014) concluded to use higher contents of HCO^{3-} and CO_3^{2-} in aerobic pre-treatment sludge as a substrate, with hydrogenotrophic methanogens, gives the best yields in aerobic digester. Whereas, lower concentrations of H_2S in biogas have been recovered, probably because of remoting air stripping at the stage of thermophilic aerobic process (Pagilla et al. 1996).

3.3 Dewatering

Post retrieval of essential outputs by gasses and by-products, the left-over sludge needs to be dewatered prior to the final disposal. Though in a solidified state, the dewater sludge generally possesses a huge portion of water, i.e. almost 70%, which demands for further drying and dewatering of the sludge, by advanced method specifically designed for the same. Various methods like chemical process, thermal methods and freeze-dewatering techniques can be used which prepare the sludge to relieve water, e.g. centrifuges or belt presses. Centrifugation is one of the most popular way because it enables easy recovery and permits easy handling of solids, at lower costs, and in shorter time periods. Other such methods are vacuum drum filter and belt filter press. Ferrous sulphate, ferric chloride, alum salts are the most opted chemicals here, and are often finalized by parameters like sludge characteristics, efficiency, price and requirements for final treatment of sludge. Compounds (chemical) employed for refrigeration refrigerators improve the process of filtering and dewatering the sludge.

The method of 'coagulation' process is completed in two parts (i) neutralization of the particle charge and (ii) binding of individual particles to the floc structure (Turovskiy and Mathai 2006). Unstable particles in the sludge are easily paired to the charged particles on high molecular weight flocculants, therefore improving the whole process of flocculation. This results in production of flocs, which ultimately releases water content form the sludge and hence desired dewatering is achieved.

3.4 *Sewage Sludge Disposal*

Post efficient dewatering of the sludge, burying in landfills or use as a fertilizer (as per the compound composition) can be done. If the sludge is toxic in nature, it is always sent for incineration and hence converted to ash (Chen et al. 2012; Werther and Ogada 1999). To choose for best suitable option for sludge treatment, some points to be considered include sewage origin, sewage to sludge reduction process, recovery of significant by-products. Such evaluations are best to optimize the treatment and prove to be cost-efficient.

Sewage sludge disposal via 'Landfill': A landfill, which is also referred to as 'sanitary landfill', is a site designed for proper disposal of sludge, which later causes no harm to the environment or to human health. The landfill set-ups have provisions of compiling wastes in compact beds by reducing their volume and the liquid and gaseous effluents are continuously monitored throughout the process (Yoshida et al. 2013; Chen et al. 2012; Werther and Ogada 1999). Along with municipal solid wastes, fecal sludge is also subjected to landfills because of its reliable operating standards. A standard, well-maintained landfill is always a better and preferable option in comparison to the open dump sites, but the best set-ups also face the problem of leakage after being piled up for many years. Hence, its always recommended to dispose sludge which is not expected to be reused further. Preventing waste remains a priority in the first place to solve current waste problems. Separation and reuse of different types of waste is much more sustainable. A standard landfill setup is basically a pit with a bottom that is seal-protected (for precluding possibility of contamination caused by groundwater) and the waste is inhumed in layers that are tightly packed together and covered (Harvey et al. 2002). Advances landfills utilize liner system at the bottom as well as on sides, a leachate remotion setup that even comprises of monitoring the groundwater, extracting gas and capping system. Planning the total capacity and site selection is based on the environmental risk assessment study (UNEP 2002). Proper supplement of nutrients and recirculation of leachate enable optimum bioreactor landfill processing (Reinhart et al. 2002). Options also include use of aerobic or anaerobic bioreactor landfills, where the techniques may accelerate biological transformation of organic contents, promoting microbial degradation of wastes and production of biogas. Waste mass should be timely supplied with pure form of liquids to keep the moisture level at

35–45% (water by weight), because moisture content is the most essential factor to enhance waste decomposition. Added liquids can be procured from: landfill leachate recovered earlier from the bottom, gas condensate, water, storm-water runoff, as well as the sludge from wastewater treatment plants (WM 2004).

Sewage sludge disposal via 'Incineration': Incineration has been used worldwide as an attractive disposal method because of its effectiveness in disposing the sludge, which also provides with some benefits like huge reduction in the final volume of sludge, thermal deconstruction of toxic contents which are organic in nature, as well as generation of minimal odor. As compared to mechanical dewatering, incineration provides up to 10% better results (Fytili and Zabaniotou 2008). The sewage sludge can be incinerated by two methods: mono-and co-combustion, where mono-combustion stands out to be much prevalent. Employing multiple-hearth with fluid bed furnaces have been reported to be the most prevalent technologies, with more efficiency because of low fuel consumption and emissions (Werle and Wilk 2010). Higher running costs and the impacts on environment are the major harmful effects for sludge combustion. Accumulation of heavy metals in ash, as well as the exhaustion of gasses are the environmental issues generated. The former is resolved by utilizing incineration ash as raw materials for cement production process during which the heavy metals get immobilized in cement (Murry et al. 2008). Co-combusting the sewage sludge with various natural resources (like coal, lignite or wood) or solid wastes of municipal origin is also an alternative for managing sewage sludge. In environmental and economic prospects, biosolid co-combustion technology meet the emission criteria of the Waste Incineration Directive and provide as essential source of energy, but lack of policy and legal clarity, supply chain insecurities and immaturity at marketing levels hinder the effectiveness of co-combustion (Cartmell et al. 2006).

3.5 *Composting*

Composting of the sewage sludge is aerobic method of stabilizing sewage sludge, inactivates pathogens and diminishes mass and moisture contents. Figure 6 represents the basic steps involved in composting (Garg 2009). The dewatered sludge and bulking agents are mixed for composting. Forced aeration and drying methods lead to bulk agent recovery, which can be again reused. After proper screening, the final products are stored and as per qualities are further disposed or sent for utilization in the market.

Biochemical decomposition of organic matter is the basic procedure for composting the mass. In an optimum environment, the process of composting is completed via four standard phases, which are characteristic of individual group of microorganisms (Moretti et al. 2015; Bień 2014; Bień and Wystalska 2011). Figure 7 describes various biochemical changes occurring throughout the process. Initial step is the mesophilic phase where temperature may rise and his may last for few days. It is followed by intensive decomposing where thermophilic (high-

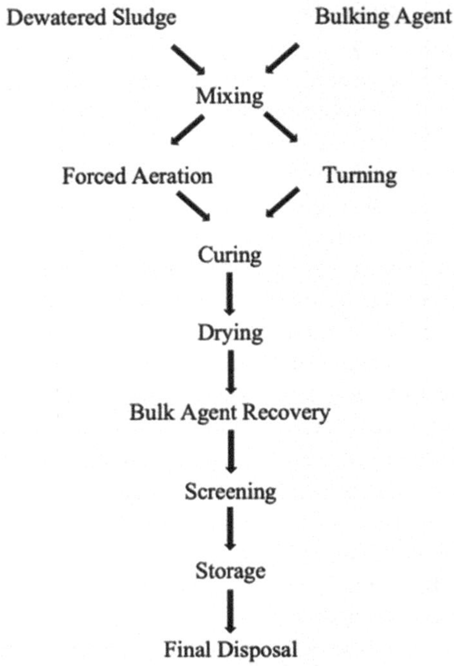

Fig. 6 The process of composting (Source: Garg 2009)

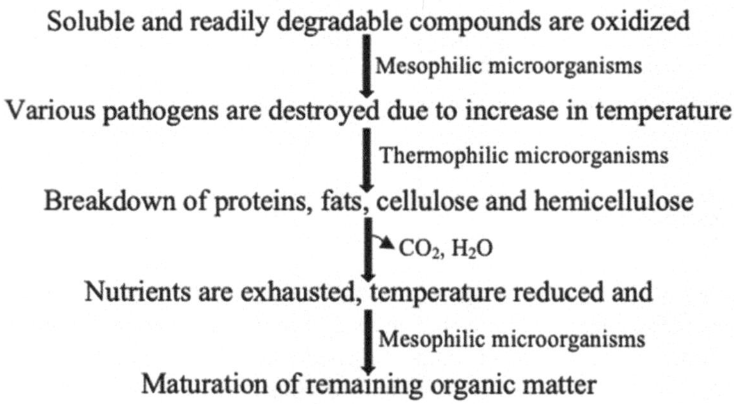

Fig. 7 Potential biochemical steps involved in sludge composting

temperature) phase lasts from som days to few days and temperature may reach up to 60–75 °C. Till such levels, the organic compounds which are biodegradable in nature get decomposed, leaving behind water, carbon dioxide and ammonia as the products. Next is the transition phase, which may be referred as composting, and start from third to fifth weeks and lasts for another 3–5 weeks, with temperature range of 30–40 °C. By the actions of selected mesophilic bacteria and fungi, hardly transformable components are decomposed here (like lignin, fat, wax, resins), with a noticeable reduction in moisture content. Secondary composting, i.e., the compost maturation phase results in cooling down of the material with a steady part of the compost (humus) and an intensive production of the macrofauna. The phase may take time of several months.

The composting process depends on a number of parameters. Temperature and reaction are particularly important. Efficiency of composting also depends on microbial strains employed, number of microorganisms, humidity, organic components, amount of oxygen etc. (Som et al. 2009; Bień 2015). The organic compounds are transformed in forms which can later be utilized as fertilizers, for structural reforming, reprocessing materials etc. The soil's physico-chemical properties, which include moisture, air as well as nutrients, are also positively altered by addition of compost, ultimately nourishing the growth of microbiome present in the soil, hence increasing the soil's biological life and the process of soil-forming. The composting process carried out at high temperatures assure the safety of compost in sanitary conditions (Tomati et al. 2010; Som et al. 2009; Eggen and Vethe 2001).

3.6 Advanced Biological Treatments

To achieve higher levels of final outcomes, advancements are proposed and employed in all the technologies worldwide. Minimization technologies can usually sum-up sewage sludge production by three distinct processes: (i) to adopt a 'water line process' which decreases over-all production of sludge; (ii) to reduce moisture content (by dewatering methodology) or (iii) to reduce contents of volatile components by stabilization. Depending upon the desired products, as well as having a stable economic back-up allows to choose the latest, and most efficient technical process. Not just the direct advantages and disadvantages (like sewage sludge production) should be considered, but also the indirect (like several other concerns related to WWTPs) should be taken into account.

3.6.1 Membrane Bioreactor (MBR)

The term membrane bioreactor (MBR) refers to a cordinated method comprising activated sludge process and filteration of sludge by membrane (Oh et al. 2012; Le-Clech et al. 2006; Chang et al. 2002). As soon the ultrafiltartion and microfiltration membranes were commercialized, late 1960s welcomed the set-up of MBR. A combined set-up of activated sludge bioreactor and cross-flow membrane filtration loop was first propsed by Dorr-Olivier Inc. which employed plymeric flat sheet membranes (of pore size ranging from 0.003–0.01µm). A major change was introduced with employing submerged membranes in bioreactor (Yamamoto et al. 1989). Earlier, MBRs were created with a separator placed outside the reactor (side stream MBR) which were dependent on high transmembrane pressure (TMP) or maintaining membrane filtration directly immersed in the bioreactor. A submerged MBR systems is generally choosen for the treatment of domestic wastewater (Berube 2010). The submerged configuration procures aeration of coarse bubbles which enables intermixing and decreases foul generation. For this set-up, aeration is the most significant feature for both hydraulic and biological process performance (Deowan et al. 2015). A proper aeration system is essential for proper suspension of solid contents, scrubbing of the membrabe surface and maintains excellent supply of oxygen to the biomass, hence enhances biodegradability and cell synthesis. Anaerobic oranoxic compartments must be intorduced to the systems for the removal of biological nutrients (Cote et al. 1998). The popularity of MBR technology over conventional processes is because high quality effluent is yielded consistently, international stringent discharge norms are followed, hydraulic and sludge retention time are controlled individually, COD is reduced and significant processes like nitrification, reduced sludge production, process intensification through high Biomass concentration with MLSS (Mixed Liquor Suspended Solids) over 8000–10,000 ppm, ability to treat high strength wastewater and reduction in post disinfection requirements.

3.6.2 Aerobic Granular Sludge Systems (AGS)

Aerobic granular sludge (AGS) refers to an exclusive microbial community that permits paralle remotion of pollutants ith C, N, P and also various other pollutants via a single sludge system. In chmical, physical and microbial characters, the AGS differs from activated sludge and it also proffers a cocised and cost-effective treatment for the removal of oxidized and reduced wastewater contaminants. AGS batch sequencing reactors are uilized for treating abattoirs, live materials, rubber, landfill leachate, dairy, breweries, textiles, sewage treatment and other effluents. But, installation procedure for AGS is time consuming when it has to be utilized for treating low-strength wastewater like sewage. Overall woking of AGS can be fastened with higher volumetric flow thorugh shorter cycles as well as mixing sewage with industrial wastewater, to uplift the formation of AGS for treating low-strength sewage (Nancharaiah and Reddy 2018). For the development of AGS, batch sequencing reactors are operated with a small sequencing time of 2–10 min (Adav et al. 2008). Wang et al. (2006) concluded that the formation of AGS was faster in reactors which had an exchange rate as high as 20 to 80%.

3.6.3 Biological Predation

The term biological predation can be defined as the system comprising higher organisms, like protozoa and metazoan with excess sludge (Atay and Akbal 2016; Semblante et al. 2014), which may be employed with water and in-line with the sludge. Procurement in the water line demands a two-stage system where the initial system has to have a minimum hydraulic retention time (HRT) to favor spread of quickly growing bacteria for treatment of wastewater, while the other one has to have an extended SRT in to favor the optimum growth of predators; (Foladori et al. 2010; Wei et al. 2003). Many researchers have worked on various applications of the sludge to treatment line and found that metazoan (especially worms and larvae) tend to be much favorable than protozoan. *Eiseniafoetida* and *Hermetiaillucens* are the most popularly employed, e.g., Kalová and Borkovcová (2013) used *Hermetiaillucens* to reduce the release of primary and secondary sewage sludge and after a treatment of 35 days, wet weight was reduced by 16%.

4 Thermo-Chemical Treatments

Different types of thermo-chemical treatment technologies are available viz., liquification; pyrolysis; gasification and wet oxidation. To carry out the process of combustion, gasification and pyrolysis, it becomes mandatory to remove moisture, else the sludge has to be dried prior, in contrary, the sludge can be directly treated by the process of liquification and wet oxidation (Singh et al. 2020; Syed-Hassan et al. 2017). Pyrolysis and liquification methods can convert organic components into bio-oil (crude), whereas gasification is known to produce the syngas (Gao et al. 2020). Wet-oxidation is not capable to generate any kind of bio-fuels. All the aforesaid techniques will be briefly explained in the subsequent sections:

4.1 Pyrolysis

Pyrolysis refers to the 'thermal degradation of biomass' for converting it into solid (charcoal-like), liquid (bio-oil) and/or gaseous products (hydrogen, methane, carbon monoxide etc.), conducted under anoxic conditions at higher temperature (ranging between 400–1000 °C) and in an inert environment with anaerobic/very low oxygen conditions (Zhang et al. 2010; Mills et al. 2014). Initial steps in pyrolysis aim at filter pressing the raw material to bring down the water content and hence the dry mass is 30%. This thickened sludge is later dried and pyrolyzed, generating the final products which vary as per the different temperatures used while pyrolysis (Hospido et al. 2005). For example, the flowchart (Fig. 8) depicts how dewatered sludge is processed to achieve the production of adsorbents (Spinosa et al. 2011). For decades, the process of pyrolysis is followed for converting wood into charcoal and this demands very slow reaction at low temperature for maximum yield of charcoal. Other types of biomass such as straw can also be decomposed to produce char like residue. The sludge is first dewatered and dried, which releases exhaust gases. This exhausted gas requires proper handling and management. The dried sludge, after pyrolysis and disintegration is finally transformed into adsorbent, which has various applications and uses.

The generation and distribution of products of pyrolysis is generally affected by few properties of sewage sludge (viz., presence of volatile content, moisture and

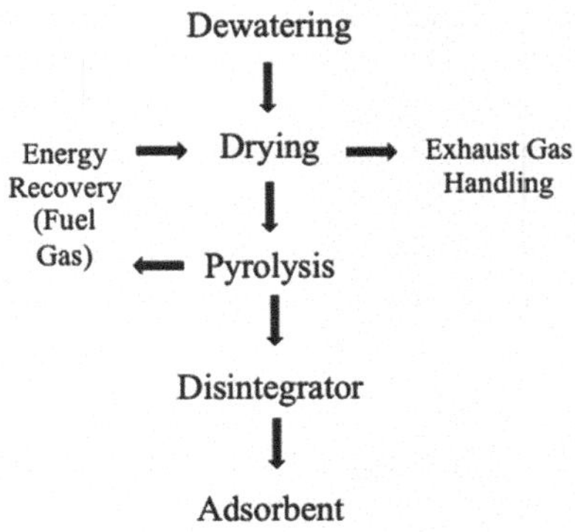

Fig. 8 Sludge pyrolysis to produce adsorbent (Source: Spinosa et al. 2011)

ash). As volatile matter elevates, it further upgrades the yield of gas and oil, along with reducing char generation (Wang et al. 2008). Products obtained at the end of the process are solid char-like residues, water, soluble organic components, various insoluble organic components and gaseous products like hydrogen, carbon monoxide, carbon dioxide, nitrogen and methane. Final outcomes of pyrolysis treatments can be of three varieties, depending upon the treatment-type: (i) one variant is the 'biochar' obtained by employing thermal degradation of sewage sludge, (ii) second is 'syngas' and 'bio-oil', (iii) production of final compounds as per the thermogravimetric analysis and reaction kinetics (Bonfiglioli et al. 2014). Whatever the desired final products and the chosen methods are, the outcomes are exclusively dependent on pyrolysis temperature, heating rate, dwell rate, atmospheric gases, pressure and raw-material (Fan et al. 2016). Undoubtedly, it has been considered that pyrolysis of sewage sludge has numerous advantages over other conventional methods such as incineration in terms of economy, recovery and with respect to controlling the emission of heavy metals as well. But overall efficiency of this method gets affected is water (moisture) is present.

Various advantages of pyrolysis are: (i) Annihilation of pathogenic microorganisms, (ii) Production of bio-char which is capable of enhancing concentrations of valuable nutrients for plant growth (like potassium, nitrogen, phosphorous, calcium, magnesium etc.) (Liu et al. 2014), (iii) Treatment conditions like temperature and resident time can be manipulated to optimize products (Bruun et al. 2012), (iv) Production of oil, char and gas certifies pyrolysis as a zero-waste process, hence minimizing environmental wastes, (v) The process employs both raw and digested sludge, (vi) The large-scale plants opting pyrolysis are economically feasible, (vii) Pyrolysis has proved to be economic and energy-efficient drying technique (Oladejo et al. 2019). But there are also some drawbacksassociated with the process of pyrolysis: (i) Elevated concentrations of heavy metals in soil, restricting its applications in agriculture (Liu et al. 2014), (ii) The whole process is complex and sludge with high water content demands compulsive dewatering, (iii) Expensive downstream treatment is necessary for disposal, reuse and storage of char, which demands high capital (Oladejo et al. 2019).

4.2 Hydrothermal Liquefaction

Hydrothermal liquefaction refers to 'low temperature with high pressure' conversion of biomass into small fragments with water and without any solvent or catalyst. The small reactive and unstable fragments can convert to a variety of oil like components through the process of re-polymerization. The transition of sludge powder from paper and pulp in the form of liquid oil by direct liquification at temperatures 200–400 °C was investigated by Xu and Lancaster (2008). Ambulant heavy metals can be easily transmuted in steady states after liquefaction, e.g. acid soluble/exchangeable and reducible contents can be converted into oxidizable and residual contents (Yuan et al. 2011; Pan 2010; Pan et al. 2009). Moreover, Li et al. (2010) removed moisture from sewage sludge and liquefied the powdered sludge in ethanol/water mixtures at different temperatures (250–400 °C) with and without the addition of catalysts. The results obtained from this study provide a promising hope to recover energy from sewage sludge in future. Figure 9 depicts the general outline of liquefaction procedure (Xu and Lancaster 2008). After liquefaction, the dried sludge produces some liquid products, which are filtered and divided on the basis of their solubility in water. The component that is soluble in water can be evaporated and finally some oil is recovered. Whereas, in case of water insoluble components, extraction is done with acetone, prior to filtration. After filtration, fractions of acetone-soluble and acetone-insoluble components are received. The acetone

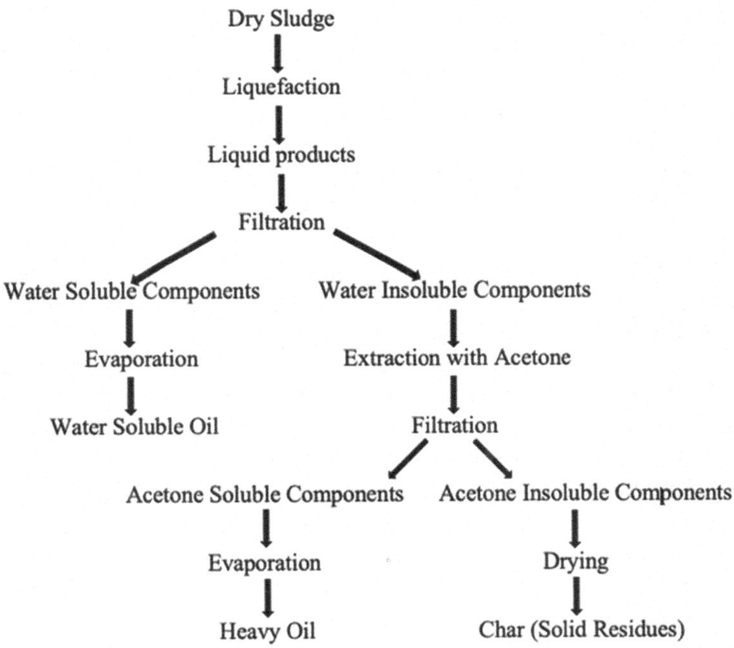

Fig. 9 Product recovery by liquefaction (Source: Xu and Lancaster 2008)

soluble components yield heavy oils after evaporation. Acetone insoluble components yield char after drying.

Advantages of the method include: (i) a major advantage of liquefaction process has been documented by Yuan et al. (2011) which records subjugated percolation of harmful heavy metals; (ii) the process is also economically feasible. But there are some drawbacks also: (i) Leng et al. (2014) have reported that higher liquefaction temperature elevates the contents of heavy metals in bio-oils; (ii) Liquefaction solvents are also able to regulate heavy metal dispersion in bio-oil as well as bio-char. Their study concluded that liquefaction with acetone can yield greener products under low temperatures.

4.3 Gasification

The method of gasification demands high temperature and pressure to process the biomass into combustible gaseous products in the presence of slighter amount of oxygen. Syngas is procured as gasification's major outcome, which is basically a type of bioenergy. It can also be utilized in the form of a natural gas substitute or may be as a raw material to procure hydrogen and synthesize chemical (Zhang et al. 2014). Now-a-days, aerobic and thermal gasification are world-wide employed treatment methods since the expenses and hazards associated with oxygen storage and use as well can be avoided. In context of thermos-chemical characters, the process of combustion, as well as gasification are similar, whereas pyrolysis may be a precursor to both the processes (Furness et al. 2000). The gasification is considered as a rate limiting step while pyrolysis occurs in a rapid manner. The process of gasification comprises of the following steps: drying of the sludge, pyrolysis (i.e. thermal decomposition), partial combustion of some gases, vapors and char, and final gasification of decomposition outcomes. Steam, air or oxygen are required to provide a gasification medium, which helps rearranging raw material's molecular structure. Some basic reactions which happen though the process of gasification include boudouard, water gas (primary and secondary), methanation, water gas shifting, steam reforming and dry reforming reactions (Buckley and Schwarz 2003). Being an alternative method of thermal treatment of sludge, gasification was opted by Werle and Dudziak (2014) to record the effects of various ways to treat wastewater and dry the sewage sludge, on the gas parameters of gasification. They recorded that the sources of wastewater and the process employed to treat wastewater had noticeable effects on sewage sludge properties. Sewage sludge with bigger oxygen required lower reaction temperatures, whereas, sewage sludge with higher contents of hydrogen had direct impact on the gas contents of gasification. They concluded that the operational parameters for gasification of sewage sludge significantly effects the profile of gasifier temperature and composition of syngas). Figure 10 depicts the steps involved in production of ceramsite by gasification of sludge (Spinosa et al. 2011). Selected agents are added in the dewatered sludge, which is later dried. Drying requires energy and produces some exhaust gases, which

Fig. 10 Sludge gasification to produce ceramsite (Source: Spinosa et al. 2011)

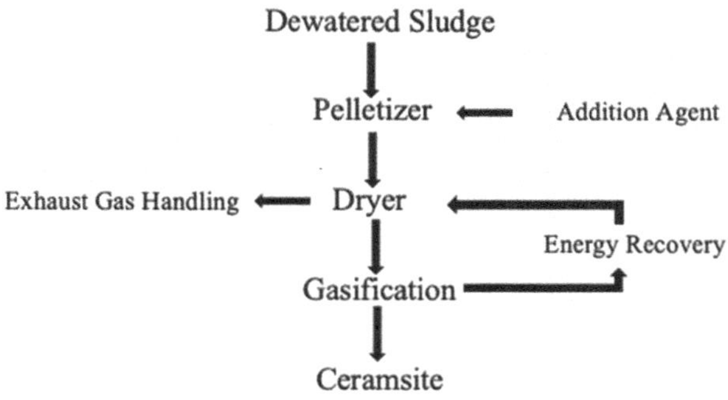

have to be handled carefully. These dried fractions finally yield ceramsite after gasification.

Advantages of gasification are: (i) the process is highly energy efficient, (ii) it is a major way of liquid fuel and chemical production from syngas, (iii) waste produced is minimal, (iv) lower emission and release of heavy metals, (v) large-scale plants are economic. Major drawbacks of the process can be listed as: (i) drying is compulsory if sludge has more than 30% moisture, (ii) reactions are complex and the technology are still not much optimized, (iii) exhaustion of organic pollutants, (iv) formation of toxic pollutants (Oladejo et al. 2019).

4.4 *Wet Oxidation*

Also known as 'wet air oxidation', this method refers to oxidation of in-organic as well as organic components in liquid phase. Wet air oxidation is known as thermo-chemical treatment methodology which is being used for treating wastewater flow from pharmaceutical as well as petrochemical industries (Lundin et al. 2004). Major parameters of the process are temperature, pressure, air supply and concentrations of solids which must be thoroughly monitored (Luduvice and Fernandes 2007). During this process, the carbon compounds of organic origin get oxidized to form carbon dioxide and into several organic compounds of lower molecular mass, at temperature (200–350 °C) and pressure (1-15Mpa). Since the process of wet oxidation requires an aqueous phase, higher pressure as well as temperature tend to be mandatory for enhancing oxygen solubility in water. Wet oxidation process cannot mineralize the biomass fully, because of the presence of soluble organic components (propionic and acetic acid) (Syed-Hassan et al. 2017). Hence, further treatment by employing biological methods is required in some cases. The flowchart (Fig. 11) depicts the steps involved in the process of wet air oxidation (Foladori et al. 2010). The sludge is first introduced to high levels of pressure for heavy recovery. This heat can be used in reactor, or can be cooled down and sludge sent to separator for obtaining gas and liquid products.

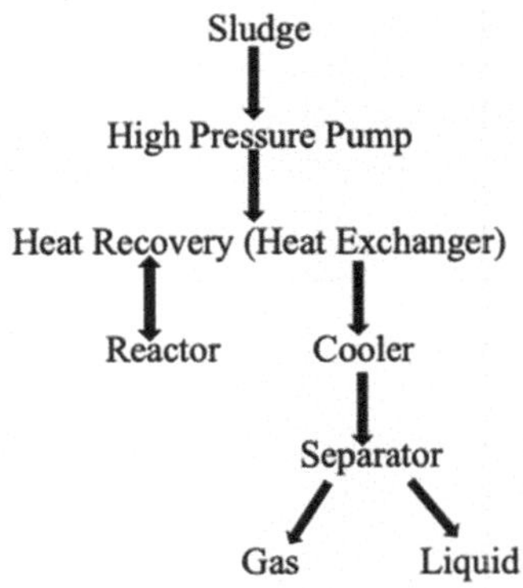

Fig. 11 Wet air oxidation process (Source: Foladori et al. 2010)

Advantages of following the process of wet air oxidation include: (i) use of phosphate and coagulants decreases sludge volume and yields >90% energy recovery (Stendahl and Jafverstrom 2004); (ii) various chemicals can be procured from the carbon-rich effluent recovered via wet oxidation method (Hii et al. 2014); (iii) no harmful by-products and the wastes produced are inert (Tungler et al. 2015). But, there are also few drawbacks of the method: (i) costly in terms of energy utilization and operation and maintenance requirements (Tyagi and Lo 2011); (ii) the process leads to corrosion of heat exchangers and reactors (Weemaes and Verstraete 1998); (iii) it demands advanced and costly construction material (Foladori et al. 2010).

5 Conclusion

Increasing human population and urbanization from last few decades have been slowly affecting the environment. This all has gradually led to the generation of pollutants in bulk, which demand for specialized technologies to transform them into harmless and/or reusable forms. Wastewater and the sludge generated, is one such problem. Proper treatment of the sludge, which emits the least harmful by-products, and ends up with production of safely disposal and/or reusable outputs is the need of hour. Various conventional methods are getting meliorated and new techniques are designed to resolve this problem. With the availability of so many options, wastewater treatment can be accomplished making smart decision regarding the method chosen. As per the characteristics and intensity of pollutants, availability of resources (economic and technical), desired outcomes and time limitations should all be evaluated to proceed further for the treatment of sludge. Also, its impact on the environment, like energy input, energy output, disposal of by-products needs to be examined thoroughly. Consideration of these parameters results in a clever decision for sludge treatment and disposal, without harming nature.

References

Adav SS, Lee DJ, Tay JH (2008) Extracellular polymeric substances and structural stability of aerobic granule. Water Res 42(6–7):1644–1650

Ahammad SZ, Graham DW, Dolfing J (2013) Water treatment: biological. In Jorgenson Se (ed) Encyclopedia of Environment Management, Taylor and Francis, p 2645–2656

Appels L, Assche AV, Willems K (2011) Peracetic acid oxidation as an alternative pre-treatment for the anaerobic digestion of waste activated sludge. Bioresour Technol 102:4124–4130

Atay S, Akbal F (2016) Classification and effects of sludge disintegration technologies integrated into sludge handling units: an overview. Clean Soil Air Water 44:1198–1213

Basim Y, Jaafarzadeh N, Farzadkia M (2016) A novel biological method for sludge volume reduction by aquatic worms. Int J Environ Sci Dev 7(4):253–256

Berube P (2010) Membrane bioreactors: theory and applications to wastewater reuse. In: Escobar IC, Sch€afer A (eds) Sustainable water for the future: water recycling versus desalination, sustainability science and engineering, 1st ed. vol. 2. Elsevier, pp 255–292

Bień J (2014) Ekspertyza, Która Będzie Stanowić Materiał Bazowy do Opracowania Strategii Postępowania z Komunalnymi Osadami Sciekowymi Na Lata 2014–2020. Politechnika Częstochowska, Częstochowa, Poland

Bień J (2015) Komunalne Osady Sciekowe—Zagospodarowanie Energetyczne i Przyrodnicze. Wydawnictwo Politechniki Częstochowskiej, Częstochowa, Poland

Bień JD, Bień B (2015) Management of municipal sewage sludge with thermal methods in the face of the landfill ban after 1 January 2016. Ecol Eng 45:36–43

Bień J, Wystalska K (2011) Sewage sludge–theory and practice. Wyd. Politech. Częstoch, Częstochowa

Bień J, Kamizela T, Kowalczyk M, Stępniak L (2004) The analysis of sedimentation and thickening of sewage sludges after ultrasonic disintegration. Environ Prot Eng 30(4):24–28

Bień JA, Kamizela T, Kowalczyk M, Mrowiec M (2009) Possibilities of gravitational and mechanical separation of sonicated activated sludge suspensions. Environ Prot Eng 35(2):67

Bonfiglioli L, Bianchini A, Pellegrini M, Saccani C (2014) Sewage sludge: characteristics and recovery options 1e21. https://doi.org/10.6092/unibo/amsacta/4027

Bora RR, Richardson RE, You F (2020) Resource recovery and waste-to-energy from wastewater sludge via thermochemical conversion technologies in support of circular economy: a comprehensive review. BMC Chem Eng 2(1):1–16

Bruun EW, Ambus P, Egsgaard H, Hauggaard-Nielsen H (2012) Effects of slow and fast pyrolysis biochar on soil C and N turnover dynamics. Soil Biol Biochem 46:73–79

Buckley JC, Schwarz PM (2003) Renewable energy from gasification of manure: an innovative technology in search of fertile policy. Environ Monit Assess 84:111–127

Callegari A, Bolognesi S, Cecconet D, Capodaglio AG (2020) Production technologies, current role, and future prospects of biofuels feedstocks: a state-of-the-art review. Crit Rev Environ Sci Technol 50(4):384–436

Carrère H, Dumas C, Battimelli A, Batstone DJ, Delgenes JP, Steyer JP, Ferrer I (2010) Pretreatment methods to improve sludge anaerobic degradability: a review. J Hazard Mater 183(1–3):1–5

Cartmell E, Gostelow P, Riddell-black D (2006) Biosolids-a fuel or a waste? An integrated appraisal of five co-combustion scenarios with policy analysis. Environ Sci Technol 40:649–658

Chang IS, Le Clech P, Jefferson B, Judd S (2002) Membrane fouling in membrane bioreactors for wastewater treatment. J Environ Eng ASCE 128:1018–1029

Chen H, Yan SH, Ye ZL, Meng HJ, Zhu YG (2012) Utilization of urban sewage sludge: Chinese perspectives. Environ Sci Pollut Res 19(5):1454–1463

Cote P, Buisson H, Praderie M (1998) Immersed membranes activated sludge process applied to the treatment of municipal wastewater. Water Sci Technol 38:437–442

Deowan SA, Bouhadjar SI, Hoinkis J, Rastogi KA (2015) Membrane bioreactors for water treatment. In: Basile A, Cassano A (eds) Advances in membrane technologies for water treatment. Elsevier, pp 155–184

Dumas C, Perez S, Paul E, Lefebvre X (2010) Combined thermophilic aerobic process and conventional anaerobic digestion: effect on sludge biodegradation and methane production. Bioresour Technol 101:2629–2636

Eggen T, Vethe O (2001) Stability indices for different composts. Compost Sci Util 9(1):19–26

Fan S, Jie T, Wang Y, Li H, Zhang H, Jun T, Wang Z, Li X (2016) Biochar prepared from co-pyrolysis of municipal sewage sludge and tea waste for the adsorption of methylene blue from aqueous solutions: kinetics, isotherm, thermodynamic and mechanism. J Mol Liq 220:432e441. https://doi.org/10.1016/j.molliq.2016.04.107

Feodorov V (2016) modern technologies for treatment and stabilization for sewage sludge from water treatment plant. Agric Agric Sci Proc 10:417–430

Foladori P, Andreottola G, Ziglio G (2010) Sludge reduction technologies in wastewater treatment plants. IWA Publishing, London

Furness DT, Hoggett LA, Judd SJ (2000) Thermochemical treatment of sewage sludge. Water Environ J 14(1):57–65

Fytili D, Zabaniotou A (2008) Utilization of sewage sludge in EU application of old and new methods-a review. Renew Sust Energ Rev 12(1):116–140

Gao N, Kamran K, Quan C, Williams PT (2020) Thermochemical conversion of sewage sludge: a critical review. Prog Energy Combust Sci 79:100843

Garg NK (2009) Multicriteria assessment of alternative sludge disposal methods. Dissertation, University of Strathclyde

Gogate PR, Pandit AB (2004) A review of imperative technologies for wastewater treatment II: hybrid methods. Adv Environ Res 8(3–4):553–597

Gray NF (2005) Water technology: an introduction for environmental scientists and engineers. Elsevier Science & Technology Books, Amsterdam

Gujer W, Zehnder AJB (1983) Conversion processes in anaerobic digestion. Water Sci Technol 15:127–167

Harvey P, Baghri S, Reed B (2002) Emergency sanitation: assessment and programme design. Loughborough University, WEDC

Hii K, Baroutian S, Parthasarathy R, Gapes DJ, Eshtiaghi N (2014) A review of wet air oxidation and thermal hydrolysis technologies in sludge treatment. Bioresour Technol 155:289–299

Hospido A, Moreira MT, Martin M, Rigola M, Feijoo G (2005) Environmental evaluation of different treatment processes for sludge from urban wastewater treatments: anaerobic digestion versus thermal processes. Int J Life Cycle Assess 10:336–345

Jang HM, Cho HU, Park SK, Ha JH, Park JM (2014) Influence of thermophilic aerobic digestion as a sludge pre-treatment and solids retention time of mesophilic anaerobic digestion on the methane production, sludge digestion and microbial communities in a sequential digestion process. Water Res 48:1–4

Kalová M, Borkovcová M (2013) Voraciouslarvae *Hermetiaillucens* and treatment of selected types of biodegradable waste. Acta Univ Agric Silvic Mendelianae Brun 61:77–83

Le-Clech P, Chen V, Fane TAG (2006) Fouling in membrane bioreactors used in wastewater treatment. J Membr Sci 284:17–53

Leng L, Yuan X, Huang H, Jiang H, Chen X, Zeng G (2014) The migration and transformation behavior of heavy metals during the liquefaction process of sewage sludge. Bioresour Technol 167:144–150

Li H, Yuan X, Zeng G, Huang D, Huang H, Tong J, Zhou M (2010) The formation of bio-oil from sludge by deoxy-liquefaction in supercritical ethanol. Bioresour Technol 101(8):2860–2866

Lin SD (2007) Water and Wastewater Calculations Manual. McGraw-Hill Companies, New York

Liu T, Liu B, Zhang W (2014) Nutrients and heavy metals in biochar produced by sewage sludge pyrolysis: its application in soil amendment. Pol J Environ Stud 23:271–275

Luduvice M, Fernandes F (2007) Sludge transformation and disposal methods. In: Andreoli CV, Sperling M, Fernandes F (eds) Biological wastewater treatment series, Sludge treatment and disposal, vol 6. IWA Publishing, London, pp 207–225

Lundin M, Olofsson M, Pettersson GJ, Zetterlund H (2004) Environmental and economic assessment of sewage sludge handling options. Resour Conserv Recycl 41(4):255–278

Miah MS, Tada C, Yang Y, Sawayama S (2005) Aerobic thermophilic bacteria enhance biogas production. J Mater Cycles Waste Manag 7:48–54

Mills N, Pearce P, Farrow J, Thorpe RB, Kirkby NF (2014) Environmental & economic life cycle assessment of current & future sewage sludge to energy technologies. Waste Manag 34 (1):185–195

Moretti SM, Bertoncini EI, Abreu-Junior CH (2015) Composting sewage sludge with green waste from tree pruning. Sci Agric 72(5):432–439

Murry A, Horvath A, Nelson K (2008) Hybrid life-cycle environmental and cost inventory of sewage sludge treatment and end-use scenarios: a case study from China. Environ Sci Technol 42:3163–3169

Nancharaiah YV, Reddy GK (2018) Aerobic granular sludge technology: mechanisms of granulation and biotechnological applications. Bioresour Technol 247:1128–1143

Nasir IM, Ghazi TI, Omar R (2012) Production of biogas from solid organic wastes through anaerobic digestion: a review. Appl Microbiol Biotechnol 95(2):321–329

Novak JT, Banjade S, Murthy SN (2011) Combined anaerobic and aerobic digestion for increased solids reduction and nitrogen removal. Water Res 45:618–624

Oh HS, Yeon KM, Yang CS, Kim SR, Lee CH, Park SY, Han JY, Lee JK (2012) Control of membrane biofouling in MBR for wastewater treatment by quorum quenching bacteria encapsulated in microporous membrane. Environ Sci Technol 46:4877–4884

Oladejo J, Shi, K, Luo, X, Yang, G, Wu, T (2019) A review of sludge-to-energy recovery methods. Energies 12(60)

Pagilla KR, Craney KC, Kido WH (1996) Aerobic thermophilic pretreatment of mixed sludge for pathogen reduction and nocardia control. Water Environ Res 68:1093–1098

Pan H (2010) Effects of liquefaction time and temperature on heavy metal removal and distribution in liquefied CCA-treated wood sludge. Chemosphere 80:438–444

Pan H, Hse CY, Gambrell R, Shupe TF (2009) Fractionation of heavy metals in liquefied chromated copper arsenate (CCA)-treated wood sludge using a modified BCR-sequential extraction procedure. Chemosphere 77:201–206

Podedworna J, Umiejewska K (2008) Sewage sludge technology. Publishing House of Warsaw University of Technology. Warsaw (in Polish)

Reinhart DR, McCreanor PT, Townsend T (2002) The bioreactor landfill: its status and future. Waste Manag Res 20(2):172–186

Ren S (2004) Assessing wastewater toxicity to activated sludge: recent research and developments. Environ Int 30(8):1151–1164

Rulkens W (2008) Sewage sludge as a biomass resource for the production of energy: overview and assessment of the various options. Energy Fuel 22(1):9–15

Samer M (2015) Biological and chemical wastewater treatment processes. In Wastewater treatment engineering Intech Open. https://doi.org/10.5772/61250

Sanin FD, Clarkson WW, Vesilind PA (2011) Sludge engineering: the treatment and disposal of wastewater sludges. DEStech Publications, Inc

Semblante GU, Hai FI, Ngo HH, Guo W, You SJ, Price WE, Nghiem LD (2014) Sludge cycling between aerobic, anoxic and anaerobic regimes to reduce sludge production during wastewater treatment: performance, mechanisms, and implications. Bioresour Technol 155:395–409

Sharma S, Bhattacharya A (2017) Drinking water contamination and treatment techniques. Appl Water Sci 7:1043–1067

Singh S, Kumar V, Dhanjal DS, Datta S, Bhatia D, Dhiman J, Singh J (2020) A sustainable paradigm of sewage sludge biochar: valorization, opportunities, challenges and future prospects. J Clean Prod 269:122259

Som MP, Lemée L, Amblès A (2009) Stability and maturity of a green waste and biowaste compost assessed on the basis of a molecular study using spectroscopy, thermal analysis, thermo-desorption and thermo-chemolysis. Bioresour Technol 100(19):4404–4416
Spinosa L (2004) From sludge to resources through biosolids. Water Sci Technol 50(9):1–9
Spinosa L, Ayol A, Baudez JC, Canziani R, Jenicek P, Leonard A, Rulkens W, Xu G, Dijk L (2011) Sustainable and innovative solutions for sewage sludge management. Water 3:702–717
Stendahl K, Jafverstrom S (2004) Recycling of sludge with the Aqua Reci-process. Water Sci Technol 49(10):233–240
Syed-Hassan SSA, Wang Y, Hu S, Su S, Xiang J (2017) Thermochemical processing of sewage sludge to energy and fuel: fundamentals, challenges and considerations. Renew Sust Energ Rev 80:888–913
Tarpani RRZ, Alfonsín C, Hospido A, Azapagic A (2020) Life cycle environmental impacts of sewage sludge treatment methods for resource recovery considering ecotoxicity of heavy metals and pharmaceutical and personal care products. J Environ Manag 260:109643
Tomati U, Madejon E, Galli E (2010) Evolution of humic acid molecular weight as an index of compost stability. Compost Sci Util 8(2):108–115
Tomei MC, Rita S, Mininni G (2011) Performance of sequential anaerobic/aerobic digestion applied to municipal sewage sludge. J Environ Manag 92:1867–1873
Tungler A, Szabados E, Hosseini AM (2015) Wet air oxidation of aqueous wastes. In: Samer M (ed) Wastewater treatment engineering, InTech Cairo, p 153–178
Turovskiy IS, Mathai PK (2006) Wastewater sludge processing. Wiley, New York
Tyagi VK, Lo SL (2011) Application of physico-chemical pretreatment methods to enhance the sludge disintegration and subsequent anaerobic digestion: an up to date review. Rev Environ Sci Biotechnol 10:215–242
UNEP (2002) A directory of environmentally sound technologies for the integrated management of solid, liquid and hazardous waste for Small Island Developing States (SIDS) in the Pacific Region. United Nations Environment Programme (UNEP), The Hague
Wang ZW, Liu Y, Tay JH (2006) The role of SBR mixed liquor volume exchange ratio in aerobic granulation. Chemosphere 62(5):767–771
Wang H, Brown SL, Magesan GN, Slade AH, Quintern M, Clinton PW, Payn TW (2008) Technological options for the management of biosolids. Environ Sci Pollut Res Int 15 (4):308–317
Weemaes M, Verstraete WH (1998) Evaluation of current wet sludge disintegration techniques. J Chem Technol Biotechnol 73:83–92
Wei Y, Van Houten RT, Borger AR, Eikelboom DH, Fan Y (2003) Minimization of excess sludge production for biological wastewater treatment. Water Res 37:4453–4467
Werle S, Dudziak M (2014) Influence of wastewater treatment and the method of sludge disposal on the gasification process. Ecol Chem Eng S 21(2):255–268
Werle S, Sobek S (2019) Gasification of sewage sludge within a circular economy perspective: a Polish case study. Environ Sci Pollut Res 26:35422–35432
Werle S, Wilk RK (2010) A review of methods for the thermal utilization of sewage sludge: the Polish perspective. Renew Energy 35:1914–1919
Werther J, Ogada T (1999) Sewage sludge combustion. Prog Energy Combust Sci 25:55–116
WM (2004): The bioreactor landfill. Cincinnati: Waste Management (WM) Bioreactor Programm
Wójtowicz A, Jędrzejewski C, Bieniowski M, Darul H (2013) Modelowe Rozwiązania w Gospodarce Osadowej. Bydgoski Dom Wydawniczy Margrafsen, Bydgoszcz, Poland
Xu C, Lancaster J (2008) Conversion of secondary pulp/paper sludge powder to liquid oil products for energy recovery by direct liquefaction in hot-compressed water. Water Res 42 (6–7):1571–1582
Yamamoto K, Hiasa M, Mahmood T, Matsua T (1989) Direct solid-liquid separation using hollow fiber membrane in an activated sludge aeration tank. Water Sci Technol 21:43–54

Yoshida H, Christensen TH, Scheutz C (2013) Life cycle assessment of sewage sludge management: a review. Waste Manag Res 31(11):1083–1101

Yuan X, Huang H, Zeng G, Li H, Wang J, Zhou C, Zhu H, Pei X, Liu Z, Liu Z (2011) Total concentrations and chemical speciation of heavy metals in liquefaction residues of sewage sludge. Bioresour Technol 102:4104–4110

Zhang LH, Xu CB, Champagne P (2010) Overview of recent advances in thermo-chemical conversion of biomass. Energy Convers Manag 51(5):969–982

Zhang LH, Xu CB, Champagne P, Mabee W (2014) Overview of current biological and thermo-chemical treatment technologies for sustainable sludge management. Waste Manag Res 32 (7):586–600

Emerging Nutrient Recovery Technologies in Sewage Sludge Management

Lalichetti Sagar, Sagar Maitra, Akbar Hossain, Ajar Nath Yadav, Sultan Singh, Deepak Kumar, Subhashisa Praharaj, Tanmoy Shankar, and Biswajit Pramanick

1 Introduction

Modern Agriculture aims mostly to increase food production to fulfill the requirement of increasing global population (Ramankutty et al. 2018), forcing the scientific community to adopt a high external input-based agricultural (HEIA) system (Abdulai and Kuhlgatz 2012; Hammoudi and Hamza 2015). Although initially, HEIA hiked global food productivity, swelling the food reserves in different countries, this has not lasted for an extended period (Mohajan 2013; Mekonnen and Leenes 2020). The unprecedented use of chemical fertilizers for the last few decades has plunged into various agricultural issues (Chandio et al. 2015). Consequently, polluting soil, water, and air, thus ensuing degradation of quality and quantity of the produce (Chakraborty et al. 2013). Now, it is high time to adopt an alternative technology that can supplement nutrients to some extent and thus reducing the

L. Sagar · S. Maitra (✉) · S. Praharaj · T. Shankar
Department of Agronomy, Centurion University of Technology and Management, Paralakhemundi, Odisha, India
e-mail: sagar.maitra@cutm.ac.in

A. Hossain
Bangladesh Wheat and Maize Research Institute, Dinajpur, Bangladesh

A. N. Yadav
Department of Biotechnology, Dr. Khem Singh Gill Akal College of Agriculture, Eternal University, Baru Sahib, Sirmour, Himachal Pradesh, India

S. Singh · D. Kumar
Department of Agronomy, Sher-e-Kashmir University of Technology and Management, Jammu, Jammu and Kashmir, India

B. Pramanick
Department of Agronomy, Dr. Rajendra Prasad Central Agricultural University, Pusa, Samastipur, Bihar, India

V. D. Rajput et al. (eds.), *Sustainable Management and Utilization of Sewage Sludge*, https://doi.org/10.1007/978-3-030-85226-9_6

dependency on chemical inputs (Savci 2012). Due to advances in technology and rapid industrialization, the quantity of sewage and sludge generated has been increasing dramatically (Muazu et al. 2017). Besides, its disposal and efficient management is a serious global concern (Usman et al. 2012). On the other side, the growing worry for fertilizer availability in the future emphasized the reclamation of nutrients from sewage to agriculture (Shaddel et al. 2019). In general, water content in human urine ranges between 90–95% in addition to some inorganic and organic salts (Shingiro et al. 2019). Human faeces consist of around three fourth water by weight and the rest one-fourth of solid material which usually consists of carbon (13%), nitrogen (14–18%) phosphorus (3.7%), and potassium (3.7%), respectively. Besides, many researchers focussed on human urine containing majorly N and constituting>50% P and K in human excreta, though faeces are rich in carbon in addition to P and K (Jensen et al. 2008). Since, animals and human being uptake nutrient-rich crops and consequently resulting in nutrient rich waste expected to be recovered through digestion (Jedrejek et al. 2016). Further, the quantity of nutrients supplied depends directly on the food intake and digestibility of the food regulates its segmentation between urine and faeces (Noziere et al. 2010; Broderick 2018).

The liquid fraction of raw sewage is called effluent or sewage, while the solid fraction is called sludge (Alrawashdeh et al. 2017). High organic matter and nutrient concentration in sewage waters widened its scope for using in agriculture and forestry for vegetation production as a fertilizer (Marinho et al. 2014). High organic matter content in the sewage and sludge improves the physicochemical and biological properties of the soil resulting in improved water holding capacity and soil aeration (Usman et al. 2012). However, the presence of heavy metals and pathogen contamination in the sludge restricted its application as a fertilizer (Behbahaninia et al. 2010). Raw sewage contains organic matter and heavy metals consisting of partly solution and suspension (Zhou et al. 2017). The heavy metal content in sludge ranges from 0.5–2% which is released into the soil upon decomposition of sludge (Elloumi et al. 2016). Landfilling with sludge without proper decontamination ultimately adds up heavy metals in the soil which ultimately may enter into the human food chain resulting in several chronic health issues (Adelekan and Abegunde 2011).

Keeping this in view, this chapter highlighted different nutrient recovery technologies available for the harmless utilization of sewage and sludge in agriculture with a goal to present the material accessible to different relevant fields. The focus of the chapter is on addressing the opportunities available through sewage and sludge management as a whole and point out the literature that narrated the detailed technologies.

2 Nutrient Recovery

There is a wide variation in use of sewage and sludge in different countries. In general, sewage sludge is comprised of all three primary nutrients, namely, nitrogen (N), phosphorus (P) and potassium (K). Dependence of chemical nutrient inputs is not a sustainable practice for crop production. For instance, the chemical P-fertilizers require raw materials from geological sources and over time these will be exhausted. Therefore, there is a need to look into the availability of nutrients from alternative sources and sewage sludge can be exploited for the purpose (Kabbe 2019). But, the presence of pollutants is restricting the direct use of sewage and sludge and that further warrants suitable technologies for proper recovery of nutrients from wastes. Nutrient recovery from sewage or sludge usually depends on the nutrient concentration in sewage/sludge (Kirchmann et al. 2016). As a known fact, nutrient concentration in sewage/sludge is comparably lower than synthetic inorganic fertilizers (Kominko et al. 2017). Nutrient recovery technologies usually cover three important steps, namely, technologies aiming at the accumulation of nutrients, the release of nutrients, and its extraction (Fig. 1).

2.1 Nutrients Accumulation

Nutrients accumulation is a key step in the recovery of nutrients from sewage sludge. Technologies involved in nutrient accumulation basically aims at sequestering nutrients from the sewage water, this, in turn, attributes towards a higher rate of recovery. Under this subheading, the role of different physicochemical and biological techniques *viz.,* prokaryotic organisms, chemical precipitants, ion exchange, algae accumulation, plant accumulation, magnetic filtration and magnetic separation in nutrient accumulation have been highlighted and the mechanisms of these techniques have been discussed.

2.1.1 Prokaryotic Organisms in Nutrients Accumulation

The process of using prokaryotic bacteria for a potential accumulation of nutrients is called as prokaryotic accumulation (Yahya et al. 2019). Purple non-sulphur bacteria, polyphosphate accumulating bacteria and blue-green algae (BGA) are most widely used in nutrient accumulation. Among these, the role of polyphosphate accumulating organism (PAO) is widely gaining importance during recent years. In general, PAOs uptake phosphorus from the sewage more than their metabolic requirement giving them a competitive advantage over other microbes but for their survival, PAOs usually require both aerobic and anaerobic environments alternatively (Fernando et al. 2019). Further, this accumulation of phosphorus can be enhanced by preferring the use of organisms based on their carbon source, thereby stressing the use of such

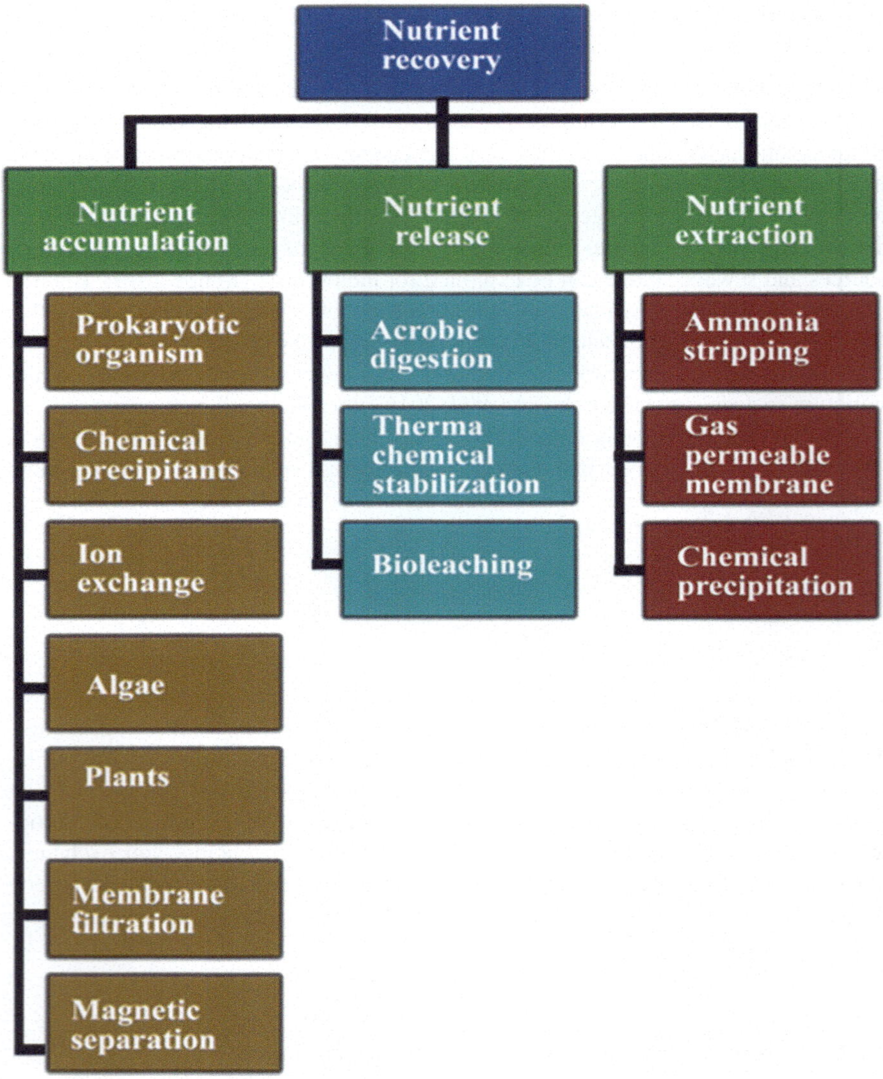

Fig. 1 Classification of nutrient recovery technologies

PAOs having volatile fatty acid as a carbon source over others, respectively (Liu et al. 2018). Similarly, the blue-green algae are reported to have the luxury in consumption of nitrogen from the sewage water accompanied by little amounts of phosphorus resulting in enhanced accumulation of primary nutrients which may release upon the further treatment, while purple nonsulphur bacteria is known to accumulate nutrients from sewage, but it also uptakes heavy metal in addition; hence raising serious concern during its recovery and usually less preferred.

2.1.2 Chemical Precipitants in Nutrient Accumulation

Accumulation of nutrients using chemical precipitants is an emerging technology in sewage water treatment. The soluble nutrients are sequestered as precipitates upon reacting with the chemical precipitants based on the principle of coagulation or flocculation (Wang et al. 2009; Achak et al. 2019). The process includes adsorption of soluble nutrients in the sewage water especially, either N or P as precipitates upon adsorption on to the colloids those are further separated in a clarifier. Besides, these chemical accumulations will also remove a significant amount of organic matter and other disease-causing organisms (Behbahaninia et al. 2010; Myint et al. 2010).

Aluminum and iron are two precipitants popularly used in the nutrient recovery from the wastewater (Peto 2010). Besides, the coagulant selection depends on the pH in the treatment plant and in general, a pH range of 6–8 is preferred for facilitating efficient functioning of the treatment plant (Kurniawan et al. 2020). On the other hand, this technique also controls the enhanced accumulation of P and N compounds in the water which would otherwise result in eutrophication, consequently protecting their adverse impact on the environment (Glibert 2017).

2.1.3 Ion Exchange in Nutrient Accumulation

This is a mass transfer process wherein the soluble nutrients of nitrogen, phosphorus and potassium present in the sewage waters are accumulated by the sorbents based on its selective preference, nutrient diffusivity, and distribution of pore size (Sanjurjo et al. 2014). In this process, the absorbing material is called sorbent while the substance that is being absorbed by the sorbent is called as absorbate (Dahri et al. 2019). The Ion exchange usually occurs at the exchange site of two phases (Bilandi and Mishra 2013).

Recently, there are several adsorbents available like an almond shell, citric acid, sawdust, charcoal etc. (Ranga and Sanghvi 2015). Among these, the use of activated charcoal is very much effective in nature (Urbanowska and Korbutowicz 2017). The porous nature of activated charcoal is attributed for its enhanced adsorption of the substance from sewage water. The large-sized particles are adsorbed by macro and mesopores while the small-sized particles are adsorbed by micropores of the charcoal (Wang et al. 2019). Nutrient solubility is directly linked to the adsorption capacity of the sorbent. More the nutrient solubility more will be the sorbent accumulation, but it depends on the nutrient concentration, solubility and its pH. At low pH condition, most of the nutrients are easily soluble, hence, marks enhanced ion exchange. Similarly, solid particle in the sewage water streams should be very low to attain higher efficiency of nutrient accumulation through ion exchange. Sorbents made of zirconium + orange waste gel reported to accumulate 57 g of phosphorus per kg of sorbent and thereby considered to be the sorbent with maximum loading capacity (Neina 2019; Ngah and Hanafiah 2008). In addition, this technology is viewed as a low-cost alternative and minimum energy involving strategy wherein biofouling using chemicals is sufficient to recover nutrients accumulated over the sorbent (Ma et al. 2020).

2.1.4 Algae in Nutrient Accumulation

Now-a-days, algae are gaining importance in sewage water treatment because of their realized ability to accrue nutrients, toxic substances, heavy metals etc. (Sen et al. 2014). Furthermore, due to this innate potential of algae and their faster growth rate; use of algae in treating wastewaters is viewed as a significant alternative to other technologies which are complex and expensive in nature (Lavrinovics and Juhna 2017). The amount and form of nutrient accumulated by algae mostly depend on the factors influencing the algal growth and development such as light intensity, pH, temperature, nutrient concentration in the effluent solution and so on (Juneja et al. 2013; Solmaz and Isik 2019).

The process involves enhanced absorption of the nutrients by algae from sewage water streams resulting in rapid algal growth with accumulated nutrients and consequently, promoting the build-up of algal biomass (Mahapatra et al. 2013). In addition, adsorption and precipitation are also involved in enhancing the nutrient recovery and accumulation by algae (Ahmed et al. 2020). Sewage treatment plants usually include three stages of treatments to remove heavy and toxic metals, thus can be used safely in agriculture as nitrogen or phosphorus supplement (Kipigroch 2018). Since the accumulation of toxic and heavy metals if allowed besides hindering the crop growth it may also enter the food chain and may cause a variety of ailments (Afshan et al. 2014). However, recent findings reported the ability of algae in accumulating toxic substances in their cell vacuoles and helped in cleaning the sewage water and thus making it utilizable (Jamuna and Noorjahan 2009).

In general, algae prefer a pH range between of 7 to 8 with an optimum temperature of about 16 to 30 °C (Aragaw and Asmare 2017). The favourable condition in the wastewater helps the algae to bloom and cover the area faster ensuring higher biomass and thus contributing towards improved carbon dioxide sequestration besides nutrient accumulation (Packer 2009). However, due to small-sized cell in algae, it is questionable on the recovery of accumulated nutrients, but in this case, advantages are dominating than the negatives associated and thus considered as a beneficial process (Mehta and Gaur 2008).

2.1.5 Plant Species in Nutrient Accumulation

Nutrient uptake by plant is a process that involves accumulation of nutrients and elimination of toxic, heavy metals and several pollutants from the sewage/ wastewater and accrual of the same in their woody cell bodies aiming to purify the entire stream which can later be used for fertigation in agriculture (Mbangi et al. 2018). This method is usually practiced in the areas dominated by wetlands where the freely floating plants accumulate more nutrients from water-logged soil. Besides, trees can also be successfully grown in these areas might also uptake volatile pollutants from the sewage water which may eject out from the leaves of the plant (Pavlineri et al. 2017; Nasr et al. 2018). These nutrients if not removed by the plant might be subjected to several losses. Even these toxic and heavy metal could leach into the groundwater, ultimately polluting the aquatic water bodies (Essien et al. 2010).

However, the efficiency of the nutrients accumulated to large extent depends on the plant biology, nutrient concentration in the sewage, temperature, pH and available dissolved oxygen (Sewwandi et al. 2010). The major concern with this practice is management since it would only be a successful approach if the harvesting is done at a regular interval (Soti et al. 2015). These harvested plants can directly use as a fertilizer in the agricultural crop field and thus ensuring recovery of nutrients.

2.1.6 Membrane Filtration in Nutrient Accumulation

Membranes include any material that acts as a barrier between any two phases and allows and restricts the movement of materials depending on the selective nature of the membrane (Cozmuta et al. 2007). Most widely applied membrane processes involved in the treatment of sewage are microfiltration, ultrafiltration, nano-filtration and reverse osmosis those separate different constituents in the wastewater based on size (Wang et al. 2011).

Microfiltration is a process that provides a barrier to suspended particulate matter of >0.1 micrometres in the sewage which is practiced efficiently subjected to a pressure of <2 bars (Gkotsis et al. 2014; Zouch et al. 2019). Ultrafiltration through specific membranes is practiced with a recommended pore size ranging from 100 to 2 nm under the influence of pressure (Polyakova and Zydney 2013). Nano-filtration includes membrane filters with a pore size from 1 to 2 nm and small-sized pores and a moderate pressure of 3 to 20 bars is essential for the process. Similarly, reverse osmosis involves some membranes with <1 nm due to further reduction in the pore size and it is usually operated under higher pressure of approximately 80 bars (Fang and Duranceau 2013). The features of different membrane processes have been presented in the following table (Table 1). Moreover, microfiltration and ultrafiltration segregate the particulates more than 0.1 micrometre while nano-filtration and reverse osmosis are generally used in segregating soluble particulate matter (Marzban et al. 2016). Passage of pre-treated sewage water through a membrane facilitates capturing of nutrients onto the surface of the membrane besides reducing

Table 1 Features of different membrane processes

Membrane process	Pressure required	Average permeability
Micro-filtration	1–3	500
Ultra-filtration	1	150
Nano-filtration	3 to 20 bars	10–20
Reverse Osmosis	>80 bars	5–10

Source: Adnan et al. (2009), Roy and Ragunath (2018)

the flow velocity directly influencing pH of the filtered water thus making it ready to use for irrigation (Dvorak et al. 2015; Shon et al. 2013).

The major issue reported with this membrane separation is subsequent retention of pathogens, heavy metals and salts limiting its direct use of concentrated nutrient mixture captured on the membrane surface (Chiama and Sarbatly 2011) which indeed, rising the involved energy costs and minimizing the membrane sustainability (Gude et al. 2011; Roy and Ragunath 2018). Comparatively, retention of pathogens on the membrane is higher while others may pass through the filter to some extent (Hava et al. 2008).

2.1.7 Magnetic Separation in Nutrient Accumulation

The magnetic separation method has been employed in sewage treatment based on its principle to separate unwanted particles having magnetic properties (Baresel et al. 2019). Sewage water gets contaminated with heavy metals as a result of rapid industrialization and that is one of the major issues harming the environment at an alarming rate (Saha and Paul 2016). Initially, this process includes adsorption of nutrients soluble in sewage water onto the surface of the carrier materials with magnetic properties (Zaidi et al. 2014; Akhter et al. 2018). Besides, it also helps in separating nonmagnetic particles and thus contributing to nutrient recovery (Naidu et al. 2020). In this regard, soluble nutrients present in the sewage water when exposed to a carrier material having magnetic properties, consequently adsorbs and retains the nutrients over it which can be recovered upon exposure to the high magnetic gradient of magnetic separators (Qi et al. 2017). Further, those accumulated or adsorbed nutrients on the surface of magnetic carriers can be recovered efficiently by passing it through high gradient magnetic separators (Issa et al. 2013). The nutrients are adhered to the magnetic field due to inherent property of the carrier where they are captured and efficiency of this method lies on the strength of nutrients with which they have adhered to the magnetic field and counteract hydrodynamic forces acting on magnetic particles (Alwani et al. 2016; Piano et al. 2019).

In general, materials like magnetite, carbonyl iron, zirconium ferrate, and iron oxide with magnetic properties are used as a carrier in this method and the magnetic field is usually based on the principle of electromagnetism (Aisida et al. 2019; Kinsler 2020). Furthermore, these high gradient magnetic separators depend mainly on the magnetic field generated by the flow of electricity thus forming a magnetic field perpendicular to the flow of electric current (Schoeffler et al. 2013). Thus, using this phenomenon, charged nutrients were sequestered from the sewage waters and strongly coagulated with the magnetic carriers facilitating nutrient recovery upon subsequent processing (Gurreri et al. 2020) (Fig. 2).

An experiment conducted in Tokyo to reduce contamination of water bodies with sewage rich in phosphorus adoption of high gradient magnetic separation (HGMS) technology using zirconium ferrite gained popularity. Zirconium ferrite is a good adsorbent of phosphate ions due to its ability to retain and high affinity for phosphate ions (Afroz et al. 2014; Ghasemian et al. 2015). These, upon exposure to the

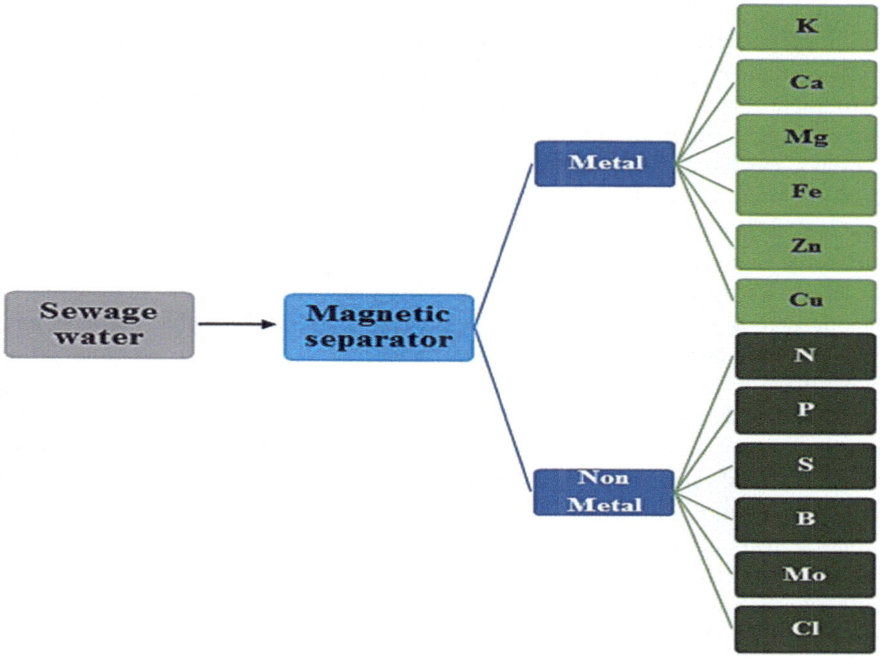

Fig. 2 Nutrient accumulation using magnetic separators

superconducting magnet, potentially sequester phosphorus which is recycled by washing the zirconium ferrite with sodium hydroxide solutions (Aigbe et al. 2017).

2.2 Nutrient Release

Nutrient release is the second step of nutrient recovery. All the nutrients accumulated are released by using different technologies. In this section, the role of anaerobic digestion technology, chemical release technology and bioleaching technology have been narrated along with mechanisms involved in the nutrient release.

2.2.1 Anaerobic Digestion in Nutrient Release

Anaerobic digestion technology involves decomposition of biodegradable waste accumulated with nutrients either through plant or algae accumulation (Ahmed et al. 2019). This process is the most common and widely used which involves waste stabilization and pathogen destruction and also helps in nutrient recovery (Wilkinson 2011). Generally, anaerobic digestion results in the conversion of an organic form of immobilized nutrients to their available forms. The digester is used for anaerobic digestion and that varies according to the characteristics of the wastes, *viz*., highly soluble solids and less soluble solids (Batstone and Jensen 2011).

The efficiency of anaerobic digestion usually depends on the availability of its favourable conditions, such as an ideal temperature ranges from 35 to 40 °C with an optimum pH ranging from 6 to 7. In general, this process completes within 20 to 30 days and the duration actually depends on the characteristics of the waste in addition to the favourable condition for the operation of this process. The nutrients released upon digestion usually bind to the digestate's solid surfaces. Furthermore, anaerobic digestion improves the soluble phosphorus and nitrogen in the sludge material (Batstone and Jensen 2011) and for better handling, the dewatering is practiced resulting in concentrating the sewage water with soluble nutrients and left-over nutrients can be used as biosolids suitable for application in agriculture (Liu et al. 2018).

Although the biosolids can be used in various agricultural activities but the nutrient content within them is low valued while that in the sewage water is having more analytical value. To improve the further solubilization of nutrients it is essential to add the complexing agent like ethylene diamine tetra-acetic acid (EDTA) that works at lower pH and thus reducing the loss of nutrients (Mehta and Damien 2012). The activated sludge usually rich in phosphorus and when it is combined with ammonia-rich digestate, further, results in struvite formation that minimizes the issues aroused during struvite preparation in nutrient extraction.

2.2.2 Thermochemical Stabilization in Nutrient Release

Thermochemical stabilization is one of the nutrient release processes which are gaining tremendous popularity in sewage sludge management in recent years. This process includes several primary stage steps, *viz*., thermal sludge drying, and incineration and enhanced stage steps, like wet oxidation, aerobic digestion or gasification and pyrolysis. The processes like pyrolysis, thermal sludge drying, gasification and incineration involve the principle of evaporation to degrade wastes with high carbon concentration. This process results in three primary stage products, namely, char, ash and oil which in turn retains potassium and phosphorus with limited amounts of nitrogen since a major portion is lost through the gas stream.

Thermal drying is the process which involves heavy use of energy to produce sufficient heat to evaporate water from the sludge resulting in decrease in volume. To improve its efficiency in certain cases, the energy requirement during the process is substituted by the utilization of biogas. Incineration is another process which involves combustion of sludge resulting in complete oxidation of organic matter present in the sludge material with some benefits like complete mineralization of sludge, pathogen-free sludge and odour free end product. Under excess oxygen concentration and high temperatures above 800 °C incineration and gasification are adopted while pyrolysis is adopted under restricted supply of oxygen and low temperatures (Bridle and Pritchard 2004). Similarly, wet oxidation operates at medium temperatures and involves thermally regulated degradation, removal of water through hydrolysis and addition of oxygen (Blocher et al. 2012).

Pyrolysis, liquefaction and gasification are considered to be thermochemical conversion technologies with several advantages over direct combustion (Thygesen et al. 2011). During direct combustion, most of the nutrients may vaporise along with excessive energy availability in the form of heat. Further, the by-products of these treatments are to be treated with some chloride salts that would help in the segregation of the heavy metals in the sludge biosolids and thereby vaporised into the air or removed as ash after converting them into heavy metal chlorides upon treatment.

2.2.3 Bioleaching in Nutrient Release

Bioleaching is a process of solubilizing heavy metals and nutrients present in the sewage waters through metabolic process undergone within the leaching micro-organisms. It usually encompasses in catalysing chemical degradation of sulphur through metabolic activities within iron and sulphur oxidizing organisms (Sannasi et al. 2010; Emmanuel et al. 2017). Different microbes that are involved in bioleaching are *Sulfobacillus* spp., *Fusarium* spp., *Acidithio bacillus* spp. and so on. Pre-requisite for micro-organisms involved in bioleaching is to adapt to extreme pH conditions and should be capable of the oxidation of iron and sulphur (Khan et al. 2014). In addition, these microbes should also release nutrients. Bioleaching is one of the cheap techniques among the other nutrient release technologies and an environmentally friendly approach. Similarly, sulphur-based bioleaching involves both direct and indirect method to convert into sulphate. In the direct method, the metal sulphides are directly converted to sulphate under the catalytic influence of any bioleaching organism. Bioleaching operation also takes place in the absence of microbes, but the presence of microbes accelerates the process without getting involved in the actual reaction. Hence, maximum bioleaching occurs when the most favourable environment for growth and development of the bacteria prevails, namely, a temperature of 20–40 °C and pH in between 1.0 to 4.5. Moreover, there should be a favourable climate for rapid solubilization of a metal (Pathak et al. 2009).

2.3 *Nutrient Extraction*

Nutrient extraction is an alternative fate to the nutrients sequestered from sewage water. This is the final most important step in nutrient recovery. This section covers the role of chemical precipitation, ammonia stripping and electro-dialysis technologies in nutrient extraction and thus facilitating the use of recovered nutrients as an alternative to the chemical fertilizers in agriculture.

2.3.1 Chemical Precipitation in Nutrient Extraction

Chemical precipitation is the most common technology used in wastewaters and it involves the transfer of nutrients present in the sewage waters into crystals or amorphous form (Shiba and Ntuli 2016). In general, struvite, iron phosphate, calcium, aluminium etc. are popular chemical precipitates used in the wastewater treatment (Alvarenga et al. 2017). Supersaturation of the wastewater is a pre-requisite for nutrient recovery from the sewage water which is usually attained by regulating temperature, changing pH or through the addition of metal ions.

2.3.2 Gas-Permeable Membranein Nutrient Extraction

The gas-permeable membrane technology basically aims to recover ammonia from the wastewaters (Kinidi et al. 2018). Further, the gas-permeable membrane is highly efficient in extracting nitrogen from wastewaters as ammonia (Priya et al. 2018). Consequently, it acts as a fertilizer supplementing the nitrogen requirement in agriculture. In this process, the sewage water rich in gaseous ammonia is passed through tubular microporous hydrophobic membranes surrounded by acidic solution (Rothrock et al. 2010). Further, upon passage through the tubular hydrophobic membrane, the ammonia will get vaporized and the volatilized ammonia passing through the gas-permeable membrane is captured either by condensation or absorbed within the acidic solution (Bae and Kim 2020).

The extraction of ammonia is more significant from wastewater with high ammonium concentration because of the dependency of this process on the partial pressure difference between liquid waste and absorbing acidic solution surrounding it (Herranz et al. 2019). This shows the significant attribution of ammonia concentration in the waste towards the efficiency of the gas-permeable membrane (Fillingham et al. 2017). The process is, further, greatly influenced by the temperature and pH wherein the higher temperature and pH above the optimum would increase the concentration of free ammonia instead of ammonium ions (Matias and Szogi 2011). This might be due to increased dissociation of ammonium ions into free ammonia which can easily permeate across the membrane and trapped by the acidic solution (Li et al. 2017).

The major limitation of this technology is that it is expensive due to the involvement of higher operational costs (Chen et al. 2015). Additionally, the efficient management of pH and temperature is very crucial which usually covers to the major percentage of operational costs (Baumann and Fuchs 2012; Kumar et al. 2020). In an experiment conducted ammoniacal nitrogen was extracted successfully from swine manure using this technology. In hydraulic retention time (HRT) for the first seven days reported the removal of 79% of total ammoniacal nitrogen (TAN) while during next five days it has been reported to remove 56% of the TAN. However, the total nutrient recovery was estimated to be 90% using these semi-continuous membranes. In this process, ammonia is directly converted into ammonium sulphate. It was further reported that dilution of the solution consequently reduced the nitrogen concentration of the final product. An enhancement in temperature around 3 °C ultimately minimised the osmotic distillation by 34% resulting in an increase the fertilizer concentration (Berta et al. 2019).

2.3.3 Ammonia Stripping in Nutrient Extraction

Ammonia stripping is the process that encompasses the mass transfer of ammonia after undergoing physio-chemical processes from the sewage waters into the gaseous phase (Leite et al. 2013). This process begins feeding the gas stripper with ammonia-rich slurry (Taşdemir et al. 2020). Usually, the ammonia in the sewage is present in the form of ammonia gas and ammonium ions. The ratio between ammonium ions and ammonia gas can be balanced by adjusting the pH in the stripper since with the increase in pH the ammonium ions concentration falls and results in increased ammonia emission. The amount of ammonia gas stripped in the stripper is directly dependent on the concentration of ammonia present in the sewage fed to the stripper tank. Before feeding the sewage to the stripper the pH of the slurry is increased by pre-treating the sewage with sodium hydroxide. This enhances the pH of the sewage effluent and fastens the stripping of ammonia gas in the stripping tank. Further, the ammonia gas is directed out of the stripping tank into the scrubber tank wherein ammonia gas is washed using sulphuric acid resulting in the formation of ammonium sulphate slurry which can be further used as a fertilizer in crop production. In addition, ammonia stripping involves several issues which include the problem of fouling, efficient emission of ammonia gas and quantitative production of sludge.

2.3.4 Electrodialysis in Nutrient Extraction

By electrodialysis, nutrient ions are separated from sewage water using anion or cation exchange membranes in the presence of electric field (Akhter et al. 2018). Electric field influences the movement of the ions present in the wastewater based on its charge (Mahmoud et al. 2010) in such a manner that the cations tend to move towards cathode while the anions incline towards the anode and simultaneously, these ions are captured using the ion exchange membranes at their respective poles (Davies and Crooks 2020). In general, the sewage water consists of both dilute and concentrate which upon activation with electric stimuli it involves the transfer of ions from dilute into the concentrate as the positively charged ions move towards the cathode and negatively charged ions towards the anode (Ghernaout et al. 2011; Tureka et al. 2017).

During this migration they tended to move across the membranes and retained over the membrane surface, *i.e.*, anion exchange membrane retained cations, while cation exchange membranes reserved anions (Fig. 3). Meanwhile, to reduce the damage to these membranes over the long run they were periodically exchanged such that anode to cathode and vice versa. This exchange process is based on the

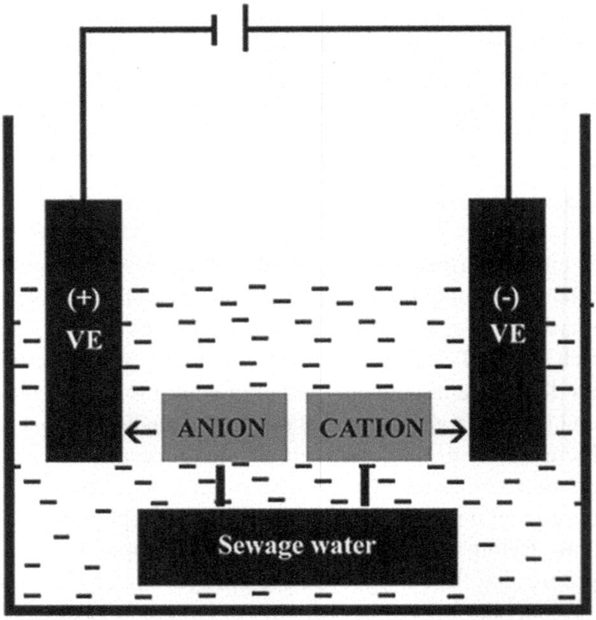

Fig. 3 Process of electrodialysis

principle of dilution resulting in longer sustainability of membranes (Hassanvand et al. 2017; Jaroszek and Dydo 2015). However, this technology is not suitable if the concentration of the salt in the sewage water streams is high (Tadimeti et al. 2015; Uqab et al. 2017). Since energy involved in reclaiming that membrane from the salt is more or less same, thus it is quite uneconomical to use electrodialysis (ED) to recover nutrients from salty sewage water (Ronan et al. 2014). At the same time, corrosive action of the chlorine ion may also affect the membrane functioning by clogging of pores thus necessitating its replacement.

3 Conclusion

Scientific management of sewage sludge is essential for making a pollution-free environment. As the sewage sludge contains nutrients, there is scope for nutrient recovery from it and using nutrients for agricultural purpose. But the nutrient recovery methods from sewage sludge are to be adopted based on suitable, economically beneficial and sustainable strategies matching with the local economy and geographical aspects. Actually, there is no single technology that can fit unanimously to all conditions. The chapter would act as an eye-opener in realizing the potential of different nutrient recovery technologies from sewage and sludge management. This additional technological alternative would supplement major nutrient demand and also focuses on reducing the use of synthetic inputs only if the all the three steps of nutrient recovery *viz.*, accumulation, release and extraction of nutrients if properly integrated. Physical, chemical and biological technologies involved in nutrient accumulation were clearly highlighted with their mechanisms significantly reporting their role in sequestration of nutrients from the sewage waters. Further, these sequestered nutrients are recovered either by nutrient release technologies or by adopting technologies involved in nutrient extraction. However, both these steps of nutrient recovery attribute to bring in soluble nutrients accumulated in physico-chemical and biological structures into a potentially used alternative to chemical fertilizers.

In sustainable waste management, three basic pillars of sustainability (3 P's, that is, planet, people and prosperity) are to be taken into consideration. Today, even the alternative sources of nutrients are not exploited properly, they can also be stored for the future. Considering the environmental and agricultural aspects, it can be concluded that proper nutrient recovery from sewage sludge can initiate a paradigm shift against over-dependence on non-renewable sources of nutrients and thus stepping forward towards sustainable agriculture.

References

Abdulai A, Kuhlgatz C (2012) Food security policy in developing countries. Oxford Handbooks Online 1–33

Achak M, Elayadi F, Boumya W (2019) Chemical coagulation/flocculation processes for removal of phenolic compounds from olive mill wastewater: a comprehensive review. Am J Appl Sci 16(3):59–91

Adelekan BA, Abegunde KD (2011) Heavy metals contamination of soil and groundwater at automobile mechanic villages in Ibadan, Nigeria. Int J Phys Sci 6(5):1045–1058

Adnan S, Hoang M, Wang H (2009) Research, development and application of membrane technology in the pulp and paper industry. Conference: In 63rd Appita Annual Conference and Exhibition, Melbourne 19–22 April Appita Inc 241

Afroz R, Masud MMM, Akhtar R, Duasa JB (2014) Challenges and future direction for water resource management policies in Malaysia. Environ Urbanization ASIA 5(1):63–81

Afshan S, Ali S, Ameen US, Farid M, Bharwana SA, Hannan F, Ahmad R (2014) Effect of different heavy metal pollution on fish. Res J Chem Environ Sci 2(1):74–79

Ahmed M, Ahmad S, Qadir G, Hayat R, Shaheen FA, Raza MA (2019) Innovative processes and technologies for nutrient recovery from wastes: a comprehensive review. Sustainability 11(18):4938

Ahmed MD, Ketife A, Momani FA, Judd S (2020) A assimilation and bioaccumulation model for the removal of heavy metals from wastewater using algae: new strategy. Proc Safety Environ Protec 144(10):52–64

Aigbe UO, Das R, Ho WH, Srinivasu V, Maity A (2017) A novel method for removal of Cr (VI) using polypyrrole magnetic nano-composite in the presence of unsteady magnetic fields. Separat Purifica Technol 194:377–384

Aisida SO, Madubuonu N, Alnasir MH, Ahmad I, Botha S, Maaza M, Ezema FI (2019) Biogenic synthesis of iron oxide nanorods using *Moringa oleifera* leaf extract for antibacterial applications. Appl Nanosci 10(1):305–315

Akhter M, Habib G, Qamar SU (2018) Application of electrodialysis in waste water treatment and impact of fouling on process performance. J Membr Sci 8(2):1–8

Alrawashdeh KA, Pugliese A, Slopiecka K, Pistolesi V, Massoli S, Bartocci P et al (2017) Co digestion of untreated and treated sewage sludge with the organic fraction of municipal solid wastes. Fermentation 3(3):1–12

Alvarenga E, Qgaard AF, Vrale L (2017) Effect of anaerobic digestion and liming on plant availability of phosphorus in iron- and aluminium precipitated sewage sludge from primary wastewater treatment plants. Water Sci Technol 75(7):1743–1752

Alwani S, Kaur R, Michel D, Chitanda JM, Verrall RE, Karunakaran C, Badea I (2016) Lysine-functionalized nanodiamonds as gene carriers: development of stable colloidal dispersion for in vitro cellular uptake studies and siRNA delivery application. Int J Nanomedicine 11(1):687

Aragaw TA, Asmare AM (2017) Experimental identifications of fresh water microalgae species and investigating the media and pH effect on the productions of microalgae. J Environ Treatment Technol 5(2):124–131

Bae W, Kim J (2020) Permeate flux and rejection behavior in submerged direct contact membrane distillation process treating a low-strength synthetic wastewater. Appl Sci 10(2):677

Baresel C, Schaller V, Jonasson C, Johansson C, Bordes R, Chauhan V, Sugunan A, Sommertune J, Welling S (2019) Functionalized magnetic particles for water treatment. Heliyon 5(8):e02325

Batstone DJ, Jensen PD (2011) Anaerobic processes. In: Wilderer P (ed) Treatise on water science. Acad Press, Oxford, pp 615–640

Baumann FW, Fuchs W (2012) The application of membrane contactors for the removal of ammonium from anaerobic digester effluent. Sep Sci Technol 47(10):1436–1442

Behbahaninia A, Mirbagheri SA, Nouri J (2010) Effects of sludge from wastewater treatment plants on heavy metals transport to soils and groundwater. Iran J Environ Health Sci Eng 7(5):401–406

Berta R, Salces BM, Vanotti MB, Gonzalez MCG (2019) Application of gas-permeable membranes for-semi continuous ammonia recovery from swine manure. Environments 6:32

Bilandi A, Mishra AK (2013) Masking of bitter drugs and sustained release formulations with their patents. Int Res J Pharm 4(8):65–74

Blocher C, Niewersch C, Melin T (2012) Phosphorus recovery from sewage sludge with a hybrid process of low-pressure wet oxidation and nano-filtration. Water Res 46:2009–2019

Bridle TR, Pritchard D (2004) Energy and nutrient recovery from sewage sludge via pyrolysis. Water Sci Technol 50:169–175

Broderick GA (2018) Review: optimizing ruminant conversion of feed protein to human food protein. Animal 12(8):1–13

Chakraborty C, Huq MM, Ahmed S, Tabassum T, Miah R (2013) Analysis of the causes and impacts of water pollution of buriganga river: a critical study. Int J Sci Technol Res 2(9):245–251

Chandio AA, Yuansheng J, Liu D, Guangshun X, Bin T (2015) Time series analysis of the impact of rising prices of inorganic fertilizers on field crops production: a case study of Pakistan. J Ecol Sust Dev 6(18):62–71

Chen XY, Thang HV, Ramirez AA, Rodrigue D, Rodrigue S (2015) Membrane gas separation technologies for biogas upgrading. RSC Adv 5(31):24399–24448

Chiama CK, Sarbatly R (2011) Purification of aquacultural water: conventional and new membrane-based techniques. Sep Purif Rev 40(2):126–160

Cozmuta I, Blanco M, Goddard WA (2007) Gas sorption and barrier properties of polymeric membranes from molecular dynamics and Monte Carlo simulations. J Physical Chem B 111(12):3151–3166

Dahri SH, Mangrio MA, Shaikh IA, Dahri SA, Steenbergen FA (2019) Effect of different forms of super absorbent polymers on soil physical & chemical properties in orchard field. World Acad J Res Eng Sci 6(2):12–20

Davies CD, Crooks RM (2020) Focusing, sorting and separating microplastics by serial faradaic ion concentration polarization†. Chem Sci 11(21):5547

Dvorak L, Gomez M, Dolina J, Cernin A (2015) Anaerobic membrane bioreactors-a mini review with emphasis on industrial wastewater treatment: applications, limitations and perspectives. Desalination Water Treat 57(41):1–15

Elloumi N, Belhaj D, Jerbi B, Zouari M, Kallel K (2016) Effects of sewage sludge on bio-accumulation of heavy metals in tomato seedlings. Span J Agric Res 14(144):13–2171

Emmanuel E, Andrew A, John OC, Emmanuel Priesley O, Ufeli D, Amina A (2017) Isolation, identification and characterization of some bacteria from soil samples of agbaja iron ore mining site of kogi state. J Bacterio Myco 4(3):79–84

Essien JP, Udofia GE, Inam E, Kim KW (2010) Effect of graded doses of heavy metals on the radial growth rate of hyphomycetous fungi from mangrove sediments of the qua iboe estuary, Nigeria. Geosyst Eng 13(4):139–146

Fang Y, Duranceau SJ (2013) Study of the effect of nanoparticles and surface morphology on reverse osmosis and nanofiltration membrane productivity. Membranes 3(3):196–225

Fernando EY, McIlroy SJ, Nierychlo M, Herbst FA, Petriglieri F, Schmid MC, Wagner M, Nielsen JL, Nielsen PH (2019) Resolving the individual contribution of key microbial populations to enhanced biological phosphorus removal with Raman-FISH. ISME J 13(8)

Fillingham M, Zaag CV, Singh J, Burtt S, Crolla A, Kinsley C, Donald DM (2017) Characterizing the performance of gas-permeable membranes as an ammonia recovery strategy from an aerobically digested dairy manure. Membranes 7(4):59

Ghasemian Z, Gahrouei DS, Manouchehri S (2015) Cobalt zinc ferrite nano-particles as a potential magnetic resonance imaging agent: an in vitro study. Avicenna J Med Biotech 7(2):64–68

Ghernaout D, Naceur MW, Ghernaout B (2011) A review of electro-coagulation as a promising coagulation process for improved organic and inorganic matters removal by electrophoresis and electroflotation. Desalination Water Treat 28(1-3):287–320

Gkotsis PK, Banti DC, Peleka EN, Zouboulis AI, Samaras PE (2014) Fouling issues in membrane bioreactors (MBRs) for wastewater treatment: major mechanisms, prevention and control strategies. PRO 2(4):795–866

Glibert PM (2017) Eutrophication, harmful algae and biodiversity-challenging paradigms in a world of complex nutrient changes. Mar Pollut Bull 124(2):591–606

Gude VG, Khandan NN, Deng S, Maganti A (2011) Energy consumption and recovery in reverse osmosis. Desalination Water Treat 36:239–260

Gurreri L, Tamburini A, Cipollina A, Micale G (2020) Electrodialysis applications in wastewater treatment for environmental protection and resources recovery: a systematic review on progress and perspectives. Membranes 10(7):146

Hammoudi A, Hamza O (2015) Food safety on domestic markets of developing countries: how to improve the contribution of export and domestic supply chains. Int J Bus Manag 10(5):20–43

Hassanvand A, Wei K, Talebi S, Chen GQ, Kentish SE (2017) The role of ion exchange membranes in membrane capacitive deionisation. Membranes 7(3):1–23

Hava DL, Wel NVD, Cohen N, Dascher CC, Houben D, Leon L et al (2008) Evasion of peptide, but not lipid antigen presentation, through pathogen-induced dendritic cell maturation. Proc Natl Acad Sci 105(32):11281–11286

Herranz MS, Bascones MS, Rodriguez JMA, Cid DC, Vanotti MB (2019) Effect of the type of gas-permeable membrane in ammonia recovery from air. Environments 6(6):70

Issa B, Obaidat MI, Albiss BA, Haik Y (2013) Magnetic nanoparticles: surface effects and properties related to biomedicine applications. Int J Mol Sci 14(11):21266–21305

Jamuna S, Noorjahan CM (2009) Treatment of sewage waste water using water hyacinth-eichhorniasp and its reuse for fish culture. Toxicol Int 16:103–106

Jaroszek H, Dydo J (2015) Ion-exchange membranes in chemical synthesis-a review. Open Chem J 14(1):1–19

Jedrejek D, Levic J, Wallace J, Oleszek W (2016) Animal by-products for feed: characteristics, European regulatory framework, and potential impacts on human and animal health and the environment. J Anim Feed Sci 25(3):189–202

Jensen PKM, Phuc PD, Knudsen LG, Dalsgaard A, Konradsen F (2008) Hygiene versus fertiliser: the use of human excreta in agriculture-a vietnamese example. Int J Hyg Environ Health 211(3-4):432–439

Juneja A, Ceballos RM, Murthy GS (2013) Effects of environmental factors and nutrient availability on the biochemical composition of algae for biofuels production: a review. Energies 6(9):4607–4638
Kabbe C (2019) Global compendium on phosphorus recovery from sweage/sludge/ash. Global Water Research Coaliation, London
Khan SA, Uddin I, Moeez S, Ahmad A (2014) Fungus-mediated preferential bioleaching of waste material such as fly-ash as a means of producing extracellular, protein capped, fluorescent and water-soluble silica nanoparticles. PLoS One 9(9):e107597
Kinidi L, Tan IVW, Wahab NBA, Tamrin KFB, Chipolito CN, Salleh SF (2018) Recent development in ammonia stripping process for industrial wastewater treatment. Int J Chem Eng 12:1–14
Kinsler P (2020) Faraday's law and magnetic induction: cause and effect, experiment and theory. Physics 2(2):148–161
Kipigroch K (2018) The use of algae in the process of heavy metal ions removal from wastewater. Desalination Water Treat 134:289–295
Kirchmann G, Borjesson H, Katterer T, Cohen Y (2016) From agricultural use of sewage sludge to nutrient extraction: a soil science outlook. Ambio 46(2):143–154
Kominko H, Gorazda K, Wzorek Z (2017) The possibility of organo-mineral fertilizer production from sewage sludge. Waste Bio Valor 8(5):1781–1791
Kumar V, Nayak SS, Katyal D (2020) Comparative evaluation of the radioactivity removal efficiency of different commercially available reverse osmosis membranes. Radiat Prot Environ 43(2):100
Kurniawan SB, Abdullah SRS, Imron MF, Said NSM, Ismail NI, Hasan HA, Othman AR, Purwanti IF (2020) Challenges and opportunities of biocoagulant/a bioflocculant application for drinking water and wastewater treatment and its potential for sludge recovery. Int J Environ Res Public Health 17(24):1–33
Lavrinovics A, Juhna T (2017) Review on challenges and limitations for algae-based wastewater treatment. J Constr Sci 20(1):17–25
Leite VD, Prasad S, Lopes WS, Sousa JTD, Barros AJM (2013) Study on ammonia stripping process of leachate from the packed towers. J Urban Environ Eng 7(2):215–222
Li T, Panther J, Qiu Y, Liu C, Huang J, Wu Y, Wong PK, An T, Zhang S, Zhao H (2017) Gas-permeable membrane-based conductivity probe capable of in situ real-time monitoring of ammonia in aquatic environments. Environ Sci Technol 51(22)
Liu Y, Li Y, Zhao J, Wang D, Yanga Q, Zeng G (2018) The feasibility of enhanced biological phosphorus removal in the novel oxic/extended idle process using fermentation liquid from sludge fermentation. RSC Adv 8(6):3321–3327
Ma G, Xu X, Tesfai M, Wang H, Xu P (2020) Developing anti-biofouling and energy-efficient cation-exchange membranes using conductive polymers and nanomaterials. J Membr Sci 603: 118034
Mahapatra DM, Chanakya HN, Ramachandra TV (2013) Treatment efficacy of algae-based sewage treatment plants. Environ Monit Assess 185(9):7145–7164
Mahmoud A, Olivier J, Vaxelaire J, Hoadley AFA (2010) Electrical field: a historical review of its application and contributions in wastewater sludge dewatering. Water Res 44(8):2381–2407
Marinho EB, Oliveira AL, Zandonadi DB, Benedito LEC, Souza RB, Figueiredo CC, Busato JG (2014) Organic matter pools and nutrient cycling in different coffee production systems in the Brazilian Cerrado. Agrofor Syst 88:767–778
Marzban R, Saberi F, Shirazi MMA (2016) Microfiltration and ultrafiltration of *Bacillus thuringiensis* fermentation broth: membrane performance and spore-crystal recovery approaches. Brazilian J Chem Eng 33(4):783–791
Matias B, Szogi AA (2011) Use of gas-permeable membranes for the removal and recovery of ammonia from high strength livestock wastewater. Proc Water Environ Feder 1:659–667
Mbangi A, Muchaonyerwa P, Zengeni R (2018) Accumulation of multiple heavy metals in plants grown on soil treated with sewage sludge for more than 50 years presents health risks and an opportunity for phyto-remediation. Water SA 44(4):569–576

Mehta CM, Damien BJ (2012) Nutrient solubilization and its availability post anaerobic digestion. Water Sci Technol 67(4):756–763

Mehta SK, Gaur JP (2008) Use of algae for removing heavy metal ions from wastewater: progress and prospects. Crit Rev Biotechnol 25(3):113–152

Mekonnen MM, Leenes WG (2020) The water footprint of global food production. Water 12(10):1–12

Mohajan HK (2013) Global food price hike is a burden to the poor. Int J Info Technol Bus Manag 19(1):1–15

Muazu ND, Jarrah N, Zubair M, Alagha O (2017) Removal of phenolic compounds from water using sewage sludge-based activated carbon adsorption: a review. Int J Environ Res Public Health 14(10):1–33

Myint AK, Yamakawa T, Kajihara Y, Zenmyo T (2010) Application of different organic and mineral fertilizers on the growth, yield and nutrient accumulation of rice in a Japanese ordinary paddy field. Sci World J 5(2):47–54

Naidu G, Tijing L, Johir MAH, Shon H, Vigneswaran S (2020) Hybrid membrane distillation: resource, nutrient and energy recovery. J Membr Sci 599:117832

Nasr FA, Shafai SA, Abdelfadil AS, Ibrahim HS, Hemdan BA (2018) Potential use of treated domestic sewage for cultivation of biofuel crops in Egypt. Int J Environ Sci Technol 16(11):7433–7442

Neina D (2019) The role of soil pH in plant nutrition and soil remediation. Appl Environ Soil Sci 2019:1–8. https://doi.org/10.1155/2019/5794869

Ngah WSW, Hanafiah MAKM (2008) Removal of heavy metal ions from wastewater by chemically modified plant wastes as adsorbents: a review. Bioresour Technol 99(10):3935–3948

Noziere P, Marty IO, Loncke C, Sauvant D (2010) Carbohydrate quantitative digestion and absorption in ruminants: from feed starch and fibre to nutrients available for tissues. Animal 4(7):1057–1074

Packer M (2009) Algal capture of carbon dioxide; biomass generation as a tool for greenhouse gas mitigation with reference to New Zealand energy strategy and policy. Energy Policy 37(9):3428–3437

Pathak A, Dastidar MG, Sreekrishnan TR (2009) Bioleaching of heavy metals from sewage sludge: a review. J Environ Manag 90(8):2343–2353

Pavlineri N, Skoulikidis NT, Tsihrintzis VA (2017) Constructed floating wetlands: a review of research, design, operation and management aspects, and data meta-analysis. Chem Eng J 308: 1120–1132

Peto MV (2010) Aluminium and iron in humans: bioaccumulation, pathology, and removal. Rejuvenation Res 13(5):589–598

Piano L, Vercelli S, Morra V, Barbero S (2019) Proceedings of the international scientific conference AIFI 2017. Therapeutic exercise: foundations, evidences and clinical reasoning in physiotherapy practice. Arch Physiother 9(1):1–59

Polyakova YS, Zydney AL (2013) Ultrafiltration membrane performance: effects of pore blockage/ constriction. J Membr Sci 434:106–120

Priya KS, Burman I, Tarafdar A, Sinha A (2018) Impact of ammonia nitrogen on COD removal efficiency in anaerobic hybrid membrane bioreactor treating synthetic leachate. Int J Environ Res 13(1):1–8

Qi Z, Joshi TP, Liu R, Li Y, Liu H, Qu J (2017) Adsorption combined with superconducting high gradient magnetic separation technique used for removal of arsenic and antimony. J Hazard Mater 343:36–48

Ramankutty N, Mehrabi Z, Jarvis KWL, Kremen C, Herrero M, Rieseberg LH (2018) Trends in global agricultural land use: implications for environmental health and food security. Annu Rev Plant Biol 69(1):1–14

Ranga SV, Sanghvi LK (2015) Selection of adsorbent for removal of dye from waste-water. Int Confer Mult Res Pract 3(1):85–89

Ronan KM, Zubair SM, Lienhard JH (2014) The benefits of hybridising electrodialysis with reverse osmosis. J Membrane Sci 469:326–335

Rothrock MJ, Szogi AA, Vanotti MB (2010) Recovery of ammonia from poultry litter using gas-permeable membranes. Am Soci of Agric and Biol Eng 53(4):1267–1275

Roy S, Ragunath S (2018) Emerging membrane technologies for water and energy sustainability: future prospects, constraints and challenges. Energies 11(11):2997

Saha P, Paul B (2016) Assessment of heavy metal pollution in water resources and their impacts: a review. J Bas Appl Eng Res 3(8):671–675

Sanjurjo MJF, Rodriguez EA, Núñez AN, Marcos MLF, Gasalla AR (2014) Nitrogen, phosphorus, potassium, calcium and magnesium release from two compressed fertilizers: column experiments. Solid Ear Discus 5(2):1351–1360

Sannasi P, Salmijah S, Kader J (2010) Effect of heavy metals to bacterial culture and the removal of heavy metals from an industrial effluent. Biosci Biotechnol Res Asia 7(2):543–557

Savci S (2012) Investigation of effect of chemical fertilizers on environment. APCBEE Procedia 1: 287–292

Schoeffler KM, Loureiro NF, Fonseca RA, Silva LO (2013) Magnetic field generation and amplification in an expanding plasma. Phys Rev Lett 182(310):1–5

Sen B, Alp MT, Sonmez F, Kocer MAT, Canpolat O (2014) Relationship of algae to water pollution and waste water treatment. In: Water Treatment Chapter: 14 (Eds., Walid Elshorbagy W, Rezaul Kabir Chowdhury RK), pp 335–354, https://doi.org/10.5772/51927

Sewwandi BGN, Weragoda SK, Mowjood MIM, Tanaka N, Sasikala S (2010) Effect of submerged and floating plants on dissolved oxygen dynamics and nitrogen removal in constructed wetlands. Trop Agric Res 21(4):353–360

Shaddel S, Davijany HB, Kabbe C, Dadgar F, Osterhus SW (2019) Sustainable sewage sludge management: from current practices to emerging nutrient recovery technologies. Sustainability 11(12):1–12

Shiba CN, Ntuli F (2016) Extraction and precipitation of phosphorus from sewage sludge. J Waste Manag 60:191–200

Shingiro C, Nyagatare G, Hirwa H, Solange U (2019) Effect of human urine and compost tea on soil physical and chemical properties case study rubirizi marshland. J Agri Sci Eng 5(3):61–66

Shon HK, Phuntsho S, Chaudhary DS, Vigneswaran S, Cho J (2013) Nanofiltration for water and wastewater treatment-a mini review. Drinking Water Eng Sci 6(1):47–53

Solmaz A, Isik M (2019) Effect of sludge retention time on biomass production and nutrient removal at an algal membrane photobioreactor. Bioenergy Res 12(1):197–204

Soti PG, Jayachandran K, Koptur S, Volin JC (2015) Effect of soil pH on growth, nutrient uptake, and mycorrhizal colonization in exotic invasive *Lygodium microphyllum*. Plant Ecol 216(7):989–998

Tadimeti JGD, Chandra A, Chattopadhyay S (2015) Optimum concentrate stream concentration in CaCl2 removal from sugar solution using electrodialysis. J Chem Eng Process Technol 6(1):1000216

Taşdemir A, Cengiz İ, Yildiz E, Bayhan YK (2020) Investigation of ammonia stripping with a hydrodynamic cavitation reactor. Ultrason sonoche 60:104741

Thygesen AM, Wernberg O, Skou E, Sommer SG (2011) Effect of incineration temperature on phosphorus availability in bio-ash from manure. Environ Technol 32(6):633–638

Tureka M, Labus K, Dydo P, Mitko K, Laskowska E, Kolon AJ (2017) A concept of hydraulic fracturing flowback treatment using electrodialysis reversal. Desalination Water Treat 64:228–232

Uqab B, Singh A, Mudasir S (2017) Impact of sewage on physico-chemical water quality of Tawi River in Jammu city. Environ Risk Assess Remediat 1(2):56–61

Urbanowska A, Korbutowicz MK (2017) The efficiency of macroporous polystyrene ion-exchange resins in natural organic matter removal from surface water. In E3S Web of Conferences. EDP Sci 22:181

Usman K, Khan S, Ghulam S, Khan MU, Khan N, Khan MA, Khalil SK (2012) Sewage sludge: an important biological resource for sustainable agriculture and its environmental implications. Am J Plant Sci 3(12):1708–1721
Wang H, Li F, Keller AA, Xu R (2009) Chemically enhanced primary treatment (CEPT) for removal of carbon and nutrients from municipal wastewater treatment plants: a case study of Shanghai. Water Sci Technol 60(7):1803–1809
Wang D, Tong F, Aerts P (2011) Application of the combined ultrafiltration and reverse osmosis for refinery wastewater reuse in sinopecyanshan plant. Desalination Water Treat 25(1-3):133–142
Wang Z, Wang Z, Xu K, Chen L, Lin Z, Liu Y (2019) Performance evaluation of adsorptive removal of Ni(II) by treated waste granular-activated carbon and new granular-activated carbon. Desalination Water Treat 161:315–326
Wilkinson KG (2011) A comparison of the drivers influencing adoption of on farm anaerobic digestion in Germany and Australia. Biomass Bioenergy 35(5):1613–1622
Yahya RZ, Arrieta JM, Cusack M, Duarte CM (2019) Airborne prokaryote and virus abundance over the red sea. Front Microbiol 10:1112
Zaidi NS, Sohaili J, Muda K, Sillanpaa M (2014) Magnetic field application and its potential in water and wastewater treatment systems. Sep Purif Rev 43(3):206–240
Zhou D, Liu D, Gao E, Li M, Luo X (2017) Effects of biochar-derived sewage sludge on heavy metal adsorption and immobilization in soils. Int J Environ Res Public Health 14(7):681
Zouch H, Cabrol L, Chifflet S, Tedetti M, Karray F, Zaghden H, Sayadi S, Quemeneur M (2019) Effect of acidic industrial effluent release on microbial diversity and trace metal dynamics during resuspension of coastal sediment. Front Microbiol 9:3103

Biostabilization of Sewage Sludge

Muhammad Ahsan, Adnan Younis, Fahad Ramzan, Usman Tariq, Muhammad Nafees, Fahim Nawaz, Gulzar Akhtar, Ahsan Akram, and Aneela Ramzan

1 Introduction

Natural form (raw sludge) of sewage sludge is rich source of pathogens, fast emerging spiteful smells and easily putrescible (Isaac and Boothroyd 1996). Processes of stabilization were established for organic matter stabilization in raw sludge by biologically degradable fraction, therefore dropping the hazard for decomposition and decreasing the pathogenic contents (Stark et al. 2015; Bhardwaj et al. 2018). The processes of stabilization can be divided into (see Fig. 1);

Bio-stabilization: Specific bacteria encourage the stabilization of biologically degradable chunk of organic matter.

Chemo-stabilization: Chemically oxidation of the organic matter achieves stabilization of sludge.

Thermo-stabilization: Heat stabilizes the volatile part of sludge in hermetically enclosed containers.

M. Ahsan (✉) · M. Nafees · A. Ramzan
Department of Horticultural Sciences, Faculty of Agriculture & Environment, The Islamia University of Bahawalpur, Bahawalpur, Pakistan
e-mail: ahsan.horti@iub.edu.pk

A. Younis · F. Ramzan · A. Akram
Institute of Horticultural Sciences, University of Agriculture, Faisalabad, Pakistan

U. Tariq
Departmrnt of Botany, GCUF, Layyah, Pakistan

F. Nawaz
Department of Agronomy, MNS, University of Agriculture, Multan, Pakistan

Institute of Crop Sciences (340h), University of Hohenheim, Stuttgart, Germany

G. Akhtar
MNS University of Agriculture, Multan, Pakistan

V. D. Rajput et al. (eds.), *Sustainable Management and Utilization of Sewage Sludge*, https://doi.org/10.1007/978-3-030-85226-9_7

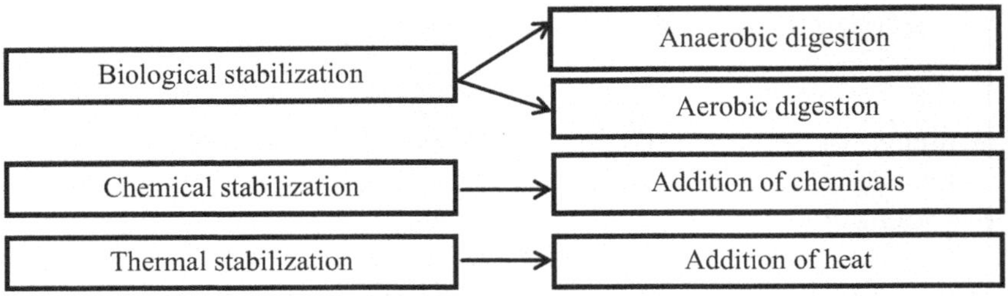

Fig. 1 Major sludge stabilization processes

Table 1 Comparison of digested sludge (anaerobically) and raw sludge

Digested sludge	Raw sludge
Less potential for odors generation	More potential for odors generation
Organic matter is stabilized	Organic matter is unstable
Lower concentration of pathogens	Higher pathogenic contents
Minor biodegradable organic matter fraction	Higher fraction of organic matter

Composting is also a common in municipal waste processing, used on limited scale by small wastewater treatment units (Tariq et al. 2012; Mouri et al. 2013). Thermal drying and alkaline treatment are also sludge stabilization processes used on limited scales.

The major focus of this chapter will be most commonly used slant of biostabilization.

All over the world, the mesophilic anaerobic digestion is the major process of sludge stabilization. The aerobic digestion of sewage sludge is not as common compared to anaerobic digestion.

1.1 Anaerobic Digestion

1.1.1 Introduction

In wastewater treatment, digestion means organic matter stabilization by bacteria in sludge which provide ideal conditions for their growth and reproduction (Sosnowski et al. 2003; USEPA 2011; Younis et al. 2014). The processes of digestion could be aerobic, anaerobic and combination of both (Abbas et al. 2011; EC 2012). Table 1 illustrates the key differences between digested sludge and raw sludge.

In anaerobic digestion, organic matter stabilization in the absence of oxygen had been identified by some workers even before the 1900s (Tyagi and Lo 2013). Because of its efficiency and heftiness, it is applicable to small and simple septic tanks as well as fully automated plants that served to larger cities/metropolitan areas (Cordell et al. 2011). In between the World War I and World War II, the process of anaerobic digestion endured significant progresses (Houtmeyers et al. 2014). In

USA, Germany and England different notions associated to this process were improved at that time, and are still being used these days in digesters design (Braun and Wellinger 2009). Table 2 shows the global sewage sludge treatment process in different parts of the world.

Anaerobic digestion is a chemical process that requires multi steps, capable of stabilizing various kinds of organic matter (Appels et al. 2011; Nasir et al. 2012). This process completes in three steps;

- Complex organic compounds like protein, cellulose, and lipid break down by enzymes into smaller soluble compounds like alcohol, fatty acids, CO_2 and NH_3.
- Different microbes alter the produce into hydrogen, CO_2 and CH_3COOH as well extra organic acids with lesser molecular weight.
- Methane-forming organisms from two groups take action: CO_2 and H_2 produced methane by one group whereas second group converts produces bicarbonates and methane from the acetates.

1.1.2 Major Rudiments for Digestion of Sludge

The constancy and efficacy of anaerobic digestion processes difference directly associated to digester environment and physiognomies of the raw sludge (Girovich 1996). The sludge which added in the anaerobic digester is a composite blend of ingredients that contains features resolute by the treatment plant area and wastewater recycling method (Jeng et al. 2006).

Commonly, the presence of micro- and macronutrients is adequate to ensure the progress of digestion if there is only sludge came from industries (Meyer et al. 2001). If macro and micronutrients are not a motive to apprehension, the sludge digester enactment might be affected by existence of foreign elements (Sonon and Gaskin 2009). Hence, following points are vital to follow:

Initial Treatment: Sludge receives after primary sedimentation comprises excessive quantity of fiber, sand, plastics and other inert materials. This material initially passes through the grit chambers and screens and settled with the main sludge, creating obstacle resulting pipe breaking, mutilating the pump rotors and other devices. Sand accretion in digester results the reduction of digester volume and efficacy.

Solid contents: Condensing of sludge aims to decrease thequantity required for digestion process. Thickening is achieved by dissolving air in major sedimentation units. It is necessary to have solid contents in sludge supply for digestion from 4% to 8%. More concentration of solids may be added if mixing and feeding tanks knob the raise of solids. Solid contents below 2.5% are commonly not advisable as excessive liquid put undesirable impact to digestion process.

Constraining elements: Microbes for anaerobic digestion are sensitive to some substances that are able to totally halt the process of digestion. An effective legislation and firm regulation on effluent release to waste stream are major ways to evade the presence of contaminated elements in wastewater. The key inhibiting agents

Table 2 Municipal sewage sludge treatment process in different countries (Source: Kelessidis and Stasinakis (2012), Lu et al. (2012), Tsagarakis et al. (1999)

	Germany	Greece	Ireland	France	Italy	Spain	Sweden	UK	Poland	USA
Stabilization										
Aerobic		**	*	*	*	*	*	*	**	*
Anaerobic	*	*	*	*	**	**	*	**	*	*
Lime		*	*	*	*	*	*	*	*	*
Composting		*	*	*	*	*	*	*	*	*
Conditioning										
Lime				*	*		*			
Inorganics		*							*	
Polymers	*	*		*	**		*		*	
Thermal		*		*	*		*			
Dewatering										
Filter press		*					*	*	*	
Centrifuges		**	*	*	*		*			
Others									*	
Thermal	**	*	*	*	*	*		*		*
Solar dry			*			*				
Long term storage					**		*	*		*
Cold fermentation		*	*	*	*	*	*	*	*	

* Common use ** most common use

include inorganic cations, non-biodegradable anionic detergents, and oxidizing agents.

Synthetic non-biodegradable cleansers are main apprehension (Bremer 2009). Though its use for manufacturing of cleansers had been stopped in different parts of regions of the world but can still be found in some areas (Lalor et al. 2012). Oxidizing agents can create a constraining act in digestion, resulting in the elimination of significant portion of organic matter (Tasker 2010) and alter the nutrient balance indigestion unit.

Inorganic cations like K, Ca, Mg and Na (even at extremely low concentration) could strongly constrain the process at elevated concentration (Lalor et al. 2012). Optimum ammonia contents range is 50–1000 mg/l, between 1000–1500 mg/l modest inhibition can occur; for 3000 mg/l and higher, solid inhibition occur (Epstein 2003).

Metals: The word *metal* in this context includes Cu, Zn, Cd, Ni, Pb and Hg (Smith and Durham 2002). The presence of these metals as metallic compound hinders the anaerobic digestion by reaction with enzyme required for processing formation of complex insoluble substances (Singh and Agrawal 2010). Except Cd and Hg, the other metals are well-thought-out micronutrients if existed in sufficient quantity (Ahsan et al. 2018a; Farooq et al. 2020).

The organic matter obliteration in digestion process of organic matter creates the contents of metals in digested sludge to form grander sludge (on dry solid basis). The poisonousness of metal varies subjected on type of metal, pH and carbonate and sulphide contents in the sludge (Wall et al. 2011; Ahsan et al. 2019).

1.1.3 Process Description

In conventional active sludge wastewater treatment plants, excess activated sludge and mixed primary sludge are stabilized by bacteria during anaerobic settings and change CO_2 and methane (Lee et al. 2014). This practice is carried on in closed reactor called anaerobic sludge digesters tank. These tanks filled with sludge and remain in it for certain time period as decided during first stage. The solids and sludge have equal detention period in digester tanks. In tanks, there are three groups of equally dependent microorganism exist which are acetogenic, hydrolytic acidogenic and methanogenic organisms (Su et al. 2009).

The colonies of these organisms remain in a vibrant symmetry and the concentration changes depends the operation environment in the tank. Denitrifying and sulphate-dropping bacteria occurred in digestion and play a major part in stabilization (Yang et al. 2012). The sulphate-dropping microbes are liable for the decrease insulphide ($S^{=}$) from sulphate (SO_4^{2-}) whereas nitrate (NO_3^-) converted to gaseous nitrogen (N_2) by denitrifying bacteria. This process of digestion occurs in pH range from 6 to 8, though pH remains almost neutral during this process by buffer ability of ammonia, sulphide and bicarbonates (Jiang et al. 2011).

The nutritious balance in digester is necessary to restrict the growing of bacteria as well as rate of organic matter stabilization. Cu, Fe, Se and Ni are major nutrients.

Among these Fe is regarded as key vital microelement in anaerobic digestion due to its main role in metabolism whereas N, P and S are less important elements (Neyens et al. 2004, Ahsan et al. 2018b).

1.1.4 Reaction Kinetics

The efficiency of slush digester is straightly connected to the contents and microorganisms population in sludge. Retention period in the digester (θ_c) should be adequate to certify the preservation of microorganisms which have a slower growing rate like methanogenic organisms, so evading their wash-out from the process (Ndegwa and Thompson 2001).

In conventionally operating anaerobic digesters, the sludge age (retention period) is equal is to hydraulic detention period and determined by following equation;

$$t = \theta_c = V/Q$$

Where;
t = detention period of hydraulic.
θ_c = retention period of solid.
V = sludge digester volume (m^3).
Q = influent flow to sludge digester (m^3/d).

Slower growing ratio of methanogenic population defines the period of reactions obligatory for digestion process to fulfilled and required retention period of sludge within the tank (Elissen et al. 2010). Other properties connected with θ_c are of highly vital in the efficiency of anaerobic digester are;

- Organic matter conservation period does not depend on volume of the sludge filled in digester on daily basis
- Detention period shorter than a critical level, the efficiency ofprocedure is abruptly decreased by washout of methanogemic organisms
- Efficiency of anaerobic digester does not upsurge as detention period elevation. As the optimal period is achieved, there is no need for more investment

Anaerobic digesters are commonly designed by considering a detention period greater than optimal to recompense infrequent operational issues like;

- Inefficient mixing of sludge system
- Changes in ambient temperature
- Variation in the sludge production rate
- Silting due to inert material accumulation in the tank

As revealed in Table 3, the anaerobic digestion kinetics varieswith methanogenic organisms. Under usual conditions, there is a perfect interface between the various organism groups and the medium. When the balance is disturbed, the process of reaction also modifies. For example, the impacts of organic overloading during an anaerobic digester are as fellows;

Table 3 Major character of anaerobic organisms

Parameter	Acetogenic and acidogenic organisms	Methanogenic organisms
pH	Less sensitive	High sensitive
Temperature	Medium sensitive	High sensitive
Growth rate	High	Slow
Volatile acids	Less sensitive	High sensitive
Toxic agents	Medium sensitive	High sensitive
Redox potential	Less sensitive	High sensitive

- methanogenic organisms are subdued by acidification of reactor
- volatile acids are produced from organic matter by acidogenic bacteria at increasing rate than methanogenic organism
- concentration of volatile acid is improved by reacting with alkalinity so preventing the buffer ability as reducing pH of medium.
- acetogenic micro-organisms are prohibited by elevating acidic medium, thus process of anaerobic digestion starts to breakdown.

1.1.5 Pathogen Reduction

Raw sewage sludge ponders greatest diversity micro-organisms (St-Hilaire et al. 2007). The quantity and kind of those organisms shows the living standard in the service area of treatment plant. The occurrence and content of some microbes in raw sludge also show the involvement of animal related areas (Seviour et al. 2009).

Digestion of sludge ominously decreases the organism population and preferring the agricultural usage of sludge (Flemming and Wingender 2001). Anaerobic stabilization acts as a limited barrier between users of sludge and pathogenic agents, thus decreasing the dangers of transmission of diseases (Morgenroth et al. 1997).

1.1.6 Biogas

Biogas produces by the process of anaerobic digestion that is produces by mixing of methane, carbon dioxide and little concentration of hydrogen sulphide, oxygen, nitrogen some traces of volatile hydrocarbons (Chen et al. 2013; Dicht et al. 2013; Ward et al. 2008). The largest producer of biogas around the word is Germany (Bodík et al. 2011; David et al. 2014).

Production of biogas in the anaerobic digester is linked with the raw sludge (Knacker and Metcalfe 2010). All around the world, there are above 1300 anaerobic digestions system in operating condition or under construction stage, which are based on sewage sludge (IEA Bioenergy 2001). Highest generation of biogas in the anaerobic digesters occurs after two hours of raw sludge supply (Lienert et al. 2007). The rate of biogas generation is expected as 0.8 m^3/kg of volatile solids wrecked that is equal to around 25 liters per occupant day (Chen et al. 2013). Thermal capability and density of biogas differ greatly with the composition. More methane contents in the biogas, more its heating level and lesser density. Approximately 23,378 kJ/m^3 (6.5 kW/m^3) heating capability is obtained from 70% of methane biogas. So, natural gas that is combinationof propane, methane and butane has heating capacity of 37,300 kJ/m^3 (10.3 kW/m^3).

Pipes for the distribution of biogas should be openly reccgnized and must be kept in well working condition and restricted routes in the treatment plants should be evaded (Lienert et al. 2013). Though frequently leakage tests are important and it is very hard to control infrequent leakage. So, extreme safety measures are necessary while using any source of ignition e.g. cutting and welding apparatus (McClellan and Halden 2010).

Explosions may occur only when an appropriate mixture of biogas and air befalls in the occurrence of a spark with a temperature more than the ignition temperature i.e. 700 °C. Naturally, both air and biogas are existed in purlieu of slush digesters, heat source cannot be entirely abolished by furnaces and control panels (Hytiriset et al. 2004). It is extremely sensible to avert air and biogas mix-upwhile gas pipeline designing. The lowest explosive limit (LEL) is lowest methane contents (about 5.0%) required to burst by ignition. Less than LEL, methane is poor for eruption to occur. The upper explosion limit (UEL) is about 15%. More than UEL, there is no oxygen to explode.

The key properties of biogas components are as follows for safety point of view;

Methane (CH_4) is colorless, odorless and inflammable between 5% and 15% LEL and UEL. It is easily dispersed contains relative density (055) lower than air. Though it is not very poisonous, but at high level might decrease the O_2 contents to suffocating level.

Carbon dioxide (CO_2) is colorless and odorless and non-ignitable. Its relative density (1.53) is greater than air while being suffocated at contents more than 2%.

Hydrogen sulphide (H_2S) is inflammable, colorless and possesses smell like rotten-egg. It is nuisance and suffocating. Its more than 1% concentration leads to unconsciousness.

The conformation of biogas produced in anaerobic digesters is presented in Table 4.

Table 4 Conformation of biogas produced in anaerobic digesters

Gas	% (volume/volume)
CH_4	62–70
CO_2	30–38
H_2S	50–3000 ppm
N_2	0.05–1.0
O_2	0.023
H_2	Less than 0.01
Water vapors	Saturation

1.1.7 Procedure and Controller of Sludge (Anaerobic) Digesters

Machinists working for a sludge treatment must identify the operational working and safely completion of a shift daily might depend on simply and easy to understand operational routine. Subsequent aspects must be taken for good performance:

- Proper frequency of sludge supply
- Simple operating circumstances of mixing system that assures homogeneity into the digestion tank.
- Greater time of detention than the growing rate of the methanogenic organisms

Alkalinity and acid contents in the digester are deeply associated to each other (Tsagarakis and Papadogiannis 2006). Volatile acid/alkaline ration is well indication of proper digestion process. Proper acid alkali ratio in the digestion process is expressed in Table 5.

Sometimes, the process of digestion may become unstable and ultimately lead to collapse of the digesters. This instability occurs in the digesters when different biochemical reactions started without proper interaction. Acid concentration elevated by outweigh of acid producing bacteria which leads to decrease in pH of the medium. Though its reasons may differ but the variability indications of digestion procedure are conjoint which contain;

- Alkalinity and pH decrease
- Production of methane decreases
- Concentration of volatile acids increases
- Elevation of CO_2 ratio in the biogas

In these conditions, the sequence of actions into the digesters is as follows,

- The digested sludge value of acid/alkalinity reaches to more than 0.3 due to increase of volatile acids;
- Alkalinity is consumed by acids results in releasing CO_2that decreases the methane contents. The ratio of volatile acid/alkali increases up to level of 0.5–0.8
- The value of pH decreases to 6.5 that prevent production of methane. In this way the digester acidified and ruin.

The ruin procedure mentioned above is not instant as it takes few days to get accomplished. Following some measure should be taken to evade its possibility to happen;

Table 5 Ratio of acids (volatile) and alkalinity

Acid/alkaline ratio	Sign
Less than 0.3	Good working of digester
0.3–0.5	Process of digestion is fail
More than 0.8	Medium becomes acidic and process may collapse is looming

- By controlling the digester's data, which is easy to know the cause of process insecurity. A steady restriction shows the existence of low-concentration toxic elements or electro-mechanical issues (defective mix up of system). An abrupt intrusion of the repairing staff is advised for this issue.
- To maintain the neutral pH, there should be added to alkali solution in the sludge is necessary.
- In case of excess metals, sodium sulphide can be added for the metal cations precipitation.
- It might be highly recommended to supply of anaerobic sludge from other under stable conditions.

Supply of raw sludge to digesters can be steady bring to standard level as recovery signs displays in the digestion process.

Rarely, there is a need of digester to pull out of services for elimination of deposited inert material. Following measures should be taken in such case;

- Sludge supply should be stopped immediately
- If possible, sludge should be shifted to other digesters
- Continuous monitoring of gas production
- Stop the mix up of heating and mixing system
- Gas outlet pipes must be isolated
- It should be confirmed that the methane contents in the gas section is less than 3%.
- Eliminate the remaining mixture from the digester
- Remove outlet and entrance projections
- Cleaning operation must be start after removal of outlet

1.1.8 Caring of the Digesters

Sampling must be done monthly or fortnightly pointing at the assessment of the internal situations in the digester. Alkalinity remains at 4000–5000 mg/L, acid contents less than 200 mg/L and pH remains almost neutral within the range of 7–7.2 under normal circumstances (Table 6).

Information of the composition of volatile acids by chromatography can also assist in the digester diagnosis. As concentration of large acid chains rises than the contents of short acid chains, the operation of digester become wobbly.

Table 7 summarizes the key reasons to failure of anaerobic digester, symptoms and suggested measures.

Table 6 Major features and suggested range of operation for anaerobic digesters

Parameter	Standard level
pH	6.9–7.3
Acids (mg/L HAc)	200
Alkalinity (mg/L $CaCO_3$)	4000–5000

Table 7 Major reasons of sludge failure (anaerobic) digesters and curative methods

Stimulating aspects of uncertainty and concerns				
Hydraulic shock	Organic shock	Toxic load	Symptoms	Suggested actions
Production of extreme sludge	Rise in sludge influent to digester	Very heavy load of heavy metals	Elevated acid contents	Add lime concentration
Very thinned supply of sludge	Increase of solids contents in the influent	Heavy detergents load	Reduction in pH and alkalinity	Decrease ratio of acid/alkali
Silting of digester	Modification in the properties of sludge	Sludge contains chlorinated organic compounds	Rise of acid/alkali ratio	Regular supply of sludge
Extreme foam	Too fast start-up of digester	Oxygen adding	Gas production deceases	Raise concentration of sludge
Methanogenic organisms wash-out	Uneven supply	Heavy sulphides	Rise of CO_2 contents in biogas	Cleanliness of the digester

Table 8 Parameters of design for mesophilic sludge digesters (Metcalf and Eddy (1991)

Item	Parameter	Value
Time of hydraulic detention (d)	Extreme active sludge Prolonged aeration Extreme active and primary sludge	10–15 12–18 15–20
Rate of organic loading (kg VS/m^3.d)	–	1.6–4.9
Demand of oxygen (kgO2/ kg VS destroy)	Endogenous respiration	2.3
Energy to keep solids in suspension	Diffused air (L/m^3. min)	20–40
DO is the digester (mg/L)	–	1–2

VS = volatile solids

2 Aerobic Digestion

2.1 Introduction

There is very much similarity with the process of activated sludge and the process of aerobic digestion (d'Antonio 1983; Duan et al. 2014). When there is regular substrate supply, then microbes are mandatory to utilize its personal monies of vigor to persist living (Adams et al. 1974). It is the endogenous phase in lack of nutrition source, the decomposable mass of cell is aerobically oxidized to CO_2, NH_3 and H_2O (Zhang et al. 2016). In this reaction, nitrate by ammonia is oxidized by following equation;

$$C_5H_7NO_2 + 7O_2 + \text{bacteria} \rightarrow 5CO_2 + NO_3 + 3H_20 + H^+$$

Extended aeration procedure of sludge digestion occurs in the aeration tanks with the oxidation of influent organic matter processing due to low food and microorganism ratio (Liu et al. 2010). Due to these conditions, the advisable process of digestion is aerobic digestion which started independently.

Presently, there are three kinds of aerobic digestion processes being used in sludge stabilization;

- Mesophilic or conventional aerobic digestion
- Digestion with oxygen
- Thermophilic digestion

2.2 *Mesophilic Aerobic Digestion*

Mesophilic aerobic digestion stabilizes and triggered extra slush in open digesters by mechanical aeration (He et al. 2007). At the range of mesophilic temperature, the digestion occurs. By flotation, the sludge is thickened to decrease the required volume of digestion.

Features to be considered in digester design are comparable to those for active sludge system like;

- Temperature
- Demand of oxygen
- Organic supply
- Requirement of power
- Detention time (t) that is equal to retention time of solids or age of sludge (θ_c)

Temperature: the solids decreasing rate of solids depends on the digester temperature. Greater rate of organic matter conversion will be under high temperature. Stabilization stops if temperature falls to 10 °C.

Demand of oxygen: supply of oxygen should meet the mass of cell respiration requirement and endorse mixing circumstances in the tanks. The level of dissolve oxygen in reactor should be in the range of 1–2 mg/L.

Organic supply: the organic supply is decreases by capacity of oxygen transfer. Concentration of solids more than 3% can lead to aerobic circumstances.

Mixing: sufficient mixing is necessary to confirm the sludge stabilization in the digester. In the diffused system of air, the mixing flow is about 30 L air/m^3.minute to reached by demand of oxygen for itself stabilization.

Detention time: Time of detention after 10–15 days, with around temperature of 20 °C, the contents of volatile solids decreased to 40% in slush. Greater temperature and time of detention must be provided to show decrease beyond 30–40% in solids.

There are few parameters which are utilized for evaluating the aerobic digester efficiency;

- Reduction of volatile solids
- Supernatant quality
- Sludge dewatering
- Odor of the sludge

2.3 Digestion with Oxygen

The digestion in the presence of oxygen in pure form is an alternative of conventional digestion during which oxygen is directly supplied to medium instead of air. The concentration of solid in digester should be 4% higher without any decrease in the rate of oxygen transferalto the biomass.

Current procedure is successful for large scale wastewater recycling, when the land is the limiting factor, and oxygen in pure form is already in use by the bio-reactor. This is efficiency elevating process and most suitable for cold climatic areas.

2.4 Thermophilic Digestion

Heat is major by product in digestion process of organic matter and temperature may rise up to 60 °C in the digester. There should be appropriate substrate to keep the microbiological activities (Suruhanjaya 2017).

This process began in 1971 for the purpose of disinfection and stabilization of sewage sludge. In those days it was suggested that thermophilic temperature will be only obtained by pure oxygen. But, later on experiments showed that plain air usage also showed effective results in achieving the higher temperatures in the process.

Major benefits of thermophilic digestion are;

- Lessening of the digester volume for the organic matter stabilization
- Formation of disinfected sludge which meets the rating of USEPA biosolids for unrestricted reuse.

There are also certain key drawbacks of this process listed below;

- Higher capital cost
- Complexity in operation
- There is build-up of foam on the surface of digester. For this issue, a freeboard with 30% digester height is highly advisable to accommodate this foam production.

2.5 Composting

It is the stabilization process of organic matter initially used by gardeners and farmers since prehistoric times. Night soil composting is conventionally used in China and considered as the utmost likely cause why the structure and nutritional status of the Chinese soil is being preserved for more than 5000 years (Ying et al. 2012).

The processes of composting may be separated into;

- Composting of static pile (aerated)
- Windrow composting (the most simple and traditional way of composting)
- In-vessel composting or composting with closed reactor

Although the composting of sewage sludge requires lot of knowledge, involvement and professionalism in designing and operation phase (Adani et al. 2000).

The major requirements for a best composting are;

- Sludge must contain adequate nutrients with C:N ratio of 25:1
- Supply of air must be delivered to keep an oxidizing environment in the windrow. For this aspect, the material types used for bulking agent is very important.
- Loss of heat control should assure 55–65 °C for the temperature in the windrow.
- Adequate humidity level must be maintained in heap. Activity of microbes decreased significantly when humidity level falls less than 34–40%.

The coreadvantages of composting are;

- Fine quality of the final product, vastly accepted in farming and gardening
- Possibility of combination with other processes of stabilization
- Cheap in cost

Major disadvantages are;

- Required for sludge with higher concentration of solids
- Its operational cost is high
- Land requirement is considerable
- High risk of producing foul-odor

References

Abbas AH, Ibrahim ABA, Nor MFM, Aris MS (2011) Characterization of Malaysian domestic sewage sludge for conversion into fuels for energy recovery plants. in National Postgraduate Conference (NPC) (Kuala Lumpur), 19–20

Adams CE, Eckenfelder WW, Stein RM (1974) Modifications to aerobic digester design. Water Res 8:213–218

Adani F, Scatigna L, Genevini P (2000) Biostabilization of mechanically separated municipal solid waste fraction. Waste Manage Res 18:471–477

Ahsan M, Younis A, Jaskani MJ, Tufail A, Riaz A, Schwinghamerd T, Tariq U, Nawaz F (2018a). Heavy metal accumulation imparts structural differences in fragrant Rosa species irrigated with marginal quality water. Ecotoxicol Environ Safe. https://doi.org/10.1016/j.ecoenv.2018.05.003

Ahsan M, Younis A, Riaz A, Jaskani MJ, Qasim M, Hameed M, Tufail A, Nafees M, Abbas H, Tariq U (2018b) Treated and untreated wastewater imparts morphological changes to scented *Rosa* species in peri-urban area. Pak J Agric Sci 55(1):111–117

Ahsan M, Younis A, Jaskani MJ, Tariq U, Shaheen MR, Tufail A, Sherani J, Nawaz F (2019) Anatomical changes in stem of scented *Rosa* spp. in response to heavy metal accumulation under wastewater treatment. Int J Agric Biol 21:1159–1165

Appels L, Joost L, Degrève J, Helsen L, Lievens B, Willems K (2011) Anaerobic digestion in global bio-energy production: potential and research challenges. Renew Sust Energ Rev 15:4295–4301

Bhardwaj L, Chauhan A, Ranjan A, Jindal T (2018). Persistentorganic pollutants in biotic and abiotic components of Antarcticpristine environment. Earth Syst Environ 2. https://doi.org/10.1007/s41748-017-0032-8

Bodík I, Sedláèek S, Kubaská M, Hutòan M (2011) Biogas production in municipal wastewater treatment plants-current status in EU with a focus on the Slovak Republic. Chem Biochem Eng Q 25:335–340

Braun R, Wellinger A (2009) Potential of co-digestion. IEA Bioenergy, Vienna

Bremer J (2009) Fertilizer prices continue to rise. http://www.hpj.com/archives/2009/jan09/jan19/Fertilizerpricescontinuetor.cfm (Accessed 14 September 2013)

Chen Y, Yu G, Cao Q, Zhang H, Lin Q, Hong Y (2013) Occurrence and environmental implications of pharmaceuticals in Chinese municipal sewage sludge. Chemosphere 93:1765–1772

Cordell D, Rosemarin A, Schröder JJ, Smit AL (2011) Towards global phosphorus security: a systems framework for phosphorus recovery and reuse options. Chemosphere 84:747–758

d'Antonio G (1983) Aerobic digestion of thickened activated sludge: reaction rate constant determination and process performance. Water Res 17:1525–1531

David H, Palanisamy K, Normanbhay S, (2014) Pre-treatment of sewage sludge to enhance biogas production to generate green energy for reduction of carbon footprint in sewage treatment plant (STP)" in International Conference and Utility Exhibition 2014 on Green Energy for Sustainable Development (Pattaya)

Dicht N, Dellbrügge RI, Morcali B (2013) Sludge treatment in Germany more than 50 years of experience current discussions-future prospects in 18th European biosolids and organic resources conference and exhibition

Duan L, Song Y, Yu H, Xia S, Hermanowicz SW (2014) The effect of solids retention times on the characterization of extracellular polymeric substances and soluble microbial products in a submerged membrane bioreactor. Bioresour Technol 163:395–398

EC (2012) Sewage sludge. http://ec.europa.eu/environment/waste/sludge/index.htm (Accessed 11 June 2014)

Elissen HJH, Mulder WJ, Hendrickx TLG, Elbersen HW, Beelen B, Temmink H, Buisman CJN (2010). Aquatic worms grown on biosolids: biomass composition and potential applications. Bioresour Technol 101ste:804–811

Epstein E (2003) Land application of sewage sludge and biosolids. Lewis, Boca Raton, Fl

Farooq A, Nadeem M, Abbas G, Shabbir A, Khalid MS, Hafiz MR, Akhtar G (2020). Cadmium partitioning, physiological and oxidative stress responses in Marigold (*Calendula calypso*) grown on contaminated soil: implications for phytoremediation. Bull Environ Contamin Toxicol https://doi.org/10.1007/s00128-020-02934-6

Flemming HC, Wingender J (2001). Relevance of microbial extracellular polymeric substances (EPSs) – Part II: Technical aspects 43:9–16

Girovich MJ (1996) Biosolids treatment and management: processes for beneficial use. Marcel Dekker Inc. Google Scholar. Huang, 1995

He SB, Xue G, Kong HN (2007) The performance of BAF using natural zeolite as filter media under conditions of low temperature and ammonium shock load. J Hazard Mater 143:291–295

Houtmeyers S, Degreve J, Willems K, Dewil R, Appels L (2014) Comparing the influence of low power ultrasonic and microwave pre-treatments on the solubilization and semi-continuous anaerobic digestion of waste activated sludge. Bioresour Technol 171:44–49

Hytiriset N, Kapellakis IE, de La Roij R, Tsagarakis KP (2004) The potential use of olive mill sludge in solidification process. Resour Conserv Recycl 40(2):129–139

IEA Bioenergy (2001) Biogas and more! Systems and Markets overview of anaerobic digestion. AEA Technology Environment, Abingdon

Isaac RA, Boothroyd Y (1996) Beneficial use of biosolids: progress in controlling metals. Water Sci Technol 34:493–497

Jeng AS, Haraldsen TK, Grønlund A, Pedersen PA (2006) Meat and bone meal as nitrogen and phosphorus fertilizer to cereals and rye grass. Nutrient Cycle Agroecosyst 76:183–191

Jiang J, Zhao Q, Wei L, Wang K, Lee DJ (2011) Degradation and characteristic changes of organic matter in sewage sludge using microbial fuel cell with ultrasound pretreatment. Bioresour Technol 102(1):272–277

Kelessidis A, Stasinakis AS (2012) Comparative study of the methods used for treatment and final disposal of sewage sludge in European countries. Waste Manag 32(6):1186–1195

Knacker T, Metcalfe C (2010) Introduction to the special issue on environmental risk assessment of pharmaceuticals. Integ Environ Assess Manag 6:511–513

Lalor STJ, Hoekstra NJ, Murphy PNC, Richards KG, Lanigan GJ (2012). Practical advice for slurry application strategies for grassland systems. Proceedings 712, International Fertiliser Society, Cambridge, 6 December 2012, UK

Lee WS, Chua ASM, Yeoh HK, Ngoh GC (2014) A review of the production and applications of waste-derived volatile fatty acids. Chem Eng J 235:83–99

Lienert J, Güdel K, Escher BI (2007) Screening method for ecotoxicological hazard assessment of 42 pharmaceuticals considering human metabolism and excretory routes. Environ Sci Technol 41:4471–4478

Liu S, Song F, Zhu N, Yuan H, Cheng J (2010) Chemical and microbial changes during autothermal thermophilic aerobic digestion (ATAD) of sewage sludge. Bioresour Technol 101:9438–9444

Lu Q, He ZL, Stoffella PJ (2012) Land application of biosolids in the USA: a review. Appl Environ Soil Sci 20:1–11

McClellan K, Halden RU (2010) Pharmaceuticals and personal care products in archived U.S. biosolids from the 2001 EPA national sewage sludge survey. Water Res 44(2):658–668

Metcalf Eddy Inc (1991) Wastewater engineering: treatment, disposal, and reuse, 3rd edn. McGraw-Hill, Singapore

Meyer VF, Redente EF, Barbarick KA, Brobst R (2001) Biosolids applications affect runoff water quality following forest fire. J Environ Qual 30:1528–1532

Morgenroth E, Sherden T, van Loosdrecht MCM, Heiinen JJ, Wilderer PA (1997) Aerobic granular sludge in sequencing batch reactors. Water Res 31:3191–3194

Mouri G, Takizawa S, Fukushi K, Oki T (2013) Estimation of the effects of chemically enhanced treatment of urban sewage system based on life-cycle management. Sustain Cities Soc 9:23–31

Nasir IM, Tinia IMG, Rozita O (2012) Production of biogas from solid organic wastes through anaerobic digestion: a review. Appl Microbiol Biotechnol 95:321–329

Ndegwa PM, Thompson SA (2001) Integrating composting and vermicomposting in the treatment and bioconversion of biosolids. Bioresour Technol 76(2):107–112

Neyens E, Baeyens J, Dewil R, De heyder B (2004) Advanced sludge treatment affects extracellular polymeric substances to improve activated sludge dewatering. J Hazard Mater 106:83–92

Seviour T, Pijuan M, Nicholson T, Keller J, Yuan Z (2009) Gel-forming exopolysaccharides explain basic differences between structures of aerobic sludge granules and floccular sludges. Water Res 43:4469–4478

Singh A, Agrawal M (2010) Effects of municipal waste water irrigation on availability of heavy metals and morpho-physiological characteristics of *Beta vulgaris* L. J Environ Biol 31:727–736

Smith SR, Durham E (2002) Nitrogen release and fertilizer value of thermally-dried biosolids. Water Environ J 16:121–126

Sonon LS, Gaskin J (2009) Metal concentration standards for land application of biosolids and other by-products in Georgia. Learn Life Bull 1353

Sosnowski P, Wieczorek A, Ledakowicz S (2003) Anaerobic co-digestion of sewage sludge and organic municipal solid wastes. Adv Environ Res 7:609–616

Stark JS, Smith J, King CK, Lindsay M, Stark S, Palmer AS Riddle M (2015) Physical, chemical, biological and ecotoxicological properties of wastewater discharged from Davis Station, Antarctica. Cold Region Sci Technol 113, https://doi.org/10.1016/j.coldregions.2015.02.006

St-Hilaire S, Sheppard C, Tomberlin JK, Irving S, Newton L, McGuire MA, Mosley EE, Hardy RW, Sealey W (2007) Fly prepupae as a feedstuff for rainbow trout, oncorhynchus mykiss. J World Aquacul Soc 38:59–67

Su H, Cheng J, Zhou J, Song W, Cen K (2009) Improving hydrogen production from cassava starch by combination of dark and photo fermentation. Int J Hydrog Energy 34:1780–1786

Suruhanjaya PAN (2017) Sewerage statistic, Malaysia. Available online at: https://www.span.gov.my/

Tariq U, Rehman S, Khan MA, Younis A, Yaseen M, Ahsan M (2012) Agricultural and municipal waste as potting media component for growth and flowering of *Dahlia hortensis* 'Figaro'. Turk J Bot 36:378–385

Tasker J (2010) Frustration at 13% fertiliser price hike. Farmers Weekly. http://www.fwi.co.uk/Articles/03/09/2010/123163/Frustration-at-13-fertiliser-price-hike.htm (Accessed 14 September 2013)

Tsagarakis KP, Papadogiannis CH (2006) Technical and economic evaluation ofthe biogas utilization for energy production at Iraklio municipality, Greece. Energy Conserv Manage 47:844–857

Tsagarakis KP, Horan NJ, Mara DD, Angelakis AN (1999). Management of biosolids from municipal wastewater treatment plants in Greece. Fourth European Biosolids and Organic Residuals Conference, Wakefield, 15–17 November 1999. Paper no 35

Tyagi VK, Lo SL (2013) Sludge: a waste or renewable source for energy and resources recovery? Renew Sust Energ Rev 25:708–728

USEPA (2011) Problem formulation for human health risk assessments of pathogens in land-applied biosolids, Cincinnatti OH. EPA/600/R-08/035F

Wall D, Jordan P, Melland AR, Mellander PE, Buckley C, Reaney SM, Shortle G (2011) Using the nutrient transfer continuum concept to evaluate the European Union Nitrates Directive National Action Programme. Environ Sci Pol 14:664–674

Ward AJ, Hobbs PJ, Holliman PJ, Jones DL (2008) Optimization of the anaerobic digestion of agricultural resources. Bioresour Technol 99:7928–7940

Yang X, Du M, Lee DJ, Wan C, Zheng L, Wan F (2012) Improved volatile fatty acids production from proteins of sewage sludge with anthraquinone-2,6-disulfonate (AQDS) under anaerobic condition. Bioresour Technol 103:494–497

Ying D, Chuanyu C, Bin H, Yueen X, Xuejuan Z, Yingxu C, Weixiang W (2012) Characterization and control of odorous gases at a landfill site: a case study in Hangzhou, China. Waste Manag 32:317–326

Younis A, Riaz A, Mustaq N, Tahir Z, Siddique MI (2014) Evaluation of the suitability of sewage and recycled water for irrigation of ornamental plants. Commun Soil Sci Plant Anal 46:62–79

Zhang Z, Zhou Y, Zhang J, Xia S, Hermanowicz SW (2016) Effects of short-time aerobic digestion on extracellular polymeric substances and sludge features of waste activated sludge. Chem Eng J 299:177–183

Sewage Sludge Treatment and Involvement of Microbes

Aryadeep Roychoudhury and Nilanjana Das

1 Introduction

Sewage sludge is the solid or semi solid slurry material left from chemical coagulation, flocculation and sedimentation during wastewater treatment processes. Domestic, municipal and industrial wastewater are the main sources of sewage sludge. Considered as an important source of pollution, sewage sludge poses a great threat to the environment and can even cause death in humans. It contains both organic and inorganic matter, plant nutrients, heavy metals like Zn, Pb, Cr, Ni, Hg, Pt, Ag, organic pollutants like polychlorinated biphenyls (PCBs) and polycyclic aromatic hydrocarbons (PAHs), and some pathogenic microorganisms (Zhang et al. 2013). The sewage sludge contains high diversity of microbial species, mainly bacteria like *Propionibacterium, Desulfobulbus, Methylobacterium, Clostridium*, etc. (Nascimento et al. 2018). The quality and quantity of sewage sludge depends on the contents, the process of stabilization, the reagents used and how much volume is reduced. The highest amount of sewage sludge production is seen in the developed countries (Krzywicka and Kwarciak-Kozłowska 2014). However, sludge treatment is an expensive process and accounts for more than 50% of the operating cost of sewage water treatment (Wang et al. 2019). There are mainly two types of sludge, viz., primary sludge and secondary sludge. While primary sludge is obtained from chemical treatments, secondary sludge is the activated biomass obtained from biological treatments. Also tertiary sludge is obtained from processes such as filtration or chemical precipitation. Three basic goals of sewage sludge processing are: (i) reducing the volume of the sludge; (ii) stabilizing the sludge so that it does not give off stale odor; (iii) proper checking so that it does not create any health hazard. The principal stages of sewage sludge processing are thickening, digestion, dewatering and disposal (Fig. 1). After wastewater treatment, the sludge contains

A. Roychoudhury (✉) · N. Das
Department of Biotechnology, St. Xavier's College (Autonomous), Kolkata, West Bengal, India

V. D. Rajput et al. (eds.), *Sustainable Management and Utilization of Sewage Sludge*, https://doi.org/10.1007/978-3-030-85226-9_8

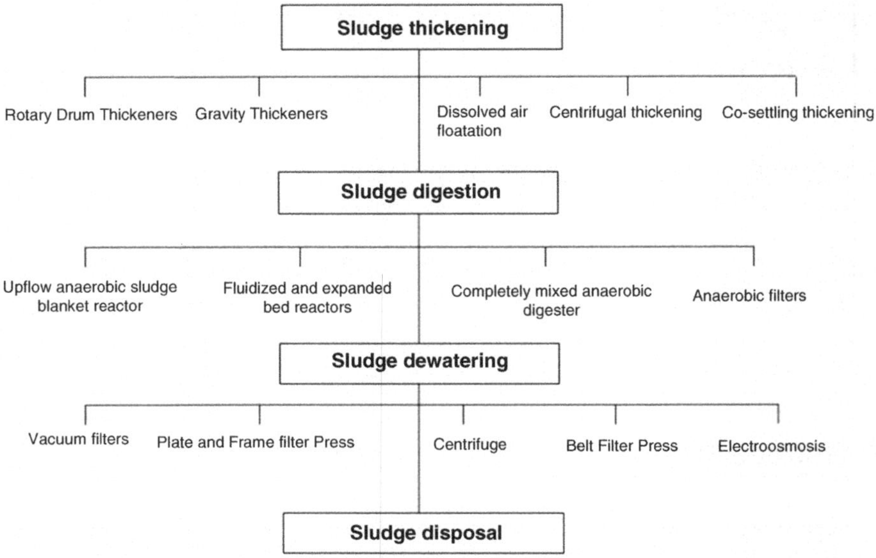

Fig. 1 The principle steps of sewage sludge processing

large volumes of water. Sludge thickening reduces the volume of water as it is difficult to handle sludge containing large amount of water. Sludge digestion is a biological process where the organic solids in the sludge are converted to liquids and gases. Dewatering is done before disposal of sludge where the disposed sludge is used as fertilizer, sanitary landfill or is incinerated. A large number of microorganisms play an important role in sewage sludge treatment. However, nowadays sewage sludge is not considered as a waste, but a source of renewable energy.

2 Types of Sludge

Sewage sludges are of two types: primary and secondary. While primary or raw sludges contain solids from wastewater treatment processes, secondary sludge contains solid as well as microorganisms produced within the treatment process. The secondary or biological sludge contains high volatile solids and low dry solids. Mixed sludge contains both primary as well as secondary sludge. Another type of sludge known as digested sludge is formed from anaerobic fermentation (Baroutian et al. 2013). Other than conventionally produced sludge, different types of sludge like liquid sludge, composted sludge, lime treated sludge, sewage cake, composted sludge etc. also exist (Usman et al. 2012). Other sludges include mineral sludge, physico chemical sludge and digested sludge.

3 Sewage Sludge Treatment

3.1 *Sludge Thickening*

The first step of sewage sludge treatment is sludge thickening. Sludge thickening reduces the water content, thereby increasing the solid content and minimizing the load for the downstream processes. The most commonly used methods for sludge thickening are rotary drum thickening, gravity or clarifier thickening, dissolved air flotation, centrifuge thickening, co-settling thickening and gravity belt thickening.

3.1.1 Rotary Drum Thickeners

Rotary drum thickener consists of a large rotating horizontal device which operates continually and automatically. The drum is differentiated into zones containing different mesh sizes. The sludge is fed through an inlet containing a finer mesh from where it enters a mixer where flocculation occurs. The sludge is then passed through the drum where centrifugal forces help to separate the solids from water. The sludge remains in the drum, while the residue water passes through a filter and is collected in a trough (Dentel and Qi 2014).

3.1.2 Gravity Thickeners

Gravity thickening is the least expensive and the most common method used for sludge thickening. It is similar to the sedimentation processes and works on the principle that gravitational force is greater on denser materials. The concept of gravity thickeners were established in 1950s by municipality for thickening of sludges (Torpey 1954). Gravity thickeners are usually large circular tanks with a collector or scraper fitted at the bottom. Sludge is fed slowly into the tank through a centre well which gets settled at the bottom by gravity and are discharged slowly by a scraper at the bottom. Gravity thickeners can be of two types: (i) plain settling where the sludge gets accumulated at the bottom by the force of gravity and scum is formed at the surface; (ii) mechanical settling where a slowly revolving sludge collector breaks the floc particles, resulting in the settling of sludge at the bottom. Gentle agitation is required to stir the sludge, allowing the water to escape through channels. The advantages of gravity thickening are simple to operate and maintain, along with low power consumption, low cost, and capacity to hold large amount of sludge. The disadvantages are that they require large area and can cause odor (Usman et al. 2012, Process Design Manual; sludge Treatment and Disposal 1979).

3.1.3 Dissolved air Floatation

Dissolved Air floatation method consists of a tank which is divided into two zones: the contact zone and the separation zone. It uses pressure to dissolve air into the wastewater that results in the formation of tiny bubbles. Collisions occur in the contact zone where the solid particles get attached with the air bubbles resulting in the formation of aggregates. These aggregates then flow to the separation zone and along with the suspended matter float to the surface (Edzwald 2010). A froth layer is formed by these float which is removed by a skimmer and the froth free water is pumped out. Often coagulants such as aluminium sulfate or ferric chloride are used for flocculation.

3.1.4 Centrifugal Thickening

Centrifugal thickening uses a large rotating cylindrical bowl which separates wastewater solids from liquid. The sludge particles are settled under the influence of the centrifugal force. The sludge containing wastewater is continuously fed to the cylindrical bowl where a conveyer continuously removes the solids and discharges the liquid (Kemp 1997). There are three types of centrifugal thickeners: basket type, solid bowl type and disc-stack type. The feed in a basket type centrifuge enters the basket and is thrown by the centrifugal forces against its wall. While the basket gets filled up with the solid, the separated liquid gets spilled out over the top of the basket. As the basket becomes full with the solid, knives attached to the basket scrap them off. The solid particles in a solid bowl centrifuge are thrown against the wall of the bowl, while the liquid gets accumulated in the centre of the cylinder. The disc-stack centrifuge that rotates about a vertical axis contains a series of cone-shaped discs with channels in between. The solids are thrown against the cone by the centrifugal force, while the liquid part exits the bottom of the discs.

3.1.5 Co-settling Thickening

Two clarifiers are used, viz., a primary clarifier where the wastewater enters and the sludge is clarified. This clarified underflow then enters into a series of thickening clarifiers from where the thickened sludge is finally discharged for dewatering. Coagulating chemicals such as ferric chloride and polymers are used to enhance the settling of the solids.

4 Sludge Digestion

Sludge digestion by microorganisms is an economic and environment friendly process. It is a two step biological process involving various aerobic and anaerobic microorganisms. In the first step, anaerobic digestion by the bacteria in a tank breaks down the complex organic molecules into simpler water soluble substances which are then fermented into fatty acids. In the second step, the sludge flows down into a second tank where it is converted into a form of biogas by methanogenic group of bacteria. Two forms of biological treatments exist in the form of attached and suspended growth. Attached growth contains a fixed biofilm containing microorganisms, biopolymers, gels and particulates fixed on a support like rock or plastic. Activated sludge digestion is a well known suspended growth process (Nelson et al. 2017). The suspended growth contains microorganisms which float freely in the mixed liquid. The biogas contains 48–65% methane which is collected and used as a biofuel for generation of power. (Ward et al. 2008). Europe is the leading producer of biogas followed by Asia. (World Bioenergy Association n.d.) Sludge digestion reduces the pathogens, odor emission, and solid content of the sludge, thus stabilizing and making it easier to dewater the sludge. China and India are among the leading producers of anaerobic sludge digestion systems among the developing countries, while Western European countries are the leading producer among the developed countries. Among the European countries, Germany is the largest biogas producer (Abbasi et al. 2012; Bodík et al. 2011) with about 10,431 plants generating 55,108 GWh/y of electricity. India having the largest number of plants (83,540) generates only 22,140 GWh/y of electricity which is mostly restricted in rural areas (Indah Water Konsortium 2013). Different types of anaerobic sludge digesters include Upflow anaerobic sludge blanket reactor, Fluidized and expanded bed reactors, Completely mixed anaerobic digester, and Anaerobic filters (Mustafa et al. 2011).

4.1 *Upflow Anaerobic Sludge Blanket Reactor*

It is mainly used by the municipality for wastewater treatment of food, paper and chemical industries. Also known as the three phase reactor, it contains three zones, the lower blanket zone, the middle dead zone and the upper gas zone. The wastewater is pumped from the bottom of the tank by a peristaltic pump. It then comes in contact with a sludge blanket containing microbial granules, mainly containing methanogens which degrade the organic compounds, resulting in the formation of gas bubbles. These gas bubbles rise up and get collected through a gas collection system. The effluent is collected in the wiers at the top of the reactor while the granules settle down at the bottom. The disadvantage with this reactor is wastewater with high solid content as these prevent the formation of the granules (Gunasekaran et al. 2019). The disadvantages of the process are low pathogen removal, low nutrient removal and requirement of post treatment of the effluent. The typical chemical oxygen demand (COD) removing efficiency of upflow anaerobic sludge blanket reactor is 70–90%.

4.2 Fluidized and Expanded Bed Reactors

The fluidized and expanded bed reactors contain microbial biofilm attached to the surface of the fluidized medium which is mainly made up of sand or granular activated carbon. This increases the catalytic efficiency of the microorganisms, resulting in higher degradation of the solid wastes. The fluidization medium at high velocities is passed through the bed of the solid particles so that these solids get suspended and behave like liquid (Gopalakrishnan et al. 2019). The solids enter the cylindrical tank from the top, while the air is circulated inside the column from the bottom causing fluidization. The advantages of fluidized reactors include large area, excellent mixing, distribution of temperatures, increased mass transfer, uniform particle distribution, less clogging, less short circuiting and low operation cost. The disadvantages include high energy to operate the bioreactor and difficulty to maintain the biofilm attached to the surface. The typical COD removing efficiency of fluidized/expanded bed reactor is 70–90%.

4.3 Completely Mixed Anaerobic Digester

It is a tank containing the wastewater mixed with microorganisms. An external heating system or an internal heating coil is placed inside the tank to adjust the temperature. The temperature should be in the mesophilic or thermophilic range and insulation is done to minimize heat loss. The tank has a head space and a rigid flexible head cover. It can be a batch or a continuous method. In the continuous process, an equal amount of the wastewater enters the tank and displaces equal amount of liquid inside continuously, while in the batch process, the wastewater comes out at the end of the process. These reactors work effectively when the solid content of the wastewater is 3–6%. One of the disadvantages of completely mixed anaerobic digestor is short circuiting which kills the microorganisms and also reduces the biogas yield.

4.4 Anaerobic Filters

Anaerobic filters are fixed bed bioreactor systems containing one or more filtration chambers which are arranged in series. A bacterial biofilm consisting of rock, gravel or stone is formed on the fixed bed reactor. The anaerobic filter bioreactor is operated in an upflow manner preventing washout of biomass, clogging and channeling (Bhattacharya et al. 2018). When the wastewater flows through the filter, the solids get trapped and the organic compounds are degraded by the bacterial biofilm attached to the fixed bed. The gas formed gets collected by a gas collector at the top and the effluent is released from the top. There are also downflow anaerobic filters where the wastewater flows in the opposite direction. The advantages include low operating cost, less use of energy, and less area requirement. The disadvantages include risk of clogging, treatment of the effluent and low pathogen and nutrient removal. The typical COD removing efficiency of anaerobic filters is 70–80%.

5 Steps of Anaerobic Digestion Process

Anaerobic digestion of sludge involves various stages like hydrolysis, acidogenesis, acetogenesis and methanogenesis (Fig. 2).

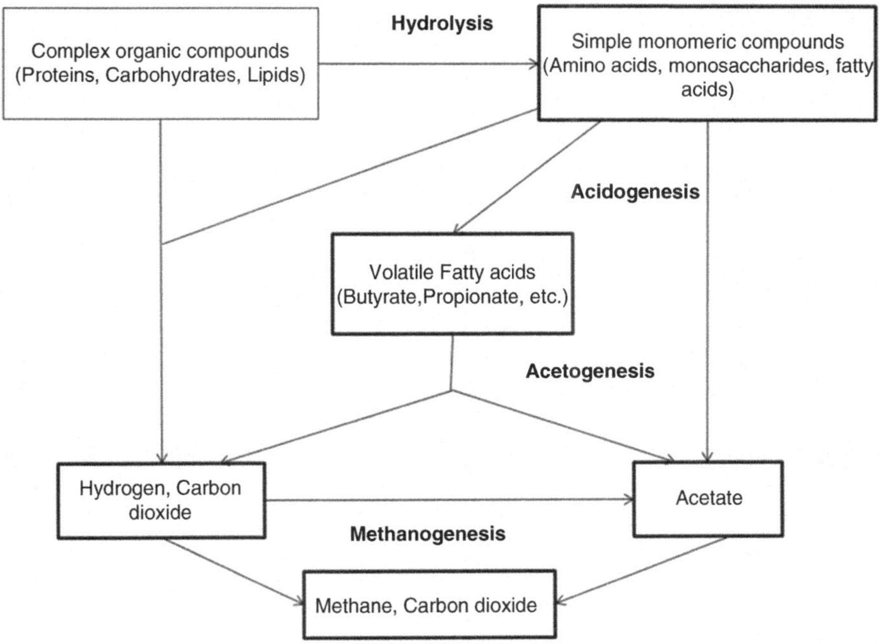

Fig. 2 Steps of anaerobic digestion process

5.1 Hydrolysis

Hydrolysis involves the breakdown of the complex insoluble organic molecules into simpler components by the bacteria in the anaerobic digestor which in turn are utilized by the acidogens. The bacteria secrete extracellular enzymes which degrade the proteins, carbohydrates and lipid polymers into their respective monomers, viz., amino acids, monosaccharides, and fatty acids. Enzymes are added to enhance the degradation of substances which are difficult to degrade like lignin, cellulose and hemicellulose (Lin et al. 2010). The enzymes produce a large surface area causing the microbial cells to attach easily, thereby enhancing the degradation process. Enzymes like protease, amylase, cellulase, endo-glycanases and glycosidases are often used for enhancing the degradation of sludge particles. Enzyme like lysozyme is often used for the destruction of the cell walls of Gram positive bacteria. For the destruction of cell walls of Gram negative bacteria, a combination of lysozyme and EDTA is used, resulting in the release of lipopolysaccharides into the solution (Luo et al. 2012). Hydrolysis is carried out specifically by the anaerobic bacteria belonging to the genera *Enterobacterium* and *Streptococcus*. Rate of hydrolysis depends upon size of particles, temperature, pH and availability of the enzymes (Shah et al. 2014). The optimal temperature for hydrolysis is 30–50 °C and the optimum pH is 5–7 (Azman 2016).

5.2 Acidogenesis

The hydrolyis products are absorbed by the cell wall of the acidogenic bacteria which convert these products into volatile fatty acids (VFAs) like butyric acid, acetic acid, formic acid, propionic acid, alcohols, aldehydes, carbon dioxide and methane. Acidogenic bacteria having a regeneration time of lesser than 36 hours, the rate of acidogenesis is believed to be faster than the other steps (Deublein and Steinhauser 2008). In protein rich wastes, amino acids breakdown to form the VFAs resulting in the formation of high amount of ammonia and sulfur dioxide which generate unpleasant smell and also result in the inhibition of the anaerobic digestion process (Park et al. 2014) The hydrogen concentration of the solution increases as a result of acidogenesis. Methanogenic bacteria cannot use the products of acidogenesis directly and relies on some obligate bacteria to convert these into acetate and hydrogen through a process called acetogenesis. Anaerobes like *Clostridium, Micrococcus, Bacillus, Pseudomonas, Flavobacterium, Acetovibrio, Propionibacterium, Butyrivibrio* are mainly responsible for carrying this step.

5.3 Acetogenesis

The higher VFAs formed by the acidogenesis are converted into acetate and hydrogen by bacteria of genera *Syntrophobacter* and *Syntrophomonas* (Schink 1997), During this process, the oxygen in the sludge is taken up by the acetogenic bacteria which create an anaerobic environment that becomes favourable for the methanogenic bacteria which are mostly obligate anaerobes. The hydrogen produced from this process sometimes exerts a toxic effect on these bacteria which carry out this process. Therefore, this hydrogen is converted into biogas by the methanogenic bacteria like *Methanobacterium propionicum* and *Methanobacterium suboxydans* resulting in syntrophism. Approximately 25% and 11% of acetate and hydrogen, respectively are produced from this step (Schink 1997; de Bok et al. 2005). While long chain fatty acids (LCFAs) containing even number of carbons will be degraded to acetate via the β-oxidation pathway directly, LCFAs containing odd number of fatty acids will be degraded first to propionate (Cirne et al. 2007). The efficiency of this acetogenesis determines how much methane will be produced at the end of the process.

5.4 Methanogenesis

In this step, the products formed from the three other steps are converted into methane by the methanogens. The acetate, carbon dioxide and hydrogen produced by acetogenesis is used up by these bacteria as energy source from which they produce mainly methane, carbon dioxide and water. The methanogenic bacteria are obligate anaerobes and are highly sensitive to oxygen. It was found that 99% of two species of methanogens, *Methanococcus vannielli* and *Methanococcus voltae* die when exposed to oxygen (Kiener and Leisinger 1983). There are two types of methanogenesis, hydrogenotrophic and acetoclastic. Hydrogenotrophic methanogenesis accounts for the production of one-third methane from the hydrogen gas produced and acetoclastic methanogenesis accounts for the production of two-third of methane from the acetate produced. *Methanosarcina* and *Methanosaeta* are the two methanogenic genera which are known to be involved in acetoclastic methanogenesis at present (Conklin et al. 2006). Hydrogenotrophic methanogenesis is performed by methanogens belonging to the orders Methanomicrobiales and Methanobacteriales (Angenent et al. 2002; Amha et al. 2017). Methanogens have a higher regeneration time and need a higher pH compared to the other bacteria involved in the other steps and a lower redox potential (Wolfe 2011; Deublein and Steinhauser 2008). Methanogenesis ends when the production of biogas stops and may take up to 40 days (Verma 2002).

6 Sludge Dewatering

After the useful gases are retrieved from sludge digestion, dewatering is performed to remove the excess water that is still left. The dewatering steps mainly depend on the type of the equipment and the type of sludge to be dewatered. Sludge dewatering mainly involves two methods, filtration and expression. Filtration is done up to a point so that no water is left and the sludge particles come in contact with each other. A solid pressure is created and the excess water within and between the solid particles is squeezed out (Novak 2006). The factors which influence sludge dewatering are rheological properties, particle size, porosity, micromorphology, surface charge and repulsive energy, and extracellular polymeric substance (EPS) (Boran et al. 2020). Methods commonly used are Vacuum filters, Plate and Frame filter Press, Centrifuge and Belt Filter Press.

6.1 Vacuum Filters

It consists of a drum with a filtering medium made of cotton, wool, mesh, nylon, plastic or mesh. The drum is submerged in the sludge and is attached to a tank. A vacuum is applied while the drum rotates and water is squeezed out of the sludge, leaving a cake on the outer surface.

6.2 Plate and Frame Filter Press

It uses filter plates to separate the solids and the liquids. The two plates are joined together forming a chamber and the sludge is squeezed out to remove the excess water. A filter cloth is attached through which the water comes out and the solids get deposited on a conveyer. This method is highly effective and beneficial as it leaves highest content of cake like solids which is easy for transport and disposal.

6.3 Centrifuge

The flocculated sludge is fed into the rotating centrifuge bowl and the sludge gets thrown against the side of the bowl. The dewatered sludge is pushed towards the end of the bowl, while the clarified liquid or centrate is ejected out of the bowl from the other end.

6.4 Belt Filter Press

Belt filter press uses a polymer flocculant that helps in the formation of stronger flocs. The sludge is dropped on a belt by a transfer pump where free water molecules are separated by gravity and the water is collected into a collection trough. As the sludge moves through a conveyer belt, water is drained out by a plough. A gravity thickener is present which repeats this process and the sludge is fed into a pressing zone. The sludge is pressed by two belt filters fitted with rollers into the pressing zone and the excess water is slowly squeezed out of the sludge and is collected in a bin.

6.5 Electroosmosis

Recently another method has been developed which uses electroosmosis with filter bags for sludge dewatering. Sludge is injected into a filter bag containing a cathode and an anode electrode, placed on a slope. The electrodes are connected to a power supply so that electroosmosis occurs and water flows down the slope (Yingchun et al. 2020). For improving sludge dewaterability, physical conditioning like sonication, freezing, thawing, adding porous substances, thermal treatment, etc. are performed. Chemical conditioning like addition of coagulating or flocculating reagents, acid/base treatment, enzymatic treatment or advanced oxidation processes are also being applied.

7 Sludge Disposal

The last stage of sewage sludge treatment is sludge disposal. The sludge obtained are either deposited on land as a landfill or disposed in the ocean which is not favourable. Some part of the sludge is also incinerated. Sewage sludge contains many essential elements like potassium, phosphorous, nitrogen and some minor elements like magnesium, sulfur, boron, and zinc needed for the proper growth of the plants making it suitable to be used as a fertilizer or a soil conditioner. It also reduces soil erosion and has a high water holding capacity. The sludge components can also be recycled and sold in the market. After sludge dewatering, the sludge must be disinfected with quicklime at a dose of 500 kg CaO/ton. Quicklime treatment increases the pH and temperature after reacting with the water, making the sludge temporarily stabilized as it gets biodegraded after the pH drops. The sludge after disinfection should not carry *Salmonella* sp., ova of helminth which is viable and not more than 2500 of fecal coliforms (Paulsrud and Nedland 1996; Ødegaard et al. 2002). Raw primary sludge needs to be composted before using it as a fertilizer. Production of compost by windrow composting using the biodegradable waste of the sludge is performed. Bio-oil having a neutral pH is also produced from the sewage sludge by pyrolysis in a fluidized bed reactor. This leads to the minimization of engine and pipeline corrosions (Arazo et al. 2017). The sludge must be disposed off properly or it will lead to environmental hazards resulting in contamination of the environment.

8 Role of Bacteria in Wastewater Treatment

The activated sludge derived after wastewater treatment contains a large number of microbes which become activated during anaerobic digestion. Almost 17 phyla were identified by high throughput sequencing of nine samples of waste activated sludge. The main bacteria found in the waste activated sludge and bioreactors include bacteroidetes, proteobacteria, and firmicutes, proteobacteria being the most abundant. The main methanogens found are Methanosarcinales and Methanomicrobiales (Shin et al. 2019). In another study, analysis of the bacterial community of activated sludge revealed that the dominant phyla are Proteobacteria (26.7–48.9%), Bacteroidetes (19.3–37.3%), Chloroflexi (2.9–17.1%), and Acidobacteria (1.5–13.8%). About 55 genera including *Dokdonella, Flavobacterium, Terrimonas, Tetrasphaera, Nitrospira* existed in almost all the samples (Xu et al. 2018). As discussed previously, large amount of ammonia is released after breakdown of protein rich wastes. This ammonia is oxidized by *Nitrosomonas* to nitrite followed by conversion of nitrite to nitrate by *Nitrobacter* (Wagner 1996). The nitrate is then reduced to dinitrogen gas by denitrifying bacteria and is liberated. The optimum temperature and pH for denitrification is around 30–35 °C and 7.0–8.0, respectively. However, it was found that during denitrification, the step of methanogenesis is completely suppressed. The nitrogen oxides formed, completely inhibited the formation of methane (Chen and Lin 1993). Though anaerobic digestion is mostly preferred, many small communities use the process of aerobic digestion which breaks down organic matter in presence of oxygen. Aerobic bacteria like *Chromobacter, Flavobacterium* and *Pseudomonas* form biofilm during secondary treatment. Formation of hydrogen sulphide by sulphate-removing bacteria becomes disadvantageous for the treatment process as it is highly corrosive and inhibits the growth of other essential microorganisms. Sulphate reducing bacteria include mesophilic δ-Proteobacteria like *Desulfobacterium*, gram-negative thermophilic bacteria like *Thermodesulfovibrio* and *Thermodesulfobium* and gram-positive bacteria like *Desulfotomaculum* and *Desulfosporomusa* (Mori et al. 2003; Muyzer and Stams 2008; Thauer et al. 2007; Thevenieau et al. 2007).

9 Role of Fungi in Wastewater Treatment

Although bacteria plays an important role in the degradation of the organic molecules into simpler substances, use of filamentous fungi has also been found to be effective during wastewater treatment. Valuable biochemical substances, large amount of fungal biomass and high value fungal proteins are formed by the conversion of the organic compounds. Fungi secrete enzymes like α-amylase which can degrade complex carbohydrates like starch (Jin et al. 1998). Yeasts and molds play an important role in the production of microbial biomass protein (MBP) during wastewater purification. Yeasts can grow at a pH less than 5, are easier to cultivate, less susceptible to contamination, can produce high energy rich biomass and have a higher growth rate in comparison to molds (Gonzalez et al. 1992; Bergmann et al. 1988; Satyawali and Balakrishnan 2007). However, the mycotoxins produced by the fungi are harmful for the growth of the other beneficial microorganisms. The vast majority of the fungi used in wastewater treatment have a few advantages over bacteria. They are mesophilic in nature, require a temperature of 20–40 °C and a pH less than 5.0 (Sankaran et al. 2010). Fungi have the capability of greater resistance to inhibitory compounds as they have more genes compared to the bacteria which confer them better reproductive selectivity, thus making them more adaptive (Guest and Smith 2002; Bennett and Lasure 1991). While fungi are cultivated in the industries for valuable products, sludge containing bacterial biomass generated after wastewater treatment is expensive to treat and is of low value. Also due to the presence of fungal hyphae, the biomass is easier to collect from the mixed liquor, thus reducing the treatment cost.

10 Role of Protozoa in Wastewater Treatment

Although microorganisms like bacteria are directly involved in wastewater treatment, there are a large number of other microorganisms which play an important role in wastewater treatment. Protozoa like amoeba, ciliates, and flagellates are present in the sludge solution and improve the quality of the effluent. They mainly feed on the suspended bacteria and keep the density of dispersed bacterial population in check. The protozoa acts as bioindicators for determining the presence of toxic heavy metals present in the wastewater. A large number of protozoa like *Euglena, Euglypha, Paramecium, Chilodonella, Trochilia, Coleps, Acineria, Aspidisca, Epistylis, Vorticella, Plagiocampa,* etc. are found in the activated sludge plants. In an activated sludge plant, the most common protozoa found are the peritrichs and hypotrichs, while cyrtophorids and testate amoebae may also be observed. Though protozoa were considered harmful for the activated sludge process earlier, it was found that the wastewater with no protozoa has a higher level of biochemical oxygen demand (BOD) (Curds et al. 1968). The protozoa helps in carbon mineralization and the excretion of these mineral nutrients are then used up by the bacteria as energy source. Release of nitrogen, phosphorous, organic carbon and some growth stimulatory compounds by the protozoa influence the growth of the bacteria (Jurgens and Matz 2002). Protozoa also helps in improving the quality of the effluent by bacterial grazing. The clearance rates of protozoa are around 4×10^{-7} to 1×10^{-6} ml medium per protozoa per hour (Bloem et al. 1988).

11 Conclusion

The purpose of sewage sludge treatment is to reduce the organic matter and the number of pathogens present in the sludge. The sludge obtained after wastewater treatment needs to undergo vigorous processing before being disposed off. It should be thoroughly checked for the presence of contaminants before using it for soil amendment or landfill. Although many different types of techniques are used, much more improvement needs to be done to maximize the reduction of the harmful contaminants present in the sludge. Since quite a large number of microorganisms are involved in the treatment process of sludge, some might be harmful for the environment and needs to be removed from the final product. Therefore, more scientific research needs to be done on sewage sludge treatment for sustainable development in the future.

References

Abbasi T, Tauseef SM, Abbasi SA (2012) Biogas energy. Springer Briefs Env Sci 2:11–23

Amha YM, Sinha P, Lagman J, Gregori M, Smith AL (2017) Elucidating microbial community adaptation to anaerobic co-digestion of fats, oils, and grease and food waste. Water Res 123:277–289

Angenent LT, Sung SW, Raskin L (2002) Methanogenic population dynamics during startup of a full-scale anaerobic sequencing batch reactor treating swine waste. Water Res 36 (18):4648–4654

Arazo RO, Genuino DAD, de Luna MDG, Capareda SC (2017) Bio-oil production from dry sewage sludge by fast pyrolysis in an electrically-heated fluidized bed reactor. Sust Env Res 27(1):7–14

Azman S (2016) Presence and role of anaerobic hydrolytic microbes in conversion of lignocellulosic biomass for biogas production. Anaerobic digestion of cellulose and hemicellulose in the presence of humic acids. Wageningen University, pp. 15–46

Baroutian S, Eshtiaghi N, Gapes DJ (2013) Rheology of a primary and secondary sewage sludge mixture: dependency on temperature and solid concentration. Bioresour Technol 140:227–233

Bennett JW, Lasure LL (1991) More gene manipulations in fungi. Academic, San Diego, CA

Bergmann FW, Abe J, Hizukuri S (1988) Selection of microorganisms which produce raw-starch degrading enzymes. Appl Microbiol Biotechnol 27:443–446

Bhattacharya J, Dev S, Das B (2018) Design of wastewater bioremediation plant and systems. Low cost wastewater bioremediation technology. Innovative treatment of sulphate and metal-rich wastewater. Butterworth Heinemann, pp. 65–313

Bloem J, Starink M, Bär-Gilissen MJB, Cappenberg TE (1988) Protozoan grazing, bacterial activity, and mineralization in two-stage continuous cultures. Appl Environ Microbiol 54:3113–3121

Bodík I, Sedláèek S, Kubaská M, Hutòan M (2011) Biogas production in municipal wastewater treatment plants-current status in EU with a focus on the Slovak Republic. Chem Biochem Eng Q 25:335–340

Boran W, Xiaohu D, Xiaoli C (2020) Critical review on dewatering of sewage sludge: influential mechanism, conditioning technologies and implications to sludge re-utilizations. Water Res 180:115912

Chen K-C, Lin Y-F (1993) The relationship between denitrifying bacteria and methanogenic bacteria in a mixed culture system of acclimated sludges. Water Res 27(12):1749–1759

Cirne DG, Paloumet X, Björnsson L, Alves MM, Mattiasson B (2007) Anaerobic digestion of lipid-rich waste—effects of lipid concentration. Renew Energy 32:965–975

Conklin A, Stensel HD, Ferguson J (2006) Growth kinetics and competition between *Methanosarcina* and *Methanosaeta* in mesophilic anaerobic digestion. Water Environ Res 78 (5):486–496

Curds CR, Cockburn A, Vandyke JM (1968) An experimental study of the role of the ciliated protozoa in the activated sludge process. Water Pollut Control 67:312–329

de Bok FAM, Harmsen HJM, Plugge CM, de Vries MC, Akkermans ADL, de Vos WM, Stams AJM (2005) The first true obligately syntrophic propionate-oxidizing bacterium, *Pelotomaculum schinkii* sp. nov., co-cultured with *Methanospirillum hungatei*, and emended description of the genus *Pelotomaculum*. Int J Syst Evol Microbiol 55(4):1697–1703

Dentel SK, Qi Y (2014) Management of sludges, biosolids, and residuals. comprehensive water quality and purification. Elsevier, pp. 223–243

Deublein D, Steinhauser A (2008) Biogas from waste and renewable resources: an introduction. Wiley: Hoboken, NJ; ISBN 978-3-527-31841-4

Edzwald JK (2010) Dissolved air flotation and me. Water Res 44(7):2077–2106

Gonzalez MP, Siso MIG, Murado MA, Pastrana L, Montemayor MI, Miron J (1992) Depuration and valuation of mussel-processing wastes: characterization of amylolytic postincubates from different species grown on an effluent. Bioresour Technol 42:133–140

Gopalakrishnan B, Khanna N, Das D (2019) Dark-fermentative biohydrogen production. Biohydrogen (2nd ed) Biomass, Biofuels, Biochemical, pp. 79–122

Guest RK, Smith DW (2002) A potential new role for fungi in a wastewater MBR biological nitrogen reduction system. J Environ Eng Sci 1:433–437

Gunasekaran M, Preethi, Usman TMM, Banu R, Kumar, G (2019) Biohydrogen production from industrial wastewater. Biofuels: Alternative Feedstocks and Conversion Processes for the Production of Liquid and Gaseous Biofuels (2nd ed), Vol 7, 100287

Indah Water Konsortium Sdn Bhd. (2013) Sustainability Report 2012–2013. Kuala Lumpurl: IWK1

Jin B, van Leeuwen J, Patel B, Yu Q (1998) Utilization of starch processing wastewater for production of microbial biomass protein and fungal α-amylase by *Aspergillus oryzae*. Bioresour Technol 66:201–206

Jurgens K, Matz C (2002) Predation as a shaping force for the phenotypic and genotypic composition of planktonic bacteria. Antoine Van Leeuwenhoek 81:413–434

Kemp JS (1997) Just the facts on dewatering systems: a review of the features of three mechanical dewatering technologies. Water Environ Technol

Kiener A, Leisinger T (1983) Oxygen sensitivity of methanogenic bacteria. Syst Appl Microbiol 4:305–312

Krzywicka A, Kwarciak-Kozłowska A (2014) Advanced oxidation processes with coke plant wastewater treatment. Water Sci Technol 69(9):1875–1878

Lin L, Yan R, Liu Y, Jiang W (2010) In-depth investigation of enzymatic hydrolysis of biomass wastes based on three major components: cellulose, hemicellulose and lignin. Bioresour Technol 101:8217–8223

Luo K, Yang Q, Li X, Yang G, Liu Y, Wang D, Zheng W, Zeng G (2012) Hydrolysis kinetics in anaerobic digestion of waste activated sludge enhanced by α-amylase. Biochem Eng J 62:17–21
Mori K, Kim H, Kakegawa T, Hanada S (2003) A novel lineage of sulfate-reducing microorganisms: *Thermodesulfobiaceae* fam. nov., *Thermodesulfobium narugense*, gen. nov., sp. nov., a new thermophilic isolate from a hot spring. Extremophiles 7:283–290
Mustafa EE, Hale O, Recep KD, Izzet O (2011) Anaerobic treatment of industrial effluents: an overview of applications, waste water – treatment and reutilization, Prof. Fernando Sebastian GarcÃa Einschlag (Ed.), ISBN: 978-953-307-249-4, InTech
Muyzer G, Stams AJM (2008) The ecology and biotechnology of sulphate-reducing bacteria. Nature Rev Microbiol 6:441–454
Nascimento AL, Souza AJ, Andrade PAM, Andreote FD, Coscione AR, Oliveira FC, Regitano JB (2018) Sewage sludge microbial structures and relations to their sources, treatments, and chemical attributes. Front Microbiol 9:1462
Nelson MJ, Nakhla G, Zhu J (2017) Fluidized-bed bioreactor applications for biological wastewater treatment: a review of research and developments. Engineering 3(3): 330–342
Novak JT (2006) Dewatering of sewage sludge. Dry Technol 24:1257–1262
Ødegaard H, Paulsrud B, Karlsson I (2002) Wastewater sludge as a resource: sludge disposal strategies and corresponding treatment technologies aimed at sustainable handling of wastewater sludge. Water Sci Technol 46(10):295–303
Park J, Park S, Kim M (2014) Anaerobic degradation of amino acids generated from the hydrolysis of sewage sludge. Environ Technol 35:1133–1139
Paulsrud B, Nedland KT (1996) Full scale experiences with processes for stabilization and disinfection of primary-chemical sewage sludge. In: Hahn HH, Hoffmann E, Ødegaard H (eds) Chemical water and wastewater treatment IV. Springer, Heidelberg, pp 368–377
Process Design Manual; sludge Treatment and Disposal, US Environmental Protection Agency, Office of Research and Development, EPA 625/1-79-011 September, 1979
Sankaran S, Khanal SK, Jasti N, Jin B, Pometto AL, Van Leeuwen JH (2010) Use of filamentous fungi for wastewater treatment and production of high value fungal byproducts: a review. Crit Rev Environ Sci Technol 40:400–449
Satyawali Y, Balakrishnan M (2007) Wastewater treatment in molasses-based alcohol distilleries for COD and color removal: a review. J Environ Manag 86:481–497
Schink B (1997) Energetics of syntrophic cooperation in methanogenic degradation. Microbiol Mol Biol Rev 61(2):262–280
Shah FA, Mahmood Q, Shah MM, Pervez A, Asad SA (2014) Microbial ecology of anaerobic digesters: the key players of anaerobiosis. Sci World J 1–21
Shin J, Cho SK, Lee J, Hwang K, Chung JW, Jang HN, Shin SG (2019) Performance and microbial community dynamics in anaerobic digestion of waste activated sludge: impact of immigration. Energies 12: 573
Thauer RK, Stakebrandt E, Hamilton WA (2007) Energy metabolism and phylogenetic diversity of sulphate reducing bacteria. In: Barton LL, Hamilton WA (eds) Sulphate-reducing bacteria: environmental and engineered systems. Cambridge University Press, Cambridge, pp 1–37
Thevenieau F, Fardeau ML, Ollivier B, Joulian C, Baena S (2007) *Desulfomicrobium thermophilum* sp nov., a novel thermophilic sulphate-reducing bacterium isolated from a terrestrial hot spring in Colombia. Extremophiles, 11: 295–303
Torpey WN (1954) Concentration of combined primary and activated sludges in separate thickening tanks. J Sanitary Eng Div Proc ASCE 80:443
Usman K, Khan S, Ghulam S, Khan M, Khan N, Khan M, Khalil S (2012) Sewage sludge: an important biological resource for sustainable agriculture and its environmental implications. Am J Plant Sci 3:1708–1721
Verma S (2002) Anaerobic digestion of biodegradable organics in municipal solid wastes. Columbia University, New York, NY
Wagner M (1996) In situ analysis of nitrifying bacteria in sewage treatment plants. Water Sci Technol 1:237–244

Wang W, Gao X, Zhang J, Yang T, Li R, Sun Y (2019) Effect of SDS and neutral protease on the release of Extracellular Polymeric Substances (EPS) from mechanical dewatered sludge. Waste Biomass Valorization 10(4):1053–1064

Ward AJ, Hobbs PJ, Holliman PJ, Jones DL (2008) Optimization of the anaerobic digestion of agricultural resources. Bioresour Technol 99:7928–7940

Wolfe RS (2011) Techniques for cultivating methanogens in methods in enzymology, vol 494. Academic Press, Cambridge, MA, pp 1–22

World Bioenergy Association (n.d.) WBA Global Bioenergy Statistics. Published by WBA

Xu S, Yao J, Ainiwaer M, Hong Y, Zhang Y (2018) Analysis of bacterial community structure of activated sludge from wastewater treatment plants in winter. BioMed Res Int 1–8

Yingchun Z, Rui S, Chen L, Dundun S, Shiming L (2020) Deep sludge dewatering method using electroosmosis with filter bags. United States. Power China. Huadong Engineering Corporation Limited (Hangzhou, CN) 20200140311

Zhang G, Jing H, Ren MZ, Zhang SK, Cheng J, Yang ZR (2013) Emission, mass balance, and distribution characteristics of PCDD/Fs and heavy metals during cocombustion of sewage sludge and coal in power plants. Environ Sci Technol 47(4):2123–2130

Role of Beneficial Microbes in Sewage Sludge Management

Someshree Mane and Pravin Khaire

1 Introduction

Because of rapid global population growth, water is expected to be one of the most dwindling resources in the twenty-first century (Day 1996). As human numbers grow, the existing resources are put under greater pressure and natural supplies are becoming more endangered. Sewage is the world's primary contributor to toxic waste to water sources and the environment. Scientists, policymakers and the public at large are increasingly aware of environmental challenges emerging in developing countries from the processing of municipal debris. In the 1970s, however, lawmakers recognised the demands of scientists by enacting regulations that guarded water sources from pollutant and organic waste disposal. According to a report conducted by the "Secretary-General of the United Nations Commission on Sustainable Development" (UNCSD 1997), existing natural water use perhaps by developing and developed countries is not sustainable as well as industrial water usage has increased by more than three times the world's population, leading to widespread public health concerns, limits and problems.

"The United States (US), which has been demanding secondary disposal of wastewater by municipalities since 1972 under the Federal Water Pollution Control Act, is the first example of this" (Stentiford 1983). "Law No. 319 (Rules for Water safety from prevention)" was adopted in Italy in 1976 and a decree was issued on January 1992 that would require it. Directive 278/CEE approved by the European Communities (EC) in 1986 calling for the guidelines for water safety to be enforced in all the EC countries. These laws have expanded the sewage sludge output, which is becoming one of organic waste's main significant sources. Animal and urban agriculture contain comparatively large amounts of organic waste. A tonne of civilization waste reaches the water supply via the discharge from domestic,

S. Mane (✉) · P. Khaire
Department of Plant Pathology and Agril. Microbiology, PGI, MPKV, Rahuri, India

V. D. Rajput et al. (eds.), *Sustainable Management and Utilization of Sewage Sludge*, https://doi.org/10.1007/978-3-030-85226-9_9

industrial and non-point sources of waterborne waste containing undesirable content (Welch 1992). Whilst wastewater treatment has been around since ancient times, it was only developed in the late 1800s & early 1900s (Chow et al. 1972). However, modern understanding of the need to sanitise and treat waste water started with John Snow's case in 1855, wherein he demonstrated that in London outbreaks of Cholera disease were caused by pollution from the Thames river (Cooper 2001). Treatment methods for waste water differ from country to country.

Untreated and/or contaminated lakes pose a major health danger because of the illness caused by sewage. Water-borne diseases continue to pose a major global threat to public health, despite major improvements in water and wastewater treatment. Around 250 million people are reportedly infected by waterborne pathogens every year, leading to 10 to 20 million dead (Anon 1996). In developed countries with lower levels or healthcare, socio-economic issues and less knowledge of public health are present for a majority of those diseases than in more industrialised countries. However, it has been reported that the occurrence of incident or waterborne diseases has also risen in the United States in the last 20 years and between 1971 and 1985 there have been more waterborne diseases relative to any recent 15 years since 1920.

Wastewater prevention policies vary across countries. In developed nations, treatment and discharge systems for urban high-income & urban low-income users can differ greatly around the world as well as within rural & urban users (Doorn et al. 2006). In most developed nations, the level of wastewater treatment varies. India emits 62 billion litres of waste water per day according to the Central Pollution Control Board (CPCB) environmental body. In certain countries across the world, public concern over dirty water is a big problem. There seems to be a possibility that the populace, influenced by the media, develops an inflated perception of the threats wastewater. Furthermore, without wastewater treatment even developing nations, there are still populations systems and in few situations, even in areas with a high level of wastewater treatment, bacteria and some contaminants, many of which have unknown ecological effects, can still be discharged into the atmosphere (LeChevallier and Au 2004).

This massive volume of organic waste, whether discarded or recycled, is a major environmental problem. Some habitats have been heavily polluted by sewage sludge, which is tainted by pathogenic organisms and also contains organic and inorganic contaminants as a result of past and ongoing waste management activities. Farmers are worried about the spread of organic waste pathogens, which could put staff and livestock feeding on drainage sludge-modified soil in risk. In addition, sludge can pollute surface and groundwater with microbes.

Sewage treatment is a method in which the contaminants are separated. The ultimate aim of the treatment of waste is to create an effluent that does not harm the environment. The findings can be catastrophic in the absence of sewage treatment, as sewage may disrupt the ecosystem. The consequences of inadequate treatment modalities resulting in microbial pathogenic contamination of the aquatic reception system, as is typical in developed countries, are discussed in this chapter.

2 Sewage Sludge Composition and Its Features

Sewage sludge is a final product of removing pollutants from municipal wastewater. It's made mainly from the sediment deposition of wastewater's organic matter, and it's located in basins designed specifically for wastewater treatment. Because of the abundance and variety of its elements, as well as the presence of xenobiotic compounds, Boyle (1990) referred to the organic fraction of sewage sludge as a "chaotic mixture." Human excreta make up the majority of the organic matter in waste sludge, which has been changed through non-biological and biological stabilisation procedures. Since the organic matter in sewage sludge is easily fermentable, it must be stabilised before it can be used in any way. The following are some examples of stabilisation procedures: (A) air or heat drying; (B) chemical treatment; (C) aerobic stabilisation (liquid state); (D) anaerobic stabilisation (biogas production) (E) composting.

"The efficacy of different sewage sludge treatment methods in reducing pathogen levels varies greatly. When water activity falls below critical levels, drying reduces the viability of most bacteria. When environmental conditions are favourable, pathogens with appropriate survival techniques (bacterial spores, cysts, etc.) can safely survive treatment and return to vegetative status. However, since drying waste organic matter does not adequately stabilise it, re-contamination may occur if the material is re-wetted, either inadvertently or on purpose, and its water content reaches a value. Salmonella has also been found to flourish in wastewater sludge of less than 10% water content" (Dumontet et al. 1999).

"The pathogen nature of the stabilising sludge and the chemical and physical modifications imposed on the sludge by the composition determine the sanitation efficiency of a physical and chemical stabilisation process. The effectiveness of biological therapy is determined by a variety of factors, including temperature, redox capability, pathogen-mineralizing microflora competition, and the susceptibility of stable organic matter to pathogen development" (Stentiford 1983). The time/temperature ratio is usually the most important factor in aerobic and anaerobic sanitation procedures, while in the composting process, both temperature and antagonistic species are involved in pathogen inactivation.

"Without any substantial volume reduction, the aerobic and anaerobic stabilisation processes produce a still liquid material with poorly stabilised organic matter and variable pathogen removal ability. Composting, on the other hand, reduces the amount of sewage sludge treated and produces a sanitary, storable solid product" (Dumontet et al. 1999; Havelaar et al. 1983). Regardless of the hygiene requirements in the countries concerned, improper sewage sludge disposal will increase oro-fecal disease transmission. Furthermore, poorly sanitised sludge dumping, either by dredging or by dumping, will increase microbial contamination of surface water.

3 Microbes in Wastewater

Microbial pathogens that are likely to be present in waste water are classified into three categories: (a) viruses (b) bacteria (c) protozoans/helmiths (LeChevallier and Au 2004).

3.1 *Viruses*

Viruses are one of the most widespread and dangerous drainage pathogens (Tree et al. 2003). According to Toze (1997), untreated waste water can contain a number of viruses with more than 103–104 viral particles per litre of waste water. Viruses, in general, are more resistant to medicine, are airborne, are more difficult to detect in environmental screening, such as waste water, and take less energy to infect than other pathogens (Gomez et al. 2006). Viruses in wastewater enter the atmosphere through infected hosts' or carriers' faeces (Leclerc et al. 2000). "Enteroviruses, which are small single-strand RNA viruses, are the most frequently encountered pathogenic viruses in wastewater and include polioviruses type 1 and 2. Others are various echovirus strains, enteroviruses and coxsackie viruses" (Tanji et al. 2002).

3.2 *Bacteria*

Among bacteria, microbial infections found in waste water are the most common. Enteric pathogens include a large number of bacterial pathogens and opportunistic pathogens that have been reported in the literature (Simpson and Charles 2000). GI infections are the most common health problems caused by bacterial pollutants in wastewater (LeChevallier and Au 2004). Diarrhoea, dysentery, leptospira interrogations, contagious diseases, typhoid, human entercolites, legionels, melioidosis, ulcers, and tumours are all common wastewater diseases, as are wastewater-induced infections (Liang et al. 2006). "Water includes recognised species producing toxin such as the *Staphylococcus aureus*, *Salmonella Spp.*, *E. coli* of food poisoning or perfering *Clostridium* can cause food toxic outbreaks" (Toze 1997) (Table 1).

3.3 *Protozoa*

"Pathogenic protozoa are found in greater abundance in sewage than in any other environmental source" (Toze 1997). "Pathogenic protozoans correlated with wastewater include, and most often are isolated from fecal matter wastewaters, such as

Table 1 The most common human pathogenic bacteria and viruses found in waste water and sewage sludge, as well as the diseases they cause

Virus		Bacteria	
Enteroviruses	Symptoms/disease caused by respective virus	Pathogen	Symptoms/ disease caused by bacteria
Polio virus	Poliomyelitis, meningitis, fever	*Salmonella* spp.	Salmonellosis, typhoid
Coxackievirus A	Herpangina, respiratory disease, meningitis, fever	*Shigella* spp.	Bacillary dysentery
Coxackievirus B	Myocarditis, congenital heart anomalies, respiratory disease, pleurodynia, rash, fever	*Escherichia coli* (enteropathogenic strains)	Gastroenteritis
Echovirus	Meningitis, respiratory disease, diarrhea, encephalitis, acute hemorrhagic conjunctivitis, fever	*Pseudomonas aeruginosa*	Otitis externa, skin infections (opportunistic pathogen)
New Enteroviruses		*Yersinia enterocolitica*	Acute gastroenteritis
Adenovirus	Respiratory disease, eye infection	*C. perfringens*	Gastroenteritis (food poisoning)
Parvovirus	Meningitis, encephalitis, respiratory disease, acute hemorrhagic conjunctivitis, fever	*C. botulinum*	Botulism
Reovirus and Astrovirus	Not clearly established	*B. anthracis*	Anthrax
Hepatitis A, C and E virus	Infectious hepatitis	*Listeria monocytogenes*	Listeriosis
Rotavirus, Calicivirus and Norwalk agent and other small round viruses	Vomiting and diarrhea	*Vibrio cholera*	Cholera
Coronavirus	Common cold	*Mycobacterium* spp.	Leprosy, tuberculosis
Adeno-associated viruses	Not clearly established, but associated with respiratory disease in children	*Leptospira* spp.	Leptospirosis
Polyomaviruses		*Campylobacter* spp.	Gastroenteritis
JC	Progressive multifocal leukoencephalopathy	*Staphylococcus*	Impetigo, wound infections, food poisoning
JC	Infections of the urinary tract	*Streptococcus*	Sore throat, necrotizing fasciitis, scarlet fever

Source: This table was constructed for this manuscript based on information obtained from Veronica Arthurson 2008; Strauch D (1991; 1998)

Entamóeba histolytica, Giadia intestinalis and Cryptoporidium parvum" (Caccio et al. 2003; Toze 1997).

3.4 Helminths

"Helminths (nematodes and tapeworms) are popular intestinal parasites that spread through human faeces, similar to enteric protozoan pathogens" (Feenstra et al. 2000). "Strongyloidiasis is caused by round worms (*Ascaris lumbricoides*), hook worms (*Ascaris duodenale* or *Nector americanus*), whip worms (*Trichuris trichiura*), and *Strongloides stercolaris*, which are commonly found in wastewater" (Feenstra et al. 2000). "It is estimated that approximately 25% of the world's population is infected with the round worm *A. lumbricoids"*. "Population growth, educational standards, rate of sanitation and irrigation, and cultural dietary patterns are responsible for the prevalence of *Ascaris* infection" (Smith et al. 2001). Gut nematodes are the most serious health concern in non—treated excreta and agricultural/aquacultural waste water, according to the World Health Organization (1998).

3.5 Yeast and Fungi

Pathogenic yeast and fungi are likely to play a secondary role in sewage loam infection of humans. Such species can cause a reasonable amount of illnesses, from allergies to severe systemic infections. There are also fungi which may develop mycotoxins when developing in particular plants and foods. *Aspergillus fumigatus* is still found in wastewater sludge and is still extensively contaminating the environment with fungal opportunists and respiratory allergens of medicinal importance. The most significant pathogenic yeast and fungi in sewage sludge are mentioned in Table 2. Up to 75% of the airborne micro-flora of the plants is composted in waste sludge *A. fumigatus*, which is still present in the atmosphere. Milner et al. (1977) stress that the health threat represented by *A. fumigatus* cannot be removed because cellulose is used as carbon source for this fungus and sometimes cellulose-rich products are used as bulking agents.

4 Microbial Indicators

It will be challenging, time intensive and exceedingly costly to locate, isolate and identify the multiple forms of microbe pathogens associated with wastewater as attempted routinely. "The indicator microorganisms are used to measure the relative risk of possible disease agents on a sample to minimise the need for these massive programmes" (Ashbolt et al. 2001). "These microorganisms must be part of the gut

Table 2 Pathogenic yeast, fungi and bacteria that are extracted from sewage sludge

		Bacterial pathogens	
Yeast	Fungi	Primary	Opportunistic
C. albicans, Trichosporon, C. neoformans, C. tropicalis, C. krusei, Candida guillermondii	*Aspergillus spp., Geotricum candidum, Epidermophyton spp., Phialophora richardsii, Trycophitum spp.*	*Motile Aeromonas, Arcobacter spp., B. anthracis, Brucella spp., C. coli, C. fetus ssp. fetus, C. jejuni, C. botulinum, C. perfringens, Escherichia coli O111: NM, Escherichia coli O157:H7, Escherichia coli O184:H21, Leptospira spp., Listeria monocytogenes, Mycobacterium spp., Pseudomonas aeruginosa, Salmonella spp., Shigella spp., Staphylococcus (coagulase positive strains), Streptococcus (beta-hemolyticus strains), Vibrio cholera, Vibrio parahaemolyticus, Vibrio vulnificus, Yersinia enterocolitica.*	*Citrobacter spp., Enterobacter spp., Escherichia coli, Klebsiella spp., Proteus spp., Providencia spp., Serratia spp.*

Source: This table was constructed for this manuscript based on information obtained from Strauch D (1991; 1998)

microbiota of warm, blood-filled animals to function effectively as markers; must be present when pathogens of infected specimens are present and missing; should be more present than pathogens; must be at least equally resistant to external factors and wastewater disinfestation as they are to pathogens" (Bitton 2005).

Escherichia coli has long been used as fecal material contamination measurement of water supply but is well known for its production and operation in the environment (Ashbolt et al. 2001). "It has been named as thermotolerant coliforms (TTCs) because *E. coli* can be grown at a high temperature (44.5 °C) and have been a key indicator in the water industry (Leclerc et al. 2000)." "Thermotolerant coliform bacteria are more likely to trigger environmental improvements and treatment systems to withstand more resistant bacterial pathogenes and almost all viruses, protozoan cysts and helminths (Ashbolt et al. 2001)." Further downside to using TTC as an indicator of faecal contamination is that certain warm-blooded bacteria in their intestines have coliform. Consequently, the detection of TTC in a water source is not necessarily confirming that the water body is contaminated by human excrement or that human pathogenic agents exist. Faecal coliforms or TTC's inappropriateness as markers of human faecal pollution of water supplies and efficacy of treatment has lead to more suitable indicator microorganisms being pursued.

"According to Ferguson et al. (1996), *Clostridium perfringes* is the most useful markers of human faecal contamination and, relative to faecal streptococci and F RNA bacteriophages, the only accurate indication for the existence of *Giardia intestinalis*. The *enterococci, bifidobacteria* and bacterioiphages are other possible bacterial markers of the existence of microbial pathogens in water (Leclerc et al. 2000). However, anaerobic indicator bacteria like the bacteroiphages and bifidobacteria are difficult to use as markers of large-scale faecal pollution due to the difficulty involved with these extreme anaerobic in handling. Recent production of DNA probes for the identification of polymerase chain reaction (PCR) alleviates the need for culture and enhances the ability of anaerobes as indicators of faecal contamination (Kreader 1995). The higher tolerance of protozoan cysts and viruses to environmental conditions and treatment processes is one of the issues associated with using bacteria as a measure of the presence of microbial pathogens in water (Tree et al. 2003; Hijnen et al. 2004; Gomez et al. 2006)."

"The low number, complexity and high cost of cultivation are especially difficult for viruses to detect in many water sources (Tanji et al. 2002)". "Bacterial viruses were investigated in order to solve these problems for the purpose of fecal matter infection and for the treatment methods to remove entry viruses (Ashbolt et al. 2001)". "Although a series of alternatives were evaluated for alternative uses of faecal coliforms, none were completely suitable. All possible markers tested to date have one or more characteristics which prevent their use as proxies for faecal coliforms" (Ashbolt et al. 2001; Bitton 2005). "Thus, despite its drawbacks, faecal pollution and productivity processes remain the most important organisms that are used to show faecal contamination" (Toze 1997). "However, progress in the detection of molecular methods used in the last 10 years may suggest that metrics may no longer be relevant" (Bitton 2005).

5 Isolation and Identification of Wastewater Pathogens

Approaches used to identify and quantify microbial populations of waste water can be divided into three categories: (a) culture, (b) immunology, & (c) nucleic acid-based.

5.1 Culture-Based Methods

In this method, selective and/or differential media are used, providing a 'presumptive identification' and may be supplemented by a number of other steps. The studies validate the characterization of isolates using biochemical, immunological, or molecular methods.

5.2 *Nucleic Acid Based Methods*

"Advances in molecular biology have transformed waste water microbiology by promoting the identification of new microbes, the detection of microbial species, and the differentiation of pathogenic and non-pathogenic bacteria that are closely associated (Persing et al. 2003)". "Nucleic acid hybridization between genes whose frequency is specific to a human or whose sequence differentiates species is the most common method of discrimination in nucleotide variation; other methods depend on chromosome restriction". "Fluorescence in situ hybridization (FISH) (Loge et al. 1999; Moter and Gobel 2000; Baudart et al. 2002; Rompre et al. 2002) and filter hybridization (Polz and Cavanaugh 1997; Jiang and Fu 2001) and polymerase chain response (PCR) are hybridization-based approaches" (von Wintzingerode et al. 1997; Polz and Cavanaugh 1998).

5.3 *Immunological Methods*

"Immunological profiling has been used to identify and, in some cases, enumerate pathogenic populations in waste water samples. These methods are focused on the innate susceptibility of immune reactions and are typically aimed at pathogen-specific antigens such as lipopolysaccharides (LPS) on cell walls, membrane and flagellar proteins, or toxins. There are three types of immunoassays: enzyme-linked immunosorbent assay (ELISA), immunofluorescent microscopy, and agglutination assays" (Besnard et al. 2000; Bitton 2005).

6 Sources of Pathogens

"The contaminants most commonly present in sewage sludge and its derivatives are determined by the state of public health as well as the presence of hospitals, tanneries, meat-processing plants, and slaughterhouses in the same area. Foodborne bacteria are one of the most serious causes of water waste in developed countries. Public health authorities around the world emphasise that only a limited proportion (10–20% of all food and waterborne disease outbreaks are actually reported, implying that the incidence of these diseases is likely to be far greater than seen in observational research. It has been calculated by the World Health Organisation that only 10 percent of European outbreaks are registered" (Bruce and Davis 1983).

In view of the pervasive dissemination of undiagnosed and unreported infectious diseases arising in single homes, contaminated people are very likely to supply oro-fecal pathogens to sanitation systems. “Barker and Bloomfield (2000) found that after household *salmonellosis, Salmonella enteritidis* lived up to 4 weeks in biofilms in the home toilet. *Salmonella spp.* was isolated from below the waterline of the toilet bowl by the same scientists, up to 50 days after experimental seeding. These results demonstrate the significance of household events of gastroenteritis in pathogenic microorganisms for the long-term enrichment of sewage”. “Sadik et al. (1998) revealed that wastewater treatment workers were at high risk of infectious diseases being contracted. This research conducted on 242 employees of various wastewater treatment plants, showed that the occurrence of gastroenteritis and gastrointestinal symptoms among these workers had risen”.

For employees exposed to water, the high prevalence of zooparasites in sewage and their low minimum infective doses are likely to pose a health danger. “Schlosser et al. (1999) found an overall intestinal zooparasite carriage of 11.8 percent after analysing 126 waste water employees in Paris. Four zooparasites were found: *Trichiurus sp, G. lamblia, Entamoeba coli,* and *Endolimax nanus*, respectively”.

7 Treatment of Wastewater

The disposal of water is a method in which the contaminants are separated. The ultimate aim of the treatment of waste is to create an effluent that does not harm the environment. The findings can be catastrophic in the absence of sewage treatment, as sewage may disrupt the ecosystem. Primary, secondary and tertiary treatment are the common water treatment systems. Primary disposal requires physical sorting of sewage using a settling basin into solids and liquid. The liquid waste is moved to secondary storage, which relies on the use of micro-organisms to extract the dissolved biological compound. To degrade the biological matter in the liquid sludge, micro-organisms typically use aerobic metabolism. In order to clean the waste, tertiary treatment is then needed so that it can be discharged into the environment.

7.1 Procedures

Sewage sludge has to be stabilised prior to land use (Fig. 1). “The stabilisation procedure typically reduces organic matter and water content, as well as the emission of unwanted odours and pathogenic microorganism concentrations. Stabilization will produce either an end product with pathogens below detection limits or sludge with reduced but detectable pathogen concentrations” (Straub et al. 1993). Common stabilisation strategies include anaerobic and aerobic digestion, lime stabilisation,

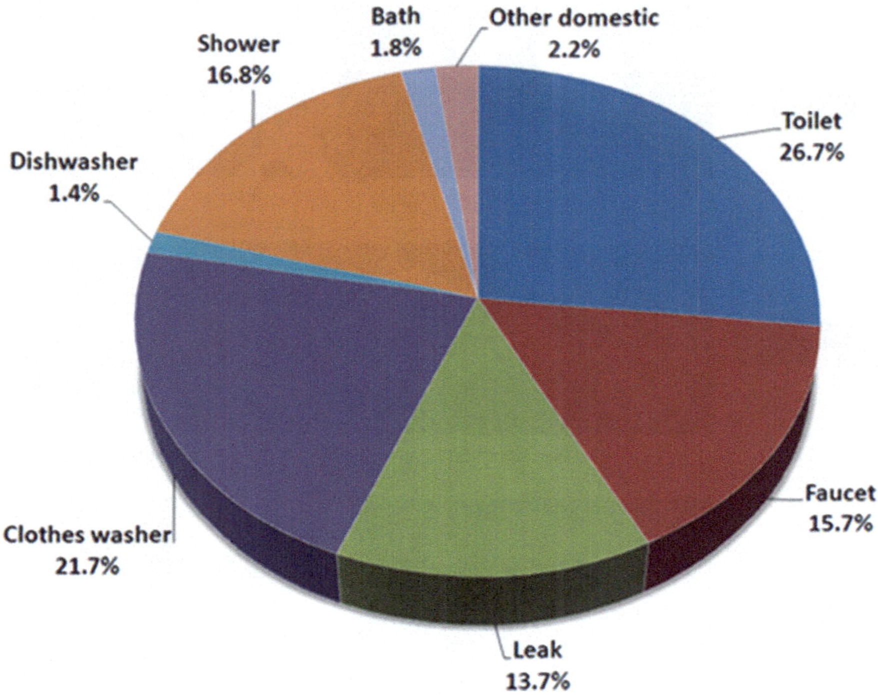

Fig. 1 Sewage Composition in Urbanized cities. (Source: This figure was obtained from Residential End Uses of Water (1999) Water Research Foundation)

composting, and heat drying. These pathways differ greatly in their capacity to reduce pathogenic microbial content in sewage sludge (Fig. 2).

7.2 *Anaerobic and Aerobic Digestion*

"Anaerobic treatment or biodegradation produces small concentrations of hydrogen, carbon dioxide, and a host of other contaminants, as well as heat and a stabilised sludge end product of a higher nitrogen content than aerobic digestion. Aerobic digestion produces small volumes of carbon dioxide, ammonia, and other contaminants, as well as vast amounts of heat and a final sludge substance. About the fact that anaerobically digested sewage sludge has higher nitrogen concentrations, after 16 weeks of incubation at 30 °C, aerobically digested sludge has higher N mineralization rates (19 to 50 percent and 16 to 41 percent, respectively)" (Douglas and Magdoff 1991). "Carbon deficiency in anaerobic sludge can explain the result and lead to insufficient C to mineralize the soil N for the decomposition of microbial biomass. Variations in N mineralization may also affect specific groups or compounds present in sludge, such as polyphenols (which prolong mineralization by

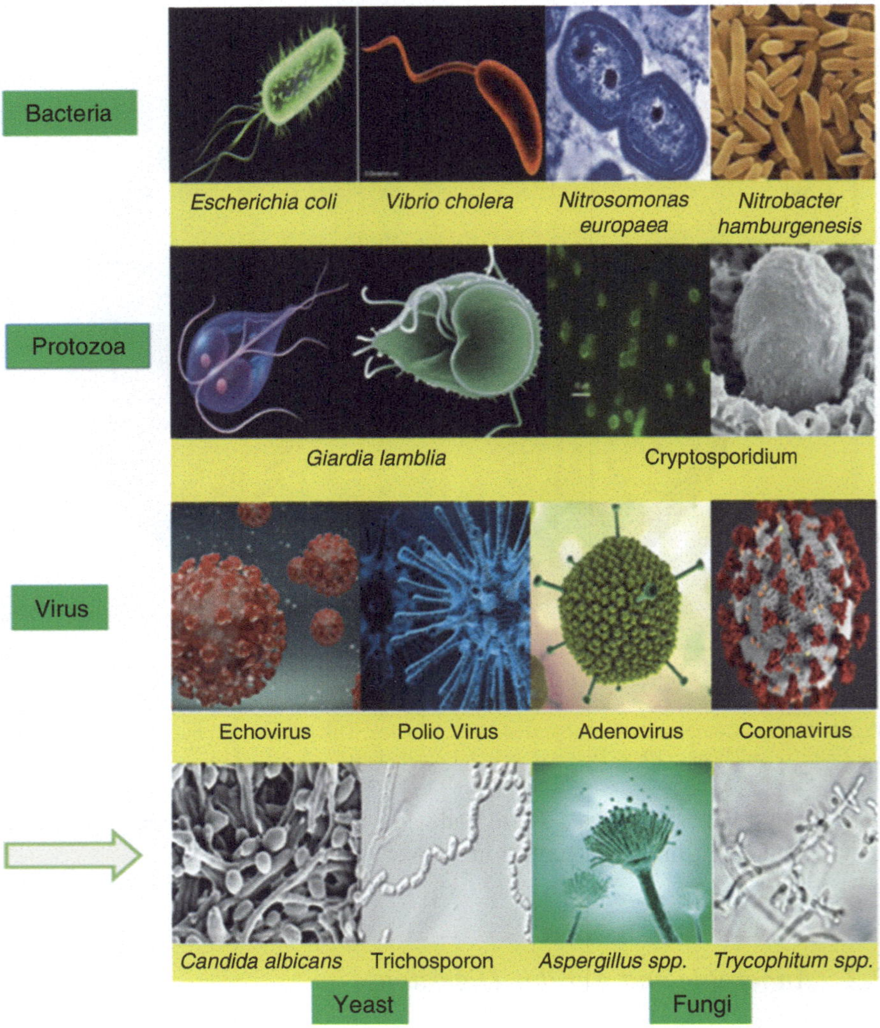

Fig. 2 The most important microorganisms in wastewater and sewage waste in urban areas. (Source: This figure was complied for this manuscript from open source Google links. https://www.slideshare.net/vaishali789/use-of-microorganisms-in-wastewater-treatment)

binding to N in proteins), soluble carbohydrates, or soil water content" (Cabrera et al. 2005).

The digestion process is either mesophilic (30–38 °C) or thermophilic, which is an important inactivation parameter (50–60 °C). Indeed, most bacteria are inactivated during heat exposure, since they are significantly above the optimal growth temperature and the duration of exposure is sufficient. Thermophilic waste treatment is clearly more useful in that the levels of vegetative pathogens and intestinal parasites than the mesophilic option (Fig. 3).

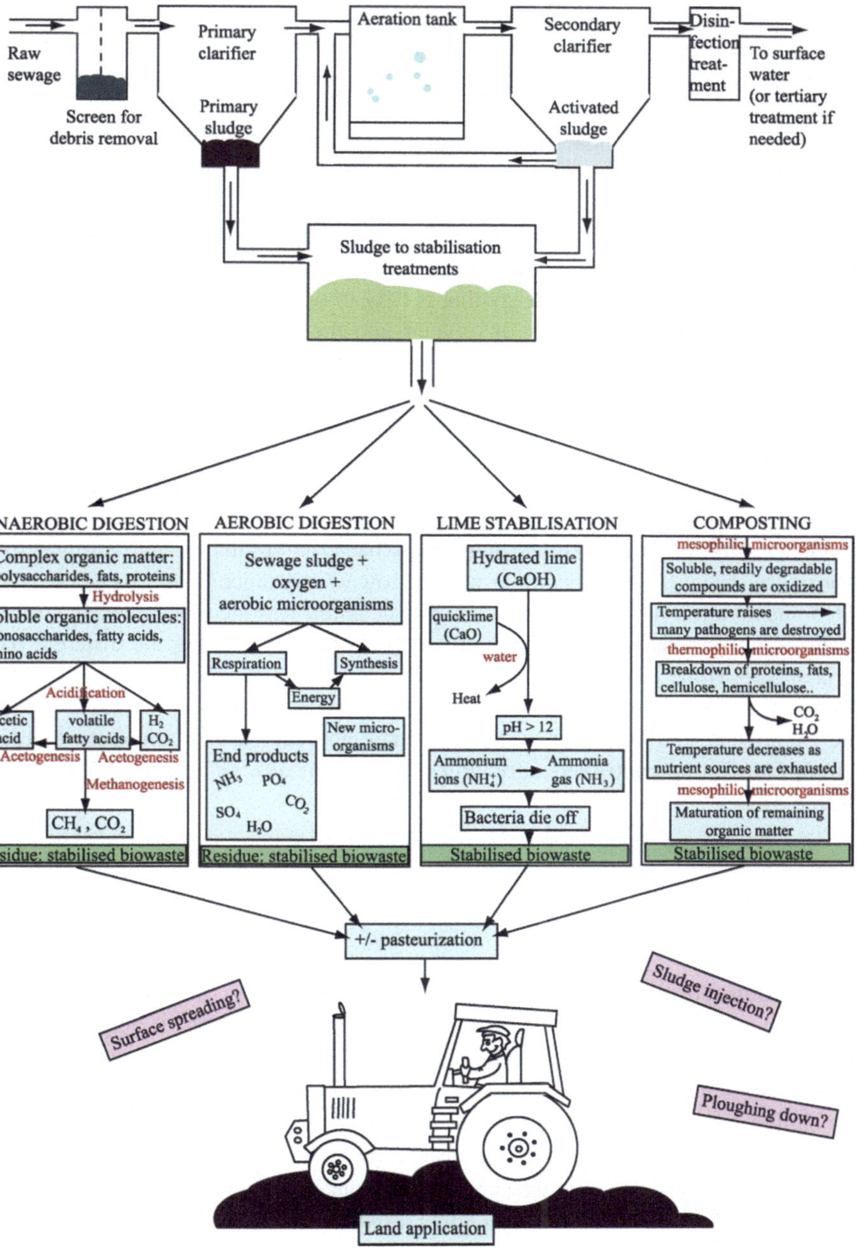

Fig. 3 Development of a Waste Sludge Flow Diagram showing potential measures for a high microbial content end product for use as a crop fertiliser. (Source: This figure used for manuscript was obtained from Veronica Arthurson 2008)

For example, "Olsen and Hammack (2000) showed that in a thermophilic anaerobic digester, *Salmonella enterica serovar Typhimurium* and *Mycobacterium paratuberculosis* are inactivated within 24 h, but the process in its mesophilic counterpart takes weeks and months.

A major benefit of anaerobic digestion over aerobic digestion is that as end products, methane and carbon dioxide (biogas) are produced, thereby providing the treatment facility with energy needs. Typically, biogas contains about 60 to 70 percent methane, 30 to 40 percent carbon dioxide, and minor quantities of other gases, including ammonia, hydrogen sulphide, and mercaptans, rendering it an incredibly useful gas rich in energy that is easy to obtain. In comparison, anaerobic digestion needs no input of air or oxygen into the system, which is highly cost-effective in contrast to oxygen-requiring sludge treatment methods" (Strauch 1991, 1998).

"Kearney et al. (1994) used mesophilic anaerobic digestion to determine the persistence of pathogenic bacteria in animal waste. The group observes that there have been decreases in viables *E. coli*, *S. enterica serovar Typhimurium*, *Yersinia enterocolitica*, *Listeria monocytogenes* and *C. jejuni* during their treatment and that, along with the previous results, indigenous bacterial strains remain stronger than laboratory strains. *Y. enterocolitica* was the lowest resistance to anaerobic digestion (90% inactivation period, 18.2 days). While *C. jejuni* was the most resistant species, it showed discrepancies in susceptibilities of disinfection among the different bacteria species (time required to inactivate 90 percent of the population, 438.6 days). These results are consistent with the findings published in Jepsen et al. (1997) that the pathogens and indicator organisms may not decrease to amounts that are appropriate for uncontrolled agricultural use by means of aerobic stabilisation" (Table 3).

7.3 Chemical Treatment

"Lime stabilisation is an interesting alternative to anaerobic and aerobic digestion, largely because of its cost efficient and functional existence" (Czechowski and Marcinkowski 2006). Hydrated lime is added in liquid waste sludge (calcium hydroxide) at a level adequate to raise pH to 12.0 for at least 2 h. The NH4 ions of the sludge are deprotonated at pH 12.0 to release ammonia gas that bactericidally acts via cell membranes of microorganisms. The elevated pH and ammonia mixture reduces the coliform bacteria by 2–7 orders of magnitude and the presence of faecal *straptococci* indicator bacteria to a small extent. Several trials have validated the need for the effective removal of *Salmonella* from sewage sludges of a robust pH of 12.0 over 20 to 60 day. Lime stabilisation was technically classified as a relatively time consuming treatment option. By contrast, Strauch confirmed that "*Salmonella* had been removed within 24 h at a steady pH of 10. The investigator found that their removal relies on a collected pH, liming period and dryness of the sludge". Bina et al. (2004) demonstrated, in line with previous results, that "the microbial content

Table 3 Pathogen-reduction of non-biological and biological sludge treatment performance

Type of Treatment		Sanitation Factor	Sanitizing Effect on				Product Stability
			Parasite Eggs	Spore	Bacteria	Viruses	
Non-biological							
Pasteurization		Heat, 30 min at 70 °C	Good	Poor	Good	Moderate	Poor
Irradiation		Ionizing radiation, 300 rad	Moderate/ good	Poor	Good	Poor	Moderate
Lime treatment	Slaked lime	High pH	Moderate	–	Good	Moderate/ good	Good, if pH remains >10
	Quick lime	High pH, 80 °C	Good	–	Good	Good	Good, if pH remains >10
Biological							
Anaerobic digestion	Mesophilic (30–35 °C)			Poor	Poor	Poor	
	Thermophilic (50–55 °C)			Good	Moderate/ good	Moderate/ good	
Aerobic digestion	Mesophilic (up to 20 °C)			Poor	Poor	Poor	
	Termophilic (50–55 °C)			Good	Good	Good	
Composting (50–60 °C)				Good	Good	–	

Source: This table was constructed for this manuscript based on information obtained from Dumontet et al. 1999; Havelaar 1983

of sewage sludge met the criteria for class B at pH 12 within 2 h, while class A sludge was collected at the same pH for *Salmonella* and faecal coliforms after 2 and 24 h respectively. Quicklime (calcium oxide), which causes an exothermic reaction with water, is an alternative to hydrated lime. The heat release normally increases the sludge temperature to 70 °C, equivalent to that achieved during pasteurisation".

7.4 Composting

Fluid sludge shall be processed before composting by a bulking agent like timber pipes, dry manure or municipal waste. Indigenous microorganisms oxidise the usable substrates present in sludge in the compost pile, leading to an extreme rise in temperature, particularly in the middle of the pile (up to 60 °C and above). The temperature of the compost pile falls to the ambient until the nitrogen sources have been exhausted and the organic content of the sludge mineralized to CO_2 and H_2O or transmitted to humic compounds. Various independent composting methods exist and the results are not necessarily the same, but the effectiveness of the process in waste sludge for the elimination of human bacteria is not determinable. This method is thus unlikely. Temperature and time are, nevertheless, the major factors that regulate pathogens inactivation, showing the value of homogenising compost in both the central pile and edge of high temperatures. In addition, the microbial inactivation process is affected by other agents including ammonia, chemical constituents, solved solids and hydroxide anions. The reduction in the amount of pathogenic bacteria in the compost can also, in theory, rely on biological control (e.g. antagonism) or rivalry among the different bacteria in the pile. "Hussong et al. (1985) observed an increase in the amount of Salmonella in the sludge treated with irradiated compost in relation to the amount treated with unirradiated compost. The authors suggested that the changes in the salmonella level can be explained by competitive composting practises between microorganisms".

7.5 Pasteurization

"The best way of removal is for biowaste pasteurisation at 70 °C for at least 1 h (Bendixen 1999). For eg, *Salmonella* is killed within 30 minutes by sludge heated at 70 °C" (Bagge et al. 2005; Bendixen 1996; Mitscherlich and Marth 1984). Pasteurization can be used either before or during the usual stabilisation process to create a material suitable for use as a fertiliser (digestion, composting, or liming). However, traditional methods of pasteurisation do not remove bacterial endospores present in sewage sludge (*Clostridium spp.* and *Bacillus spp.*) and determine the likely occurrence of those bacteria until the application of the sludge area. For the removal of endospores, an expensive procedure needs at least two different rounds of pasteurisation. Spores can be activated by primary heating into vegetative stages, which will start germinating and expanding eventually. The secondary pasteurisation phase should kill these thermal bacteria, and the incubation period between the two pasteurisation stages should be brief enough to prevent the formation of new endospores. Alternatively, "the irradiation technology of waste sludge can be more studied in order to determine whether it is an adequate solution for the elimination of pathogenic bacteria and bacterial endospores" (Cuba et al. 2003a, b). Most bacteria forming endospores are still soil endemic, so more research is required to determine whether application of treated sewage sludge could pose a higher risk.

Pasteurisation is a major choice in general for sanitization, but the procedure does not kill bacterial endospores. Regardless of the effectiveness of sanitization, it is crucial to remember the fundamental microbial principles in order to mitigate bacterial recontamination and growth while organising and handling biological products like wastewater sludge. Furthermore, pasteurisation is a detriment to costs. In particular, the heating stage usually occurs with a steam or a heat exchanger and needs considerable energy, which further illustrates the benefits of using anaerobic digestion in combination with pasteurisation, which results in energy-rich output biogas in the same plant.

8 Conclusion

Environmental protection and biodiversity are recognised at the greatest concern levels that need important worldwide intervention because it is necessary to move forward into the future. In order to ensure health, waste control, NR and ecosystem protection and toxic and pollutant treatment are the main areas that also need to be prioritised. Today, protecting the environment from deterioration isn't just the elimination of contaminants and toxins, but also the reusing and recycling of harmful substances by converting various waste into a prosperous environment.

Useful things in an aesthetic and eco-friendly way. In the current circumstances there has been an increasing trend in using certain microorganisms and a focus in terms of community seeking sustainable ways to clean up polluted habitats and waste. The potential of microorganisms with the emergence of biotechnology, increased emphasis and analysis has been given to chosen uses. In nature, microbes are specific and sometimes unpredictable. As an important agent for addressing certain environmental issues, microorganisms may be used. The experimental and reliable application of microbes is inevitably unfeeling and explores a fast development of research and novel tools to help protect our world and existing biological World Cup and environmental management methods. Finally, the use of microorganisms and microbiological techniques, especially in the form of atmosphere and other important environmental issues, has opened up new perspectives into the field of sustainable development.

References

Ashbolt NJ, Grabow WOK, Snozzi M (2001) Indicators of microbial water quality. In: Fewtrer L, Bartram J (eds) WHO water quality guidelines, standards and health. IWA Publishing, London, pp 289–316. https://www.dora.lib4ri.ch/eawag/islandora/object/eawag:11261

Bagge E, Sahlstrom L, Albihn A (2005) The effect of hygienic treatment on the microbial flora of biowaste at biogas plants. Water Res 39:4879–4886. https://doi.org/10.1016/j.watres.2005.03.016

Barker J, Bloomfield SF (2000) Survival of salmonella in bathrooms and toilets in domestic homes following salmonellosis. J Appl Microbiol 89:137–144. https://doi.org/10.1046/j.1365-2672.2000.01091.x

Baudart J, Coallier J, Laurent P, Prevost M (2002) Rapid and sensitive enumeration of viable diluted cells of members of the family enterobacteriaceae in freshwater and drinking water. Appl Environ Microbiol 68:5057–5063. https://doi.org/10.1128/AEM.68.10.5057-5063.2002

Bendixen HJ (1996) Hygienic recommendations and reduction of pathogenic microorganisms in Danish biogas plants. Dansk Vet Tidskr 11:475–482. https://www.osti.gov/etdeweb/biblio/579971

Bendixen HJ (1999) Hygienic safety—results of scientific investigations in Denmark. In Proceedings of the IEA bioenergy workshop, Hohenheim, Germany, p 27–47. Deutsche Veterinarmedizinische Gesellschaft, Giessen, Germany

Besnard V, Federighi M, Cappelier JM (2000) Evidence of viable but non-culturable state in *Listeria monocytogenes* by direct viable count and CTC-DAPI double staining. Food Microbiol 17:697–704. https://doi.org/10.1006/fmic.2000.0366

Bina B, Movahedian H, Kord I (2004) The effect of lime stabilization on the microbiological quality of sewage sludge. Iranian J Environ Health Sci Eng 1:34–38. https://www.semanticscholar.org/paper/The-Effect-of-Lime-Stabilization-on-the-Quality-of-Bina-Attar/693f2f871bf83ee83cd1f031881bc5ac883cd815

Bitton G (2005) Microbial indicators of faecal contamination: application to microbial source tracking. Report submitted to the Florida Stormwater Association 719 East Park Avenue, Tallahassee, 32301. https://www.semanticscholar.org/paper/MICROBIAL-INDICATORS-OF-FECAL-CONTAMINATION%3A-TO-Bitton/908cdc8c238477dc4317c0286d799702a4b4044d

Boyle M (1990) Biodegradation of land-applied sludge. J Environ Qual 19:640–644. https://doi.org/10.2134/jeq1990.00472425001900040002x

Bruce AM, Davis RD (1983) Utilization of sewage sludge in agriculture. Maximizing benefits and minimizing risks. In Zucconi F, De Bertoldi M, Coppola S (eds) Biological reclamation and land utilization of urban wastes, Proceedings of an International Symposium, Naples, Italy, October 11–14, 1983; pp 102–111. https://doi.org/10.1080/10473289.2001.10464313

Cabrera ML, Kissel DE, Vigil MF (2005) Nitrogen mineralization from organic residues: research opportunities. J Environ Qual 34:75–79. https://doi.org/10.2134/jeq2005.0075

Caccio SM, DeGiocome M, Aulicino F, Edoardo P (2003) Giardia cysts in wastewater treatment plants in Italy. Appl Environ Microbiol 69:3393–3398. https://doi.org/10.1128/aem.69.6.3393-3398.2003

Chow VT, Eliason R, Linsley RK (1972) Development and trends in wastewater Engineering. In Wastewater Engineering. McGraw-Hill Book Company, New York, pp 1–11. https://www.ijert.org/review-on-waste-water-treatment-technologies

Cooper PF (2001) Historical aspects of wastewater treatment. In Decentralized sanitation and reuse concepts, systems and Implementation. Edited by Lens P, Zeeman G, Lettinga G. IWAPublishing. London (UK), pp 11–38. https://www.ircwash.org/resources/decentralised-sanitation-and-reuse-concepts-systems-and-implementation

Cuba V, Pospisil M, Mucka V (2003a) Electron beam/biological processing of anaerobic and aerobic sludge. Czechoslov J Phys 53:369–374. http://biospraysehatalami.com/stabilized-sewage-sludge-intended-for-arable-land-make-use-of-needs/

Cuba V, Pospisil M, Mucka V, Jenicek P, Silber R, Dohányos M, Zábranska J (2003b) Impact of accelerated electrons on activating process and foaming potential of sludge. Rad Phys Chem 67:545–548. https://tvp.vscht.cz/department/19015?reset_infoboxy

Czechowski F, Marcinkowski T (2006) Sewage sludge stabilisation with calcium hydroxide: effect on physicochemical properties and molecular composition. Water Res 40:1895–1905. https://doi.org/10.1016/j.watres.2006.02.023

Day D (1996) How Australian social 'policy neglects environments'. Aust J Soil Water Conserv 9:3–9. https://eurekamag.com/research/002/860/002860696.php

Doorn MRJ, Towprayoon S, Maria S, Vieira M, Irving W, Palmer C, Pipatti, R, Wang C (2006) Wastewater treatment and discharge. In 2006 IPCC Guidelines for National Greenhouse Gas Inventories volume 5. (Waste). WMO, UNEP. pp 6: 1–6, 28. https://www.ipcc-nggip.iges.or.jp/public/2006gl/pdf/5_Volume5/V5_6_Ch6_Wastewater.pdf

Douglas BF, Magdoff FR (1991) An evaluation of nitrogen mineralization indices for organic residues. J Environ Qual 20:368–372. https://doi.org/10.2134/jeq1991.00472425002000020006x

Dumontet S, Dinel H, Baloda S (1999) Pathogen reduction in sewage sludge by composting and other biological treatments: a review. Biol Hort Agric 16(4):409–430. https://doi.org/10.1080/01448765.1999.9755243

Feenstra S, Hussain R, van der Hoek W (2000) Health risk of irrigation with untreated urban wastewater in the Southern Punjab, Pakistan. International Water Management Institute (IWMI) report No. 107. https://www.researchgate.net/publication/228332160_Health_Risks_of_Irrigation_with_Untreated_Urban_Wastewater_in_the_Southern_Punjab_Pakistan

Ferguson CM, Coote BG, Ashbolt NJ, Stevenson IM (1996) Relationships between indicators, pathogens and water quality in an estuarine system. Water Res 30:2045–2054. https://agris.fao.org/agris-search/search.do?recordID=GB9713089

Gomez M, de la Rua A, Garralon G, Plaza F, Hontoria E, Gomez MA (2006) Urban wastewater disinfection by filteration technologies. Desalin 190:16–28. https://doi.org/10.1016/j.desal.2005.07.014

Havelaar AH, Oosterom H, Notermans S, van Knapen F (1983) Hygienic aspects of the application of sewage sludge to land. In Zucconi F, De Bertoldi M, Coppola S (eds) Biological reclamation and land utilisation of urban wastes; Proceedings of an International Symposium, Naples, Italy, October 11–14, 1983; pp 167–200

Hijnen WAM, van der Veer AJ, Beerendonk EF, Medema GJ (2004) Increased resistance of environmental anaerobic spores to inactivation by UV. Water Sci Technol Water Suppl 4 (2):54–61. https://dspace.library.uu.nl/handle/1874/35333

Hussong D, Burge WD, Enkiri NK (1985) Occurrence, growth, and suppression of salmonellae in composted sewage sludge. Appl Environ Microbiol 50:887–893. https://doi.org/10.1128/AEM.50.4.887-893.1985

Jepsen S-E, Krause M, Gruttner H (1997) Reduction of fecal streptococcus and Salmonella by selected treatment methods for sludge and organic waste. Water Sci Technol 36:203–210. https://agris.fao.org/agris-search/search.do?recordID=GB1997039529

Kearney TE, Larkin MJ, Levett PN (1994) Metabolic activity of pathogenic bacteria during semicontinuous anaerobic digestion. Appl Environ Microbiol 60:3647–3652. https://www.ncbi.nlm.nih.gov/pmc/articles/PMC201868/

Kreader CA (1995) Design and evaluation of Bacteroides DNA probes for the specific detection of human faecal pollution. Appl Environ Microbiol 62:3259–3264. https://doi.org/10.1128/AEM.61.4.1171-1179.1995

LeChevallier MW, Au K (2004) Inactivation (disinfection) processes. In Water Treatment and Pathogen Control. IWA Publishing, pp 41–65

Leclerc H, Edberg S, Pierza V, Delattre JM (2000) Bacteriophages as indicators of enteric viruses and public health risk in groundwaters. J Appl Microbiol 88:5–21. https://doi.org/10.1046/j.1365-2672.2000.00949.x

Liang JL, Dziuban EJ, Craun GF, Hill V, Moore MR, Gelting RJ, Calderon RL, Beach MJ, Roy SL (2006) Surveillance for waterborne disease and outbreaks associated with drinking water and water not intended for drinking. CDC/MMWR 55(SS12):31–58. https://www.cdc.gov/mmwr/preview/mmwrhtml/ss5512a4.htm

Loge FJ, Emerick RD, Thompson DE, Nelson DC, Darby JL (1999) Development of a fluorescent 16s rRNA oligonucleotide probe specific to the family Enterobacteriaceae. Water Environ Res 71:75–83

Milner PD, Marsh PB, Snowden RB, Parr JF (1977) Occurrence of aspergillus fumigatus during composting of sewage sludge. Appl Environ Microbiol 34:765–772. https://www.ncbi.nlm.nih.gov/pmc/articles/PMC242745/

Mitscherlich E, Marth EH (1984) Microbial survival in the environment. Springer, Berlin. https://www.springer.com/gp/book/9783642699764

Moter A, Gobel UB (2000) Fluorescence in situ hybridization (FISH) for direct visualization of microorganisms. J Microbiol Method 41:85–112. https://doi.org/10.1016/s0167-7012(00)00152-4

Olsen AR, Hammack TS (2000) Isolation of Salmonella spp. from the Housefly, Musca domestica L., and the Dump Fly, Hydrotaea aenescens (Wiedemann) (diptera muscidae), at Caged-Layer Houses. J Food Prot 63:958–960

Persing DH, Tenover FC, Versalovic J, Tang YW, Unger ER, Relman DA, White TJE (2003) Molecular microbiology: diagnostic principles and practice. Am Soc Microbiol, Washington, DC, pp 748. https://www.journals.uchicago.edu/doi/pdf/10.1086/428262

Polz MF, Cavanaugh CM (1997) A simple method for quantification of uncultured microorganisms in the environment based on in vitro transcription of 16S rRNA. Appl Environ Microbiol 63:1028–1033. https://www.ncbi.nlm.nih.gov/pmc/articles/PMC168394/

Polz MF, Cavanaugh CM (1998) Bias in template-to-product ratios in multitemplate PCR. Appl Environ Microbiol 64:3724–3730. https://www.ncbi.nlm.nih.gov/pmc/articles/PMC106531/

Rompre A, Servais P, Baudart J, de Roubin MR, Laurent P (2002) Detection and enumeration of coliforms in drinking water: current methods and emerging approaches. J Microbiol Methods 49:31–54. https://pubmed.ncbi.nlm.nih.gov/11777581/

Sadik AK, Tammy A, Bisesi MS, Schaub EA (1998) Prevalence of infectious diseases and associated symptoms in wastewater treatment workers. Am J Ind Med 33:571–577

Schlosser O, Grall D, Laurenceau MN (1999) Intestinal parasite carriage in workers exposed to sewage. Eur J Epidemiol 15:261–265. https://link.springer.com/article/10.1023/A:1007535426462

Simpson E, Charles K (2000) The health threat posed to surrounding community by effluent discharged from rural hospital sewage treatment plants. In Proceeding of the Water Institute of Southern Africa Conference, 28 May – 1 June, Sun City, South Africa

Smith DM, Roebeling PC, Webster AJ, Kragt M, Bohnet I, Zull A (2001) Assessment of the socio-economic impacts of management options for improving water quality in Douglas shire. CSIRO Sustainable Ecosystems. Alice Springs, Atherton, Brisbane, Canberra, Darwin, Perth, Toowoomba, Townsville

Stentiford EI (1983) Recent development in composting. In: De Bertoldi M, Ferranti MP, L'Hermite P, Zucconi F (eds) Compost: production, quality and use. Elsevier Applied Science, New York, pp 52–60. https://doi.org/10.1016/0961-9534(92)90024-K

Straub TM, Pepper IL, Gerba CP (1993) Hazards from pathogenic microorganisms in land-disposed sewage sludge. Rev Environ Contam Toxicol 132:55–91. https://doi.org/10.1007/978-1-4684-7065-9_3

Strauch D (1991) Microbiological treatment of municipal sewage waste and refuse as a means of disinfection prior to recycling in agriculture. Stud Environ Sci 42:121–136

Strauch D (1998) Pathogenic microorganisms in sludge. Anaerobic digestion and disinfection methods to make sludge usable as a fertilizer. Eur Water Manage 1:12–26. https://ec.europa.eu/environment/archives/waste/sludge/pdf/sludge_eval.pdf

Tanji Y, Mizoguchi K, Yoichi M, Morita M, Hori K, Unno H (2002) Fate of coliphage in wastewater treatment process. J Biosci Bioeng 94:172–174. https://pubmed.ncbi.nlm.nih.gov/12523767/

Toze S (1997) Microbial pathogens in wastewater. CSIRO Land and Water Technical report 1/97. https://publications.csiro.au/rpr/download?pid=procite:3effd8f9-a923-4ca9-a42a-9e6ce63ec1a6&dsid=DS1

Tree JA, Adams MR, Lees DN (2003) Chlorination of indicator bacteria and viruses in primary sewage effluent. Appl Environ Microbiol 69:2038–2043. https://doi.org/10.1128/AEM.69.4.2038-2043.2003

United Nations Commission on Sustainable Development (1997) Comprehensive assessment of the fresh water resources of the world: a report of the Secretary-General. www.un.org/dpcsd/dsd/freshwat.htm

Veronica Arthurson (2008) Proper Sanitization of Sewage Sludge: a Critical Issue for a Sustainable Society, Applied and Environmental Microbiology Aug 2008, 74 (17) 5267–5275, https://doi.org/10.1128/AEM.00438-08

Welch EB (1992) Ecological effects of wastewater: applied limnology and pollutant effects. Chapman and Hall, New York https://doi.org/10.4324/9780203038499

Wintzingerode F, Gobel UB, Stackebrandt E (1997) Determination of microbial diversity in environmental samples: Pitfalls of PCR-based rRNA analysis. FEMS Microbiol Rev 21:213–229. https://pubmed.ncbi.nlm.nih.gov/9451814/

World Health Organization (1998) Cholera in 1997. Wkly Epidemiol Rec 73:201–208. https://www.who.int/cholera/statistics/en/

Sewage Sludge and Its Health Risk Assessment: Opportunities and Challenges

L. Devarishi Sharma, Indira Sarangthem, Rojeet Thangjam, Rahul Sadhukhan, Nivedita Oinam, Birjeet Yanglem, Laikhuram Banarjee, Hrangbung Jurist Anal, and Hanuman Singh Jatav

1 Introduction

It is a traditional practice to apply sewage sludge in soil, wastewater in soil, and excreta in most countries' soil. Many civilizations in Europe, the Mediterranean, and Asian countries applied excreta of human as well as animal in the soil as fertilizers and manures. To cite an example, we may indicate the use of treated wastewater between the 14th and the fifteenth century in Milanese Marcites and Valenian Huertas (Soulié and Tréméa 1991). Most of the cities of North American and

L. D. Sharma (✉)
Central Agricultural University, Imphal, India

I. Sarangthem
College of Agriculture, Imphal, Manipur, India

R. Thangjam
College of Horticulture, Thenzawl, Mizoram, India

R. Sadhukhan
Multi-Technology Testing and Vocational Training Centre, Selesih, Mizoram, India

N. Oinam
All India Coordinated Research Project (MSPE)-Imphal Centre, Imphal, Manipur, India

B. Yanglem
Agricultural Technology Management Agency, Imphal East, Manipur, India

L. Banarjee
ARYA Project, KVK, Hengbung, Senapati, Manipur, India

H. J. Anal
Department of Agricultural Chemistry & Soil Science, School of Agricultural Sciences & Rural Development, Nagaland University, Medziphema, Nagaland, India

H. S. Jatav
Department of Soil Science and Agriculture Chemistry, Sri Karan Narendra Agriculture University, Jobner, Rajasthan, India

V. D. Rajput et al. (eds.), *Sustainable Management and Utilization of Sewage Sludge*, https://doi.org/10.1007/978-3-030-85226-9_10

European nations use to dispose of the wastewater in agricultural lands to avoid the water bodies get contaminated because technologies of wastewater treatment were not yet invented (Asano et al. 2007).

In developing countries in Asia, North America, the gulf, etc., wastewater and sewage sludge are used for decades as a fertilizer source for the crops. Thus, it is evident that wastewater and sewage sludge are used as fertilizers and as a source of irrigation water in agriculture. However, developed countries have now adopted the improved technologies to treat the wastewater, and in view of the environmental and risk of the health hazard, they stop using untreated water and sewage sludge as crop nutrients in agriculture. On the contrary, the developing countries are still using untreated wastewater and sewage sludge in agriculture despite its adverse impacts on crops and the general population's health. On the basis of the available data in respect of irrigated agricultural lands, an area of 4–6 M ha approximately in the world is irrigated with wastewater for irrigation (Keraita et al. 2008). WHO (2006) estimated that 7% of the agricultural land of the world are irrigated with wastewater. Treated sewage sludge is applied on 10% of the irrigated area as compared to the area irrigated with untreated wastewater. The dose of wastewater and its application in agriculture is different from region to region. The developing countries consisting of 75% of the world's total irrigated agricultural lands are still using untreated wastewater as fertilizers and as a source of irrigation water in agriculture. However, in developed countries, the practice has become unpopular due to environmental and risk of health hazard (Jiménez and Asano 2008). They have also indicated in their review report that 46 (forty-six) countries of the world apply untreated wastewater for irrigation (Fig. 1).

WHO (2006) has published statistics concerning the effects on the use of wastewater after treatment and untreated wastewater on Gross Domestic Product and the

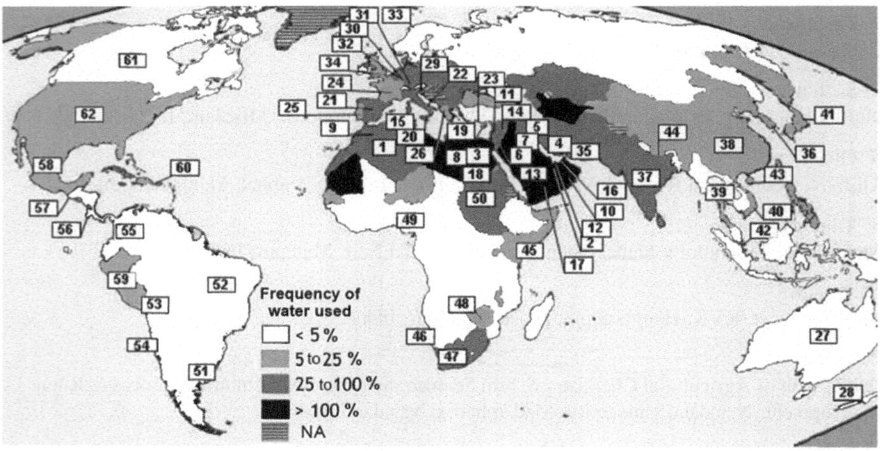

Fig. 1 Application of untreated wastewater for irrigation in various regions (Source: Jiménez and Asano 2008)

Table 1 Statistics of wastewater application in various regions for irrigation purpose

Sl. No.	Types of wastewater	Number of countries	Coverage of Sanitation	Gross Domestic/Capita (US$)
1	Untreated	23	15–65	880–4800
2	Treated and Untreated	20	41–91	1170–7800
3	Treated	20	87–100	4313–19,800

Source: WHO (2006)

improvement in sanitation (Table 1). A range between 15% and 65% of sanitation has been covered in 23 countries where untreated water is used for irrigation in agriculture, and a range between 41% and 91% of sanitation has been covered in another 20 countries which use both treated and untreated water as a source of irrigation water in agriculture whereas a range between 87% and 100% of sanitation has been covered in 20 countries where treated water is used for irrigation in agriculture.

In view of the above facts, we may conclude that due to the upsurge in population in the world has increased the production of sewage sludge, and its application in the soil is one of the options for disposal. Lack of scientific knowledge and injudicious use of sewage and sludge water in agriculture may cause several diseases like ascariasis, diarrhoea, cholera, etc.

2 Sewage Sludge

The liquid waste, which is released by houses or societies, is called sewage, whereas the sludge consists of solid waste and liquid waste generated at the time of treatment of sewage. The reutilisation of sewage sludge for cultivation is the most important option for its control and disposal adopted by the whole world (Kacprzak et al. 2017).

As per the Environment Protection Act (2001), sludge refers to

(a) The secondary product released by the treatment plants of sewage during the period when treatment of urban or domestic wastewater is carried out and other wastewater treatment plants having the same compositions in respect of urban or domestic wastewater.
(b) Secondary products released by septic tanks and similar wastewater treatment plants.
(c) Secondary products released by the treatment plants of sewage water except for the byproducts that are indicated in (a) and (b) above.

3 Sewage Sludge and Its Implication on Human Health

Exposure to sewage sludge may cause different infectious diseases to human health, but the main reasons are (i) direct contact of sewage sludge in agricultural lands by the farmers and (ii) consumption of fruits, grains, vegetables, etc., that are grown in sewage applied agricultural lands. The workers working in the sewage treatment plant are vulnerable to many infectious diseases because of their direct contact to untreated wastewater. Such occupation falls under the category of highly hazardous occupation. The other possible ways of spreading infectious diseases could be (i) peoples who reside near the sewage treatment plant, (ii) children playing near the contaminated ponds or lakes, (iii) the peoples who reside near the waste dumping site. Another way for infection of diseases is handling of sludge improperly and lack of proper hand washing and sanitization. Farmers who apply sludge in their field come to direct exposure to sludge (Westrell et al. 2004). The farmers who apply sludges in their fields and the people who work in the sewage treatment plant are frequently exposed to infectious diseases owing to their contact with sewage sludge directly for a more extended period. The people who work for the dewatering of sludge and those who work for digested sludge settlement come under the highest risk of health hazard. Many sewage sludge pathogens are reduced due to the dilution and wastewater treatment by applying different techniques. Westrell et al. (2004) indicated that exposure to sewage sludge leads to the risk of infections through various pathogens to human being according to their exposure at a different level (Table 2).

Table no. 2 shows that exposure to the virus even for a single time causes the highest risk to human health compared to protozoa and bacteria because it is removed from the sewage sludge in a lesser quantity. According to the computations, as shown in Table 2 regarding the risk of infections through various pathogens, it is seen that *Giardia* poses a severe risk to human health as compared to *Cryptosporidium.* Westrell et al. (2004) also observed that the non-viral microorganisms

Table 2 Risk of infections through various pathogens to human health in many exposure levels

Levels of exposure	*Salmonella* ($\times 10^{-10}$ level)	EHEC ($\times 10^{-10}$ level)	*Cryptosporidium* ($\times 10^{-10}$ level)	*Giardia* ($\times 10^{-10}$ level)	*Adenovirus* ($\times 10^{-10}$ level)	*Rotavirus* ($\times 10^{-10}$ level)
1	3×10^5	6×10^6	2×10^6	1×10^7	2×10^9	9×10^8
2	1×10^6	2×10^7	9×10^6	4×10^7	1	1
3	2×10^4	3×10^5	3×10^5	3×10^6	1×10^9	5×10^8
4	6×10^2	1×10^4	1×10^4	1×10^5	4×10^7	2×10^7
5	5×10^{-1}	9	5×10^2	2×10^3	6×10^5	$1 \times 10+$
6	6×10^6	1×10^8	6×10^7	2×10^8	9×10^9	$4 \times 10+$
7	3×10^6	5×10^7	2×10^7	1×10^8	1	3×10^9
8	9×10^2	2×10^4	9×10^4	2×10^4	4×10^6	2×10^6

Source: Westrell et al. (2004)

Table 3 Infection of skin diseases to the farmers who cultivate the rice field with sewage as a source of irrigation in Nam Dinh, North Vietnam

	A community of My Trung		A community of My Tan	
Diseases Diagnosed	Number = 557	Percentage	Number = 546	Percentage
Urticaria	2	0.4	6	1.1
Fungal infections in nail	18	3.2	29	5.3
Fungal infections in Skin	18	3.2	93	17
Bacterial infections in Skin	27	4.8	31	5.7
Eczema	42	7.5	117	21.4

Source: Trang et al. (2007)

(<<1) are insignificantly infectious to human health, whereas the virus causes severe infections to the people who deal with sewage sludge directly. They also observed that eating vegetables grown on sewage sludge treated soils is comparatively lesser infectious to human health than those who deal with sewage sludge directly. As for society's perspective, the level of exposures, as shown in serial 1, serial 2, and serial 7 in Table 2, must be checked. Thorn and Kerekes (2001) observed that those workers working in the wastewater treatment plant are prone to gastrointestinal ailments, headaches, tiredness, respiratory ailments, and allergies in the skin and eyes. Trang et al. (2007) investigated about the farmers affected by skin diseases for their exposure to wastewater in Nam Dinh, North Vietnam. They carried out the study only in the areas where wastewater is used as a source of irrigation and witnessed that the skin diseases varied from community to community (Table 3).

Blumenthal et al. (2000) carried out a study and reported that *Ascaris* poses skin-related diseases among the farmers along with their families for their direct exposure to wastewater for irrigation and apply sludges as a source of fertilizer on their agricultural fields than the farmer community who use rainwater as a source of irrigation. The children are more prone to disease causes diarrhoea than the adult owing to weak immunity to fight the diseases. Wastewater needs to be stored for some time and then brought to use for different purposes to lower the infestation risk of Ascaris in both adults and children than the utilization of untreated wastewater.

4 Major Pathogens Associated with Human Ailments Contained in Sewage Sludge

Sewage sludge contains many pathogens of different microbes, namely virus, protozoa, helminth worms, and bacteria, etc. cause different types of diseases in the general population (Gerba and Smith 2005), as shown in Table 4.

Table 4 Primary Pathogens associated with many diseases that are present in sewage sludge

Symptoms of the concerned diseases	Pathogens associated with human ailments
Group-Bacteria	
Gastroenteritis	*Escherichia Coli*
Gastroenteritis	*Campylobacter jejuni*
Cholera	*Vibrio cholera*
Severe gastroenteritis such as abdominal pain, diarrhoea)	*Yersinia spp.*
Dysentery	*Shigella spp.*
Typhoid, Salmonellosis	*Salmonella*
Group-Virus	
Infection in the respiratory tract and Gastroenteritis	Adenoviruses
Fever, Pneumonia, Hepatitis, Meningitis	Coxsackievirus
Group-Protozoa	
Toxoplasmosis	*Toxoplasma gondii*
Dysentry, Diarrhoea	*Balantidium coli*
Cryptosporidiosis, gastroenteritis	*Cryptosporidium*
Cramps in Abdominal and diarrhoea	*Giardia lambia*
Severe enteritis	*Entamoebahistolytica*
Group- Helminth worms	
Taeniasis	*Hymmenolepsis*
Diseases associated with Hookworm	*Necatoramericanus*
Insomnia, anorexia, anxiety	*Taeniasasginata*
Stomach ache, fever, an ache in the muscle	*Toxocaracanis*
Diarrhoea, stomach pain, loss of body weight	*Trichuristrichiura*
Pain in chest and cough	*Ascarissum*
Problems in digestion, stomach pain	*Ascarislumbricodes*

Source: Gerba and Smith (2005)

5 Sewage Sludge Application and Its Impact on Plant Pathogens

Different studies on the application of sewage sludge and its impact on plant-pathogen were carried out in the world and reported that use of sewage sludge in the soil plays a vital role to suppress different pathogens as follows:

Pathogens	Plant
1. *Sclerotinia sclerotiorum*	Tomato
2. *Ralstonia solanacearum*	Bean
3. *Sclerotium rolfsii*	Bean
4. *Rhizoctonia solanin*	Radish
5. *Pythium spp.*	Cucumber

The microbes population increases due to sewage application, which contains an increased amount of organic matter, and subsequently, it suppresses different plant pathogens. Since the microbes population is increased due to the use of sewage sludge and as a result, it causes competition between the natural populations of microbes and pathogens for water, space, and nutrients, and this is how the population of microbes suppresses the pathogens. Sewage sludge application in the soil has an antagonistic impact on the pathogen, and thus it acts as an agent to control pathogens (Chen et al. 1986). At high-temperature, pathogens that are mesophilic in nature, unable to exist. Therefore the compost, which is made of sewage sludge, lowers the risk of plant health than the untreated sewage sludge application in the soils. Verticillium, which causes wilt in pepper plant, is affected by various causes such as (1) Application of compost made of sewage sludge at various doses in the field, (2) various methods applied for the treatment of sludge, and (3) various temperatures during the period of composting. Application of sewage sludge at an optimum dose increases the yield, area of the leaf, the height of the plant, dry matter of the leaf, dry matter of the shoot in different crops. However, sewage sludge application at a high dose decreases many field crop yield and growth. The application of sewage sludge in soils has negative impacts on the physico-chemical properties, namely pH, EC and nutrients' status. Sewage sludge application in higher doses causes soil phytotoxicity, heavy metal accumulation in plants, and soil salinity (Ghini et al. 2007). Composts made of sewage sludge incapable of suppressing *Phytophthora spp.* and *Pythium ultimum.* Cellulose and lignin-rich composting materials improve the microbes' diversity in sludge compared to anaerobic composting sludge. A substrate that contains a higher percentage of oxygen and having good aeration decreases the severity of diseases such as *Phytophthora spp.* which can overgrow below the oxygen level of 15%. Compost made of sewage sludge has a lower bulk density than the raw sewage sludge. Chemical properties like EC, pH, and biological properties are conducive medium for germination of seeds and the diversity of microbes are significantly increased when it is applied with composts made of sewage sludge (Leoni and Ghini 2006). They also observed an inverse relationship between electrical conductivity and *Phytophthora nicotianae* in the plants of tobacco.

It is advantageous to apply sewage sludge in agriculture beyond doubt, but it also poses serious health hazards. Mention may be made that besides others, sewage sludge contains organic pollutants, heavy metals, and pathogens (Smith 2009; Oleszczuk 2006; Harrison et al. 2006). The standard pathogen concentrations upto the maximum levels are shown in Table 5.

The permissible limit of heavy metals concentrations in the soil and also in sludge is shown in Table 6.

In Russia, agricultural utilisation of sewage sludge is monitored by the State All-Union Regulations (17.4.3.07–2001), and in order to use as fertilizers in agricultural fields, the sewage sludge to be used should have the permissible limits of heavy metals and arsenic concentrations as shown in Table 7.

Table 5 The standard pathogen concentrations up to the maximum levels

Country	Others pathogens	Salmonella
Finland	Contains *Escherichia coli*- less than 1000 Colony-forming unit	Nil in 25 gram
France	Contains Enterovirus of about 3 MPCN/ 10 g of Dry Matter Eggs of Helminth @ 3/10gm of DM	8 Most Probable Number per 10gm of Dry Matter
Hungary	It contains Faecal coli and also contains faecal streptococci (its number less than 10%)	Nil
Italy	Nil	Contains 1000 Most Probable Number per gm of Dry Matter
Luxembourg	It contains enterobacteria @ 100 per gm, and its eggs are not transmittant	Nil
Poland	Nil	The sludge that contains salmonella is banned from use in agriculture

Source: European Commission (2009)

Table 6 The permissible limits of heavy metals concentrations in the soil and also in sludge

	Soil			
Heavy metals	pH greater than 7	pH less than 6 to pH less than 7	pH less than 5 to pH less than 6	Sewage Sludge
Zinc	200	150	100	2500
Lead	100	70	70	750
Nickel	70	50	30	300
Mercury	1	0.5	0.1	10
Copper	100	50	30	1000
Chromium	100	75	50	1000
Cadmium	1.5	1	0.5	10

Source: European Commission (2009)

Table 7 The permissible limits of heavy metals and arsenic concentrations in the sludge

	Concentration in terms of ppm	
Heavy metals	1	2
Zinc	3500	1750
Lead	500	250
Nickel	400	200
Mercury	15	7.5
Arsenic	20	10
Copper	1500	750
Chromium	1000	500
Cadmium	30	15

Source: Delibacak et al. (2020)

Table 8 Standards for sanitary sewage sludge for irrigation

Pointer	Standards	
	Group 1	Group 2
Eggs of Geohelminth and intestinal cysts consisting of pathogenic protozoa (sample per kg)	–	–
Salmonella (Cell per gram)	–	–
Group of *E. coli* bacteria (cell per gram)	1000	100

Source: Delibacak et al. (2020)

The hygiene required for wastewater used in agriculture is regulated by Sanitary and Epidemiological rules (2.1.7.573–96) in Russia, and its standards are shown in Table 8.

Group 1and Group 2 are sewage sludge.
Group 1 type of sewage sludge is applied for industrial crops, cereals, and legumes.
Group 2 type of sewage sludge is applied for crops excluding strawberries, mushrooms, and vegetables.

Both the group (1 & 2) are utilized for landfills, correction of problematic soils, aesthetic nurseries, forest, and industrial floriculture. Effluents of industries and rainwater are often run from roads to the sewage system, and as a result of which sewage sludge becomes toxic and may contain materials of organic origin as well. Many detrimental toxins may also be found in the sewage sludge-like pesticides, detergents, and various salts owing to the dumping of municipal effluents, effluents of industries, and lethal organic materials (Sommers et al. 1976). The diversity of lethal organic materials is significant and it also relates to the diverse impacts of toxins on the health of human being, i.e., mutagen, the effect of carcinogens etc. (Singh and Agrawal 2008). Modern-day advanced techniques for analysis of sewage sludge can recognize various new lethal organic materials in sewage sludge (Clarke and Smith 2011; Davis et al. 2012; Müller et al. 2006). In order to identify these lethal organic contaminants from the sewage sludge, competent professional instruments will be required. The information on lethal organic pollutants in sewage sludge is essential to prevent from dangerous health hazards. Under the circumstances, we must conduct biological tests to identify probable threats. These tests can also identify the probable interactions among specific contaminants, i.e., antagonistic and synergistic effects.

It is imperative to test the sewage sludge to detect phytotoxicity because it is often used in agriculture. Domene et al. (2010) observed in their study that the toxicity of the sewage sludge is significantly affected by different soil type, and by the arrangement of solid particles and pore space during the time of estimation (Domene et al. 2008). Suchkova et al. (2010) observed that plant species which are grown on sewage sludge applied soil affect the sewage sludge phytotoxicity.

6 Organic Pollutants

Guo et al. (2009) observed a high concentration of heavy metals toxicity and pollutant of organic origin in the sewage sludge. Industrial and domestic effluents and deposition from the atmosphere are the primary sources of the pollutants in the sewage sludge (Blanchard et al. 2001; Harrison et al. 2006; Guo et al. 2009). The pollutants as mentioned above are, therefore, built-up in the sewage sludge during the treatment of wastewater. Such pollutants may pose severe health hazard owing to their involvement in the food web of humans from the field crops, and livestock that grazes on sewage sludge applied soils (McLachlan et al. 1996). Among the contaminants, polycyclic aromatic hydrocarbons (PAHs) are the most hazardous for the atmosphere, and it originates from the smoke released by the vehicle, industrial gases etc. (Ozcan et al. 2013). Kasatikov et al. (2017) studied Moscow's sewage sludge and found 200 pollutants of organic origin produced by erroneous human activities. They belong to compounds of different chemical groups (i) unsaturated hydrocarbons, (ii) acyclic aromatic hydrocarbon, (iii) polycyclic aromatic hydrocarbon, (iv) compounds containing oxygen, and (v) aromatic cyclic hydrocarbons. The permissible limit of organic compounds in the sludge of various countries is shown in Table 9.

7 Sources of Polycyclic Aromatic Hydrocarbons

Among the exotoxicants, polycyclic aromatic hydrocarbons are the most significant which have high levels of toxicity. Polycyclic aromatic hydrocarbons are chemically inactive and are rarely affected by acids and other oxidizing agents. Zhai et al. (2011) indicated that sewage sludge contains polycyclic aromatic hydrocarbons like acenaphthene, anthracene, acenapthylene, benzo(b)fluoranthene, benzo(a)pyrene, benzo(k)fluoranthene, benzo(g,h,i)perylene, 1,2 benzanthracene, fluoranthene, pyrene and indeno (1,2,3-cd) pyrene. Polycyclic aromatic hydrocarbons in sewage are originated from the burning of petroleum, wood, kerosene, coal, and grass. Inorder to indicate the anthracenesphenathrene to anthracene ratio, the molecular mass number 178 is used as a standard. It indicates that the ratio of less than 0.10 implies polycyclic aromatic hydrocarbons from the burning of petroleum, and the ratio of greater than 0.10 implies other burning sources (Budzinski et al. 1997). Furthermore, Yunker et al. (2002) indicates that the relative proportion of [benz (a)anthraceneschrysene]: [benz(a)anthracene] less than 0.20 denotes polycyclic aromatic hydrocarbons from the burning of petroleum; the ratio between 0.20 and 0.35 denotes polycyclic aromatic hydrocarbons originated from the burning of petroleum or other burning sources, and the ratio above 0.35 denotes other burning sources. Numerous studies conducted on sewage sludge of European countries, African countries, and Asian countries showed a large concentration of polycyclic aromatic

Table 9 The permissible limit of organic compounds in the sludge of various countries

	Linear AlkylbenzeneSulfonates	Sum of Halogenated organic compounds (AOX)	Di (2-Ethylhexyl) Phthalate	Nonylphenol and Nonylphenol ethoxylates	Polycyclic Aromatic hydrocarbons	Polychlorinated biphenyls	Polychlorinated dibenzo-p-dioxins and furans (ngTEq/kg dm)
Denmark	1300	–	50	10	3[1]	–	–
European Union	2600	500	100	50	6[1]	0.8[2]	100
Germany	–	500	–	–	–	0.2[5]	100
Lower Austria	–	500	–	–	–	0.2[5]	100
Sweden	–	–	–	50	3[3]	0.4[4]	100

Source: Langenkamp et al. (2001)
[1]Summation of acenaphthene, benzo(b + j + k) fluoranthne, benzo(a)pyrene, fluoranthene, benzo(ghi)perylene, fluorine, pyrene, phenanthrene, ideno (1,2,3,-c,d) pyrene
[2]Summation of six congeners of polychlorinated biphenyls 180, 153, 138, 101, 52, 28
[3]Summation of six compounds
[4]Summation of seven congeners
[5]All the six (6) congeners polychlorinated biphenyls 180, 153, 138, 101, 52, 28

hydrocarbons ranging between 0 and 33,000 nanogram/gram dw (Poluszyńska et al. 2017; Man et al. 2016).

8 Contents of Polychlorinated Biphenyls

Sewage sludge contains polychlorinated biphenyls as well as polycyclic aromatic hydrocarbons, as reported by Wyrwicka et al. 2014; Urbaniak et al. 2014; McLachlan et al. 1996). In the environment, polychlorinated biphenyls are present owing to erroneous anthropogenic activities (Borja et al. 2011). More or less 30% of polychlorinated biphenyls have access to the environment (Benabdallah El-Hadj et al. 2007). The presence of polychlorinated biphenyls in sewage sludge reduces its value significantly for agricultural utilization. The sewage sludge must contain less than 0.8 mg/kg polychlorinated biphenyls per the European Union and Turkey regulations for utilization in agriculture.

9 Opportunities of Sewage Sludge in Agriculture

At present, India produces sewage of almost 38,354 million litres with the same quantity of sludge daily (Kaur et al. 2012). In India, the potential N, P, and K content in sewage are 350,000 tonnes year^{-1}, 150,000 tonnes year^{-1}, and 200,000 tonnes year^{-1}, respectively (Juwarkar et al. 1991). The use of sewage sludge in agriculture has become preferable since it provides recycling prospects of the organic carbon and essential plant nutrients to the soil for sustainable production of crops. Several studies have revealed that utilizing sewage sludge in agriculture has increased the yield of many vegetables and field crops. It improves the physico-chemical properties of the agricultural soil and improves microorganisms' activities in the soil because of its higher content of organic carbon. Hence, the utilization of sewage sludge provides a recycling opportunity for nutrients, and it may also be used instead of chemical fertilizer.

Application of sewage sludge in soil supplies the required primary nutrients, secondary nutrients, and micronutrients in plants which are required for the optimum growth and development of the plant. Sewage sludge serves as a conditioner of the soil, boosts its physical properties, and improves its chemical properties, and thus it reduces soil erosion. Various studies pointed out that the utilization of sewage sludge in agriculture increases the yield of the crop (Latare et al. 2014), available phosphorous (Shu et al. 2016), and its negative effect on ecology is negligible (Adair et al. 2014). Numerous studies conducted in recent years reported that the application of sewage sludge in soil not increases productivity but also improves the fertility status of the soil, water holding capacity, and soil reaction (after application of biosolids, the pH of the soil reduced, and it is considered to increase the plant uptake of maximum metals (Carvalho et al. 2013).

9.1 Status of Nutrients in Sewage Sludge

The sewage sludge of Kolkata, India, contains organic carbon between 9.05% and 14.35% with an average of 11.56%, and the content of total nitrogen ranges between 1.86% and 3.16%, with an average of 2.46%. The content of total phosphorus in sewage sludge of Kolkata, India, ranges between 1.53% and 2.42%, with an average of 1.83%, and the total potassium ranges between 0.97% and 1.64%, with an average of 1.31% (Saha 2015). The total nitrogen content of sewage sludge of different cities such as Chennai, Nagpur, Delhi, Jaipur and Ahmedabad ranges between 0.82% and 2.34% and the total phosphorus content of such sewage sludge ranges between 0.51% and 0.94% and the total potassium content of such sewage sludge ranges between 0.11% and 0.23%. The sludge of Kolkata contains total nitrogen content between 0.34% and 0.56%, total phosphorus content between 0.11% and 0.12% and total potassium content between 0.36% and 0.59% respectively (Maiti et al. 1992; Juwarkar et al. 1991). The sewage sludge of Ukkadam, Coimbatore contains 1230 mg kg^{-1} available nitrogen, 633 mg kg^{-1} available phosphorus and 380 mg kg^{-1} available potassium. It also contains 3.62% of total nitrogen, 1.46% of total phosphorus and 2.53% of total potassium (Chitdeshwari et al. 2002). The sludge of Guangzhou, China contains organic carbon between 281 gm/kg and 606 gm/kg, total nitrogen between 12.5 gm kg^{-1} and 38.3 gm kg^{-1}, total phosphorus between 11.9 gm kg^{-1} and 36.2 gm kg^{-1} and total potassium between 6.7 gm kg^{-1} and 19.1 gm kg^{-1} (Liu and Sun 2013). Comparative figures on the nutrient status of sewage sludge are indicated in Table 10. The pH of different sewage sludge is slightly acidic to alkaline in reaction. The table also indicates a drastic difference in the total organic carbon content of different sewage sludge, but there is no difference in the total nitrogen content and total phosphorus content of different sewage sludge.

Several studies were done abroad and in India revealed that utilization of sewage sludge in agriculture enhances crop yield and productivity. Epstein (2003) observed that utilizing sewage sludge in the soil increases the crop yield than those soils applied with chemical fertilizers. The application of sludge in the soils increases the amount of nutrients that are available to plants. The sludges are rich in available iron, available copper, available manganese and available zinc, organic carbon,

Table 10 Comparative data on the nutrient status of sewage sludge

Parameters	Brazil[1]	China[2]	Spain[3]	Delhi[4]	Varanasi[5]	Kolkatta[6]
Total nitrogen*	2.10	6.48	2.22	1.48	1.73	2.46
Total phosphorus*	1.0	1.65	1.66	1.61	0.72[$]	1.83
Total potassium*	–	0.49	0.47	0.20	–	1.31
Total organic carbon*	24.80	42.55	18.30	7.64	5.52	11.56
pH	6.60	7.45	7.90	6.41	7.0	6.57
EC[#]	–	–	1.20	1.39	2.28	2.60

*Expressed in percentage (%), [#]Expressed in dS m^{-1} and [$]Available phosphorus
Source: [1]Bettiol and Ghini (2011), [2]Wang et al. (2008),[3]Antolin et al. (2005),[4] Roy et al. (2013),[5]Singh and Agrawal (2010a) and [6] Saha (2015)

Table 11 Different sewage sludge doses and their impact on various crop yield

Sl. no	Type of crop	Sewage sludge doses and its impact on crop yield	Reference
1	Rice	• Sewage sludge @ 3 kg m^{-2} increases nearly 60% of the yield of the grain, sewage sludge @ 4.5 kg m^{-2} increases nearly 111% of the grain yield, sewage sludge @ 6 kg m^{-2} increases nearly 125% yield of the grain, sewage sludge of about 6.9 kg m^{-2} as well as 12 kg m^{-2} increases nearly 134% & 137% yield of the grains respectively. • Sewage sludge incorporated in alluvial soil @ 10 t ha^{-1} increases nearly 22.5% of the yield in comparison with the control, sewage sludge incorporated in alluvial soil @ 20 t ha^{-1} increases nearly 51.9% of the yield in comparison with the control, sewage sludge incorporated in alluvial soil @ 40 t ha^{-1} increases nearly 79.6% of the yield in comparison with the control and sewage sludge incorporated in alluvial soil @ 80 t ha^{-1} increases nearly 99.4% of the yield in comparison with the control	Singh and Agrawal (2010a) Saha (2015)
2	Wheat	Sewage sludge @ 3.75 kg m^{-2} increases nearly 54.1% of wheat yield.	NEPA (1992)
3	Maize	• Incorporation of sewage sludge in an increasing rate, i.e. 0 tonnes ha^{-1}, 10 tonnes ha^{-1}, 20 tonnes ha^{-1}, 30 tonnes ha^{-1}, 40 tonnes ha^{-1} and 50 tonnes ha^{-1} dry wt.) improves the germination of the seeds. Incorporation of sewage sludge @ 20 t ha^{-1} gives the highest shoot, root length, and highest area of the leaf and the lowest was observed in 0 t ha^{-1}. • Incorporation of sewage sludge @0 gm pot^{-1}, 22.7 gm pot^{-1}, 45.5 gm pot^{-1}, 30 tonnes ha^{-1} and gm pot^{-1}increases the crop yield.	Qasim et al. (2001) Chitdeshwari et al. (2002)
4	Sunflower	• Sewage sludge @ 0.7 kg m^{-2} and 1.4 m^{-2}increase nearly 30% and 31% of the yield respectively. • Incorporation of sewage sludge in an increasing rate, i.e. 0 tonnes ha^{-1}, 80 tonnes ha^{-1}, 160 tonnes ha^{-1} and 320 tonnes ha^{-1} improves the sunflower's dry weight.	Lavado (2006) Morera et al. (2002)
5	Mung bean	• Highest yield & highest weight of the pod are recorded with the application of sewage sludge @ 9.0Kgm^{-2}.	Singh and Agrawal (2010b)
6	Broad bean	• Sewage sludge @ 4.5 kg m^{-2}gives higher yield in comparison with the sewage sludge @ 0 kg m^{-2}.	Garrido et al. (2005)
7	Palak	• Use of chemical fertiliser along with sewage sludge increase the yield substantially	Roy et al. (2013)

phosphorus, and nitrogen. Hence, its utilization in agriculture improves the soil's organic carbon status and boost nutrient availability to the plants, specifically nitrogen, and, consequently, increases the crop yield. In some cases, the soils treated with sewage sludge produce a higher yield than those treated with chemical fertilizers (Table 11).

9.2 Sewage Sludge and Its Impact on Soil Physical Properties

The sewage sludge contains organic matter in higher quantity, and, therefore, it has positive impacts on soil physical properties. It also serves as a conditioner of the soil. The physical soil properties which are improved by the application sewage sludge are (i) water-holding capacity, (ii) porosity, (iii) bulk density and (iv) soil aggregate and stability (Angin and Yaganoglu 2011; Wells et al. 2000; Ramulu 2002). Usman et al. (2012) indicated that sewage sludge's application improves soil physical conditions more significantly than farm manures because the compounds of organic origin in sewage sludge are stable. In another study, the application of sewage sludge at a higher rate for wheat cultivation improves soil porosity but decreases the soil bulk density (Wortmann 2005). Studies carried out by Ojeda et al. (2003), Cogger (2005), Garcia-Orenes et al. (2005), Chambers et al. (2003) and Lindsay and Logan (1998) also reported that the sewage sludge applied at a higher rate decreases the compaction as well as the bulk density of the soil. The soil bulk density decreases when it is applied with sewage sludge, and resulting in the improvement of soil aggregates and porosity of the soil (Lindsay and Logan 1998). The soil aggregates of the sandy loam soil is improved by nearly 41% when it is applied with sewage sludge, but in the case of clay soil, there is no prominent effect on soil aggregates (Cameron et al. 1997). The soil physical properties are improved when it is applied with sewage sludge and also increases the rate of infiltration, improves better absorption of rainwater, and consequently minimizes erosion due to water (Chambers et al. 2003). Epstein (1975) investigated the effect of sewage sludge @ 0.5% on the stability of aggregates, retention of water, and hydraulic conductivity, and it showed that raw and digested sludge improves the retention of water by soil. It was also observed that raw sludge applied soil increases water retention by soil compared to digested applied soil. In an incubation study, Epstein (1975) observed a sharp increase in the hydraulic conductivity of soil that has been incubated with sludge for 27 days.

9.3 Sewage Sludge and Its Effect on Soil Chemical Properties

Sewage sludge contains a higher quantity of organic matter. When sewage sludge is applied to soils, it produces humic acid, carboxylic acid, phenolic acid, and enolic acid, which are considered as a conditioner of soil. The soil pH is increased when it is applied with sewage sludge that originates from the municipality area (Tsadilas et al. 1995). On the contrary, the soil's pH is decreased by the application of sewage sludge (Epstein et al. 1976). The soil pH variation is also interrelated with the content of $CaCO_3$ in the sludge and organic acids released from the sludge during its decomposition (Sommers 1977). Sewage sludge when applied in calcareous soils @ 100 t ha^{-1} a reduction in soil pH is observed. The initial pH was 8.2, but after application of maximum dose of sewage sludge, it decreased to pH 8.0 (Jamil et al.

2006). Sewage sludge application in soil improves the CEC, and in turn, it makes more accumulation of essential nutrients in the soil (Soon 1981). In agriculture, sewage sludge is used for its significant nitrogen content and higher nitrogen content is the best criteria for the use in agriculture (Veeken et al. 2000).

9.4 Sewage Sludge and Its Effect on Soil Biological Properties

The application of sewage sludge in soil improves productivity, improves the physical conditions, chemical properties and biological properties of the soil. In many studies conducted by Angin and Yaganoglu (2011), Göcmez and Okur (2010), Antolin et al. (2005), Saviozzi et al. (1999) observed that the use of sewage sludge in agriculture improves nutrient recycling, activities of soil enzyme such as β-glucosidase, alkaline phosphatase, dehydrogenase, BAA, protease and urease, mineralization of nitrogen, respiration rate and microbial population etc. The improvement in biomass of microorganism and activity of soil enzyme because of the presence of organic matter at a higher quantity in the sewage sludge. Banerjee et al. (1997) pointed out that the use of sludge in soil decreases the variety of microbial populations. However, it enhances the general population of the microorganisms and the conversion of the organic form of nutrients to their inorganic form along with the improvement of soil enzymes such as alkaline phosphatase, acid phosphatase, and arylsulfatase etc. Kizilkaya and Bayrakli (2005) investigated the influence of dissimilar doses of sewage sludge, i.e., 0 t ha^{-1}, 100 t ha^{-1}, 200 t $ha^{-1,}$ and 300 t ha^{-1} dry weight on soil enzymes such as urease, arylsulphatase, alkaline phosphatase, β-glucosidase. It was found that the use of sewage sludge at dissimilar doses increases the activities of soil enzymes. The activity of β-glucosidase was highest when a maximum dosage of sludge is applied in the soil containing a high C: N ratio. Urease activities, alkaline phosphatase activities, and arylsulphatase activities were highest with the application of a maximum dose of sludge in soil that possesses a low C: N ratio. A prolonged experiment of 20 years on sewage sludge found that the use of sewage sludge at dissimilar dose increases the microbial population and improves the status of soil organic carbon, i.e., 2.5 times than that of the soils which are not treated with sewage sludge. The heavy metal concentrations in the sludge decide the usefulness of sewage sludge and its impact on the soil's biological properties. Sewage sludge containing heavy metal at low concentrations are congenial for soil microbial population, soil microbial activities, and organic carbon, compared to those soils with high heavy metal concentrations showed significantly low carbon biomass (Usman et al. 2012; Knight et al. 1997; FlieBbach et al. 1994), microbial population, soil enzyme activities and soil microorganisms (Kandeler et al. 2000; Baath 1989; McGrath et al. 1988; Tyler 1981).

10 Conclusions

After careful examination of fundamental physicochemical properties, pollutant contents, harmful bacteria as well as ecotoxicity properties, sewage sludge may be considered for utilisation in agriculture. The type of soil regulates sewage sludge phytotoxicity. In due course of time, the soil incorporated with sewage sludge changes to minimise its toxicity. The level of sewage sludge toxicity is dependent on the types of crops grown. Further, its toxicity level is also dependent on kinds of sewage sludge as well as soil types. Soils applied with properly treated sewage sludge helps in reducing its toxicity. Only extract estimation of sewage sludge cannot define the risks of health hazard in using sewage sludge for agriculture.

Biosolids are rich in organic matter. They serve as a rich source of primary nutrients i.e. nitrogen, phosphorus, potassium, calcium, magnesium, sulphur, zinc, iron, manganese and copper. Biosolids can also serve as an alternative of chemical fertilizer in future.. When sewage sludge is applied to soils, it enriches the essential plant nutrient as well as also improves the soil pH. Indiscriminate utilisation of biosolids may lower soil productivity. Biosolids hold heavy metals and other residues of toxic organic materials; therefore, injudicious use of biosolids may jeopardise the food web.

Sewage sludge in colossal quantity is produced due to the rapidly increasing human population and development. Among the options for sewage sludge disposal, its application in the soil for better crop yield is the best option. In the case of incorrect and unwise dumping of biosolids often causes pollution of the river, lakes, groundwater, land degradation, and food web pollution.

References

Adair KL, Wratten S, Barnes AM, Waterhouse BR, Smith M, Lear G, Weber P, Pizey M, Boyer S (2014) Effects of biosolids on biodiesel crop yield and belowground communities. Ecol Eng 68: 270–278

Angin I, Yaganoglu AV (2011) Effects of sewage sludge application on some physical and chemical properties of a soil affected by wind erosion. J Agric Sci Technol 137:57–768

Antolin MC, Pascual I, Garcia C, Polo A, Sanchez-Diaz M (2005) Growth, yield and solute content of barley in soils treated with sewage sludge under semiarid Mediterranean conditions. Field Crop Res 94:224–237

Asano T, Burton H, Leverenz H, Tsuchihashi R, Tchobanoglous G (2007) Water reuse: issues, technologies, and applications. McGrawHill Professional, New York, p1570

Baath E (1989) Effects of heavy metals in soil on microbial processes and populations (a review). Water Air Soil Pollut 4(3–4):335–379

Banerjee MR, Burton DL, Depoe S (1997) Impact of sewage sludge application on soil biological characteristics. Agric Ecosyst Environ 66(3):241–249

Benabdallah El-Hadj T, Dosta J, Torres R, Mata-Alvarez J (2007) PCB and AOX removal in mesophilic and termophilic sewage sludge digestion. Biochem Eng J 36(3):281–287

Bettiol W, Ghini R (2011) Impacts of sewage sludge in tropical soil: a case study in Brazil. Appl Environ Soil Sci 1:11. https://doi.org/10.1155/2011/212807

Blanchard M, Teil MJ, Ollivon D, Garban B, Chestérikoff C, Chevreuil M (2001) Origin and distribution of polyaromatic hydrocarbons and polychlorobiphenyls in the urban effluents to waste water treatment plants of the Paris Area (FRANCE). Water Res 35(15):3679–3687
Blumenthal UJ, Cifuentes E, Bennett S, Quigley M, Ruiz-Palacios G (2000) The risk of enteric infections associated with wastewater reuse: the effect of season and degree of storage of wastewater. Trans R Soc Trop Med Hyg 95(2):131–137
Borja Á, Belzunce MJ, Garmendia JM, Rodríguez JG, Solaun O, Zorita I (2011) Impact of pollutants on coastal and benthic marine communities. In: Sánchez-Bayo F, van den Brink PJ, Mann RM (eds), Ecological impacts of toxic chemicals. Bentham Science Publishers Ltd. pp. 165–186
Budzinski H, Jones I, Bellocq J, Pierard C, Garrigues P (1997) Evaluation of sediment contamination by polycyclic aromatic hydrocarbons in the gironde estuary. Mar Chem 58(1–2):85–97
Cameron KC, Di HJ, McLaren RG (1997) Is soil an appropriate dumping ground for our wastes? Aust J Soil Res 35(5):995–1035. https://doi.org/10.1071/S96099
Carvalho A, Nabais C, Roiloa SR, Rodriguez-Echeverria S (2013) Revegetation of abandoned copper mines: the role of seedbanks and soil amendments. Web Ecol 13:69–77
Chambers BJ, Nicholson FA, Aitken M, Cartmell E, Rowlands C (2003) Benefits of biosolids to soil quality and fertility. Water Environ J 17(3):162–166. https://doi.org/10.1111/j.1747-6593.2003.tb00455.x
Chen W, Hoitink HAJ, Schmitthenner AF (1986) Factors affecting suppression of Pythium damping-off in container media amended with composts. Ecol Ecodemiol 77:750–755
Chitdeshwari T, Savithri P, Mahimairaja S (2002) Effect of sewage biosolid composts on the yield of crops. Indian J Environ Protect 21(10):911–912
Clarke BO, Smith SR (2011) Review of 'emerging' organic contaminants in biosolids and assessment of international research priorities for the agricultural use of biosolids. Environ Int 37(1):226–247
Cogger CG (2005) Potential compost benefits for restoration of soils disturbed by urban development. Compost Sci Util 13(4):243–251
Davis EF, Klosterhaus SL, Stapleton HM (2012) Measurement of flame retardants and triclosan in municipal sewage sludge and biosolids. Environ Int 40:1–7
Delibacak S, Voronina L, Morachevskaya E, Onguna AR (2020) Use of sewage sludge in agricultural soils: useful or harmful. Eurasian J Soil Sci 9(2):126–139
Domene X, Alcañiz JM, Andres P (2008) Comparison of solid-phase and eluate assays to gauge the eco toxicological risk of organic wastes on soil organisms. Environ Pollut 151(3):549–558
Domene X, Colón J, Uras MV, Izquierdo R, Àvila A, Alcañiz JM (2010) Role of soil properties in sewage sludge toxicity to soil collembolans. Soil Biol Biochem 42(11):1982–1990
Epstein E (1975) Effect of sewage sludge on some soil physical properties. J Environ Qual 4(1):139–142
Epstein E (2003) Land application of sewage sludge and biosolids. Lewis Publication, New York
Epstein E, Taylor JM, Chaney RL (1976) Effects of sewage sludge and sludge compost applied to soil on some soil physical and chemical properties. J Environ Qual 5(5):422–426
European Commission (2009) Environmental, economic and social impacts of the use of sewage sludge on land. Consultation Report on Options and Impacts, Report by RPA, Milieu Ltd and WRc for the European Commission, DG Environment under Study Contract DG ENV.G.4/ETU/2008/0076r
FlieBbach A, Martens R, Reber HH (1994) Soil microbial biomass and microbial activityin soils treated with heavy metal contaminated sewage sludge. Soil Biol Biochem 26(9):1201–1205. https://doi.org/10.1016/0038-0717(94)90144-9
Garcia-Orenes F, Guerrero C, Mataix-Solera J, Navarro-Pedreno J, Gomez I, Mataix-Beneyto J (2005) Factors controlling the aggregate stability and bulk density in two different degraded soils amended with biosolids. Soil Tillage Res 82(1):65–76. https://doi.org/10.1016/j.still.2004.06.004

Garrido S, Del Campo GM, Esteller MV, Vaca R, Lugo J (2005) Heavy metals in soil treated with sewage sludge composting, their effect on yield and uptake of broad bean seeds (*Viciafaba* L.). Water Air Soil Pollut 166(1):303–319

Gerba CP, Smith JES (2005) Sources of pathogenic microorganisms and their fate during land application of wastes. J Environ Qual 34:42–48

Ghini R, Rodriges F, Patricio A, Bettiol W, Gatti AIM, Holanda A, Maia N (2007) Effect of sewage sludge on suppressiveness to soil-borne plant pathogens. Soil Biol Biochem 39:2797–2805

Göcmez S, Okur N (2010) Effect of municipal sewage sludge application on microbial biomass and activity of soils. Int Sust Water Wastewater Manag 26–28 October, Konya/Turkey

Guo L, Zhang B, Xiao K, Zhang Q, Zheng M (2009) Levels and distributions of polychlorinated biphenyls in sewage sludge of urban waste water treatment plants. J Environ Sci 21(4):468–473

Harrison EZ, Oakes SR, Hysell M, Hay A (2006) Organic chemicals in sewage sludges. Sci Total Environ 367(2–3):481–497

Jamil M, Qasim M, Umar M (2006) Utilization of sewage sludge on organic fertilizer in sustainable agriculture. J Appl Sci 6(3):531–535. https://doi.org/10.3923/jas.2006.531.535

Jiménez B, Asano T (2008). Water reclamation and reuse around the world in water reuse: an international survey of current practice, issues and needs (eds. B. Jiménez and T. Asano) IWA Publishing, London, p648

Juwarkar AS, Juwarkar A, Desbhratar PB, Bal AS (1991) Exploitation of nutrient potential of sewage and sludge through land application. In: Asian experience in integrated plant nutrition, Regional office for Asia and Pacific (RAPA). Food and Agriculture Organization of the United Nations, Bangkok, pp 178–201

Kacprzak M, Neczaj E, Fijałkowski K, Grobelak A, Grosser A, Worwag M, Singh BR (2017) Sewage sludge disposal strategies for sustainable development. Environ Res 156:39–46

Kandeler E, Tscherko D, Bruce KD, Stemmer M, Hobbs PJ, Bardgett RD, Amelung W (2000) Structure and function of the soil microbial community in microhabitats of a heavy metal polluted soil. Biol Fertil Soils 32(5):390–400. https://doi.org/10.1007/s003740000268

Kasatikov VA, Shabardina NP, Raskatov VA (2017) After effect of the systematic application of urban sewage sludge on the agrobiological and ecological properties of soddy-podzolic soil. Plodorodie 1(94):43–46 [in Russian]

Kaur R, Wani SP, Singh AK and Lal K (2012) Wastewater production, treatment and use in India, presented at the 2nd Regional Workshop on Safe Use of Wastewater in Agriculture, 16–18 May, National Report, New Delhi, India [online] http://www.ais.unwater.org (Accessed 25 August 2017)

Keraita B, Jiménez B, Drechsel P (2008) Extent and implications of agricultural reuse of untreated, partly treated and diluted wastewater in developing countries. Agric Veter Sci Nutr Nat Resour 3:15–23

Kizilkaya R, Bayrakli B (2005) Effects of N-enriched sewage sludge on soil enzyme activities. Appl Soil Ecol 30(3):192–202

Knight BP, McGrath SP, Chadri AM (1997) Biomass carbon measurements substrateutilization patterns of microbial populations from soils amended with cadmium, copper orzinc. Appl Environ Microbiol 63(1):39–43

Langenkamp H, Part H, Erhardt W, Prüeß A (2001) Organic contaminants in sewage sludge for agricultural use. Joint Research Centre, Institute for Environment and Sustainability, Soil and Waste Unit, European Commission

Latare AM, Kumar O, Singh SK, Gupta A (2014) Direct and residual effect of sewage sludge on yield, heavy metals content and soil fertility under rice-wheat system. J Ecol Eng 69:17–24

Lavado RS (2006) Effects of sewage sludge application on soils and sunflower yield: quality andtoxic element accumulation. J Plant Nutr 29(6):975–984

Leoni C, Ghini R (2006) Sewage sludge effect on management of *Phytophthoranicotianae* in citrus. Crop Prot 25:10–22

Lindsay BJ, Logan T (1998) Field response of soil physical properties to sewage sludge. J Environ Qual 27(3):534–542. https://doi.org/10.2134/jeq1998.00472425002700030009x

Liu J, Sun S (2013) Total concentrations and different fractions of heavy metals in sewage sludge from Guangzhou, China. Trans Nonferrous Metals Soc China 23(8):2397–2407
Maiti PS, Sah KD, Gupta SK, Banerjee SK (1992) Evaluation of sewage sludge as a source of irrigation and manure. J Indian Soc Soil Sci 40(1):168–172
Man YB, Chow KL, Cheng Z, Mo WY, Chan YH, Lam JCW, Lau FTK, Fung WC, Wong MH (2016) Profiles and removal efficiency of polycyclic aromatic hydrocarbons by two different types of sewage treatment plants in Hong Kong. J Environ Sci 53:196–206
McGrath SP, Brookes PC, Giller KE (1988) Effects of potentially toxic metals in soil derived from past applications of sewage sludge on nitrogen fixation by *Trifoliumrepens* L. Soil Biol Biochem 20(4):415–424. https://doi.org/10.1016/0038-0717(88)90052-1
McLachlan MS, Horstmann M, Hinkel M (1996) Polychlorinated dibenzo-p-dioxins and dibenzofurans in sewage sludge: sources and fate following sludge application to land. Sci Total Environ 185(1–3):109–123
Morera MT, Echeverria J, Garrido J (2002) Bioavailability of heavy metals in soils amended with sewage sludge. Can J Soil Sci 59:433–438. https://doi.org/10.1002/ldr.580
Müller J, Böhmer W, Litz NT (2006) Occurrence of polycyclic musks in sewage sludge and their behaviour in soils and plants—Part 1: behaviour of polycyclic musks in sewage sludge of different treatment plants in summer and winter. J Soils Sediments 6(4):231–235
National Environmental Protection Agency (NEPA) (1992) Study on Municipal Wastewater Reuse. Science Press, Beijing
Ojeda G, Alcaniz JM, Ortiz O (2003) Runoff and losses by erosion in soils amended with sewage sludge. Land Degrad Dev 14(6):563–573
Oleszczuk P (2006) Characterization of Polish sewage sludges with respect to fertility and suitability for land application. J Environ Sci Health, Part A41(7):1119–1217
Ozcan S, Tor A, Aydin ME (2013) Investigation on the levels of heavy metals, polycyclic aromatic hydrocarbons, and polychlorinated biphenyls in sewage sludge samples and ecotoxicological testing. Clean Soil Air Water 41(4):411–418
Poluszyńska J, Jarosz-Krzemińska E, Helios-Rybicka E (2017) Studying the effects of two various methods of composting on the degradation levels of polycyclic aromatic hydrocarbons (PAHs) in sewage sludge. Water Air Soil Pollut 228:305
Qasim M, Javed N, Himayatullah SM, Subhan M (2001) Effect of sewage sludge on the growth of maize crop. J Biol Sci 1(2):52–54
Ramulu USS (2002) Reuse of municipal sewage and sludge in agriculture. Scientific Publishers, Jodhpur
Roy T, Singh RD, Biswas DR, Patra AK (2013) Effect of sewage sludge and inorganic fertilizers on productivity and micronutrients accumulation by palak (*Beta vulgaris*) and their availability in a TypicHaplustept. J Indian Soc Soil Sci 61(3):207–218
Saha S (2015) Remediation of heavy metals toxicity to crops in sewage-sludge contaminated soils, Published Ph.D. thesis, Bidhan Chandra KrishiViswavidyalaya, West Bengal
Saviozzi A, Biasci A, Riffaldi F, Levi-Minzi R (1999) Long-term effects of farmyardmanure and sewage sludge on some soil biochemical characteristics. Biol Fertil Soils 30(1):100–106
Shu W, Price GW, Sharifi M, Cade-Menum BJ (2016) Impact of annual and single application of alkaline treated biosolidson soil extractable phosphorus and total phosphorus. Agric Ecosyst Environ 219:111–118
Singh RP, Agrawal M (2008) Potential benefits and risks of land application of sewage sludge. Waste Manag 28(2):347–358
Singh RP, Agrawal M (2010a) Variations in heavy metal accumulation, growth and yield of rice plants grown at different sewage sludge amendment rates. Ecotoxicol Environ Saf 73(4):632–641
Singh RP, Agrawal M (2010b) Effect of different sewage sludge applications on growth and yield of *Vignaradiata* L. field crop: metal uptake by plant. Ecol Eng 36(7):969–972
Smith SR (2009) Organic contaminants in sewage sludge (biosolids) and their significance for agricultural recycling. Philos Trans R Soc A367:4005–4041

Sommers LE (1977) Chemical composition of sewage sludges and analysis of their potential use as fertilizers. J Environ Qual 6(2):225–232
Sommers LE, Nelson DW, Yost KJ (1976) Variable nature of chemical composition of sewage sludges. J Environ Qual 5(3):303–306
Soon YK (1981) Solubility and sorption of cadmium in soils amended with sewage sludge. J Soil Sci 32(1):85–95
Soulié M, Tréméa L (1991) Technologie pour le traitement et la réutilisation des eauxuséesdans le bassinméditerranéen'. In Proceedings of the 3rd meeting of the regional agency for environment, Provence – Alpes – Côte d'Azur, pp 171–255
Suchkova N, Darakas E, Ganoulis J (2010) Phytoremediation as a prospective method for rehabilitation of are as contaminated by long-term sewage sludge storage: a Ukrainian-Greek case study. Ecol Eng 36(4):373–378
Thorn J, Kerekes E (2001) Health effects among employees in sewage treatment plants: a literature survey. Am J Indust Med 40:170–179
Trang DT, Hoek WVD, Tuan ND, Cam PD, Viet VH, Luu DD, Konradsens F, Dalsgaard A (2007) Skin disease among farmers using wastewater in rice cultivation in Nam Dinh, Vietnam. Trop Med Int Health 12:51–58
Tsadilas CD, Matsi T, Barbayiannis N, Dimoyiannis D (1995) Influence of sewage sludge application on soil properties and on the distribution and availability of heavy metal fractions. Commun Soil Sci Plant Anal 26(15–16):2603–2619
Tyler G (1981) Heavy metals in soil biology and biochemistry. In Paul EA, Ladd JN (eds) Soil Biochemistry, vol 5, pp 371–414, Marcel Dekker, New York
Urbaniak M, Wyrwicka A, Serwecińska L, Zieliński M, Tołoczko W (2014) Impact of sludge originated PCDDs/PCDFs on soil contamination and Salix sp. metabolism. 14th International Multidisciplinary Scientific Geoconference and EXPO. SGEM 2014. 17–26 June 2014, Albena, Bulgaria 2(3):169–174
Usman K, Khan S, Ghulam S, Umar Khan M, Khan N, Anwar Khan M, Khalil SK (2012) Sewage sludge: an important biological resource for sustainable agriculture and its environmental implications. Am J Plant Sci 3(12):1708–1721
Veeken A, NieropK WV, Hamelers B (2000) Characterization of NaOH-extracted humic acids during composting of a biowaste. Bioresour Technol 72(1):33–41. https://doi.org/10.1016/S0960-8524(99)90096-2
Wang PF, Zhang SH, Wang C, Hou J, GuoPC LZ (2008) Study of heavy metal in sewage sludge and in Chinese cabbage grown in soil amended with sewage sludge. Afr J Biotechnol 7(9):1329–1334
Wells AT, Chan KY, Cornish PS (2000) Comparison of conventional and alternative vegetable farming systems on the properties of a yellow Earth in New South Wales. Agric Ecosyst Environ 80:47–60. https://doi.org/10.1016/S0167-8809(00)00133-X
Westrell T, Schonning C, Stenstrom TA, Ashbolt NJ (2004) QMRA (Quantitative Microbial Risk Assessment) and HACCP (Hazard Analysis and Critical Control Points) for management of pathogens in wastewater and sewage sludge treatment and reuse. Water Sci Technol 50:23–30
WHO (2006) Guidelines for the safe use of wastewater, excreta and greywater, volume 2: wastewater use in agriculture. World Health Organization, Geneva
Wortmann CS (2005) Sewage sludge utilization for crop production, Department of Agronomy and Horticulture, University of Nebraska-Lincoln, Lincoln
Wyrwicka A, Steffani S, Urbaniak M (2014) The effect of PCB-contaminated sewage sludge and sediment on metabolism of cucumber plants (*Cucumissativus* L.). Ecohydrol Hydrobiol 14(1):75–82
Yunker MB, Macdonald RW, Vingarzan R, Mitchell RH (2002) PAHs in the Fraser river basin: a critical appraisal of PAH ratios as indicators of PAH source and composition. Org Geochem 33(4):489–515
Zhai J, Tian W, Liu K (2011) Quantitative assessment of polycyclic aromatic hydrocarbons in sewage sludge from wastewater treatment plants in Qingdao, China. Environ Monit Assess 180(1–4):303–311

Scope of Antibiotic Resistance Genes in Sewage Sludge for Therapeutic Uses

Parul Chaudhary, Heena Parveen, Anuj Chaudhary, Murtaza Ali, Devika Gautam, and Parmanand

1 Introduction

Antibiotics diminish the growth of microbial population and widely use to fight against infectious diseases from the 1930s (Chee-Sanford et al. 2009). Jelic et al. (2015) observed to merely a minute quantity of antibiotics can be entirely metabolized in living beings; 20–90% is generally excreted in non metabolized form in urines and faeces. Accordingly, antibiotics are usually found in sewage waters released from industrial, veterinary and pharmaceutical industries. In response to direct contact with antimicrobial substances, the microbes mutate itself to survive under un-wanted conditions, which leads to the development of resistant behavior, failure of antibiotic treatments and causes death to humans. Antibiotic resistance bacterial pathogens is one of the mainly issues for health protection system in twenty-first century globally, which pressurize our ability to take care of infections and principally disturbing the quick worldwide increase of various and pan-resistant bacteria recognized as "superbugs" causes infections which are not curable with current antibiotics. Inappropriate/overuse of antibiotics by humans and animal farms are supposed to be a key factor for multidrug tolerance development in the bacteria (Boonyasiri et al. 2014). Therefore, it is crucial to examine the resistance pattern in bacteria from both humans and animals.

P. Chaudhary (✉) · M. Ali · D. Gautam · Parmanand
Department of Animal Biotechnology, National Dairy Research Institute, Karnal, Haryana, India

H. Parveen
Department of Dairy Microbiology, National Dairy Research Institute, Karnal, Haryana, India

A. Chaudhary
School of Agriculture and Environmental Sciences, Shobhit University, Gangoh, Uttar Pradesh, India

V. D. Rajput et al. (eds.), *Sustainable Management and Utilization of Sewage Sludge*, https://doi.org/10.1007/978-3-030-85226-9_11

The sewer environment and activated sludge system of waste water treatment plants (WWTPs) is a potential hot-spot for the propagation of antimicrobial resistance bacteria (ARB) and antibiotic resistance genes (ARGs) and their transfer events (Hembach et al. 2019). Various researches reported that ARGs are ubiquitously noticed in the soil, canal and irrigation trenches and WWTP discharges that causes serious and long-lasting threats for the living being's health worldwide (Jechalke et al. 2014). The most harmful infections are spread by multi-drug resistance bacteria: *Staphylococcus aureus* (MRSA), enlarged-spectrum β-lactamase producers (ESBL) and vancomycin-resistance enterococci (VRE). All of the abovementioned organisms are common resistant's of the healthy living beings microbiome which can develop into opportunistic pathogens. There are various pathogens involved in diseases and released from hospitals found in wastewater (*Escherichia coli, Proteus spp., Mycobacterium tuberculosis, Pseudomonas aeruginosa, Salmonella, Klebsiella pneumonia* and *Enterococci faecalis*) with clinically relevant ARGs (Alexander et al. 2020). These bacterial pathogens along with resistance towards antibiotics scheduled in the World Health Organisation global main concern listing (WHO 2017). Rising of ARB carrying ARGs in clinical, farming waste, particularly in sewage wastewater is a severe human health concern (Devarajan et al. 2015). Several ARGs reduce the susceptibility of pathogens towards antibiotics such as tetracycline (*tet*), chloramphenicol (*cml*), sulphonamide (*sul*), methicillin (*mec*), fluroquinolone (*qnr*) and b-lactam (*bla*).

Various traditional WWTPs are deliberated to eliminate high concentration of nutrients like nitrates, phosphates, carbon, ARB and ARGs from wastewater. Activated sludge is well-known technique for the biological management of waste-water. Generally, two tanks are requisite for this treatment process: in which one is used for aeration of biological reaction and other is used for settling of the sludge. Approximate 79–88% of total antibiotics and 2.0 log ARGs were reduced in activated sludge treatment, but not fully elevated from the water phase (Obayiuwana and Ibekwe 2020).

The modern development in the analytical methodology in the field of chemistry and molecular microbiology provides opportunity to identify various antibiotics and ARGs. Antibiotics residues found in low concentration in water may be detected by liquid chromatography coupled along the mass spectrometry (LC-MS) and solid phase extraction. Development of qPCR, DNA microarray, droplet digital PCR, smartChip techniques and metagenomic analysis make it likely to detect abundances of multiple ARGs present in environment. This is crucial to examine and manage of the release of antibiotics and ARGs/bacteria from WWTPs. For tackling that concern, it is prior to recognize the allocation and providence of antibiotics, ARGs and ARB in WWTPs.

2 Occurrence and Spread of ARGs in Sewage Water

Antibiotic resistance has developed serious risk to human being. In the environment, high level of ARB and ARGs comes through activities carried out by humans and major concern for public health (Fig. 1). WWTPs get the discharges from different resources and hotspots for ARGs, related to clinical pathogens (Di Cesare et al. 2016). The microbes having ability of antimicrobial resistance is to bear the effects of antimicrobial therapies. The extent of resistance evolved quickly with time and plasmid mediated resistance developed since the year 2000 towards colistin. ARGs bear major concern due to the mobility in genetic elements that may simply transfer in between microbes during horizontal gene transfer (HGT). In the environment HGT is the crucial mechanisms involved in transfer of ARGs (Yoo et al. 2020). Transfer of genetic material from one bacteria to another in various processes like transformation, conjugation, transduction and gene transfer agents. Bacterial genome harbour diverse genes encoding antimicrobial resistance. These genes can

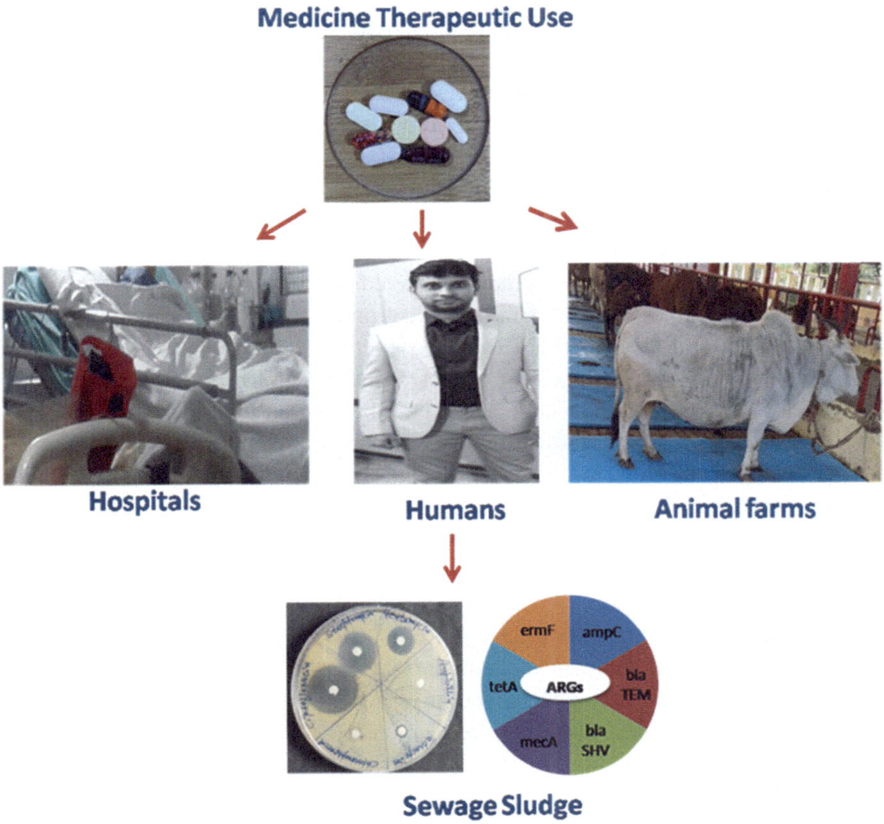

Fig. 1 Antimicrobial resistance bacteria and genes in sewage waste water

be expressed by triggers resistance mechanisms such as cell wall recycling, target protein modification, active exclusion system, porins reduction and efflux pumps (Bruchmann et al. 2013). Persistence of AMR in wastewater can be evaluated by culture dependent which expressed in CFU/ml and molecular approaches such as qPCR expressed in genes copies/ml.

3 Treatment Approaches of Wastewater

Treatment plants are key point to manage the increase of ARB from municipal waste and hospital. On the other hand, ARB in wastewater still remains high in abundance in secondary effluents of WWTPs in activated sludge (Zhang et al. 2015). Digested sludge might be possible resource of AMR for the reason that it is occasionally spread in agricultural farms like organic fertilizers. Elevated quantity of antibiotics and ARGs can control in landfills, municipal solid waste and WWTPs. To conquer the incidence and spread of ARGs, treatment method is capable to demolish ARGs by inactivating pathogens (McKinney and Pruden 2012). There are various treatment methods of wastewater like sedimentation, activated sludge, aerobic and anaerobic (combined treatment) process, disinfection, constructed wetlands and nanomaterial based treatments. Primary treatment of wastewater treatment involved in removal of large amounts of dissolved materials before discharge. Biological treatment can remove suspended solids, antibiotics, ARGs and ARB from wastewater using anaerobic, aerobic processes and combined treatment in an efficient manner. Tertiary treatment is the last treatment step in which wastewater is treated with disinfectants to inactivate the pathogens.

Biodegradation and biosorption are considered the primary mechanisms to eliminate antibiotics, ARB and ARGs in biological process. Aerobic treatment process occurs in the presence of microbes, which converts organic molecules into carbon dioxide and biomass. Under aerobic process genes like *suII, tetA, tetW* and *ermB* have been removed up to 85% at 20 °C and 13 day hydraulic residence time (Diehl and Para 2010). Anaerobic treatment process occurs in the presence of microbes, do not need oxygen and transfer organic compounds into methane and CO_2. Aanaerobic and anoxic treatment is more efficient to eliminate ARGs relatively than the aerobic process because microbes have lower activity in anaerobic condition which inhibited the propagation of resistance genes (*tet, intI* and *sul*) (Du et al. 2015). Anaerobic-aerobic sequence bioreactors (AAS) are more effective and removed 85% of ARGs over aerobic/anaerobic process (Christgen et al. 2015). Barancheshme and Munir (2018) observed that aerobic and anaerobic process when coupled in a waste water treatment performed better. The reduction of ARGs in digesters depends on the operating conditions like season of the year, reactor design, types of the ARGs targeted temperature, retention time and pH of the wastewater. Reduction in ARGs such as *suII, suIII, tetG, tetX* and *tetC* was observed with long sludge retention time (10–20 days) and pH 9–11 (Ma et al. 2011). Huang et al. (2017) found that alkaline pH inhibit the removal of ARGs by limiting the transferability of plasmids. On the other hand, ARGs resistant to erythromycin and tetracycline increased under anaerobic process (Aydin et al. 2016).

Biological treatments have proved to be well-organized for antibiotics removal at certain area within ng/L range is still estimated in biological reactors (He et al. 2015). Tetracyclines, quinolones, macrolides and sulphonamides antibiotics have been removed up to 17–100% from sewage plants in Xinjiang in anaerobic, aerobic and combined treatments (Wang et al. 2017; Liu et al. 2017).

Membrane Bioreactors are method based on mass separation of different molecular size with fine pore size membrane which act as barrier and eliminate the undesired materials from liquid or gaseous mixture. This process operates in aerobic/anaerobic conditions and recently used to remove ARB and ARGs effectively. Wang et al. (2020a, b, c) reported that ampicillin, tetracycline, ciprofloxacin and erythromycin, ARB (*E. coli, Aeromonas* and *Bacteroides*) and ARGs such as tetO, tetW, ermB and intl1 were removed efficiently in with pore size of $<0.4\mu m$.

Constructed wetlands (CWs) are constructed to stimulate normal processes to disinfect water and used to treat cattle wastewater (Abou-Elela and Hellal 2012). This process involved in reduction of BOD, nitrogen and phosphorus, COD and emerging contaminants (Matamoros et al. 2017). It represents as an alternative approach to diminish antibiotics and ARGs from wastewater (Dires et al. 2018). These systems depend on flow types, season and different plant species (Fang et al. 2017). CWs promote uptake of persistent micropollutants by plants from substrate and reduce their occurrence in wastewater effluent (Du et al. 2020). Antibiotics such as sulphonamide were easily degraded as compared to other antibiotics in the presence of *Phragmites australis* and associated microorganism (Choi et al. 2016). CWs reduced upto 97% ARGs from domestic effluents under aerobic conditions (Huang et al. 2019).

Disinfection of water is tertiary treatment frequently used at the ending of wastewater treatment procedure to diminish the pathogenic microbes. It causes inactivation of pathogens and destruction of genes involved in antibiotic resistance by using different techniques like chlorination and ozonation or combination of both, UV irradiation, activated carbon adsorption and coagulation kills a significant percentage of pathogenic microorganisms which causes bacterial, parasitic and viral disease and effective method to eliminate the ARB and ARGs from waste water (Pang et al. 2016). UV radiations can spoil the nucleic acids in microbial cells thus involved in reduction in abundance of ARGs. Sulfadiazine resistant bacteria were inactivated using chlorination (Yuan et al. 2015). Chlorination treatment significantly reduces the total ARGs (Thakali et al. 2020). Chlorination and UV in combination removed 100% of pathogenic microorganism under laboratory conditions (Zhang et al. 2019; Wang et al. 2020a, b, c). Elimination of ARB (*Enterobacteria, Enterococci* and *Staphylococci*) and ARGs (blaVIM, ermB and ampC) upto 60 to 98.9% and 18.7–99.3% during ozoniation process (Chang et al. 2017).

Concentration of antibiotics in wastewater from hospitals can be estimated as being ten times higher as compared to sewage from households. Separate treatment of hospital effluent is recommended. Release of unprocessed water into the environment has to be avoided. WHO suggests that the subsequent treatments steps must be incorporated such as mechanical, biological and tertiary treatment to reduce filterable substances less than 10 mg/L concentration and finally disinfection using chlorine and UV radiations. Sludge's from plants require appropriate digestion to attain suitable level of hygienisation. Otherwise the sludges can be dry in sludge beds for later combustion.

3.1 Nanomaterial Based Treatment of Wastewater

Metal nanoparticles have been used in developing new drugs with antibiotics and pharmaceuticals agents against AMR (Kyzioł et al. 2020). Nanomaterials can be modified to act as antimicrobial agents by their functionalized properties with existing antibiotics to boost their antimicrobial efficiency. Nanocellulose materials act as original waste water treatment as they could be modified as nanofibres having adsorption capacities to eliminate heavy metals with a wide efficiency of 63–94% (Abouzeid et al. 2019). Nanomaterials have high surface area, retention capacity, inert and have elevated energetic value and promising substitute for wastewater treatment. Among the metals such as iron and silver have antimicrobial activities (Yang et al. 2019). Silver nanomaterials inhibit the growth of *E. coli, Bacillus, S. aureus, P.aeruginosa* and *Klebsiella pneumonia*. Despite the antimicrobial properties of nanomaterials these can also show the toxic properties and promotes AMR, ARGs and develop resistance towards metals by using different mechanisms such as biofilm formation and metal resistance genes. Waseem et al. (2020) reported that pathogens can be destroyed and their genes transferred into other microbes.

4 ARBs and ARGs in Sewage Management Plants

ARBs on the behalf of several mechanism of resistance available in nature such as defence of target site, alteration of target, modification of antibiotic and reduced efflux pump permeability were found in waste water treatment plants (WWTPs). The resistance genes towards trimethoprim, macrolides and quinolones antibiotic groups are the most prevalent in the influent and effluent of treatment plants followed by tetracycline, β-lactam and lincomycins as these drugs are frequently used in medication and have firm structure (Wang et al. 2020a, b, c). Antibiotics and their metabolites into hospital and public sewage are the result of their high intake, inappropriate drug disposal practices or excreta material of animals or humans. In the climate, not only are antibiotics chemical contaminants capable of exerting lethal effects. This process resulted in the exclusion of susceptible microbes and the endurance of resistant microbiota which allocate them to conquer the undesirable effects of these antibiotics (Birosova et al. 2014). Because of their specificity and physicochemical properties, WWTPs are reservoirs of resistance places wherever the resistance spreading occurrence is especially severe, whereas antibiotics, antibiotic-resistant microbes, and resistance-determining genes are viewed as a new form of environmentally hazardous pollutants, which are so far underestimated (Rysz and Alvarez 2004; Pruden et al. 2006; Czekalski et al. 2015). In addition, WWTPs obtain waste from a variety of sources and bacteria from diverse habitats, allowing the bacteria to communicate and share genes horizontally and proving to be the primary source of antibiotics, ARB and ARG (Karkman et al. 2018). Despite the imminent need for a global framework to track antibiotic resistance and the economic and health burden it creates, knowledge concerning the proliferation of resistance in the ecosystem is still surprisingly limited. In respect of this issue, some of the common antibiotic genes are discussed below that are reported from sewage treatment plants (STPs) (Table 1).

4.1 β-Lactamase Resistance Genes

This class of antibiotic includes penicillin, cephalosporins, carbapenems and monobactam as these drugs have a common structural feature, which is beta-lactam ring that is responsible for their antimicrobial property. β-lactam antibiotics block the peptidoglycan synthesis by acylating the transpeptidase (penicillin binding proteins) that are normally involved in cross-linking of peptide chins to form peptidoglycan wall. In turn, this binding disrupts the process of terminal transpeptidation with in the bacterial cell and causes cell lysis (Eckburg et al. 2019). β-lactam antibiotics are currently one of the most frequently prescribed classes of drugs (Van Boeckel et al. 2014). Watkinson et al. (2009) measured the utmost amount of penicillin G (29 ng/L) from raw sewage, while in effluent the concentration of about 300 ng/L was detected. The cefalexin concentration (64,000 ng/L) was also found in influent, while in effluent up to 5070 ng/L concentration was estimated. As we know that WWTPs is a hostile environment for flourishing microbial growth and making them an exposure to the antibiotics waste present in the sewage. This exposure leads to the development of resistance towards different antibiotics. By the 1950s, resistance to the first antibiotic penicillin had become a significant clinical problem (Spellberg and Gilbert 2014; Lobanovska and Pilla 2017). Unfortunately, microbes developed resistance to almost all antibiotics. Bacteria are immune to β-lactam through a variety of mechanisms. Among them, first criteria to become resistant are the production of lactamase enzymes that is very common and important mechanism of resistance. Some bacteria develop resistance by lowering their affinity towards these antibiotics by altering in the active site penicillin binding proteins. Sometimes, bacteria can diminish the expression of outer membrane proteins and finally, the efflux pumps can prevent the entry of β-lactam antibiotics to the cell (Drawz and Bonomo 2010; Shaikh et al. 2015). β-lactamases are extensively scattered in the bacterial kingdom and play a major role in intrinsic and extrinsic resistance to β-lactams, which are encoded chromosally or by plasmids. Basically three types of genes (*amp, bla, oxa*) are involved in the degradation of beta-lactam ring. Abraham

Table 1 Occurrence of ARB and ARGs in sewage waste water in different regions

S. No	Bacteria/Phylum	ARGs	Genetic elements	Key Findings	References
1.	Vancomycin resistant-*Enterococci*, Methicillin resistant *Staphylococci*	*van*A, *mec*A and *amp*C	ND	They investigated drinking water quality and municipal wastewater plants for the presence of ARGs.	Schwartz et al. (2003)
2.	*Bacteroides*	*tet (A,X,W)*	Class 1 integrons (*int*1)	This study suggested that the tertiary treated municipal wastewater is a spot of disseminating ARGs into Duluth-Superior Harbor, St. Louis, Unites states which is considered as the largest freshwater port in the world	LaPara et al. (2011)
3.	*E.coli* isolates	(tetracycline, β-lactam, aminoglycosides and quinolones)	Insertion sequences elements	This study reported the prevalence of ARGs in two treatment steps of WWTPs i.e., activated sludge (AS) treatment and physico-chemical treatment. AS treatment increases in the abundance of ARGs in contrast to other treatment process	Biswal et al. (2014)
4.	ND	*tetA,tetE, qnrB, sul2, tetH, tetS, tetB, tetX and tetG*	ND	In this study, author reported different ARGs from two WWTPs in northern China. They concluded that the	Mao et al. (2015)

(continued)

Table 1 (continued)

S. No	Bacteria/Phylum	ARGs	Genetic elements	Key Findings	References
				proliferation of ARB was higher in effluent as compare to influent which is correlated with the selective pressure of antibiotics and heavy metals on ARGs.	
5.	ND	*erm*(F), *tetP* (A) and *tetP*(B)	Transposon (Tn25) and class 1 integron	This research detected several ARGs and analysed transposases in wastewater. They reported that quantity of most of the genes was highest in influent and lower in effluent water and sludge.	Karkman et al. (2016)
6.	ND	*Sul*1, *tet*(A,C,E), *qnr*S(winter) and *sul*(1,3), *tet*(A,C, E), *qnr*S	Integron (*intI*1)	Study was carried out in domestic sewage (integrated surface flow constructed wetland) in China. The occurrence of different ARGs were observed in winter and summer season and entailing that mobile genetic elements play important role in ARGs dissemination	Fang et al. (2017)
7.	*Proteobacteria, Bacteroidetes* and *Firmicutes*	*bla* (NDM, VIM, IMP,KPC and OXA-48), *sul4*,	Class 1 integrase	Hospital waste water is the point source for the distribution	Marathe et al. (2019)

(continued)

Table 1 (continued)

S. No	Bacteria/Phylum	ARGs	Genetic elements	Key Findings	References
		mphE and bla_{RSA1}		of ARB and ARGs in the environment. They explored the hospital wastewater from India and demonstrated the diversity of carbapenemases has been extended	
8.	ND	*tet* (*B,K,L,O*) and *sulIII*	Mobile genetic elements	Raw influent and final effluent containing these ARGs reported from WWTPs of Poland	Pazda et al. (2020)

*ND not discussed

and chain, 1940, reported first bacterial enzyme (AmpC β-lactamase) from *E. coli* that was responsible for the destruction of penicillin. In a study conducted by Piotrowska et al. (2017), isolated beta-lactamases genes from urban WWTPs and checked their occurrence among *Aeromonas* sp. They identified different genes bla_{OXA}, $bla_{CTX\text{-}M}$, bla_{MOX}, bla_{TEM}, bla_{ACC}, bla_{SHV} bla_{FOX}, bla_{GES}, bla_{PER}, bla_{VEB}, bla_{KPC}, *cphA, imiH,* and *cepH* belonging to 14 families. The presence of these beta lactamases suggested that treatment plants take part in spreading of ARGs.

4.2 Sulphonamide Resistance Genes

These antibiotics belong to sulpha-related group of antibiotics which contain SO_2-NH_2 moiety. In today's world, sulfonamides are not used very frequently because of their limited use. Sulfonamides are the derivatives of sulfanilamide, which obstruct the synthesis of folic acid in bacterial cell by competitively inhibiting the enzyme DHPS, dihydropteroate synthase. The condensation of p-aminobenzoic acid (PABA) and 7,8-dihydro-6-hydroxymethylpterin-pyrophosphate (DHPPP) to form dihydropteroic acid is catalyzed by DHPS enzyme and this acid play in the development of dihydrofolic acid, which is essential for the creation of DNA and proteins. The *sul*1, *sul*2 and *sul*3 genes, which encode DHPS with a low affinity for sulfonamides, are primarily responsible for sulfonamide tolerance. These genes are mostly present in bacterial species and commonly located in mobile genetic elements (transposons and mobilizable plasmids) that manifest antibiotic resistance (Byrne-Bailey et al. 2009).

4.3 Macrolides Resistance Genes

The class of macrolides consists of a large macrocyclic lactone ring to which it is possible to bind one or more deoxy sugars (cladinosis and desosamine). Macrolides are a form of polyketide natural product with antibiotic or antifungal activity that is used as a prescription medication. There are currently five macrolide antibiotics available: erythromycin, clarithromycin, azithromycin, fidaxomicin and telithromycin, the latter being a ketolide associated with it. The mechanism of their action is to inhibit the initial phases of protein synthesis by binding to the bacterial ribosome subunit 50S (Pazda et al. 2019). The widespread use of these antibiotics has led to the development of strains resistant to macrolides, especially *Streptococcus pneumoniae, Streptococcus pyogenes* and *Staphylococcus aureus.* The main mechanism of macrolide resistance is based on the synthesis of methylase enzyme, which is responsible for the methylation of 23S rRNA, which is the main site of the antibiotic's action and encoded by *erm* genes. Other genes are also involved in developing resistance for e.g. *ere* genes are responsible for the cleavage of lactone ring, *mph*(A) and *mph*(B) encoded enzyme MPH enzyme (macrolide phosphotransferases) which inactivate the antibiotic. Li et al. (2013), observed the presence of different antibiotics in sewage sludge from various treatment plants in China and the macrolide content of 3.6–69.9 ng/g was found. According to Bielen et al. (2017), high concentrations of macrolide antibiotics in waste water were recently discovered in a Croatian pharmaceutical manufacturing facility synthesising the azithromycin. In contrast, elevated levels of *mph*, *mef*, *msr* and new *erm* resistance genes and azithromycin-resistant bacteria were also identified in these wastewaters.

4.4 Tetracycline Resistance Genes

Tetracycline comes under the broad spectrum category and due to this spectrum of activity, irrelative safety and low cost, following the discovery of penicillin, tetracycline became widely used around the world. This antibiotic class contains chlortetracycline, doxycycline, minocycline, oxytetracycline, anhydrotetracycline and 6-thiatetracycline (Van Hoek et al. 2011). It is well known that this antibiotic, preventing the interaction of amino acyl tRNA with the bacterial ribosome and responsible for bacterial protein synthesis inhibition. This interaction is reversible which leads to the bacteriostatic effects of this group of antibiotics (Levy 1984; Chopra and Hawkey 1992). Various genes that encoded different proteins lead to the resistant nature in the bacterial community. Some are explained here such as *tet* [A

(C & P), B, K, L, V, Y and Z] genes caused reduction in intreacellular drug concentration and *ort*A, *tet*B(P), *tet*M, *tet*O, *tet*Q, *tet*S, *tet*T, *tet*W, *tet*X kind of genes altering the ribosome structure to prevent effective binding of antibiotic and developed resistance (Speer et al. 1992; Aminov 2001). Recently in a study conducted by Obayiuwana and Ibekwe (2020), observed that tetracycline efflux genes *tet*(A), *tet*(B) and *tet*(E) were the chief resistance genes known, but *tet* (E) genes was found in 18 of the bacterial isolates' genomes, making it the most common. *Tet* (A–C and E) efflux genes have been found in activated sludge.

5 Role of STPs in Transmission of ARGs in Bacterial World

One of the most significant challenges of the twenty-first century healthcare system is the rise of infections caused by antibiotic resistant bacteria (ARB). In several regions of the world, the Global Antimicrobial Surveillance System (GLASS) of the World Health Organization (WHO) recently showed an increased level of resistance factor in a number of severe bacterial infections. Resistant infections kill 700,000 people per year, but by 2050, the number of people killed by resistant infections would have been forecast to rise up to ten million. In low income countries, infections by AMR cause deaths in India and Thailand of 58,000 children and 38,000 adults, respectively (Fouz et al. 2020). Considering the present and potential production of antibiotic drugs, this condition is getting more serious. The ecosystem serves as a means of transmission of ARGs from animals to soil to water to sediments to waste water.

In nutrient-rich environments such as STPs, the mobilization of ARGs is improved by large bacteria cell densities, the ideal setting for ARG transfer events. Sewage as a main environmental pool of AMR has been highlighted by many studies as it represents the best environment for the persistence of ARGs (Bruchmann et al. 2013; Hembach et al. 2017). The position of ARGs on mobile genetic components like plasmids, transposons, and integrons, makes the transfer of resistance between bacteria of the same or different origins possible and easy to achieve (Allen et al. 2010). The identification of MGEs along with unique ARGs can therefore give a wide-ranging perception on antibiotic resistance and the distribution of ARGs (Lupo et al. 2012).

Mobile elements carrying ARGs have been originate in WWTP waste water and activated sludge (Zhuang et al. 2015; Guo et al. 2017). Recent research suggests that phages and phage-derived particles also take part in ARG distribution in the environment. The prevalence of ARGs poses a threat to their spread and incorporation into new bacterial contexts, especially in the bacteriophage fraction, which could result in the rise of resistant clones (Calero-Cáceres and Muniesa 2016). The dissemination of resistance nature in bacteria towards different antibiotics is generally acquired by two mechanisms: vertical (mutation) and horizontal (transformation, conjugation and transduction) evolutions. The main way for global transmission of resistance towards antibiotics is believed to be lateral gene transfer and is responsible for plasmids transfer which having ARGs (R plasmids) in gram negative bacteria (60–90%). Lawrence and Ochman, (1998) reported *E. coli* genes (17.6%) have been obtains by horizontal gene transfer. Depending on plant quality, biological sewage treatment reduces the viable count of faecal bacteria, for example coliforms and *Enterococci* (1-4 log units) (Martins da Costa et al. 2006). However in STP effluents, moderately high numbers of bacteria (103 cfu/mL) enterococci and carry antibiotic resistance genes notably in the influent, which are recorded to contain up to 103 mL coliform resistant antibiotic agent (Martins da Costa et al. 2006).

It was demonstrated that STPs are not intended to delete ARGs (Alexander et al. 2015). They found an improvement in the loads of certain ARGs (*amp*C, *bla*VIM) after traditional wastewater treatment in the bacterial community. Similarly, an effluent sample from WWTPs in Sweden, there was an increased concentration of three resistance genes (*bla*CTX-M, *tet*B and *dfr*A3) was recorded. Rahube et al. (2014) characterize different self transmissible, resistant plasmids in effluent of WWTPs (colE and IncP-1β) that carry ARGs encoding resistance towards tetracycline, chloramphenicol, beta-lactams and sulphonamide antibiotics. The IncP-1 is a group of highly promiscuous self transfer plasmids which is also capable in transferring non mobilizable plasmids. Mobile genetic elements (MGEs) facilitate the resistance towards antibiotics and quaternary ammonium compounds were originated within these plasmids (antibiotic resistance). The transmission rate of resistance genes may be influenced by external and internal factors to bacteria. External factors such as temperature, pH, detergents and other organic compounds facilitate DNA transferability to other bacteria. Internal factors like SOS response involved in transcription, metabolic changes and causes mutation which help in the survival and transfer of resistance characters. The presence of organic compounds and other physical parameters in sewage treatment plants is a key aspect for the attainment of antibiotic resistance among different bacteria.

6 Molecular Approach to Identifying ARGs

Recently, molecular approaches are exploited in studying the resistance profiles of antibiotic resistance bacteria by characterizing their genes and transfer pathways. Mainly, two types of techniques are employed in analyzing ARGs from different environmental samples i.e., cultivable dependent and culture independent approaches. Culture dependent involves antimicrobial susceptibility testing using pour plate methods. By using this, it was observed that the majority of ARB found in WWTP is typical fecal contamination markers including *E. coli*, total coliforms, and *Enterococci* and a wide range of clinically significant ARBs, such as methicillin-resistant *Staphylococcus aureus*, have been discovered (Figueira et al. 2011; Bouki et al. 2013).

Culture independent approaches i.e., q-PCR and metagenomic widely accepted to study the whole-community of ARB and ARBs from wastewater and environmental samples (Karkman et al. 2018). Genera including *Enterococcus* and *Escherichia*, which are frequently targeted using culture-dependent methods in WWTPs, were not found in nearly all abundant populations, according to metagenomic analyses. *Proteobacteria* were the most common bacteria, followed by *Actinobacteria* and genera such as *Bacillus*, *Staphylococcus*, *Lysinibacillus* and *Providencia*, were found in WWTP samples (Zhang et al. 2015). Functional metagenomic approach also play crucial role in studying functional role of ARGs. The limitations of metagenomic sequencing in detecting resistance genes can be overcome by functional metagenomics, which involves cloning and expression of DNA in experimental laboratory. Environmental DNA is cloned in fragments (10–200 kb) in a host, such as *E. coli* and the resistance of host to various antibiotics are evaluated in functional metagenomics. Subcloning, mutagenesis, or *in silico* analysis are used to screen the clones by a resistance phenotype for the determinant of antibiotic-resistance, which can be time consuming and tedious. The difficulties of cloning and expressing in host are the key drawbacks of functional metagenomics, and can be resolved to several degree via use of new hosts not *E.coli*. Proteomics in combination with functional metagenomics is a new approach to avoid the time-consuming process of identifying possible clones that include all of the DNA segments. The expressed proteins can be characterized in a high-throughput manner using proteomics along with functional metagenomics, and the putative new resistance determinants identified by comparison a strain exclusive of the cloned DNA (Fouhy et al. 2015).

7 ARGs in Therapeutic Use

AMR markers are key biological markers for the analysis of ARGs in wastewater and afford a broader perception of the resistance genes in population (Sims and Kasprzyk-Hordern 2020). It was observed that integrons having genes to increase antibiotic resistance and can be use as marker of ARG in the atmosphere (Barraud et al. 2010). Thakali et al. (2020) in USA observed that Class 1 integrase gene (*intI1*) persist during the sewage treatment which can be used as an indicator for ARGs in effluents to check the spread of AMR. Presence of *mcr-1*, *mecA*, *ermB*, *sul1*, $bla_{OXA\text{-}1}$ and t*etW* antimicrobial resistance genes were used for the treatment or indicator of methicillin, erythromycin, sulphonamide, beta-lactam and tetracycline resistant bacteria.

8 Conclusion and Future Prospects

The spreading of AMR is one of the main challenges faced by mankind in public fitness domains. The rising water insufficiency related to climate alteration requires developing modern wastewater recycle practices. An appropriate management of wastewater is crucial prior to its exonerate into aquatic system to circumvent the increase of ARB in surroundings. Direct use of sludge should be avoided to stop release of ARB to protect farmer's health and further assessed to know the pathways of ARB and ARG. In addition, understanding the transferability of resistance genes would be important to develop our capability to recognize the risk scenarios. Measures should be taken to check the utilization of antibiotic practices in human and animal medicines. It is suggested to observe the incidence of antibiotics and genes to prevent their possible ecological risk. Use of Omics approaches provides better understanding on fate of ARGs in wastewater. Further future research is required on these antibiotic substances and ARGs using molecular based approaches.

References

Abou-Elela SI, Hellal MS (2012) Municipal wastewater treatment using vertical flow constructed wetlands planted with Canna. Phragmites Cyprus Ecol Eng 47:209–213

Abouzeid RE, Khiari R, El-Wakil N, Dufresne A (2019) Current state and new trends in the use of cellulose nanomaterials for wastewater treatment. Biomacromolecules 20:573–597

Alexander J, Bollmann A, Seitz W, Schwartz T (2015) Microbiological characterization of aquatic microbiomes targeting taxonomical marker genes and antibiotic resistance genes of opportunistic bacteria. Sci Total Environ 512–513:316–325

Alexander J, Hembach N, Schwartz T (2020) Evaluation of antibiotic resistance dissemination by wastewater treatment plant effluents with different catchment areas in Germany. Sci Rep 10:8952

Allen HK, Donato J, Huimi Wang H, Cloud-Hansen KA, Davies J, Handelsman J (2010) Call of the wild: antibiotic resistance genes in natural environments. Nat Rev Microbiol 8(4):251–259

Aminov RI (2001) Molecular ecology of tetracycline resistance: development and validation of primers for detection of tetracycline resistance genes encoding ribosomal protection proteins. Appl Environ Microbiol 67:22–32

Aydin S, Ince B, Ince O (2016) Assessment of anaerobic bacterial diversity and its effects on anaerobic system stability and the occurrence of antibiotic resistance genes. Bioresour Technol 207:332–338

Barancheshme F, Munir M (2018) Strategies to combat antibiotic resistance in the wastewater treatment plants. Front Microbiol 8:2603

Barraud O, Baclet MC, Denis F, Ploy MC (2010) Quantitative multiplex real-time PCR for detecting class 1, 2 and 3 integrons. J Antimicrob Chemother 65:1642–1645

Bielen A, Šimatović A, Kosić-Vukšić J, Senta I, Ahel M, Babić S, Udiković-Kolić N (2017) Negative environmental impacts of antibiotic-contaminated effluents from pharmaceutical industries. Water Res 126:79–87

Birosova L, Mackulak T, Bodik I, Ryba J, Skubak J, Grabic R (2014) Pilot study of seasonal occurrence and distribution of antibiotics and drug resistant bacteria in wastewater treatment plants in Slovakia. Sci Total Environ 490:440–444

Biswal BK, Mazza A, Masson L, Gehr R, Frigon D (2014) Impact of wastewater treatment processes on antimicrobial resistance genes and their co-occurrence with virulence genes in *Escherichia coli*. Water Res 50:245–253

Boonyasiri A, Tangkoskul T, Seenama C, Saiyarin J, Tiengrim S, Thamlikitkul V (2014) Prevalence of antibiotic resistant bacteria in healthy adults, foods, food animals, and the environment in selected areas in Thailand. Pathog Glob Health 108(5):235–245

Bouki C, Venieri D, Diamadopoulos E (2013) Detection and fate of antibiotic resistant bacteria in wastewater treatment plants: a review. Ecotoxicol Environ Saf 91:1–9

Bruchmann J, Kirchen S, Schwartz T (2013) Sub-inhibitory concentrations of antibiotics and wastewater influencing biofilm formation and gene expression of multi-resistant *Pseudomonas aeruginosa* wastewater isolates. Environ Sci Pollut Res Int 20:3539–3549

Byrne-Bailey KG, Gaze WH, Kay P, Boxall AB, Hawkey PM (2009) Prevalence of sulfonamide resistance genes in bacterial isolates from manured agricultural soils and pig slurry in the United Kingdom. Antimicrob Agents Chemother 53:696–702

Calero-Cáceres W, Muniesa M (2016) Persistence of naturally occurring antibiotic resistance genes in the bacteria and bacteriophage fractions of wastewater. Water Res 95:11–18

Chang PH, Juhrend B, Olson TM, Marrs CF, Wigginton KR (2017) Degradation of extracellular antibiotic resistance genes with UV254 treatment. Environ Sci Technol 51:6185–6192

Chee-Sanford JC, Mackie RI, Koike S, Krapac IG, Lin Y-F, Yannarell AC, Maxwell S, Aminov RI (2009) Fate and transport of antibiotic residues and antibiotic resistance genes following land application of manure waste. J Environ Qual 38(3):1086–1108

Choi YJ, Kim LH, Zoh KD (2016) Removal characteristics and mechanism of antibiotics using constructed wetlands. Ecol Eng 91:85–92

Chopra IPM, Hawkey HM (1992) Tetracyclines, molecular and clinical aspects. J Antimicrob Chemother 29:245–277

Christgen B, Yang Y, Ahammad SZ, Li B, Rodriquez DC, Zhang T (2015) Metagenomics shows that low-energy anaerobic-aerobic treatment reactors reduce antibiotic resistance gene levels from domestic wastewater. Environ Sci Technol 49:2577–2584

Czekalski N, Sigdel R, Birtel J, Matthews B, Bürgmann H (2015) Does human activity impact the natural antibiotic resistance background? Abundance of antibiotic resistance genes in 21 Swiss lakes. Environ Int 81:45–55

Devarajan N, Laffite A, Graham ND, Meijer M, Prabakar K, Mubedi JI (2015) Accumulation of clinically relevant antibiotic-resistance genes, bacterial load, and metals in freshwater lake sediments in Central Europe. Environ Sci Technol 49:6528–6537

Di Cesare A, Eckert EM, D'Urso S, Bertoni R, Gillan DC, Wattiez R (2016) Co-occurrence of integrase 1, antibiotic and heavy metal resistance genes in municipal wastewater treatment plants. Water Res 94:208–214

Diehl D, LaPara T (2010) Effect of temperature on the fate of genes encoding tetracycline resistance and the integrase of class 1 integrons within anaerobic and aerobic digesters treating municipal wastewater solids. Environ Sci Technol 44:9128–9133

Dires S, Birhanu T, Ambelu A, Sahilu G (2018) Antibiotic resistant bacteria removal of subsurface flow constructed wetlands from hospital wastewater. J Environ Chem Eng 6:4265–4272

Drawz SM, Bonomo RA (2010) Three decades of β-lactamase inhibitors. Clin Microbiol Rev 23 (1):160–201

Du J, Geng J, Ren H, Ding L, Xu K, Zhang Y (2015) Variation of antibiotic resistance genes in municipal wastewater treatment plant with A2O-MBR system. Environ Sci Pollut Res 22:3715–3726

Du L, Zhao Y, Wang C, Zhang H, Chen Q, Zhang X, Zhang L, Wu J, Wu Z, Zhou Q (2020) Removal performance of antibiotics and antibiotic resistance genes in swine wastewater by integrated vertical-flow constructed wetlands with zeolite substrate. Sci Total Environ 721:137765

Eckburg PB, Lister T, Walpole S, Keutzer T, Utley L, Tomayko J, Kopp E, Farinola N, Coleman S (2019) Safety, tolerability, pharmacokinetics, and drug interaction potential of SPR741, an intravenous potentiator, after single and multiple ascending doses and when combined with β-lactam antibiotics in healthy subjects. Antimicrob Agents Chemother 63(9)

Fang H, Zhang Q, Nie X, Chen B, Xiao Y, Zhou Q, Liang X (2017) Occurrence and elimination of antibiotic resistance genes in a long-term operation integrated surface flow constructed wetland. Chemosphere 173:99–106

Figueira V, Serra E, Manaia CM (2011) Differential patterns of antimicrobial resistance in population subsets of *Escherichia coli* isolated from waste- and surface waters. Sci Total Environ 409:1017–1023

Fouhy F, Stanton C, Cotter PD, Hill C, Walsh F (2015) Proteomics as the final step in the functional metagenomics study of antimicrobial resistance. Front Microbiol 6:172

Fouz N, Pangesti KNA, Yasir M, Al-Malki AL, Azhar EI, Hill-Cawthorne GA, Abd El Ghany M (2020) The contribution of wastewater to the transmission of antimicrobial resistance in the environment: implications of mass gathering settings. Trop Med Infect Dis 5(1):33

Guo J, Li J, Chen H, Bond PL, Yuan Z (2017) Metagenomic analysis reveals wastewater treatment plants as hotspots of antibiotic resistance genes and mobile genetic elements. Water Res 123:468–478

He K, Soares AD, Adejumo H, McDiarmid M, Squibb K, Laney L (2015) Detection of a wide variety of human and veterinary fluoroquinolone antibiotics in municipal wastewater and wastewater-impacted surface water. J Pharm Biomed Anal 106:136–143

Hembach N, Schmid F, Alexander J, Hiller C, Rogall ET, Schwartz T (2017) Occurrence of the mcr-1 colistin resistance gene and other clinically relevant antibiotic resistance genes in microbial populations at different municipal wastewater treatment plants in Germany. Front Microbiol 8:1–11

Hembach N, Alexander J, Hiller C, Wieland A, Schwartz T (2019) Dissemination prevention of antibiotic resistant and facultative pathogenic bacteria by ultrafiltration and ozone treatment at an urban wastewater treatment plant. Sci Rep 9:1–12

Huang H, Zheng X, Chen Y, Liu H, Wan R, Su Y (2017) Alkaline fermentation of waste sludge causes a significant reduction of antibiotic resistance genes in anaerobic reactors. Sci Total Environ 580:380–387

Huang XF, Ye GY, Yi NK, Lu LJ, Zhang L, Yang LY, Xiao L, Liu J (2019) Effect of plant physiological characteristics on the removal of conventional and emerging pollutants from aquaculture wastewater by constructed wetlands. Ecol Eng 135:45–53

Jechalke S, Heuer H, Siemens J, Amelung W, Smalla K (2014) Fate and effects of veterinary antibiotics in soil. Trends Microbiol 22:536–545

Jelic A, Rodriguez-Mozaz S, Barceló D, Gutierrez O (2015) Impact of in-sewer transformation on 43 pharmaceuticals in a pressurized sewer under anaerobic conditions. Water Res 68:98–108

Karkman A, Johnson TA, Lyra C, Stedtfeld RD, Tamminen M, Tiedje JM, Virta M (2016) High-throughput quantification of antibiotic resistance genes from an urban wastewater treatment plant. FEMS Microbiol Ecol 92:014

Karkman A, Do TT, Walsh F, Virta MPJ (2018) Antibiotic-resistance genes in waste water. Trends Microbiol 26:220–228

Kyzioł A, Khan W, Sebastian V, Kyzioł K (2020) Tackling microbial infections and increasing resistance involving formulations based on antimicrobial polymers. Chem Eng J 385:123888

LaPara TM, Burch TR, McNamara PJ, Tan DT, Yan M, Eichmiller JJ (2011) Tertiary-treated municipal wastewater is a significant point source of antibiotic resistance genes into Duluth-Superior Harbor. Environ Sci Technol 45(22):9543–9549

Levy SB (1984) Resistance to the tetracyclines. In: Bryan LE (ed) Antimicrobial drug resistance. Academic, Orlando, FL, pp 191–240

Li W, Shi Y, Gao L, Liu J, Cai Y (2013) Occurrence, distribution and potential affecting factors of antibiotics in sewage sludge of wastewater treatment plants in China. Sci Total Environ 445:306–313

Liu J, Lu J, Tong Y, Li C (2017) Occurrence and elimination of antibiotics in three sewage treatment plants with different treatment technologies in Urumqi and Shihezi, Xinjiang. Water Sci Technol 75:1474–1484

Lobanovska M, Pilla G (2017) Focus: drug development: penicillin's discovery and antibiotic resistance: lessons for the future? Yale J Biol Med 90(1):135

Lupo A, Coyne S, Berendonk T (2012) Origin and evolution of antibiotic resistance: the common mechanism of emergence and spread in water bodies. Front Microbiol 3

Ma Y, Wilson CA, Novak JT, Riffat R, Aynur S, Murthy S, Pruden A (2011) Effect of various sludge digestion conditions on sulfonamide, macrolide, and tetracycline resistance genes and class I integrons. Environ Sci Technol 45:7855–7861

Mao D, Yu S, Rysz M, Luo Y, Yang F, Li F, Alvarez PJJ (2015) Prevalence and proliferation of antibiotic resistance genes in two municipal wastewater treatment plants. Water Res 85:458–466

Marathe NP, Berglund F, Razavi M (2019) Sewage effluent from an Indian hospital harbors novel carbapenemases and integron-borne antibiotic resistance genes. Microbiome 7:97

Martins da Costa P, Vaz-Pires P, Bernardo F (2006) Antimicrobial resistance in *Enterococcus* spp. isolated in inflow, effluent and sludge from municipal sewage water treatment plants. Water Res 40(8):1735–1740

Matamoros V, Rodríguez Y, Bayona JM (2017) Mitigation of emerging contaminants by full-scale horizontal flow constructed wetlands fed with secondary treated wastewater. Ecol Eng 99:222–227

McKinney CW, Pruden A (2012) Ultraviolet disinfection of antibiotic resistant bacteria and their antibiotic resistance genes in water and wastewater. Environ Sci Technol 46:13393–13400

Obayiuwana A, Ibekwe AM (2020) Antibiotic resistance genes occurrence in wastewaters from selected pharmaceutical facilities in Nigeria. Water 12:1897

Pang Y, Huang J, Xi J, Hu H, Zhu Y (2016) Effect of ultraviolet irradiation and chlorination on ampicillin-resistant *Escherichia coli* and its ampicillin resistance gene. Front Environ Sci Eng 10:522–530

Pazda M, Kumirska J, Stepnowski P, Mulkiewicz E (2019) Antibiotic resistance genes identified in wastewater treatment plant systems – a review. Sci Total Environ 134023

Pazda M, Rybicka M, Stolte S, Piotr Bielawski K, Stepnowski P, Kumirska J, Mulkiewicz E (2020) Identification of selected antibiotic resistance genes in two different wastewater treatment plant systems in Poland: a preliminary study. Molecules 25(12):2851

Piotrowska M, Przygodzińska D, Matyjewicz K, Popowska M (2017) Occurrence and variety of β-lactamase genes among aeromonas spp isolated from Urban wastewater treatment plant. Front Microbiol 8

Pruden A, Pei R, Storteboom H, Carlson KH (2006) Antibiotic resistance genes as emerging contaminants: studies in northern Colorado. Environ Sci Technol 40:7445–7450

Rahube TO, Viana LS, Koraimann G, Yost CK (2014) Characterization and comparative analysis of antibiotic resistance plasmids isolated from a wastewater treatment plant. Front Microbiol 5:558

Rysz M, Alvarez PJJ (2004) Amplification and attenuation of tetracycline resistance in soil bacteria: aquifer column experiments. Water Res 38:3705–3712

Schwartz T, Kohnen W, Jansen B, Obst U (2003) Detection of antibiotic-resistant bacteria and their resistance genes in wastewater, surface water, and drinking water biofilms. FEMS Microbiol Ecol 43(3):325–335

Shaikh S, Fatima J, Shakil S, Rizvi SMD, Kamal MA (2015) Antibiotic resistance and extended spectrum beta-lactamases: types, epidemiology and treatment. Saudi J Biological Sci 22:90–101

Sims N, Kasprzyk-Hordern B (2020) Future perspectives of wastewater-based epidemiology: Monitoring infectious disease spread and resistance to the community level. Environ Int 139:105689

Speer BS, Shoemaker NB, Salyers AA (1992) Bacterial resistance to tetracycline: mechanisms, transfer, and clinical significance. Clin Microbiol Rev 5:387–399

Spellberg B, Gilbert DN (2014) The future of antibiotics and resistance: a tribute to a career of leadership by John Bartlett. Clin Infect Dis 59:71–75

Thakali O, Brooks JP, Shahin S, Sherchan SP, Haramoto E (2020) Removal of antibiotic resistance genes at two conventional wastewater treatment plants of Louisiana, USA. Water 12:1729

Van Boeckel TP, Gandra S, Ashok A, Caudron Q, Grenfell BT, Levin SA, Laxminarayan R (2014) Global antibiotic consumption 2000 to 2010: an analysis of national pharmaceutical sales data. Lancet Infect Dis 14:742–750

Van Hoek A, Mevius D, Guerra B, Mullany P, Roberts A, Aarts H (2011) Acquired antibiotic resistance genes: an overview. Front Microbiol 2:203

Wang M, Shen W, Yan L, Wang XH, Xu H (2017) Stepwise impact of urban wastewater treatment on the bacterial community structure, antibiotic contents, and prevalence of antimicrobial resistance. Environ Pollut 231:1578–1585

Wang J, Chu L, Wojnárovits L, Takács E (2020a) Occurrence and fate of antibiotics, antibiotic resistant genes (ARGs) and antibiotic resistant bacteria (ARB) in municipal wastewater treatment plant: an overview. Sci Total Environ 140997

Wang S, Ma X, Liu Y, Yi X, Du G, Li J (2020b) Fate of antibiotics, antibiotic-resistant bacteria, and cell-free antibiotic-resistant genes in full-scale membrane bioreactor wastewater treatment plants. Bioresour Technol 302:122825

Wang H, Wang J, Li S, Ding G, Wang K, Zhuang T, Huang X, Wang X (2020c) Synergistic effect of UV/chlorine in bacterial inactivation, resistance gene removal, and gene conjugative transfer blocking. Water Res 185:116290

Waseem H, Jameel S, Ali J, Jamal A, Ali MI (2020) Recent advances in treatment technologies for antibiotics and antimicrobial resistance genes. In Antibiotics and antimicrobial resistance genes, 1st ed.; Springer: Berlin, vol 2020, pp 395–413

Watkinson AJ, Murby EJ, Kolpin DW, Costanzo SD (2009) The occurrence of antibiotics in an urban watershed: from wastewater to drinking water. Sci Total Environ 407:2711–2723

WHO (2017) Global priority list of antibiotic-resistant bacteria to guide research, discovery, and development of new antibiotics. WHO 2017:1–7

Yang X, Wei Q, Shao H, Jiang X (2019) Multivalent aminosaccharide-based gold nanoparticles as narrow-spectrum antibiotics in vivo. ACS Appl Mater Interfaces 11:7725–7730

Yoo K, Yoo H, Lee J, Choi EJ, Park J (2020) Exploring the antibiotic resistome in activated sludge and anaerobic digestion sludge in an urban wastewater treatment plant via metagenomic analysis. J Microbiol 58:123–130

Yuan QB, Guo MT, Yang J (2015) Fate of antibiotic resistant bacteria and genes during wastewater chlorination: implication for antibiotic resistance control. PLoS One 10:e0119403

Zhang S, Han B, Gu J, Wang C, Wang P, Ma Y, Cao J, He Z (2015) Fate of antibiotic resistant cultivable heterotrophic bacteria and antibiotic resistance genes in wastewater treatment processes. Chemosphere 135:138–145

Zhang T, Hu Y, Jiang L, Yao S, Lin K, Zhou Y, Cui C (2019) Removal of antibiotic resistance genes and control of horizontal transfer risk by UV, chlorination and UV/chlorination treatments of drinking water. Chem Eng J 358:589–597

Zhuang Y, Ren H, Geng J, Zhang Y, Zhang Y, Ding L, Xu K (2015) Inactivation of antibiotic resistance genes in municipal wastewater by chlorination, ultraviolet, and ozonation disinfection. Environ Sci Pollut Res 22:7037–7044

Required Quality of Sewage Sludge as an Agricultural Soil Amendment

Muhammad Ashar Ayub, Muhammad Adnan, Wajid Umar, Zia Ur Rahman Farooqi, Muhammad Awais, Hamaad Raza Ahmad, Bisma Imran Ch, and Ayesha Siddique

1 Introduction

The wastewater produced from communities or households is commonly called sewage where a blend of water and solid produced during its treatment is called as sludge (Kacprzak et al. 2017). Sewage sludge (SS) is the basic product of wastewater which includes nutrients, organic matter and harmful constituents i.e., micro-pollutants, pathogens, and heavy metals (European Commission 2001; Laura et al. 2020). The application of wastewater, SS and excreta to the soil is a common method all over the world. For many decades in northern Europe, China and many Asian countries are utilizing animal and human excreta as fertilizers or manures (Meena et al. 2020), it has been proven beneficial for soil properties to some extent (Roig et al. 2012). In countries like Brazil, the use of SS in the agriculture sector is expanding because it provides required limestone (88%), nitrogen (74%), P_2O_5 (73%) and K_2O_5 (35%) to crops (Bittencourt et al. 2014). In Tunisia Mediterranean regions, succeeding sewage sludge amendment improved the enzyme activities, microbial biomass and fertility of the soil (Hamdi et al. 2019). In Britain, it is used to enhance the forest areas fertility (Moffat 2006). Moreover, about 26.4% of sewage sludge was utilized for the cultivation of land and composting in agriculture in Poland (GUS (Poland)—Central Statistical Office 2018). The sludge is a great source to improve yields and can potentially provide nutrients to enhance the fertility

M. A. Ayub (✉) · Z. U. R. Farooqi · M. Awais · H. R. Ahmad · B. Imran Ch · A. Siddique
Institute of Soil and Environmental Sciences, University of Agriculture Faisalabad, Faisalabad, Pakistan

M. Adnan
Department of Agronomy, College of Agriculture, University of Sargodha, Sargodha, Pakistan

W. Umar
Institute of Environmental Science, Hungarian University of Agriculture and Life Sciences, Gödöllő, Hungary

V. D. Rajput et al. (eds.), *Sustainable Management and Utilization of Sewage Sludge*, https://doi.org/10.1007/978-3-030-85226-9_12

of the soil (Belaid 2015). Eid et al. (2019) stated that the application of sewage to soil is a safe practice and improve *Corchorus olitorius* plant growth and no heavy metal hazard was observed. It is reported that the dried sewage sludge contains organic matter (50–70%), mineral constituents (30–50%) including inorganic carbon (1–4%), P (0.5–2.5%), N (3.4–4.0%) and considerable amount of micronutrients (Samolada and Zabaniotou 2014). The reported work by Asgari and coworkers show that the crop yield was significantly increased in soils having a sufficient amount of amended sludge as compared to control (Asgari et al. 2019). Furthermore, the application of SS is economically and ecologically sustainable than other strategies of disposal (Kacprzak et al. 2017). However, they also contained a high concentration of numerous contaminants like heavy metals including Zn, Cu, Cd, Pb, As, Ni, Hg, Se and Cr that cause severe threats to the crop and food (Kacprzak et al. 2017; Raheem et al. 2018). So, the safer limit to use sewage sludge in agriculture crop production system have varied from country to country in all over the world (He et al. 2005).

Wastewater treatment-derived sewage sludge is very difficult to manage due to the presence of pathogenic organisms and heavy metals. Therefore, society and government have a keen interest in the safe and beneficial use of SS. The process of thermal processing of wastewater before its use provides a promising disposal option to manage this waste. The production of biochar from the wastewater SS is the beneficial and safe method of SS disposal/recycling as compared to the direct SS application to farmland (Zong et al. 2018). The formation of larger quantities of SS poses many disposal and environmental issues all over the world. However, their proper treatment and application to farmland improved soil and plant health along with a reduction in negative impacts on the environment. The biological, chemical and physical properties of soil as well as supply of nutrients improved by the utilization of sludge as organic amendment which has also increased soil fertility and crop production (Angin et al. 2017). The nutritious values of organic waste materials like sludge are endorsed as benefits and it may reduce the use of synthetic chemicals. Beside few beneficial aspects, poor quality sewage sludge can be a hazard for agriculture crops as well as human health as can be a sources of pollutants and pathogenic microbes. Due to this, some treatment forms should be adopted to make them appropriate for application on land (Yue et al. 2017).

The SS is a suspension of concentrated solids, mainly comprising of OM, usually rich in minerals. Moreover, it contains many trace elements (heavy metals) that may or may not be essential for the growth of plants. So, pressure is increasing to apply safe and purrified SS to agricultural lands because of ecological concerns allied with disposal of all these materials in land as this material is being produced in high volumes. The positive impact of SS on plant and soil has been documented by many researchers whereas SS contains soluble salts and HMs, which may cause phytotoxicity impacts on crops (Mohamed et al. 2018). The Table 1 summarizes the heavy metal permissible limits of various metals in SS categorized by different authorities.

Keeping in mind the importance and hazardous effects of sewage sludge in agriculture, present draft is compiled keeping in mind the advantages, limitation, and quality assurance of sewage sludge prior to using as a soil amendment.

Table 1 Permissible limits of heavy metals in sewage sludge extracted with permission from information compiled by Wiśniowska and Włodarczyk-Makuła (2018)

Legal act	Concentration limit (mg/kg d.m)							
	Cu	Ni	Cr	Pb	Zn	Hg	Cd	References
US legislation (Part 503)	4300	420	3000	840	7500	57	85	Inglezakis et al. (2014)
EU law	1000/ 1750	300–400	No limited	750/ 1200	2500/ 4000	16/ 25	20/ 40	Council Directive (1986)
Polish ministry of environment	1000	300	500	750	2500	16	20	Ordinance on sewage sludge (2015)
New proposal for a directive	1000	300	1000	500	2500	10	10	Working document on the bio-waste and sludge (2010)

2 Role of Sewage Sludge in Agricultural Production

Sewage sludge can only be defined as residual material (semi-solid) which is unavoidably leftover from the industrial or municipal processes of wastewater treatment. The quick increase in the population tied with industrialization has increased the production of SS manifolds, which may be speculated to rise in future. Design and engineering of the wastewater treatment plants play a significant role in the safe disposal, processing and reuse of sludge.

Wastewater is mainly treated via different methods (i) physically (sedimentation and flotation) (ii) chemically (flocculation) (iii) biologically (microbial treatment process). Moreover, wastewater treatment processes are grouped into the subsystems like primary, secondary and tertiary wastewater treatment. Usually, SS is utilized for energy generation or resource recovery. A typical SS contains 59–88% w/v O.M, 50–55% of carbon, 25–30% of oxygen, 10–15% of nitrogen, 6–10% of hydrogen with a slight amount of sulfur and phosphorus (Rehman et al. 2018). Although, the use of SS could be beneficial for agricultural soils to improve the organic matter and nutrients status of soils, but it also signifies risk due to contaminants i.e. organic compounds, pathogens and heavy metals (Lamastra et al. 2018). The SS derived amendments are commonly organic amendments that have many beneficial impacts on the biological, chemical and physical properties of agricultural soils. The SS derived compost enhances the structure of soil and preventing from the formation of crust, erosion and surface runoff. It also increases the porosity of soil, retention of water and hydraulic conductivity. Moreover, its application results in a complex volume of bonding, storage and residual pores. The collective impact of these changes in the differential porosity is very high than total porosity. Worldwide it is expected that about four billion tons of solid wastes are being produced per annum (Vaish et al. 2016) and it contains various nutrients which can be effectively reused (Kirchmann et al. 2017). It was already reported that crop yield in SS amended soils is higher than control (Asgari et al. 2019; Singh and Agrawal 2008) as it contains organic and inorganic matter, and microorganisms in suspended or dissolved forms (Raheem et al. 2018). Overall, dried SS contains organic matter (50–70%), mineral constituents (30–50%) including inorganic carbon (1–4%), P (0.5–2.5%), N (3.4–4.0%) and considerable micronutrients amounts which can be benificial for plants (Samolada and Zabaniotou 2014).

The application of SS in the agriculture production system is becoming common and accepted worldwide due to nutrients and organic matter presence (Sharma et al. 2017). It also improves the biological, chemical and physical properties of the soil (Singh and Agrawal 2008). It contains about 50% OM which is well recognised to enhance soil physical health like improving soil porosity, water-holding capacity and bulk density. Its application significantly improved organic matter contents from 1.38% to 4.83% (Eid et al. 2020). Moreover, the positive impact of SS application on plant growth are also well reported by many scientists (Eid et al. 2017a, b, 2018, 2019; Singh and Agrawal 2008). Similarly, Sharma et al. (2017) stated that SS improves crop yield and nutritional value. Furthermore, it was already documented that barley grain yield increased due to repeated application of SS (Antolín et al. 2005). In another work, SS was utilized as an alternative substrate for *Ailanthus altissima* growth as compared to normal soil because the SS led to more nutrient which results in better *A. altissima* growth (Liu et al. 2019). Moreover, its stable complexes organic structure also decreased the availability of heavy metals (Kominko et al. 2017). Belhaj et al. (2016) reported that the application of SS in *Helianthus annuus* improved the total N, P, organic matter and exchangeable Ca, K and Na as compared to un-amended ones. Moreover, they also observed an increase in shoot and root length, leaves biomass and antioxidant activities. Furthermore, according to the previous experiments, the application of SS improved fertility of soil and crop production of many plant species such as cucumber (Eid et al. 2017b), spinach (Eid et al. 2017a), wheat (Eid et al. 2019) and broad bean (Eid et al. 2018). The overall fate of SS application is diverse and can be depicted by a layout presented in (Fig. 1).

3 Limitations of Sewage Sludge as an Agricultural Soil Amendment

The application of sewage sludge to soil also poses some threats that need to be addressed depending on wastewater type (combined, industrial or municipal), as it can contain some biologically active pollutants, heavy metals, specifically Zn and some organic pollutants (Zennegg et al. 2013; Farsang et al. 2020). The metal pollutants present in untreated SS can be a health hazard as these metals can be uptaken by plants and deposit in their edible portions making their way into human food chain. Similarly, organic pollutants as well as pathogens present in the SS can also be a health hazard for human health. But in the literature, no specific data can be observed regarding adverse effects caused by sludge application on land but Ashekuzzaman et al. (2019) reported the following characteristics of SS: increasing levels of persistent toxic compounds in wildlife, vegetation and soil,

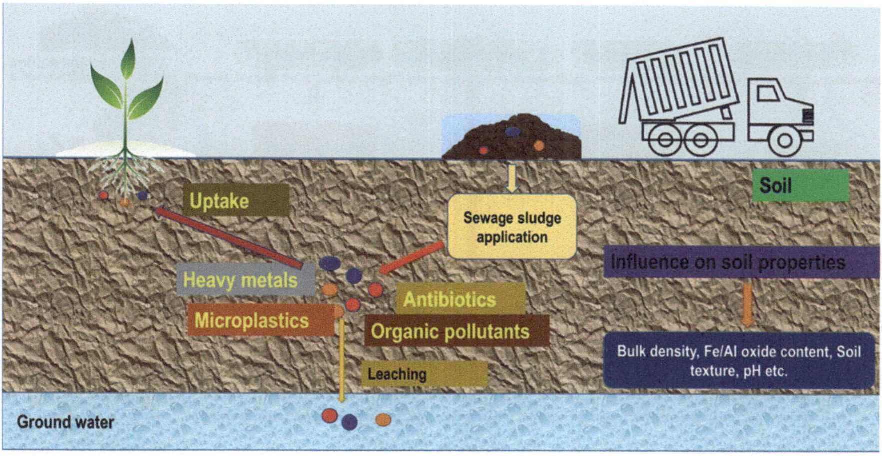

Fig. 1 Fate of sewage sludge in agricultural soils

increased emission of greenhouse gas i.e. methane and nitrous oxide, and release of odorous complexes.

The common chemicals present in wastewater are di-hydrogen cations, hydrocarbons, detergents, pharmaceutical residues, phosphorus and nitrogenous compounds. Moreover, the human or animal wastes contain various kinds of bacteria, viruses and protozoa which is the source of microbial contamination (Ohoro et al. 2019) and SS derived from these OMs source can also contain same pathogens. The degree of hazard depends on applied SS quantity, structure, crop species and conventional controls (Latare et al. 2014). The accumulation and persistence of inorganic and organic pollutants harm human health, crop growth and microbial ecosystem (Iglesias et al. 2018). Bondarczuk et al. (2016) stated that the application of SS in land causing antibiotic resistance, which is the linkage among human and antibiotic resistance, assessment of ecological risk and antibiotic resistant genes and antibiotic resistant bacteria produced during soil fertilisation. So, the European and national legislations strictly monitor the application of industrial sludge, the frequency and dose SS applications must be checked to overcome soil contamination on large scale. In addition to agronomic restrictions, environmental limitations should also play an important role in its practical application (Roig et al. 2012). Furthermore, SS also contain toxic elements like polycyclic aromatic hydrocarbons (PAHs), dibenzo-p-furans (PCDD/Fs), polybrominated diphenyl ethers (PBDEs), di (2-ethylhexyl) phthalate (DEHP), polychlorinated dibenzo-p-dioxins, polychlorinated biphenyls (PCBs), heavy metals, synthetic steroids, detergent residues, endogenous hormones, personal care products and pharmaceuticals (Singh and Agrawal 2008). The excessive use of SS in soils increased the bioavailability of many heavy metals which ultimately cause negative impact on agricultural soils (Usman et al. 2012).

The results from long-term experiments have revealed that the application of SS raise the concentration of heavy metals in soils (Nogueira et al. 2008). The reports

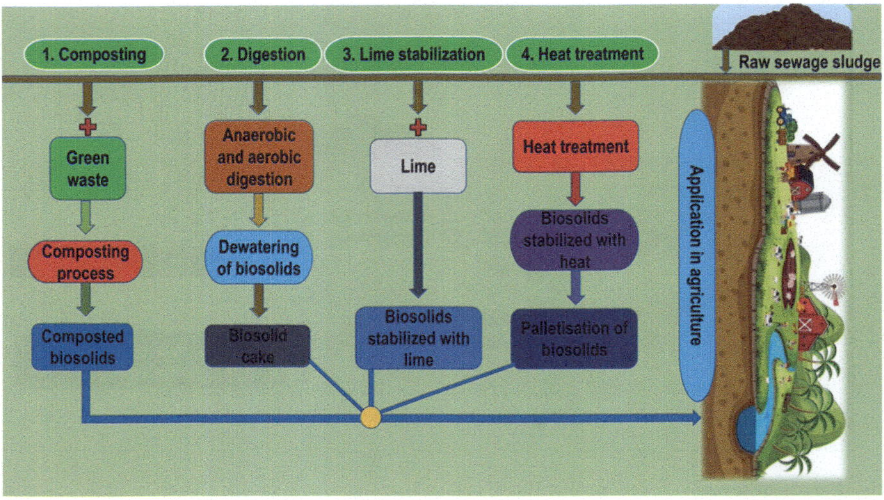

Fig. 2 Transformation of sewage sludge for soil applications

showed that the soils of tropical areas are contaminated with heavy metals after the addition of SS (Marques et al. 2007). Its uptake by the crops depends on the variety and structure of soils. Likewise, most vegetative plant parts, particularly leaves, transport much more heavy metals than nuts, fruits or seeds (Melo et al. 2007). The nutrients like phosphorus and nitrogen in SS may leach down and contaminate the groundwater. The SS being biodegradable need severe attention because they may cause putrefaction on disposal. It is imperative to pass raw SS through different treatment processes before soil application.

So, the treatment of SS is essential before the disposal of SS (Pathak et al. 2009). The higher concentration of Cd in soluble and exchangeable forms was observed in SS amended soil even after twenty-three years of interruption by McGrath et al. (2000), which evaluates the impact of SS on Mn, Zn, Cu, Cd, Fe in two wheat varieties, i.e., Roshan (*T. aestivum* cv. Roshan) and Sivand (*T. aestivum* cv. Sivand). This research summarized that treated stem and seed of Sivand, poses the concentrations of Cd, Mn, Fe and Zn lower than the standard limit of toxicity while Pb and Cu concentration were above the limit. On the other hand, in Roshan, Mn, Cd, Fe, Pb and Zn concentrations were lower than the standard limit while the concentration of Cu was higher than the standard limit (Fig. 2).

4 Quality Assurance of Sewage Sludge Prior to Using as a Soil Amendment

4.1 *Physic-chemical Characteristics*

The SS composition is highly different and depends on many factors like applied technology to treat wastewater, seasons and influent source area. Normally, the dewatered SS contains organic matter (50–70%), mineral components (inorganic carbon about 1–4%), nitrogen (3.4–4.0%), phosphorus (0.5–2.5%) and other nutrients including some micronutrients. For instance, phosphorus is predicted to become more exhausted or scarce in coming 50–100 years, So, its recovery from the SS is becoming a great alternative (Connor et al. 2017).

4.2 *Heavy Metal Concentration*

The SS is composed of micronutrients, macronutrients, inorganic and organic pollutants, trace metals/metalloid pollutants, microorganisms and organic compounds (Singh and Agrawal 2008). It is used as agricultural fertilizer because it contains minerals (P and N) and OM, and its proper utilization enables us to recycle P, N and OM (Fernández et al. 2009). However, if managed improperly, it may be easily decomposed by anaerobic processes which finally cause environmental pollution. The SS also contains some heavy metals like Cr, Hg, Cu, Pb, Cd, Ni, Zn and many organic pollutants i.e. polycyclic aromatic and polychlorinated biphenyls hydrocarbons (Hung et al. 2015). These all contaminants derived from industrial wastewater. So, the conversion of all these hazardous chemicals into non-toxic forms, various techniques have been developed (Rizzardini and Goi 2014) (Table 2).

4.3 *Microbiological Parameters*

The influence of contaminants on microorganisms has also been studied under controlled environmental conditions and by giving special attention to the physiology and abundance of certain species and strains (Brynhildsen et al. 1988). On the other hand, little attention is paid to the changes made due to metal stress in microbial communities of soil.

Individual factors like microbial biomass N (Nmic) and C (Cmic), enzyme activities and basic respiration is widely utilized for the measurement of soil management, which includes areas where SS application was carried out (Armenta et al. 2012). The reduction in biomass and enzymatic activities of microbes were recorded in different studies due to the application of SS in agricultural soil, however SS amendment increased enzyme activities of microbes (Banerjee et al. 1997).

Table 2 Concentration of heavy metals (mg/kg), Microbes (CFU g^{-1}) and Physic-chemical characteristics of sludge

Heavy metals (mg/kg)							Physic-chemical characteristics				Microbes (CFU g^{-1})		
Cu	Pb	Cr	Ni	Hg	Cd	Zn	pH	EC mS/cm	OM	Moisture %	*E. coli*	Salmonella sp.	Ref.
140.8	<5.6	<5.6	22.6	<1.3	1	–	7.1	3.49	67.5%	83.9	$8.9*10^2$	Absent	Alvarenga et al. (2015)
124.37	73.77	52.07	17.39	–	6.17	2610	–	–	–	–	–	–	Praspaliauskas et al. (2020)
103.6	32.8	21.1	9.5	0.9	0.7	650	9.1	4.2	71.5%	–	–	–	Praspaliauskas et al. (2020)
10–20	<1	0.5	10–20	0.6–2	0.2	40–50	8–8.5	5–10	35–40%		Absent	Absent	Cucina et al. (2019)
196	47	73	40		<3	936	–	–	–	–	–	–	Cucina et al. (2019)
587	48.2	170	–	–	3.47	1062	5.91	10.8	501 g/kg	–	–	–	Zuo et al. (2019)
174.4	35	–	22.2	–	4.04	342	7.7	1.7	18.5%	–	–	–	Hamdi et al. (2019)
11	6	5	7	–	<0.2	21	8.5	0.41	1%	34.91	–	–	Delgado et al. (2019)
6.20	3.80	1.10	<1	<0.02	<0.03	76	3.8	–	25%	40.55	–	Absent	Delgado et al. (2019)
–	–	–	–	–	–	–	7.7	1.7	31.8%	–	–	–	Delgado et al. (2019)
115	207	100	–	–	–	306	4.8	–	21%	17	–	–	de Figueiredo et al. (2019)
162.8	87.79	–	–	–	2.73	766.4	6.39	–	10.3%	–	–	–	de Figueiredo et al. (2019)
312	40	206	102	–	2.7	1590	8.4	–	33.8	74	–	–	Nascimento et al. (2020)
174.88	42.94	46.21	32.42	–	0.67	–	–	–	28.15	–	–	–	Bastida et al. (2019)
174.4	–	–	22.2	–	4.04	342	7.7	1.7	18.5	–	–	–	Bastida et al. (2019)
90						534	7.05	2.61	52.7	80			Koutroubas et al. (2020)

Cu	Pb	Cr	Ni	Hg	Cd	Zn	pH	EC mS/cm	OM	Moisture %	*E. coli*	Salmonella sp.	Ref.
876.7	20.1	–	–	–	1.8	429.5	5.7	–	83.5	–	–	–	Mohamed et al. (2018)
115	207	100	–	–	–	306	4.8	–	21	17	–	–	de Figueiredo et al. (2019)
110.9	42.8	25.4	14.8	–	2.3	1004	6.22	–	59.45	–	–	–	Skowrońska et al. (2020)
10.40	0.58	0.88	1.24	<0.5	<0.5	20.30	–	–	–	–	–	2	Skowrońska et al. (2020)
116–127	110–213	65–435	27–79	0.15–1.62	8–13	363–592	–	–	–	–	–	–	Belmeskine et al. (2020)
8608	–	9623	1410	–	2.32	8293	8.13	2.68	7.81	0.00	–	–	Xie et al. (2020)
200–300	100–150	100–150	40–60	<0.1	<0.1	800–900	7.5–8	–	50–60	–	–	–	Roig et al. (2012)
270	132	42.9	19.2	–	1.1	1360	8.2	4.1	58.1	–	–	–	Madrid et al. (2020)
89.03	20.75	49.27	11.5	0.51	0.49	505.7	6.5	2.25	63.55	–	>100	Absent	Romanos et al. (2019)
654	36	85	42	–	<2	1940	–	–	–	–	–	–	Romanos et al. (2019)
240	–	90	80	–	1.80	1500	6.09	3.87	45.9	–	–	–	Bozkurt and Yarılgaç (2010)
136	145	121	39	ND	3.2	1731	6.57	–	74.9	–	–	–	Natal-da-Luz et al. (2009)
205.3	29.03	27.12	14.73	0.77	0.37	429.5	7.54	1.04	61.97	76.26	–	–	Carbonell et al. (2009)
1121.9	–	155.7	52.8	–	3.3	2127.3	6.32	–	37.6	–	–	–	Bai et al. (2017)
317.7	60	35.5	47.17	–	154.5	785.3	7.0	2.28	–	–	–	–	Singh and Agrawal (2008)
189	47	–	–	–	–	1290	6.43	1.24	7.19	–	–	–	Roy et al. (2019)
1121.9	–	155.4	52.8	–	3.3	2127	6.32	–	21.62	–	–	–	Bai et al. (2017)
100	10	–	15	–	2.6	–	6.67	1.2	22.45	–	–	–	Rehman et al. (2018)

(continued)

Table 2 (continued)

Heavy metals (mg/kg)							Physic-chemical characteristics				Microbes (CFU g^{-1})		
Cu	Pb	Cr	Ni	Hg	Cd	Zn	pH	EC mS/cm	OM	Moisture %	*E. coli*	Salmonella sp.	Ref.
62.3	26	27.1	2.09	–	ND	5.89	7.3	1.7	–	–	–	–	Singh and Kumar (2020)
749.2	61.3	–	28.7	1.04	2.91	838.5	7.05	3.8	53.4	–	–	–	Arriagada et al. (2009)
343	106	–	37.7	–	–	563	7.55	2.85	32.8	–	–	–	Gwenzi et al. (2016)
17.9	81	32.8	20.9	0.44	1.15	215	6.1	2.83	83.2	–	–	–	Mohamed et al. (2018)
230	69	115	35	–	<5	500	12	–	19	–	–	–	Kidd et al. (2007)

4.4 *Phytotoxicity Evaluation*

The impacts of SS on crop growth and production result as the action of many factors like the existence of organic matter content, the bioavailability of dangerous pollutants (PCBs, PAHs and heavy metals), soil structure, pH of soil, climatic conditions, microbial charge, nutrient balance, microbial charge and humidity (Puiu et al. 2019). The biodegradation of organic compounds produced secondary compounds and these compounds cause toxicity because of mobility and bioavailability. These chemicals are mostly not monitored due to their uptake by plants and leaching in groundwater (Oleszczuk 2008). The growth and development mechanism of plants are influenced by metals. Moreover, all the heavy metals present in SS cause cell membranes damage in plants. Further, these heavy metals reduce the rate of transpiration by damaging photosynthetic organelles, the destruction of protein synthesis and the production of lipid peroxidation. Many heavy metals Cu, Ni, Pb, Cr, Cd and Zn are much toxic metals produced by the application of SS (Khan et al. 2018). Prior to use any kind of SS as an agricultural amendment, a pilot investigation must be conducted to screen its organic and inorganic contents and phytotoxicity to specific crops.

5 Future Perspectives

The application of SS to agricultural soil denotes the most appropriate economic and environmental option but this practice needs careful control due to the hazardous chemicals in SS like organic and heavy metal contaminants. Keeping in mind the importance and drawbacks, chemical analysis should be carried out to assess the ecological safety of raw and processed SS. Moreover, the installations of smaller treatment plants for collecting SS from small industrial units is also very effective. Furthermore, the management process of SS must be developed independently for every treatment plant, then it will be economically and ecologically acceptable.

References

Alvarenga P, Mourinha C, Farto M, Santos T, Palma P, Sengo J, Cunha-Queda C (2015) Sewage sludge, compost and other representative organic wastes as agricultural soil amendments: benefits versus limiting factors. Waste Manag 40:44–52. https://doi.org/10.1016/j.wasman.2015.01.027

Angin I, Aslantas R, Gunes A, Kose M, Ozkan G (2017) Effects of sewage sludge amendment on some soil properties, growth, yield and nutrient content of raspberry (*Rubus idaeus* L.). Erwerbs-obstbau 59(2):93–99. https://doi.org/10.1007/s10341-016-0303-9

Antolín MC, Pascaul I, Garcia C, Polo A, Sanchez-Diaz M (2005) Growth, yield and solute content of barley in soils treated with sewage sludge under semiarid Mediterranean conditions. Field Crops Res 94:224–237. https://doi.org/10.1016/j.fcr.2005.01.009

Armenta R, Vaca R, Lugo J, Del-Aguila P (2012) Microbial and biochemical properties of an agricultural Mexican soil amended with sewage sludge. Rev Bras Ciênc Solo 36:1646–1655. https://doi.org/10.1590/S0100-06832012000500029

Asgari LB, Najafi N, Moghiseh E, Mosaferi M, Hadian J (2019) Effects of gamma irradiated and non-irradiated sewage sludge on growth characteristics, leaf chlorophyll index, and macronutrients concentrations in Basil. J Soil Sci Plant Nutr 19(3):580–591. https://doi.org/10.1007/s42729-019-00057-4

Ashekuzzaman SM, Kwapinska M, Leahy JJ, Richards K, Fenton O (2019) Novel use of Dairy Processing Sludge Derived Pyrogenic Char (DPS-PC) to remove phosphorus in discharge effluents. In: Waste and biomass valorization. Springer, Berlin

Bai Y, Zang C, Gu M, Gu C, Shao H, Guan Y, Feng K (2017) Sewage sludge as an initial fertility driver for rapid improvement of mudflat salt-soils. Sci Total Environ 578:47–55. https://doi.org/10.1016/j.scitotenv.2016.06.083

Banerjee MR, Burton DL, Depoe S (1997) Impact of sewage sludge application on soil biological characteristics. Agric Ecosyst Environ 66:241–249. https://doi.org/10.1016/S0167-8809(97)00129-1

Bastida F, Jehmlich N, Martínez-Navarro J, Bayona V, García C, Moreno JL (2019) The effects of struvite and sewage sludge on plant yield and the microbial community of a semiarid Mediterranean soil. Geoderma 337:1051–1057. https://doi.org/10.1016/j.geoderma.2018.10.046

Belaid D (2015) Utilisation des boues résiduelles de station d'épuration en Algérie

Belhaj D, Elloumi N, Jerbi B, Zouari M, Abdallah FB, Ayadi H, Kallel M (2016) Effects of sewage sludge fertilizer on heavy metal accumulation and consequent responses of sunflower (*Helianthus annuus*). Environ Sci Pollut Res 23:20168–20177. https://doi.org/10.1007/s11356-016-7193-0

Belmeskine H, Ouameur WA, Dilmi N, Aouabed A (2020) The vermicomposting for agricultural valorization of sludge from Algerian wastewater treatment plant: impact on growth of snap bean *Phaseolus vulgaris* L. Heliyon 6(8):e04679. https://doi.org/10.1016/j.heliyon.2020.e04679

Bittencourt S, Serrat BM, Aisse MM, Gomes D (2014) Sewage sludgeusage in agriculture: a case study of its destination in the Curitiba Metropolitan Region, Paraná, Brazil. Water Air Soil Pollut 225:2074. https://doi.org/10.1007/s11270-014-2074-y

Bondarczuk K, Markowicz A, Piotrowska-Seget Z (2016) The urgent need for risk assessment on the antibiotic resistance spread via sewage sludge land application. Environ Int 87:49–55. https://doi.org/10.1016/j.envint.2015.11.011

Bozkurt MA, Yarılgaç T (2010) The use of sewage sludge as an organic matter source in apple trees. Pol J Environ Stud 19(2):267–274

Brynhildsen L, Lundgren BV, Allard B, Rosswall T (1988) Effects of glucose nitration's on cadmium, copper, mercury and zinc toxicity to a Kebsie IIa sp. Appl Environ Microbiol 55:1375–1379

Carbonell G, Pro J, Gómez N, Babín MM, Fernández C, Alonso E, Tarazona JV (2009) Sewage sludge applied to agricultural soil: ecotoxicological effects on representative soil organisms. Ecotoxicol Environ Saf 72(4):1309–1319. https://doi.org/10.1016/j.ecoenv.2009.01.007

Central Statistical Office, Environmental Protection 2018. (in Polish)

Connor R, Renata A, Ortigara C, Koncagul E, Uhlenbrook S, Lamizana-Diallo BM (2017) The United Nations world water development report 2017. Wastewater: The untapped resource. The United Nations World Water Development Report

Cucina M, Ricci A, Zadra C, Pezzolla D, Tacconi C, Sordi S, Gigliotti G (2019) Benefitsand risks of long-term recycling of pharmaceutical sewage sludge on agricultural soil. Sci Total Environ 695:133762. https://doi.org/10.1016/j.scitotenv.2019.133762

Delgado M, Maeso FJ, Martín JV, Gonzalez MI, Martinez S (2019) Valorization of sludge from the quartz industry as soil amendment and crop production. Soil Tillage Res 194:104320. https://doi.org/10.1016/j.still.2019.104320

Directive C (1986) Council directive on the protection of the environment, and in particular of the soil, when sewage sludge is used in agriculture. Off J Eur Commun 181:6–12

Eid EM, El-Bebany AF, Alrumman SA, Abd El-latif H, Taher MA, Fawy KF (2017a) Effects of different sewage sludge applications on heavymetal accumulation, growth and yield of spinach (*Spinacia oleracea* L.). Int J Phytoremediation 19:340–347. https://doi.org/10.1080/15226514.2016.1225286

Eid EM, Alrumman SA, El-Bebany AF, Hesham A, Taher MA, Fawy KF (2017b) The effects of different sewage sludge amendment rates onthe heavymetal bioaccumulation, growth and biomass of cucumbers (*Cucumis sativus* L.). Environ Sci Pollut Res 24:16371–16382. https://doi.org/10.1007/s11356-017-9289-6

Eid EM, Alrumman SA, El-Bebany AF, Fawy KF, Taher MA, Hesham A, El-Shaboury GA, Ahmed MT (2018) The evaluation of sewage sludge application as a fertilizer for broad bean (*Faba sativaBernh.*) crops. Food Energy Secur 7:e00142. https://doi.org/10.1002/fes3.142

Eid EM, Alrumman SA, El-Bebany AF, Fawy KF, Taher MA, Hesham A, El-Shaboury GA, Ahmed MT (2019) Evaluation of the potential of sewage sludge as a valuable fertilizer for wheat (*Triticum aestivum* L.) crops. Environ Sci Pollut Res 26:392–401. https://doi.org/10.1007/s11356-018-3617-3

Eid EM, Hussain AA, Taher MA, Galal TM, Shaltout KH, Sewelam N (2020) Sewage sludge application enhances the growth of *Corchorus olitorius* plants and provides a sustainable practice for nutrient recirculation in agricultural soils. J Soil Sci Plant Nutr 20(1):149–159

European Commission (2001) Disposal and recycling routes for sewage sludge part 3 e scientific and technical report. European Commission DG Environment. http://ec.europa.eu/environment/archives/waste/sludge/pdf/sludge_disposal3.pdf. (Accessed 10 June 2019)

Farsang A, Babcsányi I, Ladányi Z, Perei K, Bodor A, Csányi KT, Barta K (2020) Evaluating the effects of sewage sludge compost applications on the microbial activity, the nutrient and heavy metal content of a Chernozem soil in a field survey. Arab J Geosci 13(19):1–9

Fernández J, Palza C, García-Gil JC, Polo A (2009) Biochemical properties and barley yield in a semiarid Mediterranean soil amended with two kinds of sewage sludge. Appl Soil Ecol 42:18–24. https://doi.org/10.1016/j.apsoil.2009.01.006

Figueiredo CC, Farias WM, Coser TR, Paula AM, Silva MRS, Paz-Ferreiro J (2019) Sewage sludge biochar alters root colonization of mycorrhizal fungi in a soil cultivated with corn. Eur J Soil Biol 93:103092. https://doi.org/10.1016/j.ejsobi.2019.103092

Gwenzi W, Muzava M, Mapanda F, Tauro TP (2016) Comparative short-term effects of sewage sludge and its biochar on soil properties, maize growth and uptake of nutrients on a tropical clay soil in Zimbabwe. J Integr Agric 15(6):1395–1406. https://doi.org/10.1016/S2095-3119(15)61154-6

Hamdi H, Hechmi S, Khelil MN, Zoghlami IR, Benzarti S, Mokni-Tlili S, Hassen A, Jedidi N (2019) Repetitive land application of urban sewage sludge: effect of amendment rates and soil texture on fertility and degradation parameters. Catena 172:11–20. https://doi.org/10.1016/j.catena.2018.08.015

He ZL, Yang XE, Stoffella PJ (2005) Trace elements in agroecosystems and impacts on the environment. J Trace Elem Med Biol 19:125–140. https://doi.org/10.1016/j.jtemb.2005.02.010

Hung CV, Cam BD, Mai PT, Dzung BQ (2015) Heavy metals and polycyclic aromatic hydrocarbons in municipal sewage sludge from a river in highly urbanized metropolitan area in Hanoi, Vietnam: levels, accumulation pattern and assessment of land application. Environ Geochem Health 37:133–146. https://doi.org/10.1007/s10653-014-9635-2

Iglesias M, Marguí E, Camps F, Hidalgo M (2018) Extractability and crop transfer of potentially toxic elements from mediterranean agricultural soils following long-term sewage sludge applications as a fertilizer replacement to barley and maize crops. Waste Manag 75:312–318. https://doi.org/10.1016/j.wasman.2018.01.024

Inglezakis VJ, Zorpas AA, Karagiannidis A, Samaras P, Voukkali I, Sklari S (2014) European Union legislation on sewage sludge management. Fresenius Environ Bull 23(2):635–639

Kacprzak M, Neczaj E, Fijałkowski K, Grobelak A, Grosser A, Worwag M, Singh BR (2017) Sewage sludge disposal strategies for sustainable development. Environ Res 156:39–46. https://doi.org/10.1016/j.envres.2017.03.010

Khan ZI, Ahmad K, Saddar H, Ugulu I, Wajid K, Bashir H, Dogan Y (2018) Manganese bioaccumulation and translocation of in forages grown in soil irrigated with city effluent: An evaluation on health risk. Res J Pharm Biol Chem Sci 9(5):759–770

Kidd PS, Domínguez-Rodríguez MJ, Díez J, Monterroso C (2007) Bioavailability and plant accumulation of heavy metals and phosphorus in agricultural soils amended by long-term application of sewage sludge. Chemosphere 66(8):1458–1467. https://doi.org/10.1016/j.chemosphere.2006.09.007

Kirchmann H, Börjesson G, Kätterer T, Cohen Y (2017) From agricultural use of sewage sludge to nutrient extraction: a soil science outlook. Ambio 46(2):143–154

Kominko H, Gorazda K, Wzorek Z (2017) The possibility of organ mineral fertilizer production from sewage sludge. Waste Biomass Valor 8(5):1781–1791

Koutroubas SD, Antoniadis V, Damalas CA, Fotiadis S (2020) Sunflower growth and yield response to sewage sludge application under contrasting water availability conditions. Ind Crop Prod 154:112670. https://doi.org/10.1016/j.indcrop.2020.112670

Lamastra L, Suciu NA, Trevisan M (2018) Sewage sludge for sustainable agriculture: contaminants' contents and potential use as fertilizer. Chem Biol Technol Agric 5(1):10–22

Latare AM, Kumar O, Singh SK, Gupta A (2014) Direct and residual effect of sewage sludge on yield, heavy metals content and soil fertility under rice-wheat system. Ecol Eng 69:17–24. https://doi.org/10.1016/j.ecoleng.2014.03.066

Laura F, Tamara A, Müller A, Hiroshan H, Christina D, Serena C (2020) Selecting sustainable sewage sludge reuse options through a systematic assessment framework: methodology and case study in Latin America. J Clean Prod 242:118389. https://doi.org/10.1016/j.jclepro.2019.118389

Liu X, Liu L, Leng P, Hu Z (2019) Feasible and effective reuse of municipal sludge for vegetation restoration: physiochemical characteristics and microbial diversity. Sci Rep 9(1):879

Madrid F, Rubio-Bellido M, Morillo E (2020) Extraction of nonylphenol, pyrene and phenanthrene from sewage sludge and composted biosolids by cyclodextrins and rhamnolipids. Sci Total Environ 715:136986. https://doi.org/10.1016/j.scitotenv.2020.136986

Marques MO, Nogueira TAR, Fonseca IM, Marques TA (2007) Contents of Cr, Ni, Pb and Zn treated with sewage sludge and cultivated with sugarcane. Rev Biol Ciênc Terra 7:133–143

McGrath SP, Zhao FJ, Dunham SJ, Crosland AR, Coleman K (2000) Long-term changes in the extractability and bioavailability of zinc and cadmium after sludge application. J Environ Qual 29(3):875–883. https://doi.org/10.2134/jeq2000.00472425002900030025x

Meena AL, Kumari S, Lekshmi N, Kumar CG, Kumar S (2020) Sewage sludge application in agriculture: impact on crops and human health. Food Sci Rep 1:53–58

Melo WJ, Aguiar OS, Melo GMP, Melo VP (2007) Nickel in a tropical soil treated with sewage sludge and cropped with maize in a longterm field study. Soil Biol Biochem 39:1341–1347. https://doi.org/10.1016/j.soilbio.2006.12.010

Moffat A (2006) Use of sewage sludges and composts in forestry. Forcommission https://doi.org/10.1177/1088357608329173

Mohamed B, Mounia K, Aziz A, Ahmed H, Rachid B, Lotfi A (2018) Sewage sludge used as organic manure in Moroccan sunflower culture: effects on certain soil properties, growth and yield components. Sci Total Environ 627:681–688. https://doi.org/10.1016/j.scitotenv.2018.01.258

Nascimento AL, Souza AJ, Oliveira FC, Coscione AR, Viana DG, Regitano JB (2020) Chemical attributes of sewage sludge: relationships to sources and treatments, and implications for sludge usage in agriculture. J Clean Prod 258:120746. https://doi.org/10.1016/j.jclepro.2020.120746

Natal-da-Luz T, Tidona S, Jesus B, Morais PV, Sousa JP (2009) The use of sewage sludge as soil amendment. The need for an ecotoxicological evaluation. J Soils Sediments 9(3):246–257

Nogueira TAR, Oliveira LR, Melo WJ, Fonseca IM, Melo GMP, Melo VW, Marques MO (2008) Cadmium, chromium, lead and zinc in maize plants and oxisol after nine annual applications of sewage sludge. Rev Bras Ciênc Solo 32:2195–2207. https://doi.org/10.1590/S01006832008000500040

Ohoro CR, Adeniji AO, Okoh AI, Okoh OO (2019) Distribution and chemical analysis of pharmaceuticals and personal care products (PPCPS) in the aquatic systems: a review. Intl J Environ Res Public Health 16:3026. https://doi.org/10.3390/ijerph16173026

Oleszczuk P (2008) The toxicity of compost from sewage sludge evaluated by the direct contact tests phytotoxic and ostracodtoxkit. Waste Manag 28(9):1645–1653

Ordinance of the Minister of the Environment on sewage sludge 06. 02 (2015)

Pathak A, Dastidar MG, Sreekrishnan TR (2009) Bioleaching of heavy metals from sewage sludge: a review. J Environ Manag 90(8):2343–2353. https://doi.org/10.1016/j.jenvman.2008.11.005

Praspaliauskas M, Žaltauskaitė J, Pedišius N, Striūgas N (2020) Comprehensive evaluation of sewage sludge and sewage sludge char soil amendment impact on the industrialhemp growth performance and heavy metal accumulation. Ind Crop Prod 150:112396. https://doi.org/10.1016/j.indcrop.2020.112396

Puiu D, Popescu M, Niculescu M, Pascu FL, Galaon T, Postolache C (2019) Mobility of some high persistent organochlorine compounds from soil to Mentha Piperita. Rev Chim 70(1):278–282

Raheem A, Sikarwar VS, He J, Dastyar W, Dionysiou DD, Wang W, Zhao M (2018) Opportunities and challenges in sustainable treatment and resource reuse of sewage sludge: a review. Chem Eng J 337:616–641. https://doi.org/10.1016/j.cej.2017.12.149

Rehman R, Rizwan M, Qayyum MF, Ali S, Zia-ur-Rehman M, Zafar-ul-Hye M, Iqbal MF (2018) Efficiency of various sewage sludges and their biochars in improving selected soil properties and growth of wheat (*Triticum aestivum*). Environ Manag 223:607–613. https://doi.org/10.1016/j.jenvman.2018.06.081

Rizzardini CB, Goi D (2014) Sustainability of domestic sewage sludge disposal. Sustainability 6:2424–2434. https://doi.org/10.3390/su6052424

Roig N, Sierra J, Martí E, Nadal M, Schuhmacher M, Domingo JL (2012) Long-term amendment of Spanish soils with sewagesludge: effects on soil functioning. Agric Ecosyst Environ 158:41–48. https://doi.org/10.1016/j.agee.2012.05.016

Romanos D, Nemer N, Khairallah Y, Abi-Saab MT (2019) Assessing the quality of sewage sludge as an agricultural soil amendment in Mediterranean habitats. Int J Recycl Org Waste Agric 8 (1):377–383

Roy T, Biswas DR, Ghosh A, Patra AK, Singh RD, Sarkar A, Biswas SS (2019) Dynamics of culturable microbial fraction in an Inceptisol under short-term amendment with municipal sludge from different sources. Appl Soil Ecol 136:116–121. https://doi.org/10.1016/j.apsoil.2018.12.024

Samolada M, Zabaniotou A (2014) Comparative assessment of municipal sewage sludge incineration, gasification and pyrolysis for a sustainable sludge-to-energy management in Greece. Waste Manag 34:411–420. https://doi.org/10.1016/j.wasman.2013.11.003

Sharma B, Sarkar A, Singh P, Singh RP (2017) Agricultural utilization of biosolids: a review on potential effects on soil and plant grown. Waste Manag 64:117–132. https://doi.org/10.1016/j.wasman.2017.03.002

Singh RP, Agrawal M (2008) Potential benefits and risks of land application of sewage sludge. Waste Manag 28:347–358. https://doi.org/10.1016/j.wasman.2006.12.010

Singh D, Kumar A (2020) Binary mixture of nanoparticles in sewage sludge: impact on spinach growth. Chemosphere 66:126794. https://doi.org/10.1016/j.chemosphere.2020.126794

Skowrońska M, Bielińska EJ, Szymański K, Futa B, Antonkiewicz J, Kołodziej B (2020) An integrated assessment of the long-term impact of municipal sewage sludge on the chemical and biological properties of soil. Catena 189:104484. https://doi.org/10.1016/j.catena.2020.104484

Usman K, Khan S, Ghulam S, Khan MU, Khan N, Khan MA, Khalil SK (2012) Sewage sludge: an important biological resource for sustainable agriculture and its environmental implications. Am J Plant Sci 3:1708–1721. https://doi.org/10.4236/ajps.2012.312209

Vaish B, Sarkar A, Singh P, Singh PK, Sengupta C, Singh RP (2016) Prospects of biomethanation in Indian urban solid waste: stepping towards a sustainable future. In: Recycling of solid waste for biofuels and bio-chemicals. Springer, Singapore, pp 1–29

Wiśniowska E, Włodarczyk-Makuła M (2018) The effect of selected acidic or alkaline chemical agents amendment on leachability of selected heavy metals from sewage sludge. Sci Total Environ 633:463–469. https://doi.org/10.1016/j.scitotenv.2018.03.163

Working document on sludge and biowaste, Brussels 21. 09, 2010.

Xie S, Yu G, Ma J, Wang G, Wang Q, You F, Li C (2020) Chemical speciation and distribution of potentially toxic elements in soilless cultivation of cucumber with sewage sludge biochar addition. Environ Res 191:110188. https://doi.org/10.1016/j.envres.2020.110188

Yue Y, Cui L, Lin Q, Li G, Zhao X (2017) Efficiency of sewage sludge biochar in improving urban soil properties and promoting grass growth. Chemosphere 173:551–556. https://doi.org/10.1016/j.chemosphere.2017.01.096

Zennegg M, Munoz M, Schmid P, Gerecke AC (2013) Temporal trends of persistent organic pollutants in digested sewage sludge (1993–2012). Environ Int 60:202–208. https://doi.org/10.1016/j.envint.2013.08.020

Zong Y, Wang Y, Sheng Y, WuC LS (2018) Ameliorating soil acidity and physical properties of two contrasting texture Ultisols with wastewater sludge biochar. Environ Sci Pollut Res 25 (26):25726–25733

Zuo W, Gu C, Zhang W, Xu K, Wang Y, Bai Y, Dai Q (2019) Sewage sludge amendment improved soil properties and sweet sorghum yield and quality in a newly reclaimed mudflat land. Sci Total Environ 654:541–549. https://doi.org/10.1016/j.scitotenv.2018.11.127

Detoxification of Sewage Sludge by Natural Attenuation and Application as a Fertilizer

Ambika Chaturvedi, Pooja Saraswat, Anamika Gupta, Mrinalini Prasad, and Rajiv Ranjan

1 Introduction

Sewage sludge is actually the derivative or the byproduct, which is obtained during the waste water treatment plants. It has become an issue due to the increasing volume day by day and also due to its adverse impact. Sewage sludge is a semisolid material which remains suspended during wastewater treatment. Sludge is generally divided into two main types i.e. primary and secondary sludge (Zhang et al. 2017). Primary sludge is generated as suspended solid via gravitational sedimentation and secondary sludge generated via microbial activities. Sewage sludge contains inorganic material, organic material and toxic substances. Sludge contains potentially harmful tracer metals due to which its management is necessary. There are many management strategies which work on energy production such as heat, electricity or biofuel (Rulkens 2008; Sommers 1977). Due to the continuous rising of sewage sludge amount, also the need of the theoretical and particle research knowledge for use of sludge as nutritive source also increases (Antonkiewicz et al. 2020). Due to the presence of a high number of micronutrients and organic matter, sewage sludge can be used as fertilizers, however, it also possesses some adverse effect due to the presence of heavy metals and pathogens (Usman et al. 2012). Some methods and processes are also there for the use of sewage as fertilizer (Moore 1985). The use of sewage sludge as a substitute to fertilizer is a best way to recycle the essential nutrients present within the sewage sludge which also contribute in enhancing the chemical, physical as well as biological property of soil (Saha et al. 2017). If the sewage sludge used as a fertilizer there are some factors, which are necessary to consider to predict the effect on soil such as the high organic content of sewage sludge must have a positive impact on soil. It is also necessary that the microbial

A. Chaturvedi · P. Saraswat · A. Gupta · M. Prasad · R. Ranjan (✉)
Plant Molecular Biology Lab, Department of Botany, Dayalbagh Educational Institute (Deemed University), Dayalbagh, Agra, UP, India

V. D. Rajput et al. (eds.), *Sustainable Management and Utilization of Sewage Sludge*, https://doi.org/10.1007/978-3-030-85226-9_13

environment present within the soil also get stimulated. If the sewage sludges used as fertilizer, it not always enhance the fertility of soil by substitute chemical fertilizer it can also generate some adverse impact on the soil due to the accumulation of heavy metal such as Ni, Cd, Zn and Pb in plant tissues (Saha et al. 2014).

2 Sewage Sludge (SS): A Worrying Component of Total Waste

Due to the continuous increase in urbanization, new environmental and social concern generated. The waste generation is one of the main concerns (Gutberlet 2018). Sewage sludge also known as biosolid is produced as residue or byproduct, which is generated during wastewater treatment (Pathak et al. 2009). Sewage sludge is produced due to the deposition of solids from chemical coagulation during treatment of wastewater (Baily 2009). Properties or the characterization of sewage sludge content decides on the basis of its origin. The central source for production of sewage sludge is generally municipal, domestic and industrial wastewater (Grosser and Neczaj 2017). Sewage sludge can be divided into two categories on the bases of their source. If the sludge is obtained from suspended solids and organics through gravitational sedimentation, then it is known as Primary Sludge. If the sludge produced by the process of organic matter decomposition by microorganism than it is known as secondary sludge.

Mismanagement of solid waste is one of the major environmental concerns. Sewage sludge is a complex mixture of large size of solids for instance sand, toxic and nontoxic, organic and inorganic substances (Da Silva et al. 2020; Bresters 1998; Gupta et al. 2015). Physical, biological as well as chemical parameter help in the characterization of sewage sludge. Physical parameter are described by processing of sewage sludge, chemical parameters are described by the presence of nutrients and toxic compound quantity and the biological parameter are described by the microbial activity and organic content (Grosser 2017). The presence of a number of chemicals in sludge is impossible to measure (Rank and Nielsen 1998). These toxic chemicals can be divided into potential and organic pollutants. Potential pollutant such as Al (Aluminium), Cd (Cadmium), Co (Cobalt), Cu (Copper), Cr (Chromium), Fe (Iron), Mn (Manganese), Hg (Mercury), Mo (Molybdenum), Ni (Nickle), Pb (Lead), Ti (Titanium) and Zn (Zinc) and organic pollutants includes, nonyl phenol (NP), PAHs, PCBs, di-(2-ethylhexyl) phthalate (DEHP), linear alkyl benzene sulphonates (LAS), NPE, dioxins (PCDD) and furans (PCDF) (Álvarez et al. 2002). Heavy metal within the sewage sludge namely Zn, Cu, Ni, Cd, Cr, Hg are big concern not only on environment but also negatively impact on health (Krogmann et al. 1999). The aggregation of heavy metals in sewage is generally from industries such as chemical, pharmaceuticals, pigment industries, etc. (Andreoli et al. 2007). Due to the large amount of sewage sludge production throughout the globe the proper management of sewage sludge is necessary. There are two main strategies which can be used for sewage sludge management one is reuse or another is final disposal via different routes such as landspreading, incineration, landfilling, other such as silviculture land reclamation, wet oxidation, pyrolysis, gasification etc. (Aubain et al. 2002; Kacprzak et al. 2017). Recently, "NEW" paradigm (Nutriety-Energy-Water) was adopted by the World Water Environment Federation (WEF) for wastewater treatment plants.

2.1 SS Contains: Source of Nutrients in Agricultural

For intensive farming the input of organic matter is desired to maintain the fertility of soil, so the crop yield can be enhanced. As we know from above discussion large amount of organic matter, micronutrients and macronutrient are present within the sewage sludge (Usman et al. 2013). Macronutrients of sewage sludge are considered as an amazing source of nutrients and organic constituents for plants (Logan and Harrison 1995) Sewage sludge is also a rich source of nitrogen, potassium and phosphorous and these all nutrients are very useful in agriculture and the availability of nutrient contain of sewage sludge based upon sludge treatment. In Indian sewage sludge the percent of organic matter is high which is about 50% while inorganic carbon is about 2–4% (Sommers et al. 1976).

The sewage sludge application in agricultural soil is a way to recycle the nutrient content in it. The use of sludge effects on PH of soil, and enhance organic matter content, electrical conductivity of agricultural land (Tsadilas et al. 1995). When sewage sludge is used as an organic fertilizer it is favourable for plant yield (Tester 1990). Uses of sewage sludge in agricultural land also enhance the microbial activity, urease activity and phosphate synthesis in soil (Sastre et al. 1996). Sludge is generally treated before recycle or dispose, pre-treatment is necessary to reduce the water content, pathogens present in it and enhance properties as well (Usman et al. 2012). Belhaj & colleagues used processed sewage sludge for *Helianthus annuus* and observed that the pH of soil decreased with increase in the electrical conductivity, organic matter contents, inorganic contents (available N and P, exchangeable Ca, Na, and K) in soil pre-treated with sewage sludge lead to increased root and shoot length, number of leaves, plant biomass, as well as antioxidant activities of *Helianthus annuus* (Belhaj et al. 2016). Casado-Vela & colleagues research result suggest that particular amount of sewage compost on per-meter square cauliflower land show favourable impact on physical, biological properties and nutrient supply of that land (Casado-Vela et al. 2006). Direct use of sewage sludge on soil is restricted due presence of heavy metal as mention earlier but, it is more effective on agricultural soil if the organo-mineral fertilizer are sewage sludge derivative and use to modify by adding require supplements of minerals. These sorts of fertilizer shows efficient effect on soil in comparison to conventional fertilizers (Kominko et al. 2017). Seleiman and colleagues conclude in their research that sewage sludge can also be used as an alimentary source when it is about bioenergy crops such as oilseed rape, fibre hemp, and white lupin (Seleiman et al. 2010). Sohaili and colleagues result suggest that the sewage sludge provide favourable effect on growth of *Abelmoschus esculentus* (Sohaili et al. 2012). Median concentrations of anaerobically digests sewage sludges have N content 4.2%, P, 3% and k also 3%, Pb, 300 mg/kg, Zn about 1890 mg/kg, Cu, 1000 mg/kg, Cd 16 mg/kg (Sommers 1977). Sewage sludge has sufficient nitrogen and phosphorus for plants. Total nitrogen content in SS is about 41–50 kg/t but the available nitrogen content is comparatively small for plants. Although, the mineralization converts this nitrogen into such forms which plant can uptake. Nyamangara and Mzezewa results suggest that the sewage sludge use in agricultural land significantly increases the phosphorous content about 19- and 57-folds (Nyamangara and Mzezewa 2001).

Sewage sludge is rich in phosphorus with heavy metal contamination. Franz uses sewage sludge for formation of phosphorous base fertilizer with the help of some processing steps and also prove that they are equally efficient for plant uptake as other fertilizers by testing them on green house plant (Franz 2008). Organo-mineral fertilizer is formed by using sewage sludge and theses fertilizers modify via input of mineral fertilizer, provide better crop yield in comparison to conventional fertilizers (Kominko et al. 2017). The impact of domestic sewage sludge upon soil and on wheat yield examine by the Jamil and Qasim conclude that the soil pH, organic matter contains of soil trace metals and other important compound of soil fertility get increases and the wheat production also favourably affect with increasing sewage sludge level in soil (Jamil et al. 2006). The acclimated sewage sludge and spent mushroom substrate of *L. edodes* combination prove to improve soil and reduce multi-polycyclic aromatic hydrocarbon pollutant from soil. This combination of ASS and SMS also effects the fungal and bacterial population in soil (Wang et al. 2016).

3 Processing of Sewage Sludge: In General

To use sewage sludge as fertilizer on agricultural land its processing is necessary. Sewage sludge consists large amount of toxic substance such as furans, chlorine derivates, aromatic hydrocarbon, etc. Therefore, the processing of sludge is considered as mandatory step. In general, the processing of sewage sludge can be divided into steps which are represented in Fig. 1 (Demirbas et al. 2017).

As Fig. 1 indicates the different steps of processing of sewage sludge in which each step has particular step or requirement such as preliminary treatment include screening and comminuting. Preliminary treatment initially removes screening and grit, but the it does not show very much effect on pathogens, which may be released in bioaerosols in preliminary treatment of sludge. Specifically, it works on removing suspended solid in liquid sludge and scum (floating organic material).

Primary thickening includes floating drainage, gravity, belt, centrifuges, liquid sludge sublimation includes lime addition, anaerobic and aerobic digestion. Secondary thickening includes same processes as primary thickening where conditioning of waste water is done by chemical treatment, elutriation and thermal treatment. Dewatering is the major step which is used to remove all moisture content in SS. This

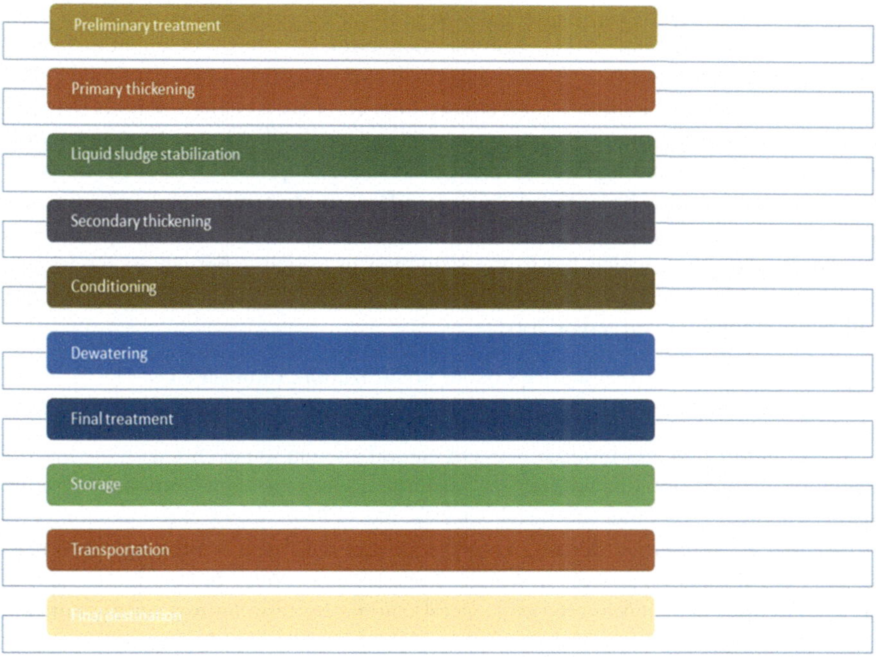

Fig. 1 The different step which generally uses to process sewage sludge to using it as fertilizer

process can be done by using composting, drying, line addition and also wet oxidation. After all these processes sludge is stored as dry sludge, liquid sludge or compost. During the time of storage this stored sludge is then transported via different routes such as pipelines, road and reaches to final destination such as agricultural land, forest etc. After complete processing the processed sludge can be further used for organic recycling process so it can be safely applicable on agricultural land (Metcalf et al. 2014).

4 Natural Attenuation

The definition of natural attenuation varies according to objectives; different groups give different definitions regarding natural attenuation. For instance, according to Environmental Protecting Agency natural attenuation is a process which includes physical, chemical as well as biological processes without human input led to the reduction of toxicity, volume and concentration of contaminants in ground water (Rittmann 2004). Natural attenuation is site specific process. Some biological process of natural attenuation involves dispersion, biodegradation, sorption, dilution, volatilisation, and radioactive decay. While, some group consider natural attenuation as a method used to decrease the pollutant content by natural mean, such as with the help of microorganisms and their products at a specific site (Kouzuma and Watanabe 2011). Natural attenuation is considered as an emerging cost saving technology. It is an in- situ process of reduction of contaminants. The process of natural attenuation requires more time compare to other methods. Natural attenuation has different ways such as one method can be the destruction of contaminants through the biodegradable process or abiotic process. Concentration of contaminants can also be reduced through diffusion, dispersion and volatilization like processes. On the other hand, mobilization of contaminants can stop which is known as adsorption which led to the reduction in toxicity (Fernandez Rodríguez et al. 2014). The process of natural attenuation can be divided into two categories which are (Mulligan and Yong 2004)

- Abiotic process (Diffusion, dispersion, volatilization, dilution etc.)
- Biotic process (through action of microorganism or microorganism product)

As the name indicates biotic process in natural attenuation occurs with the help of 'specialized microorganisms' such as bacteria, fungus etc. are known to remove major impurities of other bacterial species like *Vibrio logei* and *Pseudomonas nitroreducens* to degrade the toxic content of azo dye from wastewater treatment plant by decolorization process in which *Vibrio* species proved more effective at 6–7 pH and around 30 °C (Adedayo et al. 2004). Some bacteria known for the conversion of insoluble metallic sulfides into the soluble metallic sulfides in sludge such as *Thiobacillus ferrooxidans* and *T. thiooxidans* (Bosecker 1997). Fungus is also proven effective for the detoxification of SS, in an experiment conducted by Dhouib and colleagues in which they conclude that white-rot fungus effectively detoxifies the olive mill wastewater and degrades its polyphenolic compound. Figure 2 demonstrates how white-rot fungi is used for OMW detoxification (Dhouib et al. 2006).

Fungi like *Penicillium ochrochloron* prove their effectiveness in detoxification of malachite green impurities and convert them to N, N-dimethyl-aniline hydrochloride and p-benzyl-N, N-dimethylaniline which are harmless to plants (Shedbalkar and Jadhav 2011). Fungus *U. isabellina* is known to remove heavy metals as well as xenobiotics (Janicki et al. 2018). A term i.e., vermicomposting which defines the mass production of earthworm in waste, earthworm's secretes specific enzymes which work as biological stimulators for soil and possess biodegradation capacity. Sinha and colleagues studied the action of some species of earthworm on biodegradation of some community wastes of India (Sinha et al. 2002).

Now a days for the best result of detoxification of SS, natural attenuations combine with different other approaches as Hayes and Jewell disclose a method which works on principle of biological and chemical methods of aggregation for the detoxification of sewage sludge (Hayes and Jewell 1981).

There is another term which is also related to natural attenuation that is monitored natural attenuation. Monitored natural attenuation consider as an in-situ remediation approach and works on to diminish the mass of contaminants in soil and ground water (Jorgensen et al. 2009). Natural attenuation process do not require continuous input of human so it requires verification and monitoring to assure the effectiveness of the process (Fernandez Rodríguez et al. 2014). Monitored natural attenuation can

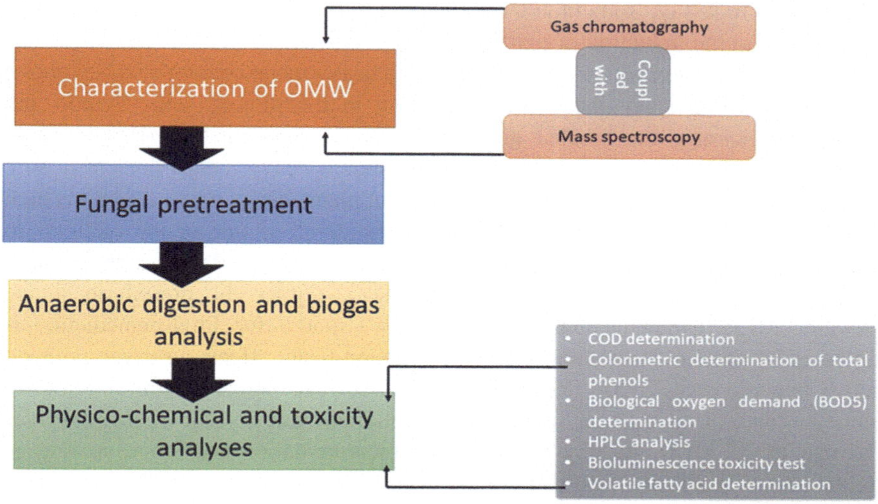

Fig. 2 How fungi use nowadays for detoxification of OMW

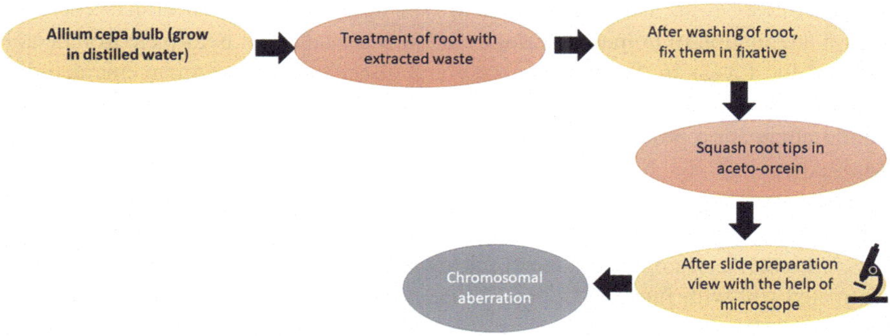

Fig. 3 Demonstration of A. cepa test use for natural attenuation monitoring

also consider as a viable remediation approach on soil with high native microbial population (Sarkar et al. 2005). Monitored natural attenuation can also be performed using *Allium cepa*, because of the induces DNA damage in *A. cepa* by SS samples and decrease in this effect also noticed during attenuation time period (Mazzeo et al. 2015). Figure 3 indicates the *A. cepa* test for MNA.

The monitoring on NA can also be done via *Viciafaba* test (Bhat et al. 2018). Sewage sludge is very beneficial for soil fertility. To test the potential of sewage sludge, it is generally used to bury in holes and for different time period these holes are free of pollutants. After each cycle of natural attenuation aqueous and organic extract are obtained and then later zebrafish embryo and yeast-based bioassay are used to test different toxicity. Estrogenic and dioxin-like activity tested via yeast-based bioassay, sometime zebrafishes are also used to check dioxin-like activity due to the induction of marker gene *cyp1a* (Mazzeo et al. 2016a, b).

If the soil is treated with sewage sludge, then heavy metal content also becomes a big issue for the heavy metal immobilization in the mining soils. Penido and colleagues examine the biochar and sewage sludge combination (Penido et al. 2019). Tai and Mcbride found that the large amount heavy metals present in soil, which previously treated with sewage and also in plants heavy metal content is high (Tai et al. 2016). Phytoremediation is also considered as one of the methods to reduce impurities for instance *Jatrophacurcas* can be used to decrease the contamination of heavy metals mainly chromium, lead, copper etc. present in sludge (Awalla 2013). Clemente and colleagues prove that if the Phytoremediation is followed by natural attenuation, then it shows significant effects on pyrite-polluted soil (Clemente et al. 2006). Some impurities of sewage sludge cannot be removed alone by natural attenuation such as polycyclic aromatic hydrocarbon impurities (Kosnar et al. 2018). As phytoremediation, bioremediation also improve sewage treated soil, specifically with addition of some stimulant agents such as sugarcane bagasse (Sommaggio et al. 2018). Ornamental plants use for phytoextraction of heavy metals due to the different detoxification pathways and their phytoextraction efficiency of heavy metals can also enhance through different strategies (AsgariLajayer et al. 2019). Mazzeo and colleagues examine the effectiveness of natural attenuation of domestic sewage sludge and evaluate it before environmental disposal for 1 year by Salmonella assay and also examine the mutagenic activity from around 6 months of natural attenuation. Mazzeo and coworkers also monitored natural attenuation process by genotypic assays in detoxification of sewage sludge (Mazzeo et al. 2016a, b).

If sewage sludge is expected to apply on agricultural land, then it should be pollutant free so it will not harm plant physiology. Sometime waste water treatment does not remove all impurities one of the approaches of natural attenuation is to treat SS treated soil with fugus. A fungus *Trametesversicolor* prove to reduce the toxic contain of micropollutant present in sewage sludge (Rodriguez-Rodriguez et al. 2011). SS detoxification via natural attenuation process also has some drawbacks such as long-term performance monitoring as it also requires long term institution controls (Forstner and Gerth 2001)

4.1 Processing of Sewage by Natural Attenuation

The rough idea of natural attenuation process is provided in Fig. 4.

Anaerobic sewage sludge sample firstly collects from the domestic plants of waste water treatment, then the sample of sewage sludge is used to dewater using centrifugation or belt filter process (Novak 2006). According to Mazzo and colleague's protocol, the specific amount of sewage sludge collected in perforated plastic bags and buried in wholes for different period of time till 1 year (Mazzeo et al.

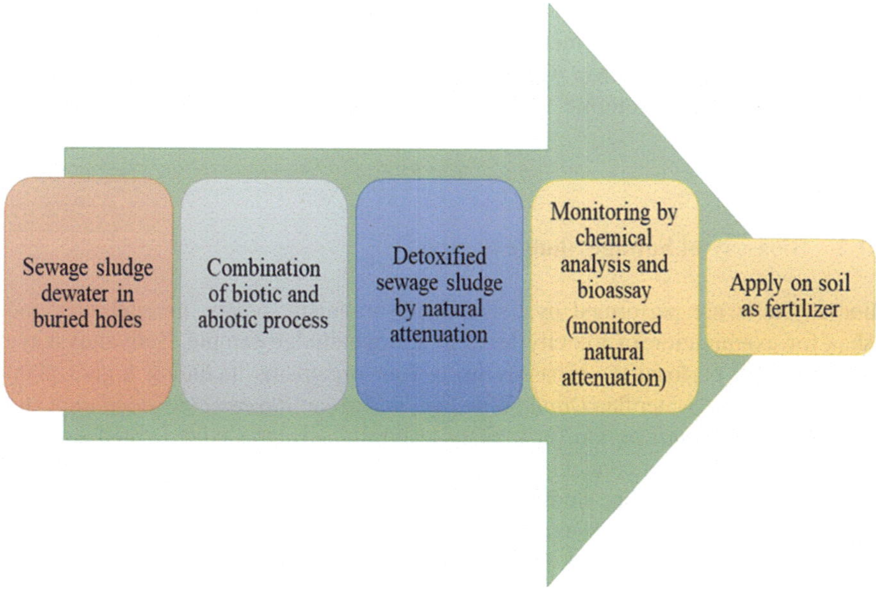

Fig. 4 Diagram shows how natural attenuation process occurs

2015). Using different protocols aqueous extract and organic extract of sludge, sample are prepared and then these extracts are further used for chemical analysis.

4.1.1 Chemical Analysis of Sludge Sample

Management of processed sewage sludge and its chemical analysis after each cycle of natural attenuation is necessary so the environmental hazards can be minimised (Cieslik et al. 2015). Chemical analysis of solid sludge (100% SS) will be necessary to perform after each cycle of natural attenuation that can be done by using different method such as yeast estrogen screen (YES) which provide the idea for the estimation of estrogenic potential (Pawlowski et al. 2004). Techniques such as liquid chromatography, group analogue internal standards and mass chromatography have been proven effective for the successful determination of antibiotic substances such as sulfonamide, b-lactams, trimethoprim tetracyclines (Lindberg et al. 2004). Toxicity testing can be done by using solid phase extraction which is followed by the chromatographic approach (gas chromatography-mass spectrometry or liquid chromatography-mass spectrometry) (Farre and Barcelo 2003). Where gas chromatography coupled with mass spectroscopy is used to determine ibuprofen, naproxen, ketoprofen, triclosan, phenolic endocrine, etc. (Samaras et al. 2011). High-resolution gas chromatography/mass spectrometry are used to determine alkyl benzenesulfonates (McEvoy and Giger 1986). Veenaas and colleagues used two-dimensional gas Chromatography-High Resolution Mass Spectroscopy for non-target screening analysis and time-trend analysis of sewage sludge contaminants (Veenaas et al. 2018). Trace metals such as Mn, Ni, Zn, Pb, Cr, Cd, Cu and V, and major elements including Mg, Ca, Fe and Al that can be determined in the sludge using coupled plasma-atomic emission spectroscopy (ICP-AES) (Sandroni and Smith 2002).

4.1.2 Bioassay of Sewage Sludge Sample

Bioassay tests are performed by using micro-organism, plants, invertebrates and fishes for examination of toxicity level in sewage sludge sample. Selivanovskaya and Latypova, perform the bioassay using four organisms including higher plant *Raphanussativus*, water flea *Daphnia magna*, protozoan *Paramecium caudatum* and bacteria *Pseudomonas putida* for toxicity testing in sewage sludge treated soil and sewage sludge sample, conclude that these kind of toxicity tests are very useful in metal contaminated sewage sludge (Selivanovskaya and Latypova 2003). For micro-organism conducted bioassay, bacteria such as *Vibrio fischeri* and rotifer, *Brachionusplicatilis* are used for ecotoxicological evaluation of sewage sludge (Park et al. 2005). Terrestrial bioassay can be conducted to test toxicity using plants such as Barley seed germination for 14 days and sprout growth for 5 days, lettuce seed germination and worm mortality, etc. Another term is Liquid-phase bioassays use also for toxic reduction (Renoux et al. 2001). Plants such as Wheat, Soybean are also used for seed germination bioassay or plant growth bioassay for sewage sludge compost phytotoxicity, relative seed germination and root elongation percentage (Araujo and Monteiro 2005).

It is necessary to test the bioassay activity of sewage sludge treated soil. If the high amount of heavy metallic contamination is there in sewage treated soil, then soil microbial biomass decreases (Fliesbach et al. 1994). There is a term i.e., bioluminescence bioassay which may be considered as an appropriate tool for toxicity threshold test. Chaudari and colleagues conduct an experiment to determined the toxicity of Zn in sewage sludge treated water via bioluminescence assay for which they used bacterial species like *Escherichia coli* and *Pseudomonas fluorescens* (Chaudri et al. 1999). Microorganism such as yeast and *E. coli* can be used for microorganism based bioassay. Two yeast based bioassay namely ER-RYA and AhR-RYA have been used by Mazzeo and colleagues to evaluate the biological activity and estrogenic activities of sewage sludge. For detection of estrogenic activity due to the presence of inserted estrogenic sensitive intein at specific site of lacZ gene. A strain of *Escherichia coli* i.e., DIER was constructed (Liang et al. 2011). Free living Rhizobium can also be used as a bio-indicator to detect heavy metal toxicity (Horswell et al. 2003). Recombinant yeast assay chemical analysis can be combinedly use to evaluate for removal of estrogenic substances (Onda et al. 2002). Cyst based bioassay can be used to detect chronic toxicity using fresh water rotifer *Brachionuscalciflorus,* which determine the toxicity of copper, pentachlorophenol, DC and lindane. Invertebrate Daphnia based bioassays also provide large range of advantages such as short reduction cycle, high range of sensitivity (Janssen et al. 1994).

From past two decades biosensors are actively used to test sewage sludge toxicity, like bioluminescence based-biosensors are considered as an inexpensive and fast technique for bioavailability of metal evaluation (McGrath et al. 1999). Bacterial and fungal bioluminescence-based biosensors are generally used for the indication of heavy metal toxicity in sewage sludge sample (Horswell et al. 2006). Nematode *Caenorhabditiselegans is* used as a bioluminescence biosensor which can indicates the impact of sublethal level of environmental pollutant (McLaggan et al. 2012).

5 Impact of Sewage Application on Soil

The use of sewage sludge in the form of fertilizer limits due to the presence of toxic contaminants. Thus it affects the physical, biological as well as the chemical behavior of agricultural land in various ways (Singh and Agrawal 2008). Organic matter which present in sewage sludge effect on physical property such as porosity, bulk density, pH etc., chemical properties such as cation exchange capacity, electric conductivity and biological properties like micro and macro-biological population etc. (Clapp et al. 1986). Physical property of soil such as pH of soil decrease in sewage amended soil (Wong 1998), bulk density shows negative effect (Hemmat et al. 2010). Ojeda and colleagues conducted an experiment and concluded that sewage sludge treatment enhances soil infiltration by decrease in erosion (Ojeda et al. 2003). Humus content is an another physical property of sewage sludge which increases in the soil containing sewage sludge (Koskela 1983). Sewage sludge is nutrient rich and affects soils chemical property as well such as N, P content increases in soil treated with SS and the continuous application of sludge increases the availability of Zn, Cd and Ni to the plant (Soon et al. 1980). The application of sewage sludge for a long period of time lead to the rise in organic matter content and soluble phenolic compounds of soil (Roig et al. 2012) and electric conductivity (one of the chemical properties of soil show an increase) (Wong 1998). Applications of sludge on agricultural land also affects its biological activity such as microbial biomass of soil and enzyme activity increases (Banerjee et al. 1997). An experiment conducted by Paz-Ferreiro and colleagues for determination of biochemical property of soil and geometric mean analysis of enzyme activity in sewage sludge applied soil. Their conclusion suggested that the geometrical mean shows a decrease, whether it shows an increase when sewage sludge biochar is applied to the soil (Paz-Ferreiro et al. 2011). Lakhdar and colleagues analyse the impact of sewage sludge and compost on salt affected soil biological activity and conclude in their research that due to the presence of sewage sludge in the soil had lead to the increase in arylsluphatase activity of soil (Lakhdar et al. 2010). Kizilkaya and Bayrakli conduct an experiment to determine the soil enzyme activity after appling

N-enriched sewage sludge and found that some sort of heavy metals are there in sewage sludge which adversely influence the enzymatic activity of soil during sludge decomposition (Kızılkaya and Bayraklı 2005). Some research also claim that the humic acid content of soil also increases with long term sewage sludge application in soil and also show significant increase in enzyme activity and micro-organism count (Sastre et al. 1996) where presents of toxic metal as Zn, copper, lead, Cadmium in food crops due to sewage application to agricultural soil create health risk concern to human and animal population, the health risk index high value which is greater than 1 indicates their potential to health risk (Chaoua et al. 2018). The different forms of copper and zinc in sewage sludge can be ranked as follows (Shrivastava and Banerjee 1998).

$$\begin{aligned}\text{Cu} &: \text{acid soluble} > \text{residual} > \text{reducible} > \text{oxidisable} > \text{exchangeableZn}\\ &: \text{acid soluble} > \text{reducible} > \text{residual} > \text{oxidisable} > \text{exchangeable}\end{aligned}$$

If the large amount of heavy metal contamination are present within sewage sludge sample then they create harmful effects on human as well as on animal population such as Cd can impact on heart and kidney where Cr, Ni, and Pb are mutagenic and carcinogenic (Satarug 2018; Casalegno et al. 2015). Da Silva Souza and colleagues perform an experiment to determine the toxic potential of sewage sludge by expose of Diplopod (*Rhinocricuspadbergi*) and fish (*Xiphophorusmaculatus*) in sewage sludge and lime treated sewage sludge and concluded that SS can cause tissue damage (Da Silva et al. 2020). Table 1 indicates the impact of sewage sludge on different plant species.

6 Conclusion and Future Prospects

From the above discussion here, we can conclude that sewage management is necessary to reduce the adverse effect of sewage sludge on environment or health. We can utilize sewage sludge (SS) in various ways, but it has high nutritive qualities which inspire SS use as fertilizer. The main concerns of SS used as fertilizer is that it has toxic substances so its detoxification also a necessity. Natural attenuation is that process which occurs at the original site of sewage sludge without human input and detoxifies the SS via a series of naturally occurring abiotic and biotic process. Monitoring of natural attenuation reduces the chances of toxic substance present. Natural attenuation provides pure sewage sludge as fertilizer with low (less than harmful ranges) or no harmful substances, however its long-term application as fertilizer can be harmful for agricultural soil. Sewage sludge detoxification via natural attenuation provides cheap solution for the management of sewage sludge.

Table 1 Different plant species affected by sewage sludge application on soil

Plants	Impact	Sewage sludge amendment rate	Reference
Mung bean (*Vignaradiata)*	Decline of protein content but nutrients content high in seed also high heavy metal contents	6, 9 and 12 kg m^{-2}	
Beta vulgaris var. Allgreen H-1	Biomass of plant reduce with decrease in length and area of leaf, shoot biomass and yield decreased increment seen in lipid peroxidation, protein, oxidation activity and proline contents	20% and 40% (w/w)	Singh and Agrawal (2007)
(*Lepidiumsativum, Sorghum saccharatum, Sinapis alba*	Root growth inhibition	0.1–6.4%, 0.03–9.4% and 6.6–22.1% (% of sewage sludge kg^{-1} soil	Oleszczuk and Hollert (2011)
maize (Zea Mays L)	Copper and manganese, Increase in germination contents increase w	200 kg N/ha 10–30-ton ha^{-1}	Qasim et al. (2001)
Festucaarundinacea Schreb	No toxic or detrimental effects on fescue, 30% higher yield with increased heavy metal control in fescue, however K, Ca, and Mg content low compare to increased P content	5.6 Mt./ha dry wt.	Boswell (1975)
Flax (*Linum usitatissimum*)	Faste development and high bio-mass production	2:1 and 10:1 (v/v soil	Tsakou et al. (2002)

Acknowledgement We would like to thank Director, Dayalbagh Educational Institute and Head, Department of Botany, Dayalbagh, Agra for providing support and encouragement.

References

Adedayo O, Javadpour S, Taylor C, Anderson WA, Moo-Young M (2004) Decolourization and detoxification of methyl red by aerobic bacteria from a wastewater treatment plant. World J Microbiol Biotechnol 20:545–550

Álvarez EA, Mochon MC, Sanchez JCJ, Rodniguez MT (2002) Heavy metal extractable forms in sludge from wastewater treatment plants. Chemosphere 47:765–775

Andreoli CV, Von Sperling M, Fernandes F, Ronteltap M (2007) Sludge treatment and disposal. IWA Publishing

Antonkiewicz J, Popławska A, Kołodziej B, Ciarkowska K, Gambus F, Bryk M, Babula J (2020) Application of ash and municipal sewage sludge as macronutrient sources in sustainable plant biomass production. J Environ Manag 264:110450

Araujo ASF, Monteiro RTR (2005) Plant bioassays to assess toxicity of textile sludge compost. Sci Agric 62:286–290

AsgariLajayer B, KhademMoghadam N, Maghsoodi MR, Ghorbanpour M, Kariman K (2019) Phytoextraction of heavy metals from contaminated soil, water and atmosphere using ornamental plants: mechanisms and efficiency improvement strategies. Environ Sci Pol 26:8468–8484
Aubain P, Gazzo A, le Moux J, Mugnier E (2002) Disposal and recycling routes for sewage sludge. Synthesis report 22 February 2002. Arthur Andersen, EC DG Environment–B/2
Awalla CO (2013) Phytoremediation of sewage sludge in soils contaminated with heavy metals. Glob J Environ Sci 12:13–19
Baily RE (2009) Sludge: types, treatment processes and disposal. Nova Science Publishers
Banerjee MR, Burton DL, Depoe S (1997) Impact of sewage sludge application on soil biological characteristics. Agric Ecosyst Environ 66:241–249
Belhaj D, Elloumi N, Jerbi B, Zouari M, Abdallah FB, Ayadi H, Kallel M (2016) Effects of sewage sludge fertilizer on heavy metal accumulation and consequent responses of sunflower (*Helianthus annuus*). Environ Sci Pollut Res 23:20168–20177
Bhat SA, Singh S, Singh J, Kumar S, Bhawana, Vig AP (2018) Bioremediation and detoxification of industrial wastes by earthworms: Vermicompost as powerful crop nutrient in sustainable agriculture. Bioresour Technol 252:172–179
Bosecker K (1997) Bioleaching: metal solubilization by microorganisms. FEMS Microbiol Rev 20:591–604
Boswell FC (1975) Municipal sewage sludge and selected element applications to soil: effect on soil and fescue1. J Environ Qual 4:267
Bresters AR (1998) Sludge treatment and disposal: management approaches and experiences (No. 7). EUR-OP
Casado-Vela J, Selles S, Navarro J, Bustamante MA, Mataix J, Guerrero C, Gomez I (2006) Evaluation of composted sewage sludge as nutritional source for horticultural soils. Waste Manag 26:946–952
Casalegno C, Schifanella O, Zennaro E, Marroncelli S, Briant R (2015) Collate literature data on toxicity of Chromium (Cr) and Nickel (Ni) in experimental animals and humans. EFSA Support Publ 12:478E
Chaoua S, Boussaa S, El Gharmali A, Boumezzough A (2018) Impact of irrigation with wastewater on accumulation of heavy metals in soil and crops in the region of Marrakech in Morocco. J Saudi Soc Agric Sci 18:429–436
Chaudri AM, Knight BP, Barbosa-Jefferson VL, Preston S, Paton GI, Killham K, McGrath SP (1999) Determination of acute Zn toxicity in pore water from soils previously treated with sewage sludge using bioluminescence assays. Environ Sci Technol 33:1880–1885
Cieslik BM, Namiesnik J, Konieczka P (2015) Review of sewage sludge management: standards, regulations and analytical methods. J Clean Prod 90:1–15
Clapp CE, Stark SA, Clay DE, Larson WE (1986) Sewage sludge organic matter and soil properties. The Role of Organic Matter in Modern Agriculture 209–253
Clemente R, Almela C, Bernal MP (2006) A remediation strategy based on active phytoremediation followed by natural attenuation in a soil contaminated by pyrite waste. Environ Pollut 143:397–406
Da Silva Souza T, Lacerda D, Aguiar LL, Martins MNC, Augusto de Oliveira David J (2020) Toxic potential of sewage sludge: histopathological effects on soil and aquatic bioindicators. Ecol Indic 111:105980
Demirbas A, Edris G, Alalayah WM (2017) Sludge production from municipal wastewater treatment in sewage treatment plant. Energy Sour A Recov Util Environ Effects 39:999–1006
Dhouib A, Ellouz M, AlouiF SS (2006) Effect of bioaugmentation of activated sludge with white-rot fungi on olive mill wastewater detoxification. Lett Appl Microbiol 42:405–411
Farre M, Barcelo D (2003) Toxicity testing of wastewater and sewage sludge by biosensors, bioassays and chemical analysis. Trends Analyt Chem 22:299–310
Fernandez Rodríguez MD, García Gómez MC, Alonso Blazquez N, Tarazona JV (2014) Soil pollution remediation. Encycl Toxicol:344–355

Fliesbach A, Martens R, Reber H (1994) Soil microbial biomass and microbial activity in soils treated with heavy metal contaminated sewage sludge. Soil Biol Biochem 26:1201–1205

Forstner U, Gerth J (2001) Natural attenuation—non-destructive processes. In Treatment of contaminated soil, Springer, Berlin, pp 567–586

Franz M (2008) Phosphate fertilizer from sewage sludge ash (SSA). Waste Manag 28:1809–1818

Grosser A (2017) The influence of decreased hydraulic retention time on the performance and stability of co-digestion of sewage sludge with grease trap sludge and organic fraction of municipal waste. J Environ Manag 203:1143–1157

Grosser A, Neczaj E (2017) Pretreatment methods as a means of boosting methane production from sewage sludge and its mixtures with grease trap sludge. E3S Web Conf 22:00058

Gupta N, Yadav KK, Kumar V (2015) A review on current status of municipal solid waste management in India. J Environ Sci 37:206–217

Gutberlet J (2018) Waste in the city: challenges and opportunities for urban agglomerations. Urban Agglomeration 21:191

Hayes TD, Jewell WJ, inventors (1981) Combined biological-chemical detoxification of organics. United States patent US 4,277,342, 7 Jul 1981

Hemmat A, Aghilinategh N, RezainejadY Sadeghi M (2010) Long-term impacts of municipal solid waste compost, sewage sludge and farmyard manure application on organic carbon, bulk density and consistency limits of a calcareous soil in central Iran. Soil Tillage Res 108:43–50

Horswell J, Speir TW, van Schaik AP (2003) Bio-indicators to assess impacts of heavy metals in land-applied sewage sludge. Soil Biol Biochem 35:1501–1505

Horswell J, Weitz HJ, Percival HJ, Speir TW (2006) Impact of heavy metal amended sewage sludge on forest soils as assessed by bacterial and fungal biosensors. Biol Fertil Soils 42:569–576

Jamil M, Qacim M, Umar M (2006) Utilization of sewage sludge as organic fertilizer in sustainable agriculture. J Appl Sci 6:531–535

Janicki T, Długoński J, Krupiński M (2018) Detoxification and simultaneous removal of phenolic xenobiotics and heavy metals with endocrine-disrupting activity by the non-ligninolytic fungus Umbelopsisisabellina. J Hazard Mater 360:661–669

Janssen CR, Persoone G, Snell TW (1994) Cyst-based toxicity tests. VIII. Short-chronic toxicity tests with the freshwater rotifer Brachionuscalyciflorus. Aquat Toxicol 28:243–258

Jorgensen KS, Salminen JM, Bjorklof K (2009). Monitored Natural Attenuation. Bioremediation 217–233

Kacprzak M, Neczaj E, Fijałkowski K, Grobelak A, Grosser A, Worwag M, Singh BR (2017) Sewage sludge disposal strategies for sustainable development. Environ Res 156:39–46

Kızılkaya R, Bayraklı B (2005) Effects of N-enriched sewage sludge on soil enzyme activities. Appl Soil Ecol 30:192–202

Kominko H, Gorazda K, Wzorek Z (2017) The possibility of organo-mineral fertilizer production from sewage sludge. Waste Biomass Valorization 8:1781–1791

Koskela I, (1983) Effect of sewage sludge on soil humus content. In The influence of sewage sludge application on physical and biological properties of soils. Springer, Dordrecht, p 136–140

Kosnar Z, Mercl F, Tlustos P (2018) Ability of natural attenuation and phytoremediation using maize (*Zea mays* L.) to decrease soil contents of polycyclic aromatic hydrocarbons (PAHs) derived from biomass fly ash in comparison with PAHs–spiked soil. Ecotoxicol Environ Saf 153:16–22

Kouzuma A, Watanabe K (2011) Molecular approaches for the analysis of natural attenuation and bioremediation. Compreh Biotechnol:25–36

Krogmann U, Boyles LS, Bamka WJ, Chaiprapat S, Martel CJ (1999) Biosolids and sludge management. Water Environ Res 71:692–714

Lakhdar A, Scelza R, Scotti R, Rao MA, Jedidi N, Gianfreda L, Abdelly C (2010) The effect of compost and sewage sludge on soil biologic activities in salt affected soil. Revista de La Ciencia Del Suelo y Nutrición Vegetal 10:40–47

Liang R, Zhou J, Liu J (2011) Construction of a bacterial assay for estrogen detection based on an estrogen-sensitive intein. AEM 77:2488–2495

Lindberg R, Jarnheimer PA, Olsen B, Johansson M, Tysklind M (2004) Determination of antibiotic substances in hospital sewage water using solid phase extraction and liquid chromatography/mass spectrometry and group analogue internal standards. Chemosphere 57:1479–1488
Logan TJ, Harrison BJ (1995) Physical characteristics of alkaline stabilized sewage sludge (N-viro soil) and their effects on soil physical properties. J Environ Qual 24:153
Mazzeo DEC, Fernandes TCC, Levy CE, Fontanetti CS, Marin-Morales MA (2015) Monitoring the natural attenuation of a sewage sludge toxicity using the *Allium cepa* test. Ecol Indic 56:60–69
Mazzeo DEC, Casado M, Piña B, Marin-Morales MA (2016a) Detoxification of sewage sludge by natural attenuation and implications for its use as a fertilizer on agricultural soils. Sci Total Environ 572:978–985
Mazzeo DEC, Fernandes TCC, Marin-Morales MA (2016b) Attesting the efficiency of monitored natural attenuation in the detoxification of sewage sludge by means of genotoxic and mutagenic bioassays. Chemosphere 163:508–515
McEvoy J, Giger W (1986) Determination of linear alkylbenzenesulfonates in sewage sludge by high-resolution gas chromatography/mass spectrometry. Environ Sci Technol 20:376–383
McGrath SP, Knight B, Killham K, Preston S, Paton GI (1999) Assessment of the toxicity of metals in soils amended with sewage sludge using a chemical speciation technique and alux-based biosensor. Environ Toxicol Chem 18:659–663
McLaggan D, Amezaga MR, Petra E, Frost A, Duff EI, Rhind SM, Lagido C (2012) Impact of sublethal levels of environmental pollutants found in sewage sludge on a novel Caenorhabditis elegans model biosensor. PLoS ONE
Metcalf & Eddy, Abu-Orf M, Bowden G, Burton FL, Pfrang W, Stensel HD, AECOM (Firm) (2014) Wastewater engineering: treatment and resource recovery. McGraw Hill Education
Moore WP (1985) Method of converting sewage sludge to fertilizer US Patent 4,519,831, 28 may 1985
Mulligan CN, Yong RN (2004) Natural attenuation of contaminated soils. Environ Int 30:587–601
Novak JT (2006) Dewatering of sewage sludge. Dry Technol 24:1257–1262
Nyamangara J, Mzezewa J (2001) Effect of long-term application of sewage sludge to a grazed grass pasture on organic carbon and nutrients of a clay soil in Zimbabwe. Nutr Cycl Agroecosyst 59:13–18
Ojeda G, Alcañiz JM, Ortiz O (2003) Runoff and losses by erosion in soils amended with sewage sludge. Land Degrad Dev 14:563–573
Oleszczuk P, Hollert H (2011) Comparison of sewage sludge toxicity to plants and invertebrates in three different soils. Chemosphere 83:502–509
Onda K, Yang SY, Miya A, Tanaka T (2002) Evaluation of estrogen-like activity on sewage treatment processes using recombinant yeast. Water Sci Technol 46:367–373
Park GS, Lee SH, Park SY, Yoon SJ, Lee SM, Chung CS, Kim SH (2005) Ecotoxicological evaluation of sewage sludge using bioluminescent marine bacteria and rotifer. Ocean Sci J 40:91–100
Pathak A, Dastidar MG, Sreekrishnan TR (2009) Bioleaching of heavy metals from sewage sludge: a review. J Environ Manag 90:2343–2353
Pawlowski S, Ternes TA, Bonerz M, Rastall AC, Erdinger L, Braunbeck T (2004) Estrogenicity of solid phase-extracted water samples from two municipal sewage treatment plant effluents and river Rhine water using the yeast estrogen screen. Toxicol In Vitro 18:129–138
Paz-Ferreiro J, Gasco G, Gutiérrez B, Mendez A (2011) Soil biochemical activities and the geometric mean of enzyme activities after application of sewage sludge and sewage sludge biochar to soil. Biol Fertil Soils 48:511–517
Penido ES, Martins GC, Mendes TBM, Melo LCA, do Rosario Guimarães I, Guilherme LRG (2019) Combining biochar and sewage sludge for immobilization of heavy metals in mining soils. Ecotoxicol Environ Saf 172:326–333
Qasim M, Javed N, Subhan M (2001) Effect of sewage sludge on the growth of maize crop

Rank J, Nielsen M (1998) Genotoxicity testing of wastewater sludge using the Allium cepa anaphase-telophase chromosome aberration assay. Mutat Res Genet Toxicol Environ Mutagen 418:113–119

Renoux AY, Tyagi RD, Samson R (2001) Assessment of toxicity reduction after metal removal in bioleached sewage sludge. Water Res 35:1415–1424

Rittmann BE (2004) Definition, objectives, and evaluation of natural attenuation. Biodegradation 15:349–357

Rodriguez-Rodriguez CE, Jelic A, Llorca M, Farre M, Caminal G, Petrovic M, Vicent T (2011) Solid-phase treatment with the fungus Trametesversicolor substantially reduces pharmaceutical concentrations and toxicity from sewage sludge. Bioresour Technol 102:5602–5608

Roig N, Sierra J, Martí E, Nadal M, Schuhmacher M, Domingo JL (2012) Long-term amendment of Spanish soils with sewage sludge: effects on soil functioning. Agric Ecosyst Environ 158:41–48

Rulkens W (2008) Sewage sludge as a biomass resource for the production of energy: overview and assessment of the various options†. Energy Fuel 22:9–15

Saha S, Hazra GC, Saha B, Mandal B (2014) Assessment of heavy metals contamination in different crops grown in long-term sewage-irrigated areas of Kolkata, West Bengal, India. Environ Monit Assess 187

Saha S, Saha BN, Pati S, Pal B, Hazra GC (2017) Agricultural use of sewage sludge in India: benefits and potential risk of heavy metals contamination and possible remediation options–a review. Int J Environ Technol Manag 20:183–199

Samaras VG, Thomaidis NS, Stasinakis AS, Lekkas TD (2011) An analytical method for the simultaneous trace determination of acidic pharmaceuticals and phenolic endocrine disrupting chemicals in wastewater and sewage sludge by gas chromatography-mass spectrometry. Anal Bioanal Chem 399:2549–2561

Sandroni V, Smith CM (2002) Microwave digestion of sludge, soil and sediment samples for metal analysis by inductively coupled plasma–atomic emission spectrometry. Anal Chim Acta 468:335–344

Sarkar D, Ferguson M, Datta R, Birnbaum S (2005) Bioremediation of petroleum hydrocarbons in contaminated soils: comparison of biosolids addition, carbon supplementation, and monitored natural attenuation. Environ Pollut 136:187–195

Sastre I, Vicente MA, & Lobo MC (1996) Influence of the application of sewage sludges on soil microbial activity. Bioresour Technol 57:19–23

Satarug S (2018) Dietary cadmium intake and its effects on kidneys. Toxics 6:15

Seleiman M, Santanen A, Manninen-Egilmez P, Stoddard F, Makela P (2010) Sewage sludge as nutrient source for bioenergy crops. Suomen Maataloustieteellisen Seuran Tiedote 31:1–5

Selivanovskaya SY, Latypova VZ (2003) The use of bioassays for evaluating the toxicity of sewage sludge and sewage sludge-amended soil. J Soils Sediments 3:85–92

Shedbalkar U, Jadhav JP (2011) Detoxification of malachite green and textile industrial effluent by Penicilliumochrochloron. Biotechnol Bioprocess Eng 16:196–204

Shrivastava S, Banerjee DK (1998) Operationally determined speciation of copper and zinc in sewage sludge. Chem Speciat Bioavailab 10:137–143

Singh RP, Agrawal M (2007) Effects of sewage sludge amendment on heavy metal accumulation and consequent responses of *Beta vulgaris* plants. Chemosphere 67:2229–2240

Singh RP, Agrawal M (2008) Potential benefits and risks of land application of sewage sludge. Waste Manag 28:347–358

Sinha RK, Herat S, Agarwal S, Asadi R, Carretero E (2002) Environmentalist 22:261–268

Sohaili J, Zaidi NS, Loon SC (2012) Nutrients content of sewage sludge and its utilization towards horticulture plant. J Emerg Trends Eng Appl Sci 3:81–85

Sommaggio LRD, Mazzeo DEC, Sant' Anna D d A e S, Levy CE, Marin-Morales MA (2018) Ecotoxicological and microbiological assessment of sewage sludge associated with sugarcane bagasse. Ecotoxicol Environ Saf 147:550–557

Sommers LE (1977) Chemical composition of sewage sludges and analysis of their potential use as fertilizers1. J Environ Qual 6:225

Sommers LE, Nelson DW & Yost KJ (1976) Variable nature of chemical composition of sewage sludges1. J Environ Qual 5:303

Soon YK, Bates TE, Moyer JR (1980) Land application of chemically treated sewage sludge: III. Effects on soil and plant heavy metal content. Am Soc Agron Crop Sci Soc Am Soil Sci Soc Am 9:497–504

Tai Y, Li Z, Mcbride MB (2016) Natural attenuation of toxic metal phytoavailability in 35-year-old sewage sludge-amended soil. Environ Monit Assess 188:241

Tester CF (1990) Organic amendment effects on physical and chemical properties of a sandy soil. Sci Soc Am J 54:827

Tsadilas CD, Matsi T, Barbayiannis N, Dimoyiannis D (1995) Influence of sewage sludge application on soil properties and on the distribution and availability of heavy metal fractions. Commun Soil Sci Plant Anal 26:2603–2619

Tsakou A, Roulia M, Christodoulakis NS (2002) Growth of flax plants (linumusitatissimum) as affected by water and sludge from a sewage treatment plant. Bull Environ Contam Toxicol 68:56–63

Usman K, Khan S, Ghulam S, Khan MU, Khan N, Khan MA, Khalil SK (2012) Sewage sludge: an important biological resource for sustainable agriculture and its environmental implications. Am J Plant Sci 3:1708–1721

Usman AR, Alkredaa RS, Al-Wabel MI (2013) Heavy metal contamination in sediments and mangroves from the coast of Red Sea: Avicennia marina as potential metal bioaccumulator. Ecotoxicol Environ Saf 97:263–270

Veenaas C, Bignert A, Liljelind P, Haglund P (2018) Nontarget screening and time-trend analysis of sewage sludge contaminants via two-dimensional gas chromatography–high resolution mass spectrometry. Environ Sci Technol 52:7813–7822

Wang C, Yu D, Shi W, Jiao K, Wu B, Xu H (2016) Application of spent mushroom (Lentinula edodes) substrate and acclimated sewage sludge on the bioremediation of polycyclic aromatic hydrocarbon polluted soil. RSC Adv 6:37274–37285

Wong J (1998) Effect of sewage sludge amendment on soil microbial activity and nutrient mineralization. Environ Int 24:935–943

Zhang Q, Hu J, Lee DJ, Chang Y, Lee YJ (2017) Sludge treatment: current research trends. Bioresour Technol 243:1159–1172

Socio Economic Aspects of Sewage Sludge Use in Agriculture

Shabana Gulzar, Shafiqa Gul, Abhik Patra, Kiran Kumar Mohapatra, Hanuman Singh Jatav, and Vishnu D. Rajput

1 Introduction

The universal water calamity, the scarcity of clean water and the production of large quantities of waste water have led to the use of agricultural waste water. Waste water is a multifaceted reserve, with certain benefits and even some drawbacks to its use. It can have enormous benefits for farmers in particular and society in general. It supplies growers with a secure supply of crop water, preserves nutrients, decreases the need for synthetic fertilizers, raises agricultural productivity and farmland yields, and is an inexpensive method for the cleaner discharge of industrial waste water. The development and reuse of waste water has risen steadily with growing water shortage in many areas of the planet as a consequence of growing population, urbanization and industrialization growth. As industries move into urban areas, the derivation of waste water has also distorted from being primarily organic (from

S. Gulzar (✉)
Department of Botany, Cluster University, Srinagar, J&K, India

S. Gul
Department of Management Studies, University of Kashmir, Srinagar, J&K, India

A. Patra
Department of Soil Science and Agricultural Chemistry, Institute of Agricultural Sciences, Banaras Hindu University, Varanasi, Uttar Pradesh, India

K. K. Mohapatra
Department of Soil Science and Agricultural Chemistry, College of Agriculture, Odisha University of Agriculture and Technology, Bhubaneswar, Odisha, India

H. S. Jatav
Department of Soil Science and Agriculture Chemistry, Sri Karan Narendra Agriculture University, Jobner, Rajasthan, India

V. D. Rajput
Southern Federal University, Rostov-on-Don, Russia

V. D. Rajput et al. (eds.), *Sustainable Management and Utilization of Sewage Sludge*, https://doi.org/10.1007/978-3-030-85226-9_14

human waste) to containing more perilous machinery such as heavy metals and other pollutants. Therefore, the use of waste water may have detrimental impacts on the communities and habitats that use this resource as well. An rise in the prevalence of wastewater-borne illnesses, rapid environmental degradation, as well as societal problems like inconvenience, poor ecological standards, bad sanitation, foul smell, etc. are likely to be caused by the systematic use of toxic waste water. Schemes for waste water treatment have usually been motivated by class, fortification and cost-effectiveness concerns. However, it is critical to monitor a structural approach that combines both anthropological and particular functions and can lead to improvements in waste water resource planning and management. As a result, a conceptual change is necessary not just to decrease ecological harm, but also to emphasis the sustainable use of waste water as a resource that must be managed properly in order to conserve water in the future. Extensive research must be conducted from an economic perspective before irrigation of waste water can be allowed as a means of increasing the availability of agricultural water. The structural risks and benefits of such waste water re-use should be measured in this regard. Standard cost benefit analysis also struggles to measure and commercialize wastewater reuse-related external costs. In order to enable a good outcome, ecological assessment methods and other relevant instruments should also be used. In addition, not only from a societal, financial and environmental point of view, but also from a sustainable viewpoint, the economic consequences of the drainage of waste water need to be calculated.

2 Sewage Sludge: An Essential Resource

Sewage sludge is a rich resource of nutrients, inorganic and organic compounds that make recycling and reuse worthwhile. Any compound and microbial contaminants can also be found and the disproportionate use of unregulated waste water can negatively impact people's wellbeing and the climate (United Nations 2015). There are also some reimbursements in terms of enhancing food security, creating source of revenue prospects, adjusting to weather alter and safe habitats for efficient waste water management (Corcoran et al. 2010). There have been several instances in the world where purified water is used successfully for consumption purposes; in Namibia, for example, 35% of all intake water is treated (Lazarova et al. 2013).

Wastewater can also be extracted and converted into various energy sources. Waste water and its bio-solids can be derived from various sources of energy, the most common being biogas. It can be combusted for the production of heat or electricity and used as a fuel for vehicles (Oki and Kanae 2006; Zimmerman and Mihelcic Smith 2008; Conley et al. 2009). Wastewater treatment stations are increasingly producing electricity, which is very important because energy consumption for wastewater treatment plants is an indispensable necessity. The waste water treatment in some developing countries requires the refining of faecal sludge into dry fuel *viz.*, briquettes (Funamizu et al. 2001; Logan et al. 2006). Water for irrigation and drinking purposes is however the most essential resources extracted from wastewater treatment. Fecal slots are widely used as fertilizer, especially in septic plants, due to low pollution in household sanitation systems compared with wastewater wastewater treatment plant bio-solids. Certain treating methods extract nutrients from wastewater during treatment, *i.e.,* the rich N and P structure, rather than from treatment materials (Bashan and Bashan 2004; Guest 2009; Larsen et al. 2009).

2.1 Usage of Sewage Sludge in Agriculture

In agriculture, waste water is commonly used as it is an abundant reservoir of nutrient and offers all the moisture content needed for crop growth enhancement (Fig. 1). It has been observed that some crops generate higher yields with wastewater irrigation, with a minimum requirement for chemical fertilizers, which is economical for farmers. A systematic research by researchers has resulted in a description of the impact of processed and unregulated waste water under various agricultural programmes on a range of efficiency and output metrics.

These experiments indicate that processed wastewater can be exploited to achieve greater yields for better crops than would normally be practicable. In many nations, there is a proliferation of the use of untreated wastewater, which is undoubtedly a threat to the climate, human health and crop yields in particular. Although higher nutrient levels do not inherently raise crop yields, the large volume of vital minerals for plant growth and the production of untreated sewage are encouraged by the farmers to use it for crop irrigation as it lowers input costs. Many crops demand specific quantities of NPK for optimal production, even those produced in urban agriculture. Once the prescribed quantity of NPK has been met, it can significantly impact crop growth and yield. For instance, total N accessible to the plants by irrigation via waste water suggests high dose of nitrogen for optimum yields, vegetative growth may be stimulated, but there are many development disruptions, and in severe circumstances, may lead to a loss of yield (Singh and Mishra 1987).

Consequently, structure and configuration of untreated waste water is a very important parameter which must be taken into consideration prior to being used in agriculture. The paucity of agricultural waste results in chemical pollution that, at greater quantities, may be toxic to plants. There are few contaminants that can causes food poisoning, although most studies demonstrate that they are present at levels safe for human consumption. The proliferation of household waste water, on the other hand, may contribute to high levels of salinity that can influence the yield of salt-sensitive crops. In the stretched run, overload nutrients and salts institute in waste water leaching under the root zone of the plant will persuade the superiority of groundwater stores. The actual result, nevertheless, relies on a variety of considerations, such as the height of the water table, the reliability of groundwater, surface drainage and the reach of drainage of waste water. For many residents in developed nations, groundwater is a major supply of clean water. Consequently, before a large waste water irrigation scheme is designed, the possibility of groundwater leakage

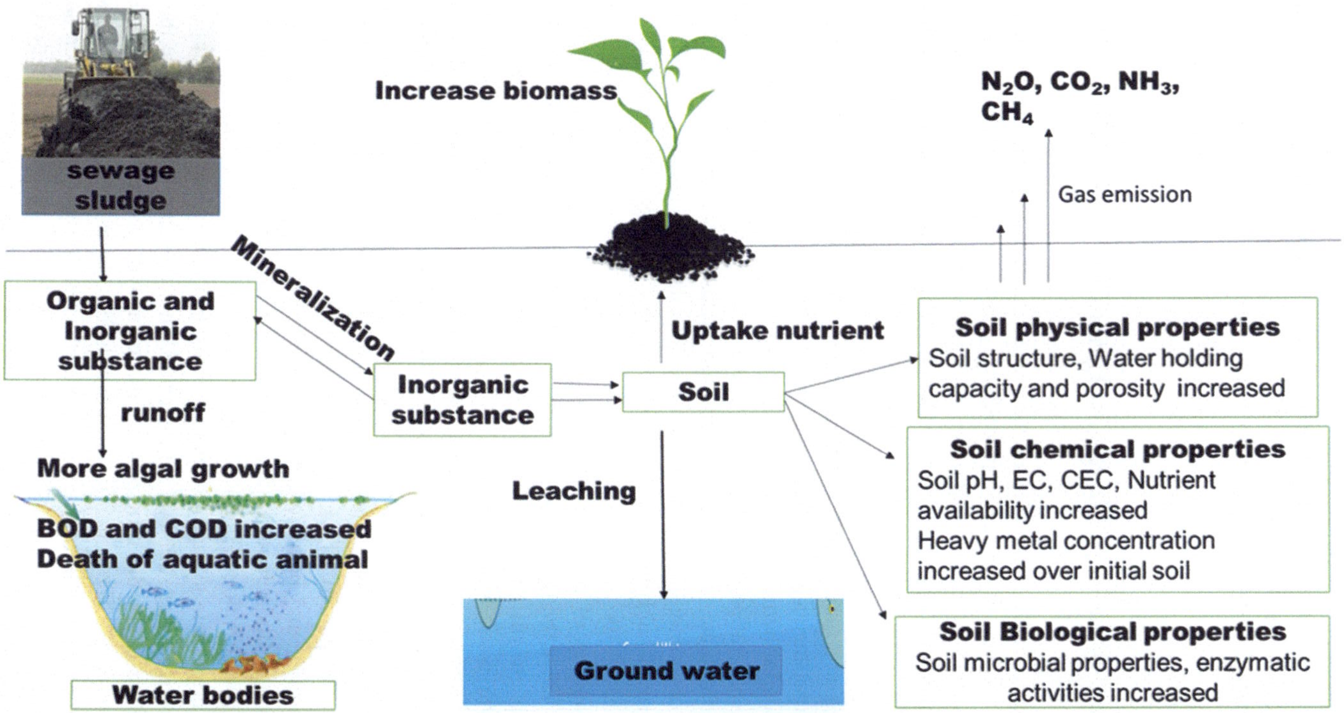

Fig. 1 Environmental aspects of sewage sludge application in agriculture

requires to be measured. Waste water irrigation has the ability to change harmful bacteria and viruses into groundwater in terms of salt and nitrate buildup, albeit only under certain conditions (NRC report 1996). Farid et al. (1993) stated that long-run waste water irrigation for crop irrigation at fields in the Greater Cairo area, where untreated or primary treated wastewater has been applied for irrigation since 1915, has resulted in an unusual rise in groundwater salinity. Evidence of coli contamination from groundwater, which was also discovered in Mexico, corroborated with Downs et al. (1999) and Gallegos et al. (1999). A related analysis (Rashed et al. 1995) reveals that chloride, sulphate, TDS and soluble osmotic pressure in groundwater in the arid wastewater field of Gabar el as far in Greater Cairo are much greater than average sewage liquid waste amounts. The leaching and draining of waste water, used for field irrigation, will act as an outlet for groundwater reservoirs to refill wells. 50–70 percent of irrigation water will flow into groundwater aquifers in certain regions (Rashed et al. 1995). Consequently, the impact of poured waste water on the safety and regeneration of groundwater is very relevant. Given the negative efficiency, in locations where freshwater supplies are limited, contamination of groundwater by waste water may be a crucial ecological and financial test. It can, in this situation, be used as a benefit in certain circumstances. There is also a direct equilibrium between the gains of recharging reservoirs and the danger of contaminating groundwater.

2.2 *Disadvantage of the Use of Sewage Sludge in Agriculture*

Wastewater can have detrimental impacts on agriculture as well (Fig. 2). Its practice in agriculture can amplify the vulnerability to transmittable diseases of producers, consumers and surrounding communities; lead to groundwater pollution; long-term utilize of wastewater can have adverse effects on soil resources-salt upsurge, grave metals in soils that can decrease the capacity for soil production in the elongated period; adverse effects on in close proximity property prices and when these are widespread dimensions. These involve wastewater quantity and origin (housing, industrial, automotive); wastewater structure; magnitude or level of pre-use diagnosis; administration matters relating to wastewater decommissioning at secondary level; management considerations relevant to the consumption of wastewater at household level, namely measures of operation, at tertiary level. Nutritional compounds can trigger eutrophication as water from drainage irrigation systems flows, in fact, into small restricted lakes and water sources and groundwater. In plant microbiological water body societies, this generates disparities (Smith et al. 1999). In exchange, this can influence other upper forms of marine life and may have an effect on the existence of water species, thus reducing diversity. Local people benefit from these bodies of water, and the biological implications may be translated into measurable potential costs.

Congesting of organic matter, for instance, culminating in a reduction in dissolved oxygen, can contribute to adjustments in the structure of marine

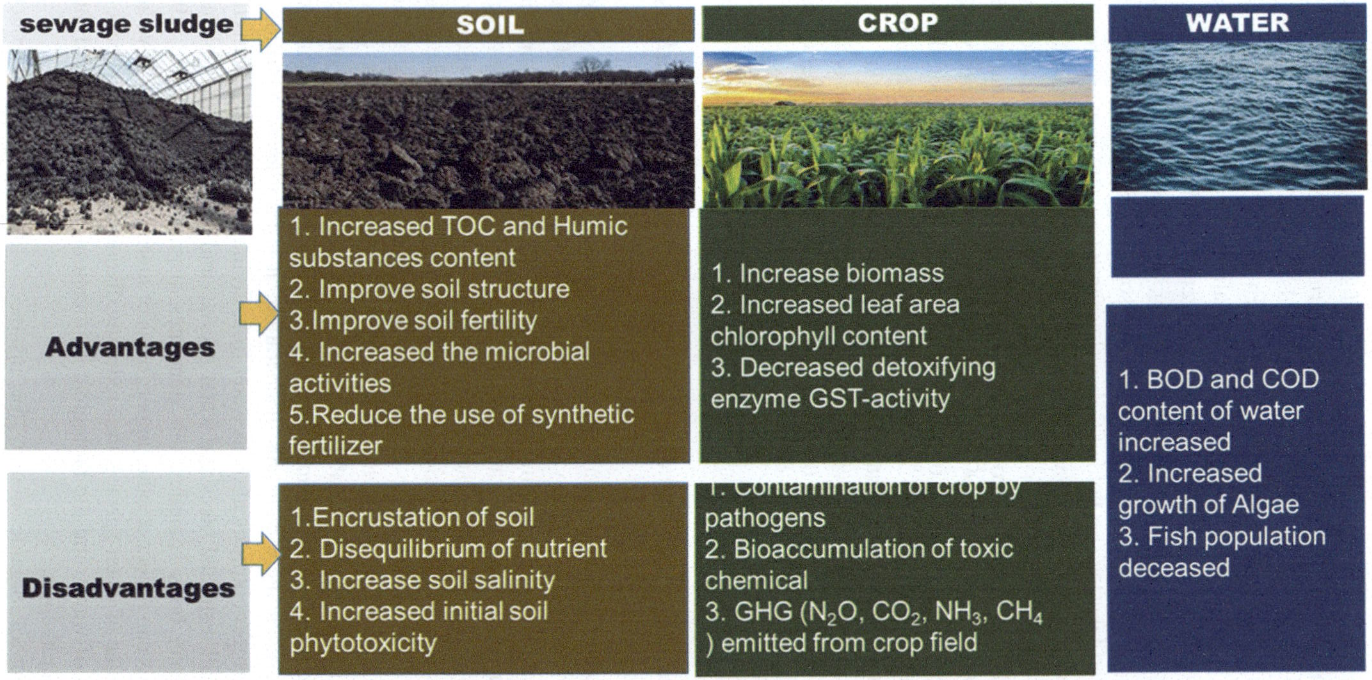

Fig. 2 Advantages and disadvantages of uses of sewage sludge in agriculture

organisms, *i.e.,* fish mortality and a reduction in fisheries. It is possible to calculate the eutrophication ability of drainage irrigation utilizing biological indexes or biomarkers, which can be defined through acceptable financial measurement methods in currency terms. Alkalinity and salt (high sodium content in soil) caused by waste water can also have detrimental impacts on soil viability, which in turn can impact land values and lease income. On the contrary, considering the relevance of the waste water supply, it is also possible to bring value to land irrigated with waste water. As a result, based on the situation, we can infer that waste water irrigation has the potential to influence land prices which can have a substantial effect on real estate values. In the analysis of the implications of waste water drainage, it should also be viewed as a profitability item. In many contexts, waste water impacts the ecosystem: waste water, for example, can comprise high nutritional value such as nitrogen and phosphate. If water sources acquire substantial amounts of these resources, they can promote unsustainable plant growth, which can release pollutants into water bodies, cause oxygen loss, and result in de-oxygenated hotspots. This process lowers species and changes the structure and superiority of organisms and limits the consistency of water for recycling (UN 2015; World Water Council 2012). An illustration is the effects of wastewater on the atmosphere across the globe: methane and nitrous oxide (powerful world heating gases) pollution linked to wastewater processing could rise by 50 percent and 25 percent between 1990 and 2020, respectively (Corcoran et al. 2010).

As per the fourth Global Water Development Report by UNESCO in 2012, just 20 percent of wastewater handled worldwide is handled right. Processing efficiency typically relies on the nation's income level; therefore in high-income nations the processing facility is up to 70 percent of wastewater generated opposed to 8 percent in low nations (Sato et al. 2013). Consequently, it is not a right to react to the problem of waste water disposal, but a cautious, practical and innovative act worthy of protecting public health and maintaining the protection of the ecosystem. A philosophical change against its recognition as a resource for wealth creation is developing, whilst the bulk of wastewater has so far been treated as a disposal challenge. In this way, waste water can be converted from a challenge to a resource of social and economic importance (Drechsel et al. 2015). Wastewater contains bacterial microbes that have the potential to cause infection, such as bacteria, disease vectors. In this respect, human parasites, such as protozoa and helminth larvae, in specific, are of distinct relevance as they tend to be the most complicated to eliminate by treatment options and have been involved in a number of digestive bacterial infections in both industrialized and evolving economies. Even so, the actual risk of individuals getting ill must be measured through the calculation of clinical outcomes, not the incidence of toxins in water. While, possible hazard is very extraordinary, a variety of other considerations result in a corresponding risk. The unregulated waste water for agriculture in all demographic ranges presents a significant hazard to public health. The element of uncertainty can, nevertheless, vary amongst people of different ages. Undiagnosed drainage of waste water leads to a relatively greater proportion of diseases of crochet worms (Feenstra et al. 2000) and Ascariasis in children (Cifuentes et al. 2000; Habbari et al. 2000).

If ingested vast concentrations and can be toxic, heavy metals in waste water pose a health danger. Because plants cannot tolerate and perish from high amounts of these poisons until they become a hazard to people, there may be little doubt regarding the ingestion of heavy metals by crops and the risk presented to customers. These research findings have significant significance for identifying environmental health concerns associated with waste water irrigation. Second, they propose that the risk evaluation of environmental health is an vital judgment predictor for irrigation of waste water and that all adolescents should be viewed as a potential group of risk.

2.3 Sociological Impacts of Wastewater Reuse

Public effects are concerns or doubts posed by the public over waste water irrigation. This can entail many issues such as annoyance, bad quality of the atmosphere, poor sanitation, smell, disturbance, increased risk of injuries, etc.; socioeconomic challenges such as food security, health and safety, impairment, degradation of property values, and land use restoration; natural capital aspects *i.e.,* vital water supply pollution, fish loss, wildlife, tropical plants, etc. In an attempt to prevent exploitation by advocacy organizations, public fears about future or real threats of drainage irrigation will create market hazards that need to be treated properly. By the availability of appropriate amounts of coverage, company liabilities and potential liabilities can be covered. As most emerging economies, like Pakistan, don't have insurance expertise in the agriculture sector, the overall liability insurance payment for waste water irrigation is considered high at the onset. In fact, discount and compensation structures are expected to vary considerably between crops and territories. Regardless of scientific data, social awareness tends to be the driving element underpinning the effectiveness or failure of wastewater reuse programmes, particularly if wastewater is cleaned using contemporary technology and health hazards are thoroughly addressed and monitored. The introduction of a waste water scheme may be aided or restricted, based on public preferences, perceptions and actions. Negative public standards would hinder the moving ahead of well projects. On the other hand, strong civic consciousness is a central constituent in the triumphant espousal of waste water recycling, which leads to cognizance (Drechsel et al. 2015; Friedler et al. 2006). As a result of insufficient citizen feedback, urban populations have rejected a number of wastewater treatment initiatives by municipalities and water sources across the world, contributing to false public perceptions (PMSEIC 2003).

Ethnic, societal, educational and/or social and economic influences involve other variables that differ on the area and the situation, (Drechsel et al. 2015). As per Bruvold, when people spoke about specific use options, the level of human contact had a bigger effect, while when the basic use method was used, other factors, such as well-being, environment, treatment, delivery and sustainability, had a stronger influence on the attitudes of people (Bruvold 1988). Thus it proposes that, in conjunction with the appropriateness and desire of people/users, it is important to consider the multiple goals of the sustainability alternatives and to choose the reuse initiatives that are most probable to be approved by the group or somehow make the investment worthwhile.

2.4 Economic Perspective of Waste Water Irrigation

Based on the degree of care and the quality of the crop, the consequences of waste water irrigation on crops can differ greatly (Fig. 2). From an economics perspective, under effective horticultural and water management methods, the optimal solutions can be gained from waste water irrigation of crops: (1) greater outputs, (2) extra irrigation water and (3) saved fertilizer value. Conversely, it can be badly influenced if plant food nutrients provided by waste-water irrigation contribute to nutritional over-supply outputs. Until that time, financial studies on drainage irrigation had been conducted from specialized perspectives, such as municipalities lowering treatment costs or farmers or regional bodies increasing income. As a moderate wastewater dumping technique, land treatment of partly treated wastewater has been around for a very lot longer. Young and Epp (1980) conducted a simulated analysis of the costs of urban wastewater land usage and their impact on crop choices. According to their findings, a range of factors impact the cost of land treatment, including the quantity of pre-treatment, pumping costs, land expenses, yearly application rate, crop type, and wastewater control. They find that, via the sales effect and productivity of the land treatment scheme, crop choice has a significant effect on the profitability. If the water delta is big, it is feasible to use waste water more efficiently while increasing agricultural production and preserving the scheme's ability to be renovated. The authors looked examined the impact of crop choices on income and expenditure flows, as well as system efficiency, using three crop patterns: Canary grass, alfalfa, maize, and forest crops. Their analysis reveals that, because reed canary grass uses waste water over the year, it is a more sustainable and expense process. If waste water can be utilized for extended periods of time, alfalfa and maize may become more expense than reed canary grass. Plant plantations have a poorer rate of nutrient elimination (extended growing period and poor harvest) and lower earnings, but they are more productive since they can utilize water all year and are more appropriate for the general population than crop irrigation. This outcome has significant policy ramifications. These trees may be cultivated every 8 to 10 years to alter the severely polluted urban environment and boost revenues, in addition to functioning as natural air conditioners and greenhouse gas sinks.

A lengthy numerical scheduling algorithm was used by Dinar and Yaron (1986) in order to optimize regional profits, relating to limitations *viz.*, wastewater treatment emerging technologies, agricultural production innovations, prices and labor legislation. The data demonstrate that granting a wastewater irrigation rebate, which had a significant transportation cost, increased the farmers' revenue. When all waste water was treated and all farmers participated in waste water irrigation, the local benefit was optimum at a reimbursement level of 50%. Participation in a local systematic strategy has supported all contributing stakeholders, both internal and external, such as producers, the economy, the ecosystem, and water ecosystems. The analysis presumed that farmers could not trade their water rights or, more simply,' inter-farm fresh water allocation transfer is not permitted.' Thus, in the lieu of water markets, the study assumes only a short and mid resolution. Important quality gains can typically be made by the commercialization of water in a dynamic environment (since that might not be the case for Israel) and, in exchange, the need for incentives can be decreased. Segarra et al. (1996) also utilizes a complex simulation model to estimate the optimal crop rotation process worthy of using both sewage sludge, recovering resources, and maximizing income in the agronomic and temperature changes of Lubbock, Texas. In order to raise net profits, they predict that alfalfa, wheat-corn, wheat-grain sorghum, and cotton are acceptable varieties of crops. Choosing commercially efficient cultivation practices reduces the maintenance and disposal routinely obtained by communities. Therefore, this ensures that towns can gain from joint ventures with nearby producers for the drainage of waste water. Darwish et al. (1999), employs a proposed framework in the Tyre region, Lebanon, to determine the best crop production to maximize the crop yields. The findings demonstrate that the drainage of the sea without agricultural production (least profitable), the use of waste water irrigation for traditional cultivation trends, and the application of new plants to suitable crop patterns are value maximization alternatives in sorted array (highly profitable). For concept of project trends, extra irrigation and fertilizers are required to optimize the income of farmers. This ensures, sufficient plant food nutrients and the relative humidity of waste water are successfully reused as new crops are planted. Consequently, this review reveals that changes in crop varieties are a critical aspect for the efficient usage of crop irrigation waste water supplies. The results indicate that the key advantages of waste water irrigation are essential water and fertilizer recovery, stronger agricultural production, a diversification work constructively, and efficiency gains for processing. It is essential to emphasize that waste water irrigation gives massive benefit income in the case of sea-based disposal, opposed to zero revenue in the existing crops grown.

An additional research has demonstrated the economic advantages and dangers involved with the continuing use of town waste water for field irrigation in Guanajuato, Mexico (Scott et al. 2000). The sophisticated River Aquifer Modeling Model was used in the study to anticipate changes in water quality under various wastewater supervision scenarios. Farm survey and modeling findings showed that the use of untreated waste water on land caused in significantly higher amounts of salinity and coli. The investigation used a chance cost or substitute benefit method to calculate dollar values for water and waste water nutrient quality. Local water value added calculations supplied by supplemental studies are used to assess the water value of waste water. As nutrients are given in abundance of plant needs, the true economic value of resources would be overestimated by the result of a strategy to nutritional quality value. The writers then used the savings on fertilizer bills and fertilizer application expenses as a more accurate indication of nutrient profit. The outcomes recommended that waste water is a good asset for the ecosystem and that waste water from agriculture is an appropriate option to expensive care. The report, nevertheless, recognizes that the use of waste water could have harmful effect on people and the atmosphere and that these implications should be assessed. The report, furthermore, recognizes that the use of waste water can have damaging consequences on health and wellbeing, and these implications should be measured.

3 Emerging Developments in Agricultural Wastewater Control

Shuval et al. (1997) created a quantitative risk analysis and a risk management method to assess the financial feasibility of WHO and USEPA microbiological health guidelines. Their cost estimation for the two cases of wastewater treatment (WHO requirements and USEPA guidelines) suggests that, for very expensive disease control, achieving USEPA guidelines would involve an extra $3–30 million per case. On contrary, WHO requirements can be satisfied by utilizing low-cost, efficient, land-based processing systems such as waste stabilization ponds, which can dependably achieve high microbial levels and efficiency. Consequently, the WHO recommendations are a more technically realistic and economically feasible choice for industrialized economies than the USEPA rules. The majority of European countries have not documented standards for the usage of waste water for irrigation, excluding Germany and France. The EU Guidelines, as established, recommend covering all sustainable agriculture dimensions, conservation of soil and groundwater, maximization of yields and hygiene standards refers to the safety of public health.

The degree of wastewater treatment used to irrigate crops relies on the nature of the field, regional constraints and compliance standards. Cost studies of wastewater treatment reveal that, at significantly higher stages, average prices are ridiculous (Schleich et al. 1996). Even so, provided the crop value the magnitude of water scarcity, and public interest, this relatively higher processing expense may also be justifiable. In the absence of any contractual constraints, such as environmental standards requirements, cost reduction should remain the primary aim of wastewater treatment facilities. Even so, experiments suggest that water quality is favored in order to reduce costs (Schwarz and McConnell 1993). In fact, for a variety of purposes, most developed countries use solid wastes water for irrigation, namely disposal costs and the scarcity of vital nutrients.

Nevertheless, until agricultural use, wastewater treatment is deemed crucial: first, from the perspective of public health welfare, and secondly, from the perspective of local religious and social values (Mara 2000). A large amount of research and innovation has been undertaken in Israel, notably in the field of waste water reuse, in light of these needs, water shortages, dry land agriculture, hot weather patterns, and the high commercial benefit of water sources. Sludge treatment disposal is the most well engineering technology and there are various technologies and providing encouragement for effective waste management (Asano et al. 1985; NRC report

1996). In the existence of an unsustainable accumulation of toxic waste, the use of activated sludge accompanied by supplementary treatment processes using elevated physiological systems is an appropriate treatment alternative for conventional wastewater treatment. But in most developed nations, high electricity prices, infrastructure needs, and scheduled maintenance problems render it unsuccessful.

4 Cost-Effective Treatment of Wastewater

When laws require waste water to be processed before re-use for crop production in developing countries, cost factors are critical in selecting an acceptable method (Fig. 3). Land-based systems have worked hard to establish themselves as one of the best wastewater treatment methods, particularly in arid and semi-arid areas (Young and Epp 1980), since they give that land is widely accessible at cheap rates, they are capable of producing similar levels of nutrient removal at a significantly low cost. Recycling and reusing waste water, as well as plant food ingredients for crop production, are additional advantages. A classic design of waste stabilization ponds includes anaerobic, optional, and maturation ponds. Studies express numerous variations and usefulness of these land-based systems in Egypt (Shereif et al. 1995), Morocco (Yagoubi et al. 2000), and Israel (Juanico et al. 1995).

These data definitely demonstrate that waste water treatment utilizing waste stabilization ponds is highly competent and cost-effective in terms of both capital and operational expenses. More handling in maturation ponds requires unfettered irrigation (or chlorination). Mara (2000) indicates that for tolerant irrigation, the land stipulates treatment levels that are more than double those sought for limited irrigation. It is also projected that unrestricted irrigation be chosen only if it is monetarily realistic. Simply put, unrestricted irrigation can only be preferred if the difference between the net present value of loose crops and the net present value of restricted crops harvested surpasses the discounted expenditure of maturing ponds. An alternate to the utilization of vocational disposal for conventional municipal waste processing is the application of floating aquatic organisms in developed wetlands (non-land-based systems). They can be used just as single species or as wetlands equipped with different organisms feeding on waste water full of nutrients. Lately, numerous significant papers have been proposed in this area. Two recent experiments, one concentrating on nitrogen undervaluing (Bramwell and Prasad 1995) and the other concentrating on the microbial productivity of these tropical animals (Karpiscak Martin et al. 1996), show that the system is low-cost and requires little land (0.27 m^2 per head for single-species systems) and is ideally compatible with the needs of small populations. Such constructed environments virtually remove, on a broad scale, BOD concentrations, nitrogen content and pathogen densities in secondary wastewater treatment. Additionally, many species of birds are drawn to these wetland areas. As a result, this aquatic, multi-species wetland system has a big influence on both economic sustainability and biological variety.

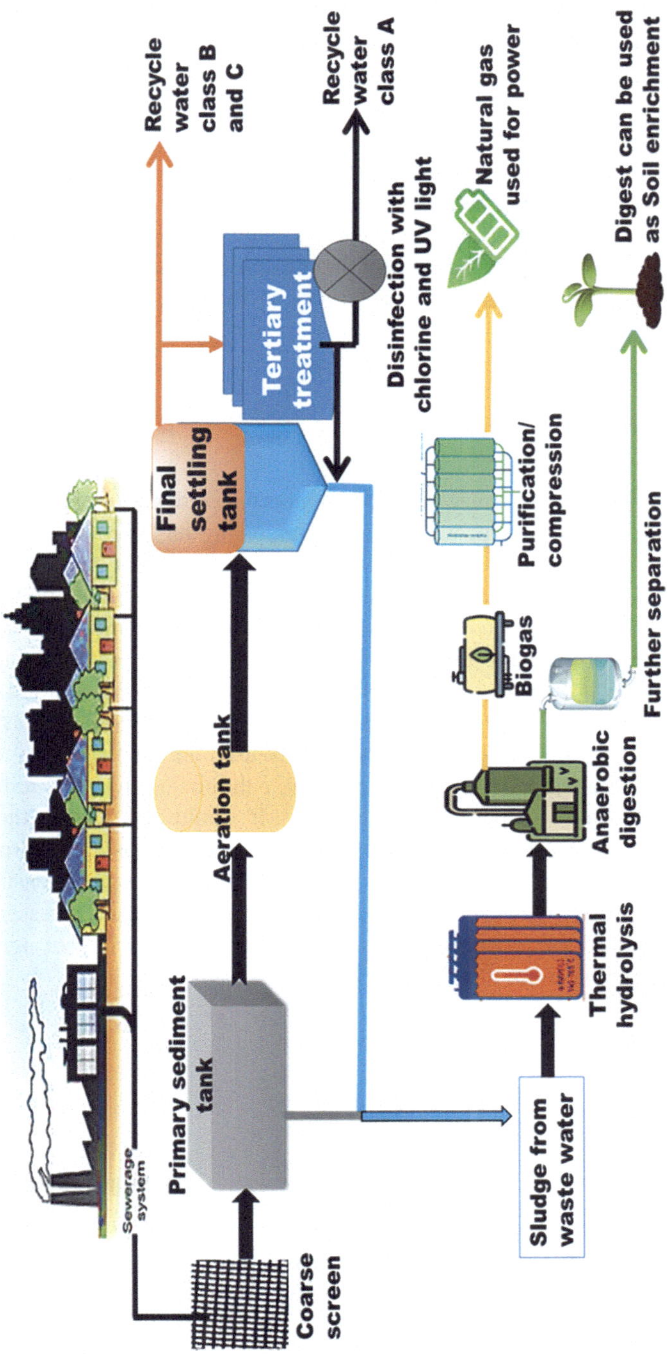

Fig. 3 Processes of waste water treatments

The application of anoxic, discretionary and development ponds for waste stabilization ponds is a substitution for a packed vocational surgical procedure, namely decontamination, which should be emphasized here. However, de-chlorination can be accomplished using chlorine, ozone, or ultraviolet (UV) light. Chlorine is an element that depletes ozone and is known to have significant detrimental ecological effects. Consequently, in order to enable ecologically minded consumers to patronize items grown with sewage processed by evolution and ecologically responsible treatment methods such as waste stabilization ponds, this downside of the conventional treatment process may be used while calculating prices. Even so, a very greater standard of integrity in treatment and customer education will be required, which in industrialized economies is sadly very costly.

5 Conclusions

With growing demands for water resources to meet the demands of expanding public populations around the world, waste water composting has become progressively essential. Thus, the comprehensive plans for water conservation need to be set up rapidly. That being said, social considerations such as public opinion, public acceptance and the demographic component, including the advancement of wastewater treatment technologies, have important ramifications for the positive outcome of wastewater reuse. By introducing and incorporating a behavioral solution, considering both social and technical considerations and their expense in waste water processing, the facilities will be more commonly used and appropriately operated by all groups of users, namely women and men, culminating in dramatic improvements in terms of consistency and productivity. It would also lead to the greater development of society by providing subsidies, processing, housing, environmental stewardship, aspects of health wellbeing of the family, as well as ensuring the availability of water whereas addressing the social issues also. In essence, this would contribute to the environmental sustainability and recycling of agricultural wastewater treatment technologies.

References

Asano TR, Smit G, Tchobanoglous G (1985) Municipal wastewater: treatment and reclaimed water characteristics. In: Pettygrove GS, Asano T (eds) Irrigation with reclaimed municipal wastewater-a guidance manual. Chelsea, Mich: Lewise Publishers, Inc.

Bashan L, Bashan Y (2004) Recent advances in removing phosphorus from wastewater and its future use as a fertilizer (1997–2003). Water Res 38:4222–4246

Bramwell SA, Prasad PVD (1995) Performance of a Small aquatic plant wastewater treatment system under Caribbean conditions. J Environ Manag 44:213–222

Bruvold WH (1988) Public opinion on water reuse options. J Water Pollut Control Fed 60:45–49

Cifuentes EM, Gomez U, Blumenthal MM, Tellez-Rojo I, Romieu G, Ruiz-Palacios, Ruiz-Velazco S (2000) Risk factors for Giardia intestinalis infection in agricultural villages practicing wastewater irrigation in Mexico. Am J Trop Med Hyg 62:388–392

Conley D, Paerl H, Howarth R, Boesch D, Seitzinger S, Havens K, Lancelot C, Likens G (2009) Controlling eutrophication: nitrogen and phosphorus. Science 323:1014–1015

Corcoran E, Nellemann C, Baker E, Bos R, Osborn D, Savelli H (2010) Sickwater? The central role of wastewater management in sustainable development. A rapid response assessment. United Nations Environment Programme, UN-HABITAT, GRID-Arendal; www.grida.no. ISBN: 978-82-7701-075-5

Darwish MR, El-Awar FA, Sharara M, Hamdar B (1999) Economic-environmental approach for optimum wastewater utilization in irrigation: a case study in Lebanon. Appl Eng Agric 15:41–48

Dinar A, Yaron D (1986) Treatment optimization of municipal wastewater and reuse for regional irrigation. Water Resour Res 22:331–338

Downs TJ, Cifuentes-Garcia E, Suffet IM (1999) Risk screening for exposure to groundwater pollution in a wastewater irrigation district of the Mexico City region. Environ Health Perspect 107:553–561

Drechsel P, Mahjoub O, Keraita B (2015) Social and cultural dimensions in wastewater use. In: Drechsel P, Qadir M, Wichelns D (eds) Wastewater: economic asset in an urbanizing world. Dordrecht: Springer, pp 75–92

Farid MSM, Atta S, Rashid M, Munnink JO, Platenburg R (1993) The impact of reuse of domestic wastewater from irrigation on groundwater quality. Water Sci Technol 27:147–157

Feenstra S, Hussain R, Van der Hoek W (2000) Health risks of irrigation with untreated urban wastewater in the Southern Punjab, Pakistan, IWMI Pakistan Report no. 107. International Water Management Institute, and Institute of Public Health, Lahore

Friedler E, Lahav O, Jizhaki H (2006) Study of urban population attitudes towards various wastewater reuse options: Israel as a case study. Environ Manag 81:360–370

Funamizu N, Iida M, Sakakura Y, Takakuwa T (2001) Reuse of heat energy in wastewater: implementation examples in Japan. Water Sci Technol 43:277–285

Gallegos E, Warren A, Robles E, Campoy E, Calderon A, Sainz G, Bonilla P, Escolero O (1999) The effects of wastewater irrigation on groundwater quality in Mexico. Water Sci Technol 40: 45–52

Guest EA (2009) A new planning and design paradigm to achieves stainable resource recovery from wastewater. Environ Sci Technol 43:6126–6130

Habbari K, Tifnouti A, Bitton B, Mandil A (2000) Geohelminthic infections associated with raw wastewater reuse for agricultural purposes in Beni-Mellal, Morocco. Parasitol Int 48:249–254

Juanico M, Yossi A, Beny T, Gedaliah S (1995) Effect of effluent addition to a freshwater reservoir on the filter clogging capacity of irrigation water. Water Res 29:1695–1702

Karpiscak Martin M, Charles P, Gerba Pamela M, Watt Kennith E, Jeanne Falabi A (1996) Multi-species plant systems for wastewater quality improvements and habitat enhancement. Water Sci Technol 33:231–236

Larsen T, Alder A, Eggen R, Maurer M, Lienert J (2009) Source separation: will we see a paradigm shift in wastewater handling? Environ Sci Technol 43:124–129

Lazarova V, Asano T, Bahri A, Anderson J (2013) Milestones in water reuse: the best success stories. IWA, pp 408

Logan B, Hamelers B, Rozendal R, Schröder U, Keller J, Freguia S, Aelterman P, Verstraete W, Rabaey K (2006) Microbial fuel cells: methodology and technology. Environ Sci Technol 40: 5181–5192

Mara DD (2000) The production of microbiologically safe effluents for wastewater reuse in the Middle East and North Africa. Water Air Soil Pollut 123:595–603

National Research Council (NRC) (1996) Use of reclaimed water and sludge in food crop production. National Academy Press, Washington, DC

Oki T, Kanae S (2006) Global hydrological cycle and world water resources. Science 313:1068–1072

PMSEIC (Prime Minister's Science, Engineering and Innovation Council) (2003) Recycling water for our cities. Paper prepared by an independent working group for PMSEIC, pp. 1–45

Rashed M, Awad SR, Salam MA, Smidt E (1995) Monitoring of groundwater in Gabal el Asfar wastewater irrigated area (Greater Cairo). Water Sci Technol 32:163–169

Sato T, Qadir M, Yamamoto S, Endo T, Zahoor A (2013) Global, regional, a country level need for data on wastewater generation, treatment, and use. Agric Water Manage 130:1–13

Schleich J, White D, Stephenson K (1996) Cost implications in achieving alternative water quality targets. Water Res Rech 32:2879–2884

Schwarz GE, McConnell VD (1993) Local choice and wastewater treatment plant performance. Water Res Rech 29:1589–1600

Scott CA, Zarazua JA, Levine G (2000). Urban-wastewater reuse for crop production in the water-short Guanajuato River Basin, Mexico. IWMI Research Report No. 41. International Water Management Institute, Colombo: Sri Lanka

Segarra E, Ragy Darwish M, Don Ethridge E (1996) Returns to municipalities from integrating crop production with wastewater disposal. Resour Conserv Recy 17:97–107

Shereif MM, El-S Easa M, El-Samra MI, Mancy KH (1995) A demonstration of wastewater treatment for reuse applications in fish production and irrigation in Suez, Egypt. Water Sci Technol 32:137–144

Shuval H, Lampert Y, Fattal B (1997) Development of a risk assessment approach for evaluating wastewater reuse standards for agriculture. Water Sci Technol 53:15–20

Singh KK, Mishra LC (1987) Effect of fertilizer factory effluent on soil and crop productivity. Water Air Soil Pollut 33:309–320

Smith VH, Tilman GD, Nekola JC (1999) Eutrophication: impacts of excess nutrient inputs on freshwater, marine, and terrestrial ecosystems. Environ Pollut 100:179–196

UNESCO. Managing water under uncertainty and risk (2012) The United Nations World Water Development Report 4. United Nations Educational, Scientific and Cultural Organization, Paris

UN-Water. Wastewater Management (2015) A UN-Water Analytical Brief

World Water Council. 6th World Water Forum, Marseille, 12–17 March 2012. http://www.solutionsforwater.org/objectifs/1-2-8-operator-efficiency-and-effectiveness-in-urban-wastewater-collection-and-treatment. Accessed 31/3/14

Yagoubi MA, Foutlane L, Bourchich J, Jellal C, Wittland Yachioui MEL (2000) Study on the performance of the wastewater stabilisation pond of Boujaad, Morocco. Aqua (Oxford) 49:203–209

Young CE, Epp DJ (1980) Land treatment of municipal wastewater in small communities. Am J Agric Econ 62:238–243

Zimmerman J, Mihelcic Smith J (2008) Global stressors on water quality and quantity. Environ Sci Technol 42:4247–4254

Sustainable Use of Sewage Sludge in Soil Fertility and Crop Production

Majid Abdoli

Abbreviations

As	Arsenic
AWC	Available water capacity
BY	Biological yield
$C_3H_6O_2$	Propionic acid
$C_4H_6O_5$	Malic acid
$C_6H_8O_7$	Citric acid
Ca	Calcium
Cd	Cadmium
CEC	Cation exchange capacity
CF	Chemical fertilizers
Cr	Chromium
Ct	Cattle
Cu	Copper
EC	Electrical conductivity
FC	Field capacity
Fe	Iron
GY	Grain yield
H_2CO_3	Carbonic acid
Hg	Mercury
K	Potassium
kGy	Kilograys
Ks	Saturated hydraulic conductivity

M. Abdoli (✉)
Department of Plant Production and Genetics, Faculty of Agriculture, University of Maragheh, Maragheh, Iran

Nomads Affairs Organization of Iran (NAOI), Kermanshah, Iran

V. D. Rajput et al. (eds.), *Sustainable Management and Utilization of Sewage Sludge*, https://doi.org/10.1007/978-3-030-85226-9_15

LAI	Leaf area index
LCI	Leaf chlorophyll index
Mg	Magnesium
Mn	Manganese
Mo	Molybdenum
MWD	Mean weight diameter of aggregates
N	Nitrogen
Na	Sodium
Ni	Nickel
OC	Organic carbon
OCFS	Complex organic fertilizer of sludge
OM	Organic matter
P	Phosphorus
Pb	Lead
pH	Soil acidity
Pl	Poultry
PWP	Permanent wilting point
qCO_2	Metabolic quotient
S	Sulfate
SAR	Sodium adsorption ratio
Se	Selenium
Sh	Sheep
SS	Sewage sludge
total N	Total nitrogen
TSP	Triple superphosphate
Zn	Zinc

1 Introduction

Population growth and industrialization are among the many factors that have led to the production of escalating volumes of wastewater and industrial effluents beyond the self-purifying capacity of the surrounding limited land areas where they are produced. However, this devastating challenge may be transformed into an opportunity by reusing the wastewater and SS as a source of materials, water, and energy.

Sewage sludge (SS) is the residual, semi-solid material that is produced as a by-product during wastewater treatment of municipal or industrial (Mokhtari et al. 2017). Sewage sludge management is one of the most challenging sections of wastewater treatment in terms of economic and environmental issues. With the increase of human societies, the production of SS will also increase, and appropriate solutions must be found for its proper disposal. There are various solutions for SS disposal (Fig. 1), the most important of which are: (1) incineration, (2) sanitary burial, and (3) use in agriculture. Today, environmentalists recommend the use of SS

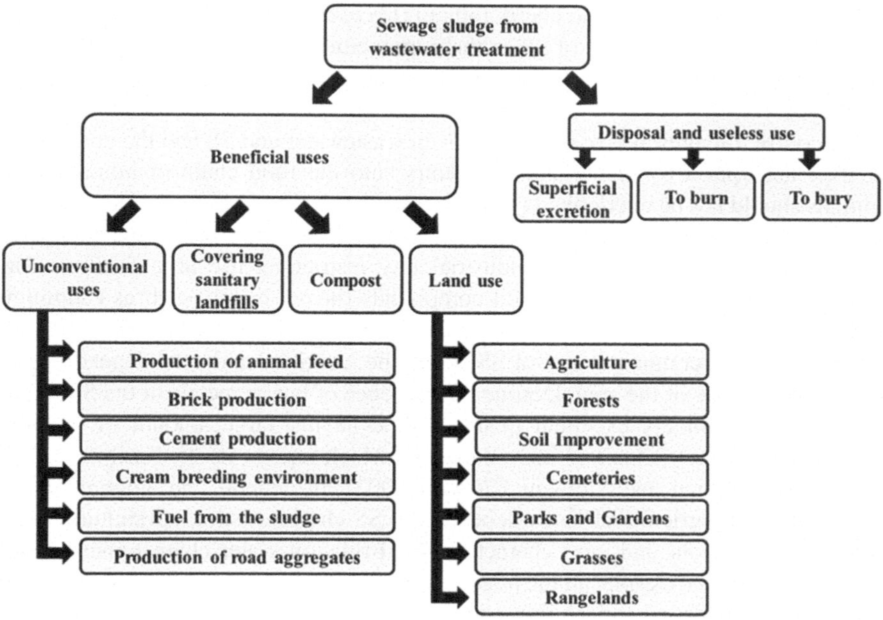

Fig. 1 Types of uses of sewage sludge

as fertilizer in agriculture due to legal restrictions on methods of burning and burial of SS on land and oceans. Soil use of SS in crop production offers an alternative technique for its management and disposal. Due to the presence of various contaminants and pathogenic organisms in SS can pose a high risk to public health, its safe and effective use in agricultural land requires the development of special guidance (Rahmani et al. 2014; Barati Rashvanlou et al. 2018). In this connection, Vaseghi et al. (2005) stated that SS is potentially a valuable fertilizer. However, the SS effect on chemical characteristics of the soil and heavy metals should be taken into consideration before its widespread use in agriculture.

Sewage sludge as an organic fertilizer has economic benefits. In barren areas, due to poor fertility and costly soil transfer, the use of SS as a cheap fertilizer improves plant growth (Shafieepour et al. 2011). Wastewater and SS can meet the water and nutritional needs such as nitrogen (N), phosphorus (P), and potassium (K) of the plant (Shafieepour et al. 2011), so they are considered as cheap sources of water and fertilizer (Nazari et al. 2006). Application of SS as a cheap and waste material is increasing for supporting plant essential elements especially in the soils of arid and semi-arid regions due to lack of organic matter (OM) in the soil. The results of various researches have shown that the application of SS increased the amount of organic carbon (OC) in soil and N, P, and K in soil and plant, and the straw and grain yields in wheat (Fathololomi et al. 2015; Shahbazi et al. 2018), barley (Chorom and Aghaei Foroushani 2007), rice (Latare et al. 2014), corn (Karimpour et al. 2010; Saadat et al. 2012; Hoshyar and Baghaie 2017), eggplant (*Solanum melongena*;

Kumar and Chopra 2016), and cherry tomato (*Lycopersicon esculentum* L.; Hossain et al. 2015) plants. The effect of SS on the improvement of the above-cited attributes mostly depends on the origin (urban or industrial) and used the rate of SS and also plant type.

Of course, the presence of heavy metals in wastewater and SS and the possibility of their absorption by plants and their entry into the food chain of humans and animals should not be overlooked (Mirhosseini et al. 2007). In this case, Salehi et al. (2013) reported that the high heavy metals such as lead (Pb) levels in the sludge of wastewater treatment of Kaveh industrial city restrict its use as a fertilizer for agriculture. Therefore, these harmful compounds (heavy metals such as cadmium (Cd) and Pb) should be removed from the SS during the physical and chemical process, and/or a certain amount of SS should be identified and used depending on the soil conditions of the area. Despite the presence of heavy metals in the SS but it bears an exceptionally excellent quality, while having large amount of organic materials, macronutrients, and micronutrients, which represents the fertilizer value of sludge (Torabian and Momeni Farahani 2002). In general, the amount of SS consumption in agricultural lands depends on: SS characteristics, agricultural land characteristics, plant and crop characteristics, irrigation water characteristics, and climatic conditions. Despite all the positive impacts associated with the consumption of SS on the physico-chemical properties of soil, there are still many concerns about health and environmental issues.

2 Effect of Sewage Sludge on Soil Fertility and Soil Physico-Chemical and Biological Characteristics

The application of OM such as urban SS may help sustainable soil fertility via improving the biological and physico-chemical soil properties. Wastewater sludge has relatively high salts and amounts of OM which favorably affect the different properties of the receiving soil. Hence, it can be used as a cheap fertilizer if properly utilized (Shirani et al. 2010). In the following, the effect of SS on physico-chemical and biological soil characteristics is discussed separately.

2.1 Soil Physical Characteristics

The use of SS in fine-textured soils (such as clay loam, silty clay loam, sandy clay, silty clay, and clay) can improve granulation, porosity, permeability, and soil aeration as well as in coarse-textured soils (like sand, loamy sand, and sandy loam) can improve water and nutrient retention. Coarse-textured soils have generally unstable structure and low OM. Using SS is one of the solutions for their associated problems. In this regard, Fathololomi and Asghari (2015) reported that all rates of SS from 30 to 180 ton ha^{-1} considerably decreased saturated hydraulic conductivity (Ks), particle density macropores (>75μm), soil particle density, and soil bulk density, relative to the control, in the ranges of 12.9 to 44.1, 9.12 to 146, 1.32 to 4.43, and 3.34 to 9.88%, respectively. Also, use of SS at rates 30 to 180 ton ha^{-1}, increased total porosity, mean weight diameter of aggregates (MWD), field capacity (FC), permanent wilting point (PWP), available water capacity (AWC), mesopores (30–75μm), and micropores (<30μm), as compared with the control, within the ranges of 1.24 to 6.0, 64.8 to 233, 1.91 to 15.1, 1.30 to 15.9, 2.22 to 15.0, 10.7 to 29.0, and 1.56 to 24.2%, respectively. Therefore, they stated that the Ardabil municipal SS can be used to improve the physical quality of coarse-textured soils. On the other hand, Shirani et al. (2010) during 3 years of research stated that the increasing amounts of SS enhanced soil infiltration, Ks, MWD, and electrical conductivity (EC) but decreased bulk density. They stated a considerable correlation was observed between the amount of SS and soil physical properties, which indicating the high effect of SS on physical properties of soil.

The reason for the diminution in bulk density is due to the high content of OC in the SS (Angin and Yaganoglu 2011; Fathololomi and Asghari 2015; Shahbazi et al. 2018). Due to the lower specific gravity of OM than soil, it is expected that the bulk density of the soil + OM mixture is less than the bulk density of the original soil (even if it has no effect on soil structure). On the other hand, it is likely that part of the reduction in soil bulk density is due to increased aggregate stability (Fathololomi and Asghari 2015) and thus improved soil structure due to the use of SS. Porosity is the volume of soil occupied by empty pores and is inversely related to the bulk density. Increased OM due to the use of SS reduces the bulk density and reciprocally increases soil porosity. Organic matter provides the materials needed for aggregate formation and stability during its conversion to humus compounds. Because SS contains a lot of OM, it will be effective in increasing weight diameter of aggregates. Other researchers such as Bahremand et al. (2003) and Shirani et al. (2010) have noted this case that can be seen in Table 1.

According to the capillary relationship ($h = 0.3/d$; h: soil suction and d: pore diameter), with the narrowing of soil pores, the suction of water in the soil increases and a large amount of water is prevented from leaving the soil due to the force of gravity, resulting in an increase of FC (Warrick 2002). Due to the fact that SS increases the micro and meso pores (Fathololomi and Asghari 2015), so it indirectly increases FC. Angin and Yaganoglu (2011) also expressed a 26.42, 30.59, and 34.47% increase in FC at the treatment of SS application with 40, 80, and 120 ton ha^{-1} compared to the control, respectively. In another study by Bahremand et al. (2003) reported that the SS use considerably increased the MWD, Ks, final infiltration rate, FC, PWP, and AWC, while it significantly decreased soil bulk density. They stated that SS can help to improve soil physical conditions and this impact persists over long periods. Also, Angin and Yaganoglu (2011) showed that in all amounts of SS use (40, 80, and 120 ton ha^{-1}), the particle density and bulk density of the soil was significantly reduced but porosity, permeability coefficient, FC, PWP, and AWC of the soil was significantly increased compared to the control (no use of

Table 1 Effect of sewage sludge on the changes of soil physical properties

Application of sewage sludge at the rates (ton ha^{-1})	Condition	Region	Increase in soil physical properties over untreated control (%)											References
			Soil bulk density	Ks	Final soil infiltration rate	MWD	FC	PWP	AWC	Total porosity	Macropores (>75μm)	Mesopores (75–30μm)	Micropores (<30μm)	
25	Farm	Najafabad, Iran	1.6	119.1	300.0	189.5	13.5	4.5	27.2	–	–	–	–	Bahremand et al. (2003)
50	Farm	Najafabad, Iran	−3.1	139.0	393.2	299.4	13.6	2.9	29.9	–	–	–	–	Bahremand et al. (2003)
100	Farm	Najafabad, Iran	−11.0	321.3	886.5	334.8	19.3	9.2	34.7	–	–	–	–	Bahremand et al. (2003)
22.5	Farm	Najafabad, Iran	−10.0	100.0	218.2	46.9	–	–	–	–	–	–	–	Shirani et al. (2010)
45	Farm	Najafabad, Iran	−18.5	427.3	263.6	44.9	–	–	–	–	–	–	–	Shirani et al. (2010)
30	Pot and greenhouse	Ardabil, Iran	–	−129.3	–	–	–	–	–	–	−9.12	–	–	Fathololomi and Asghari (2015)
60	Pot and greenhouse	Ardabil, Iran	–	−241.4	–	–	–	–	13.0	3.62	−30.27	–	–	Fathololomi and Asghari (2015)
120	Pot and greenhouse	Ardabil, Iran	–	−332.4	–	183.8	14.93	15.8	14.5	4.08	−93.36	20.44	20.6	Fathololomi and Asghari (2015)
180	Pot and greenhouse	Ardabil, Iran	−9.9	−440.5	–	233.09	15.14	–	14.98	6.08	−145.79	29.0	24.2	Fathololomi and Asghari (2015)

Ks saturated hydraulic conductivity, *MWD* mean weight diameter of aggregates, *FC* field capacity, *PWP* permanent wilting point, *AWC* available water capacity

SS). However, the effectiveness of SS decreased over time (after 3 years) due to the mineralization of OM from the sludge.

2.2 *Soil Chemical Characteristics*

2.2.1 Soil Acidity (pH)

Land use of SS changes some chemical soil properties including soil acidity (pH). In this case, Vaseghi et al. (2005) reported that the SS application significantly increased total nitrogen (total N), plant-available P, K, OM, and cation exchange capacity (CEC) in the four areas soils in Iran. But, soil pH considerably decreased as a result of SS application. Zare et al. (2015) stated that soil nutrients concentrations subjected to SS increased, and soil pH significantly decreased. They stated that the SS decreased pH by 0.4 units compared with blank treatment. Najafi and Mardomi (2013) reported the pH of soil solution in uncultivated pots decreased after the SS application. Other researchers also have reported reduced pH due to the use of different amount of SS (Ghamari and Danesh 2007; Rahimi Alashty et al. 2011; Saadat et al. 2012; Hoshyar and Baghaie 2017; Marjovvi and Mashayekhi 2018). However, a study reported the application of 15 ton ha^{-1} SS compost increase pH by 0.4 units compared to soil without SS compost (Baghaie 2018). The rate of change in pH depends on the properties of the soil, including its texture and buffering capacity. Table 2 shows the changes in pH due to different amounts of SS. Of the reasons for the decrease in pH due to the application of SS are (1) the formation of more carbonic acid (H_2CO_3) and organic acids such as citric acid ($C_6H_8O_7$), malic acid ($C_4H_6O_5$), and propionic acid ($C_3H_6O_2$) during the decomposition of OM, (2) nitrification, sulfurization, and oxidation of OM, and also (3) due to the mineralization of N in OM, H+ ions are produced which reduce the acidity of soil (Veeresh et al. 2003; Cheng et al. 2007; Angin and Yaganoglu 2011; Akbarnejad et al. 2013). Soil pH is a critical soil parameter in SS applied lands, because it considerably affects the bioavailable forms of metals. The result of reduced pH is an increase in the solubility of some trace elements (such as zinc (Zn) and Fe) and these elements are more available to the plant especially in relatively calcareous and alkaline soils. It should be noted that in highly calcareous soils with high buffering capacity, the use of SS may not have an effect on reducing pH.

2.2.2 Electrical Conductivity (EC)

Sewage sludge contains a large amount of salts (such as salts of sodium (Na), calcium (Ca), and K) and minerals that increase the EC of the soil. In this regard, Akbarnejad et al. (2013) reported that with increasing amounts of municipal solid waste compost and SS from 15 to 30 ton ha^{-1}, OC and EC of soil increased. Furthermore, Saadat et al. (2012) stated that the use of SS leads to significant

Table 2 Effect of sewage sludge on the changes of chemical properties and concentrations of macro and micronutrients in soil

Application of sewage sludge at the rates (ton ha^{-1})	Condition	Region	Increase in soil chemical properties and nutrients concentrations over untreated control (%)																References
			OC	pH	EC	SAR	N	P	K	Ca	Mg	Na	Fe	Zn	Cu	Mn	Pb	Cd	
15	Farm	Esfahan, Iran	45.9	−0.8	25.5	–	33.3	334.8	−4.3	–	–	–	45.5	2330.0	200.0	–	92.0	–	Marjovvi and Mashayekhi (2018)
30	Farm	Esfahan, Iran	65.6	−2.9	39.4	–	66.7	749.8	−3.6	–	–	–	67.3	5700.0	353.9	–	112.8	–	Marjovvi and Mashayekhi (2018)
15	Farm	Esfahan (Roudasht), Iran	39.6	–	85.2	–	–	456.6	10.6	–	–	–	65.1	1066.7	233.7	1.9	34.7	–	Marjovvi and Mashayekhi (2019)
30	Farm	Esfahan (Roudasht), Iran	60.4	–	90.1	–	–	844.1	48.5	–	–	–	95.6	1663.6	453.5	8.8	72.6	–	Marjovvi and Mashayekhi (2019)
50	Farm	Rasht, Iran	12.7	−2.4	4.8	–	130.0	43.5	−12.9	–	–	–	–	–	–	–	–	–	Vaseghi et al. (2005)
100	Farm	Rasht, Iran	22.8	−4.3	60.3	–	180.0	47.8	0.0	–	–	–	–	–	–	–	–	–	Vaseghi et al. (2005)
200	Farm	Rasht, Iran	1.8	−5.9	64.0	–	180.0	85.2	2.7	–	–	–	–	–	–	–	–	–	Vaseghi et al. (2005)
50	Farm	Langroud, Iran	11.2	0.0	19.3	–	172.7	60.0	19.6	–	–	–	–	–	–	–	–	–	Vaseghi et al. (2005)
100	Farm	Langroud, Iran	39.1	0.0	103.7	–	263.6	66.0	51.4	–	–	–	–	–	–	–	–	–	Vaseghi et al. (2005)
200	Farm	Langroud, Iran	58.7	−4.2	147.7	–	281.8	238.0	55.1	–	–	–	–	–	–	–	–	–	Vaseghi et al. (2005)
50	Farm	Lahijan, Iran	56.8	−1.9	27.4	–	125.0	51.4	8.1	–	–	–	–	–	–	–	–	–	Vaseghi et al. (2005)
100	Farm	Lahijan, Iran	97.9	−1.7	86.3	–	168.8	128.6	12.7	–	–	–	–	–	–	–	–	–	Vaseghi et al. (2005)

200	Farm	Lahijan, Iran	118.9	−3.6	154.7	–	218.8	214.3	16.3	–	–	–	–	–	–	–	–	–	Vaseghi et al. (2005)
50	Farm	Isfahan, Iran	76.0	−2.5	20.7	–	212.5	103.1	2.7	–	–	–	–	–	–	–	–	–	Vaseghi et al. (2005)
100	Farm	Isfahan, Iran	140.0	−5.0	37.9	–	337.5	181.3	21.5	–	–	–	–	–	–	–	–	–	Vaseghi et al. (2005)
200	Farm	Isfahan, Iran	238.0	−6.3	69.0	–	400.0	253.1	23.8	–	–	–	–	–	–	–	–	–	Vaseghi et al. (2005)
20	Farm	Sari, Iran	41.3	−0.3	2.6	–	–	–	–	–	–	–	–	–	–	–	321.5	212.5	Rahimi Alashty et al. (2011)
40	Farm	Sari, Iran	47.8	−1.8	66.7	–	–	–	–	–	–	–	–	–	–	–	468.4	225.0	Rahimi Alashty et al. (2011)
22.5	Farm	Najafabad, Iran	–	0.0	89.1	–	–	–	–	–	–	–	–	–	–	–	–	–	Shirani et al. (2010)
45	Farm	Najafabad, Iran	–	−1.3	143.4	–	–	–	–	–	–	–	–	–	–	–	–	–	Shirani et al. (2010)
20	Farm	Sari, Iran	–	–	–	–	–	–	–	–	–	–	11.9	309.7	30.6	22.4	–	–	Hosseinpour and Ghajar Sepanlou (2013)
40	Farm	Sari, Iran	–	–	–	–	–	–	–	–	–	–	16.1	633.0	63.8	14.5	–	–	Hosseinpour and Ghajar Sepanlou (2013)
20	Farm	Sari, Iran	–	–	–	–	–	–	–	–	–	–	62.2	64.9	247.9	208.1	–	–	Ahmad Abadi et al. (2012)
40	Farm	Sari, Iran	–	–	–	–	–	–	–	–	–	–	78.8	80.1	356.3	344.4	–	–	Ahmad Abadi et al. (2012)
20	Farm	Sari, Iran	–	–	–	–	–	–	–	–	–	–	0.3	313.7	4.9	42.8	–	–	Hosseinpour et al. (2016)
40	Farm	Sari, Iran	–	–	–	–	–	–	–	–	–	–	4.7	640.2	40.2	36.6	–	–	Hosseinpour et al. (2016)

(continued)

Table 2 (continued)

Application of sewage sludge at the rates (ton ha^{-1})	Condition	Region	Increase in soil chemical properties and nutrients concentrations over untreated control (%)																References
			OC	pH	EC	SAR	N	P	K	Ca	Mg	Na	Fe	Zn	Cu	Mn	Pb	Cd	
10	Pot	Ahar, Iran	62.6	−2.6	24.3	–	37.3	–	–	–	–	–	–	–	–	–	–	–	Kassray et al. (2008)
20	Pot	Ahar, Iran	112.3	3.9	39.3	–	128.4	–	–	–	–	–	–	–	–	–	–	–	Kassray et al. (2008)
30	Pot	Ahar, Iran	112.8	0.1	58.7	–	50.7	–	–	–	–	–	–	–	–	–	–	–	Kassray et al. (2008)
40	Pot	Ahar, Iran	212.3	−2.3	69.4	–	109.0	–	–	–	–	–	–	–	–	–	–	–	Kassray et al. (2008)
50	Pot	Ahar, Iran	237.9	1.7	71.8	–	198.5	–	–	–	–	–	–	–	–	–	–	–	Kassray et al. (2008)
60	Pot	Ahar, Iran	275.4	−3.1	93.2	–	267.2	–	–	–	–	–	–	–	–	–	–	–	Kassray et al. (2008)
70	Pot	Ahar, Iran	303.6	−4.4	102.9	–	295.5	–	–	–	–	–	–	–	–	–	–	–	Kassray et al. (2008)
50	Pot and greenhouse	Ahvaz, Iran	100	−1.3	33.3	–	28.6	106.9	52.9	61.0	–	42.9	–	–	–	–	–	–	Zare et al. (2015)
100	Pot and greenhouse	Ahvaz, Iran	200	−3.1	54.2	–	42.9	247.9	85.7	137.3	–	85.7	–	–	–	–	–	–	Zare et al. (2015)
150	Pot and greenhouse	Ahvaz, Iran	260	−3.8	91.7	–	42.9	286.2	128.6	198.3	–	128.6	–	–	–	–	–	–	Zare et al. (2015)
60	Pot and greenhouse	Urmia (Nazlou), Iran	–	2.1	8.2	–	55.6	−62.0	1.9	–	–	–	–	–	–	–	–	–	Rasouli Sadaghiani and Sepehr (2011)
5	Pot and greenhouse	Tabriz (Oskou), Iran	24.5	5.3	74.6	–	–	–	–	–	–	–	–	–	–	–	–	–	Kassray and Saedi (2010)
10	Pot and greenhouse	Tabriz (Oskou), Iran	40.8	6.7	83.0	–	–	–	–	–	–	–	–	–	–	–	–	–	Kassray and Saedi (2010)

20	Pot and greenhouse	Tabriz (Oskou), Iran	64.6	5.3	74.6	–	–	–	–	–	–	–	–	–	–	–	–	–	Kassray and Saedi (2010)
30	Pot and greenhouse	Tabriz (Oskou), Iran	82.7	6.7	163.4	–	–	–	–	–	–	–	–	–	–	–	–	–	Kassray and Saedi (2010)
40	Pot and greenhouse	Tabriz (Oskou), Iran	107.1	8.0	181.7	–	–	–	–	–	–	–	–	–	–	–	–	–	Kassray and Saedi (2010)
15 g kg^{-1} soil	Pot and greenhouse	Gorgan, Iran	92.3	−2.9	112.6	–	–	–	–	–	–	–	–	–	–	–	–	–	Saadat et al. (2012)
20 g kg^{-1} soil	Pot and greenhouse	Gorgan, Iran	111.4	−3.1	113.8	–	–	–	–	–	–	–	–	–	–	–	–	–	Saadat et al. (2012)
25 g kg^{-1} soil	Pot and greenhouse	Gorgan, Iran	127.2	−4.4	195.4	–	–	–	–	–	–	–	–	–	–	–	–	–	Saadat et al. (2012)
50	Greenhouse	Mashhad, Iran	80.0	−2.0	63.2	15.7	15.2	124.7	−12.2	88.5	80.1	55.8	–	–	–	–	59.5	700.0	Ghamari and Danesh (2007)
100	Greenhouse	Mashhad, Iran	140.0	−2.9	89.5	16.4	18.6	150.6	−5.8	150.0	89.8	67.3	–	–	–	–	100.0	900.0	Ghamari and Danesh (2007)
200	Greenhouse	Mashhad, Iran	320.0	−5.2	157.9	96.4	162.1	510.2	−1.6	69.2	223.7	215.4	–	–	–	–	264.9	2100.0	Ghamari and Danesh (2007)
400	Greenhouse	Mashhad, Iran	780.0	−8.0	342.1	393.6	946.9	835.5	11.6	90.4	230.5	700.0	–	–	–	–	529.7	4400.0	Ghamari and Danesh (2007)
30	Greenhouse	Ardabil, Iran	21.9	–	56.3	–	81.0	16.7	5.34	–	–	–	–	–	–	–	–	–	Fathololomi et al. (2015)
180	Greenhouse	Ardabil, Iran	176.0	–	175.0	–	310.3	123.0	117.8	–	–	–	–	–	–	–	–	–	Fathololomi et al. (2015)

OC organic carbon, *pH* soil acidity, *EC* electrical conductivity, *SAR* sodium adsorption ratio, *N* nitrogen, *P* phosphorus, *K* potassium, *Ca* calcium, *Mg* magnesium, *Na* sodium, *Fe* Iron, *Zn* zinc, *Cu* copper, *Mn* manganese, *Pb* lead, *Cd* cadmium

increase of soil OM and EC. Also, Ghamari and Danesh (2007) reported that an enhancement in SS application caused a significant increase in the properties of soil such as EC and OC except for the pH which showed a significant decrease. On the other hand, Hoshyar and Baghaie (2017) stated that increasing the loading rates of SS from 0 to 30 ton ha^{-1} caused a significant increase in CEC by 3.5 units. Shahbazi et al. (2018) reported that with increasing the level of SS, soil EC increased. While soil pH and soil bulk density at treatment with 60 and 120 g kg^{-1} of SS showed a significant decrease as compared to the control treatment. Najafi and Mardomi (2013) reported the soil EC in uncultivated pots increased after the SS application. Fathololomi et al. (2015) reported that the best-improving effect of SS on soil and plant attributes was obtained for 120 and 180 ton ha^{-1} treatments and a significant difference were not observed between these treatments. Therefore, with consideration of the high EC of used sludge, the 120 ton ha^{-1} treatment can be recommended as a suitable use rate for improving the soil chemical quality. Increased soil EC due to the use of SS has been reported in other studies (Vaseghi et al. 2005; Casado-Vela et al. 2006; Singh and Agrawal 2007; Baghaie 2018). One of the reasons for the effect of SS application on increasing the EC of the soil is due to the high amount of solutes in the SS. With proper irrigation management and control of conditions, can be prevented the adverse effects of increasing soil salinity and disrupting plant growth due to the use of SS. Angin and Yaganoglu (2011) stated that soil EC increased with SS application; but, this impact decreased with time. The decrease of soil EC could be attributed to the loss of SS effectiveness and irrigation, which helped leaching (Perez-Murcia et al. 2006; Gascó and Lobo 2007).

2.2.3 Organic Carbon (OC)

More than 24% of the dry matter of SS is organic carbon (OC), which is a significant amount. Organic carbon is the most important parameter and indicator for enhancement of soil quality and fertility (Reeves 1997; Ghamari and Danesh 2007). Increasing the amount of soil OM (through SS) has favorable effects on the soil physico-chemical properties. The presence of this compound causes a lot of pores in the soil, reduces the soil bulk density, and instead increases the stability of aggregates and soil fertility. Zare et al. (2015) stated that SS improved some soil chemical properties. 150 ton ha^{-1} SS without fertilizer treatment had the best effect on the soil properties. Sewage sludge increased the OM content about twice times comparing with blank treatment. In recent years, Shahbazi et al. (2018) reported that the treatments of SS in comparison with control treatment increased the OC, total N, and absorbable K and P of the soil. Hojati et al. (2006) stated that the use of SS increased soil OC and total N compared with control treatment. An increasing trend was observed in OC and total N, as the rates and times of applications increased. Also, Marjovvi and Mashayekhi (2019) reported that the organic fertilizers (municipal compost fertilizer and SS) application increased soil OM. Fathololomi and Asghari (2015) reported that application of the SS at amounts 30 to 180 ton ha^{-1}, increased OC as compared with the control, within the ranges of 21.9 to 176%, respectively. Recent evidence demonstrates that SS significantly increased OC in the soil (Hemmat et al. 2010; Angin and Yaganoglu 2011; Fathololomi and Asghari 2015; Kumar and Chopra 2016; Baghaie 2018). Rahimi Alashty et al. (2011) indicated that the use of SS was a significant effect on pH, EC, OC, Pb, and Cd concentrations in soil (available and total). They reported that the application of 40 ton ha^{-1} SS in 3 continuous years decreased the amount of pH and increased the amount of OC and EC in soil. Ahmadpoor et al. (2011) reported that with increasing application periods of SS from 1 to 3 years, the amount of soil OM significantly enhanced. Also an increasing trend was observed in total N. While mean, with increasing the rates and application periods, the amount of available K and P increased significantly.

2.2.4 Total Nitrogen (Total N)

More than 97% of soil N is organic form and the addition of good quality organic fertilizers (such as SS) provides the N needed by the plants (Balkcom et al. 2001; Shahbazi et al. 2018). Angin and Yaganoglu (2011) and Zare et al. (2015) reported an increase in soil N due to the use of SS. Also, Kassray et al. (2008) reported that an increase of SS application from 0 to 70 ton ha^{-1} led to increased EC, OM, and total N. But, the use of sludge on the amount of soil carbonates and C/N ratio doesn't have significant impacts; on the other hand, the effect of SS use on soil pH alterations has shown the decline of pH in high levels applications in comparison to 20 ton ha^{-1} is been remarkable. Chemical and biological changes in the rhizosphere following organic residues as well as manures application are important processes which influence N mineralization and nutrients bioavailibity in soils. Rasouli Sadaghiani and Sepehr (2011) reported that the highest mineralized N in the rhizosphere and non-rhizosphere soil were achieved in SS (227.5 mg kg^{-1}) and poultry manure (Pl; 214.8 mg kg^{-1}), respectively. At the rhizosphere, net N mineralization was occurred in SS and Pl treatments whereas cattle (Ct) and sheep (Sh) applied soil showed N net immobilization. Except for SS, all applied residues showed net immobilization in non-rhizosphere soil. The highest total mineralized N (mineralized N exist in soil and absorbed N by plants) were as fallows in treatments: Pl $>$ SS $>$ Sh $>$ Ct.

2.2.5 Macro and Micronutrients

In connection with the impact of SS on the macro and micronutrients, Ghamari and Danesh (2007) reported that an increment in SS use caused a significant enhancement of macronutrients and micronutrients in the soil such as N, P, K, magnesium (Mg), Ca, availability of Pb, and availability of Cd. Recent evidence demonstrates that SS significantly increased the K, Ca, Mg, total N, phosphate (P), sulfate (S), Zn, copper (Cu), manganese (Mn), chromium (Cr), and Cd in the soil (Kumar and Chopra 2016; Shahbazi et al. 2018). Latare et al. (2014) reported that the enhancement in available nutrients content of soil was also recorded with increasing levels of

SS use after harvest of wheat and rice plants. Zare et al. (2015) stated that the impact of SS on soil K content was fewer than P and N. This can be due to the small amount of K in the SS. While, few researchers have stated that the effect of SS is high on all high-consumption elements like N, P, and K (Vaseghi et al. 2005; Zare et al. 2015; Fathololomi et al. 2015; Shahbazi et al. 2018). The changes of macro and microelements in the soil due to different amounts of SS are shown in Table 2.

In comparison between SS with other organic fertilizers, Rasouli Sadaghiani and Sepehr (2011) reported plants acquired considerably more N, P, K, Ca, Mg, Fe, Zn, Cu, and Mn in SS treatment than other manures (such as Pl, Ct, and Sh manures). Also, Marjovvi and Mashayekhi (2019) reported that the use of both types of organic fertilizers (SS and municipal compost fertilizer), especially at higher levels, caused a considerable increment in K, P, Cu, Zn, Fe, Mn, and Pb in soil. Moreover, increase in Cu, Mn, Fe, and Zn in soil has been reported with increasing consumption of SS and more times (Hosseinpour et al. 2016).

The soils of Asia are calcareous to varying degrees and it is very difficult to retain P in calcareous soils. Therefore, the use of SS is suitable for these soils. Marjovvi and Mashayekhi (2018) reported that as a result of organic fertilizers consumption (25 and 50 ton ha^{-1} municipal compost and also 15 and 30 ton ha^{-1} SS), the pH decreases and concentrations of some essential elements concordant with the Pb increased considerably in the soil. On the other hand, Shafieepour et al. (2011) stated that in sludge soil treatments, physical (increasing percentage of saturated moisture) and chemical (reducing EC) properties of amended soil are improved. Also in most cases, the concentrations of micronutrients (such as Zn, Fe, Mn, and Cu) and nutrients in the soil and leaves samples with the proportionality of increasing in the rate of SS loading increased. However, in some special cases, negligible changes in concentrations of Cd and Pb in plant have been observed.

2.2.6 Sodium (Na) and Sodium Adsorption Ratio (SAR)

Some researchers have observed an increment in the anion and cation concentrations in soil solution due to increased application of SS (Baker and Mathews 1983). Sodium-ion toxicity occurs when its concentration in soil solution increases. The direct effects of high sodium (Na) concentration on plants are the accumulation of this element in plants, as well as its indirect effects are nutritional imbalances and deterioration of soil properties. Ghamari and Danesh (2007) reported that an enhancement in SS use caused a significant increment in Na and sodium adsorption ratio (SAR) at soil. Recent evidence demonstrates that SS considerably increased the Na in soil (Kumar and Chopra 2016). The SS application in short periods and in small quantities may not be a problem in terms of soil Na, but to have a sustainable agricultural system, it is better to use SS cautiously. Shafieepour et al. (2011) stated that for SS application on Kish Island, it is necessary to check agronomic nutrients amounts and accumulation of heavy metals. But, it's recommended to use any kind of salt resisted native plants for landscapes. In the case of sensitive plants, SS should be added at least 6 months in advance to ensure that condition of soil physico-chemical are achieved.

2.3 Soil Biological Characteristics

Sewage sludge, in addition to changes in soil physico-chemical properties, also affects the biological properties of soil (Hargreaves et al. 2008). The use of SS due to the increase in carbon and nutrients available, can stimulate microbial activity and/or prevent microbial activity due to the presence of heavy metals and other contaminants (Baath 1989). Therefore, the behavior of the microbial population depends on the amount and quality of SS added to the soil. In this case, Kassray et al. (2008) reported that with an enhancement of SS use, the number of microorganisms and respiration rate has increased significantly. Ahmadpoor et al. (2011) reported that activities of urease and alkaline phosphates improved with increasing the organic fertilizers and application periods. Also, Ghorbani et al. (2017) stated that the impact of urban SS on chemical and biological properties was significant. So that the use of urban SS led to increased microbial biomass carbon (15.2 and 26.5 times) and basal respiration (16 and 27 times) in the water repellency soils ($S_{50} = 50{:}50$ and $S_{80} = 80{:}20$; sewage sludge weight: soil ratio) than in control soil. They reported that the positive impact of SS might due to a high content of nutrients and OC in urban SS and reduction the labile OM and nutrients during the incubation period. In contrast, Hernandez et al. (2002) showed that soil biological activities such as enzymatic activity and microbial biomass were reduced by the use of SS.

The release of OC by plant roots (Herman et al. 2006) and the increase of organic resources such as sludge and various organic fertilizers increase the population and activity of microbes (Hadas et al. 2004). Dehghan Manshadi et al. (2012) reported that the use of SS at all levels (included 20 and 40 ton of SS, and also 20 and 40 ton of SS + 50% of chemical fertilizers (CF) ha^{-1}), increased soil OC and soil microbial respiration and enzyme activity were compared with the controls, but the trend of increasing among the treatments was not similar. In both two levels of treatment enriched with fertilizer (20 and 40 ton of SS + 50% of CF ha^{-1}), the traits increased. Meanwhile, the amount of enzyme activity and respiration in high amounts of using SS has a decreasing trend. The maximum amount of OC, in 40 ton SS + 50% CF ha^{-1} with 4 years application treatment, and soil microbial respiration in 20 ton SS + 50% CF ha^{-1} with 4 years application treatment were measured. But, the maximum amount of acid and alkaline phosphatase was observed in 40 ton ha^{-1} SS with 2 years of application. Based on their results, the 20 ton SS + 50% CF ha^{-1} in 4 years application treatments were best. Fernandes et al. (2005) also showed that the use of SS increases the rate of soil basal respiration. So that with the consumption of SS, the soil basal respiration rate increased from 15.8 to 123.6 and also from 3.15 to 91.6 mg CO_2 kg^{-1} soil per hour in layers of 0–10 and 10–20 cm, respectively. Also, they showed that the C and N microbial biomass, enzymatic activity, and metabolic quotient (qCO_2) in the soil increased as SS was added, and their values were positively correlated with SS rates. The activities of soil urease and amylase enhancement as SS amounts increased (1, 2, 4, and 8 times the recommended rate based) and were considerably correlated with soil microbial biomass. On the other hand, Rasouli Sadaghiani and Sepehr (2011) stated that the treatments applied with SS showed the highest microbial activity compared to other organic residues (Ct, Sh, and Pl manures).

Transformation of essential elements from organic form under the influences of intra- and extra-cellular enzymes and other microbially mediated processes can increase plant growth. In this case, Hojati et al. (2006) reported that increasing the times and rates of SS application enhanced L-glutaminase, alkaline phosphatase, arylsulfatase, ß-glucosidase activities, microbial biomass index, and corn yield, significantly. They stated that four consecutive applications of 100 Mg ha^{-1} SS were associated with the highest levels of the bio-indicators. Increased enzymatic activity with use of SS has also been reported by other researchers (Fernandes et al. 2005; Ros et al. 2006). In contrast, some researchers stated that heavy metals (especially in very large amounts of SS) reduce the activity of soil enzymes by inactivating alkaline and acidic phosphatase enzymes (Hernandez et al. 2002; Dehghan Manshadi et al. 2012). In the following, the impacts of SS use on the quantity and quality of the product are discussed.

3 Effect of Sewage Sludge on Plant and Crop

3.1 Yield, Agronomic Traits and Some Morphological Characteristics

3.1.1 Positive Effects

Proper nutrition of the plant is one of the momentous factors in improving the quantity and quality of the product. The use of SS in agricultural lands should be such that it can provide the elements needed by the plant, and this depends on the chemical composition of the soil and the SS used. Increased yield and agronomic traits of various crops with the consumption of SS has been reported by different researchers (Gardio et al. 2005; Bozkurt et al. 2006; Kumar and Chopra 2016; Marjovvi and Mashayekhi 2018; Shakarami and Maroufi 2019). In the following, the effect of SS on some agricultural products is mentioned.

Various studies have been conducted on the effect of SS on the cereals. For example, Fathololomi et al. (2015) reported that the SS application increased grain yield (GY) and wheat growth indices such as shoot dry weight, height and grain weight, and leaf area. Similarly, Shahbazi et al. (2018) reported that the application of SS comparing control treatment caused a significant increment in the leaf surface, dry weight of aerial parts, grain number, and total grain weight in the Sivand and Roshan the two studied varieties of wheat. In another study on barley, Chorom and Aghaei Foroushani (2007) stated that the grass yield considerably increased in the plot treated with 100 ton ha^{-1} SS. Also, GY in the two treatments (50 and 100 ton ha^{-1} SS) outstandingly increased. On the other hand, study of Latare et al. (2014) showed that there was significant increment in straw and grain yields of rice and wheat with use of SS. So that the GY of rice increased 45% at 40 ton ha^{-1} SS use over no sludge (Latare et al. 2014). In another study, the yield of rice was increased by 18–19% when complex organic fertilizer of sludge (OCFS) was used (Xie et al. 2001). Nazari et al. (2006) stated that dry matter of shoot and root of wheat, barley, and corn in well water + SS treatment was higher than other treatments. Grain yield of corn increased significantly with SS application and this fertilizer impact was visible 5 years after a single SS use (Karimpour et al. 2010). Also, Saadat et al. (2012) reported that the SS application increased significantly maize yield. Al Zoubi et al. (2008) reported an enhancement in the yield of corn and vetch as a result of the addition of SS. Karimpour and Afyuni (2007) reported that the addition of SS to the soil for 5 years increased yield of maize plant. Also, Marjovvi and Mashayekhi (2018) stated that the GY of corn was increased considerably by 5 years of organic fertilizers (SS and municipal compost) usage in comparison with control treatment. The reason for this increase in performance is the presence of macro and microelements in SS. In addition, the application of SS improves the physico-chemical and biological properties of the soil, which results in improved soil fertility and suitable conditions for plant growth (Chiba et al. 2008; Hargreaves et al. 2008; Alcantara et al. 2009; Franco et al. 2010; Ghorbani et al. 2017; Marjovvi and Mashayekhi 2019). Effects of SS application on different plants are in shown Table 3.

In addition to the positive effect of SS use on cereals, its positive impacts are evident on summer crops and vegetables. For example, the Sohrabi et al. (2017) stated that the application of 100 ton ha^{-1} SS considerably increased dry weight of biomass and root, plant height, root length, and leaf area index (LAI) of lettuce, but had no significant effect on chlorophyll content of lettuce. Similarly, Afyuni et al. (1998) reported that the SS increased the crop yields of spinach (*Spinacia oleracea* L.) and lettuce (*Lactuca sativa* L.) significantly. Also in another study, the effect of SS on spinach (*Spinacea oleracea*) plant yield was significant and increased this trait (Vaseghi et al. 2005). Boostani and Ronaghi (2011) showed that the addition of all levels of SS (10, 20, 40, and 80 g kg^{-1} soil) caused a significant enhancement in the spinach shoots weight in clay loam, sandy loam, and sandy of soil textures. The effect of fertilizer in increasing nutrients concentrations and yield was less than the 40 and 80 g SS kg^{-1} soil. The main reason for increasing the yield of spinach with the use of SS can be considered the presence of relatively large amounts of OM and essential elements in SS (Boostani and Ronaghi 2011). Also, Kumar and Chopra (2016) reported that the maximum agronomic performance in the form of dry weight and length of plant and root, the content of chlorophyll, LAI, and yield of plant as well as biochemical components like crude protein, dietary fiber, and total carbohydrate of eggplant (*Solanum melongena*) were observed at 50% concentration of SS in rainy and summer seasons. On the other hand, Ghanavati et al. (2013) reported that application of SS considerably increased growth parameters such as dry weight and length of root and weight of nodes in berseem clover (*Trifolium alexandrum* L.), but the roots colonization percentage was significantly reduced.

Table 3 Effects of sewage sludge application on different plants

Sewage sludge application rate	Plant type	Impacts	References
50 and 100 ton ha^{-1}	Barley	• Increased N, P, K, and Cd in vegetative parts compared to control. • Increased N, P, K, Fe, and Zn in grain compared to control. • Increased grass yield and grain yield compared to control.	Chorom and Aghaei Foroushani (2007)
50, 100, 200, and 400 ton ha^{-1}	Barley	• No changed dry yield of shoot and 100-grain weight compared to control.	Ghamari and Danesh (2007)
60 and 120 g kg^{-1} soil	Wheat	• Increased leaf surface, dry weight of aerial parts, grain number, and total grain weight compared to control.	Shahbazi et al. (2018)
15, 20, and 25 g kg^{-1} soil	Maize	• Increased dry weight of shoot and root compared to control. • Increased leaf number, plant height, width of end leaf, and chlorophyll compared to control. • Increased concentrations of Pb and Cd in shoot and root compared to control.	Saadat et al. (2012)
15 and 30 ton ha^{-1}	Maize	• Increased yield compared to control.	Hoshyar and Baghaie (2017)
14.2, 28.4, and 56.7 ton ha^{-1}	Sunflower	• Increased stem diameter, plant height, biological yield (BY), leaf area index (LAI), and leaf chlorophyll index (LCI) compared to control.	Kazemalilou et al. (2018)
20 and 40 ton ha^{-1}	*Haloxylon pp* (Ciahtagh in Persian)	• Decreased fresh weight, plant height, and number of branch compared to control.	Rezaii et al. (2016)
20 and 40 ton ha^{-1}	*Nitraria schoberi* (Garadagh in Persian)	• Increased fresh weight, plant height, and number of branch compared to control.	Rezaii et al. (2016)
25, 50, and 100 ton ha^{-1}	Berseem clover	• Increased dry weight of shoot, root weight, and root length compared to control.	Ghanavati et al. (2013)

In connection with medicinal and rangeland plants, Karimzadeh et al. (2016) showed that the highest rate of increase in height and canopy, basal diameter, and canopy of Garadagh plant (*Nitraria schoberi*) in prairie soil treatment sludge 100% and non-sludge effluent level for height, basal area, and canopy diameter was 100, 50, and 75%, respectively. Furthermore, Shakarami and Maroufi (2019) reported that the addition of wastewater and SS increased the fresh and dry weights of mint (*Mentha spicata*). As, in comparison control, the combination of wastewater +100 ton ha^{-1} SS treatment increased the fresh and dry weights of the plant to

257 and 239%, respectively. Also, the performance of plants was increased during the next harvesting. Many researchers reported that providing balance nutrients and gradual nutrient release from organic sources during the period of growth can be a positive role in enhancing the growth of mint. The reasons for increasing the yield of plants include: having the OM, increasing soil water holding capacity, strengthening the plant hormone-like activities, increasing nutrient uptake by plants, and generally improve the chemical and physical soil structure, noted (Angin and Yaganoglu 2011; Hojati et al. 2006; Shahbazi et al. 2018; Marjovvi and Mashayekhi 2019).

About garden products, Zare et al. (2015) stated that physiological function of olive (*Olea europaea*) plants subjected to SS increased. The impact of SS and CF on the vegetative components of the olive plant including the number of lateral branches, number of new leaves, height and dry weight of leaf, significantly increased. So that, 150 ton ha^{-1} SS without fertilizer treatment had the best effect on the properties of olive plants. Sewage sludge increased the number of leaf and leaf area about twice times comparing with blank treatment.

3.1.2 Negative Effects

Against all the positive benefits of using SS on quantitative performance but excessive increase in SS may have a negative impact on yield and lead to reduced yield by causing toxicity of elements (including Pb and Cd) in the plant. Determining this range depends on different soil and plant conditions. For example, in calcareous soils with high acidity, the use of larger amounts of SS does not reduce yield. However, the application of SS reduces yield in acidic soils due to toxicity and excessive acidification of the soil. In this regard, Kassray and Saedi (2010) stated that the application of SS higher than 10 ton ha^{-1}, due to its toxic effect, decreased the fresh and dry weights of above-ground tomato plant parts drastically. For example, the use of 30 ton ha^{-1} SS decreased dry and fresh weight by 20 and 44% as compared with that of control, respectively. They found that the water content of tomato did not differ considerably in comparison with that of control when 10 to 20 ton ha^{-1} SS were applied; however its difference was significant when 30 to 40 ton ha^{-1} SS were applied. In another study, Kassray and Saedi (2010) reported that the dry, fresh, and water content of fruit differences with control were non-significant when 5 and 10 ton ha^{-1} SS were used to the pots. But, applying SS beyond 10 ton ha^{-1} reduced all these three attributes significantly. A study stated that applying 20, 30, and 40 ton ha^{-1} SS delayed growth and development, and caused manifest necrosis in the leaf margins of mature and young plants in first month, but the toxic impact of SS on necrosis decreased at later periods of growth. Higher availability of Fe at higher SS addition rates along with concentrations of soluble anion and cation in soil solution, especially chlorine and probably other side effects caused by SS use are the factors that may have toxic impacts on plants (Kassray and Saedi 2010). Also, Ghamari and Danesh (2007) stated that the sign of chlorosis was observed on leaves of barley in application of 200 and 400 ton ha^{-1} SS. Furthermore, a significant increment in the Pb and Cd concentrations in the barley leaves and stem was indicated due to the enhancement in SS use rate. While, the impact of SS addition on the plant aerials dry weight and grain weight was not significant.

Recent evidence demonstrates that in the *Haloxylon pp* (Ciahtagh in Persian) plant increasing levels of wastewater and SS reduced plant height. Some of the branches in this plant by increasing SS rates are decreasing and it shows different effects in different wastewater levels. However, in the *Nitraria schoberi* (Garadagh in Persian) plant different wastewater rates significant effect in both soil types. Increase SS amounts in soil increment the number of branches in this plant. The impact of different SS on the weight of fresh in *Haloxylon pp* plant is decreasing and different effects in various wastewater amounts. While in the *Nitraria schoberi* plant, the effluent impact is not significant (Rezaii et al. 2016). Some researchers believe that the addition of SS to agricultural land in the first year does not significantly increase yields and the reason is the lack of soil microbial balance in the release of plant nutrients from the added material. In addition, if the SS contains large amounts of heavy metals and salts, it may cause toxicity to the plant or in some cases lead to reduced yields due to increased soil salinity and disruption of the balance of nutrients (Boostani and Ronaghi 2011). In view of the above, the necessary care should be taken in the use of SS in agricultural lands.

3.2 Product Quality

Producing a high-quality crop requires soil with favorable physical, biological, and chemical conditions. Among the OM used in the agricultural soils, SS has a significant impact on soil properties and, as mentioned, improves the physical, biological, and chemical specification of soil as well as soil fertility. So in addition to increasing yield, it will improve product quality.

According to the purpose of plant cultivation (grain for human nutrition and/or forage for animal feed), various results on the effect of SS use on the quality and nutrients of the product have been presented by various researchers in plants. For example in a study, the Marjovvi and Mashayekhi (2019) reported that the application of organic fertilizers (municipal compost fertilizer and SS) increased the macronutrients (e.g. N, P, and K) and micronutrients (e.g. Zn, Mn, Fe, and Cu) in the onion (*A. cepa* L.) shoot. The highest contents of nutrient elements in onion shoot were observed in the treatment of 50 ton ha^{-1} municipal compost. They stated that in most cases, there was no significant difference between 50 ton ha^{-1} municipal compost and 30 ton ha^{-1} SS treatments. In this study the crop rotations were onion (first stage: after 1 year), wheat, sugar beet, forage corn, and onion (second stage: after 5 years); so there were two stages of onion cultivation. The concentrations of macro and micronutrients increased considerably with the application of organic fertilizers by 5 years. In the case of Cu, this trend was reversed, and the amount of absorption of this element in the second stage was lower than that of the first stage due to the enhancement in the other nutrients concentrations such as P and the competition among the different elements in the absorption by the plant. Reciprocally, the amount of Cu was reduced uptake in the second stage. This was due to an increment of concentrations of other nutrients, such as P, and the competition between various elements to be absorbed by the plant. The application of various fertilizer treatments in the first stage only increased the rate of Zn in the onion bulb, while in the second stage, as a result of several years of waste compost and SS consumption, the N, K, Cu, Fe, and Zn concentrations in the onion bulb increased considerably. The concentrations of Pb and Cd in the plant were low in different treatments (Marjovvi and Mashayekhi 2019). In another study, Chorom and Aghaei Foroushani (2007) reported that the addition of SS increased N, P, K, and Ca in vegetative parts of barley compared to control. Also, the use of SS considerably increased N, P, K, Fe, and Zn in grain of this plant. The extent of changes of product quality (including macro and microelements in the different parts of plant such as root, shoot, leaf, flower, and grain) on different plants due to different amounts of SS is shown in Table 4.

Shakarami and Maroufi (2019) reported that total N concentration in the mint (*Mentha spicata*) plant at the combination of raw wastewater +100 ton ha^{-1} SS treatment was 3.12 times greater than that in control. Also, the highest amount of P (0.67% dry weight) and K (3.85 mg in dry weight) was observed in this treatment. The reasons for the enhancement in the amount of N, P, and K in the plant can be noted rich wastewater and sludge in elements. Saadat et al. (2012) reported that the SS application increased significantly N, P, and Na concentrations in maize. They also revealed that the application of SS increased the concentration of Pb in maize above-ground biomass, but this increase was statistically significant only in the 20% rate of SS in comparison with control treatment. Whalen and Chang (2002) found in their experiments that long-term use of OM (such as SS) causes P to be stored with lower energy bonds and its usability increases. Najafi and Mohammadnejad (2016) also reached similar results in connection with the increment in concentration of P in the shoots of maize as a result of the consumption of SS. Najafi et al. (2012) reported that by application of SS and manure, and increasing their amounts from 0 to 15 and 30 g kg^{-1} soil, the uptake and concentration of P in shoot and root, the uptake of Ca in shoot and root, and the uptake and concentration of Na in the shoot of sunflower (*Helianthus annuus* L.) increased. However, the K uptake and concentration in shoot increased by application of manure and increasing its level while decreased by the application of SS. In contrast, Nazari et al. (2006) stated that effluent and SS treatments had no effect on increasing the concentration of P in wheat, barley, and corn plants.

On the other hand, Asgari Lajayer et al. (2020) stated that adding 15 and 30 g SS kg^{-1} soil increases the uptake of P, K, Ca, Mg, and Na in the root and shoot, as well as the N of the shoot of basil (*Ocimum basilicum* L.). The use of 60 g irradiated and non-irradiated SS with the irradiation doses used in this research per kg of soil did not cause significant impacts on P, K, Mg, and Na absorption in the root and shoot, Ca in the root, and N in the shoot. But, it considerably decreased the Ca absorption of shoots. The maximum uptake of P, K, Ca, and Mg in the root and shoot and N in the shoot was obtained in 30 g kg^{-1} SS irradiated with 20 kilograys (kGy), as well as the

Table 4 Effect of sewage sludge on the changes of concentrations of nutrients in root, shoot, leaf, flower, grain, and plant

Application of sewage sludge at the rates (ton ha^{-1})	Condition	Region	Increase of nutrients concentrations over untreated control (%)												Section of plant	Crop type	References
			N	P	K	Ca	Mg	Na	Fe	Zn	Cu	Mn	Pb	Cd			
20	Farm	Sari, Iran	–	–	–	–	–	–	−0.2	18.2	41.8	−9.7	–	–	Shoot	Radish	Hosseinpour and Ghajar Sepanlou (2013)
40	Farm	Sari, Iran	–	–	–	–	–	–	1.2	41.7	80.1	3.0	–	–	Shoot	Radish	Hosseinpour and Ghajar Sepanlou (2013)
20	Farm	Sari, Iran	–	–	–	–	–	–	36.1	74.3	63.2	32.6	–	–	Root	Radish	Hosseinpour and Ghajar Sepanlou (2013)
40	Farm	Sari, Iran	–	–	–	–	–	–	43.0	148.2	98.0	35.3	–	–	Root	Radish	Hosseinpour and Ghajar Sepanlou (2013)
20	Farm	Sari, Iran	–	–	–	–	–	–	75.5	130.1	151.9	48.6	–	–	Leaf	Borage	Ahmad Abadi et al. (2012)
40	Farm	Sari, Iran	–	–	–	–	–	–	158.8	139.9	192.7	157.0	–	–	Leaf	Borage	Ahmad Abadi et al. (2012)
20	Farm	Sari, Iran	–	–	–	–	–	–	60.9	102.0	70.8	80.8	–	–	Flower	Borage	Ahmad Abadi et al. (2012)
40	Farm	Sari, Iran	–	–	–	–	–	–	85.1	189.5	133.9	283.3	–	–	Flower	Borage	Ahmad Abadi et al. (2012)
20	Farm	Sari, Iran	–	–	–	–	–	–	−8.0	−17.6	−20.2	8.6	–	–	Shoot	Lettuce	Hosseinpour et al. (2016)

40	Farm	Sari, Iran	–	–	–	–	–	–	3.1	2.5	6.9	14.4	–	–	Shoot	Lettuce	Hosseinpour et al. (2016)
20	Farm	Sari, Iran	–	–	–	–	–	–	−23.9	2.6	32.9	−37.5	–	–	Root	Lettuce	Hosseinpour et al. (2016)
40	Farm	Sari, Iran	–	–	–	–	–	–	24.3	44.7	78.1	1.8	–	–	Root	Lettuce	Hosseinpour et al. (2016)
20	Farm	Sari, Iran	–	–	–	–	–	–	–	–	–	–	157.0	160.6	Shoot	Lettuce	Rahimi Alashty et al. (2011)
40	Farm	Sari, Iran	–	–	–	–	–	–	–	–	–	–	256.0	164.8	Shoot	Lettuce	Rahimi Alashty et al. (2011)
20	Farm	Sari, Iran	–	–	–	–	–	–	–	–	–	–	18.1	344.6	Root	Lettuce	Rahimi Alashty et al. (2011)
40	Farm	Sari, Iran	–	–	–	–	–	–	–	–	–	–	70.6	357.8	Root	Lettuce	Rahimi Alashty et al. (2011)
20	Farm	Sari, Iran	–	–	–	–	–	–	–	–	–	–	268.1	183.1	Shoot	Radish	Rahimi Alashty et al. (2011)
40	Farm	Sari, Iran	–	–	–	–	–	–	–	–	–	–	294.4	364.4	Shoot	Radish	Rahimi Alashty et al. (2011)
20	Farm	Sari, Iran	–	–	–	–	–	–	–	–	–	–	134.2	126.4	Root	Radish	Rahimi Alashty et al. (2011)
40	Farm	Sari, Iran	–	–	–	–	–	–	–	–	–	–	137.9	213.2	Root	Radish	Rahimi Alashty et al. (2011)
15 g kg^{-1} soil	Pot and greenhouse	Tabriz, Iran	–	61.1	−17.1	17.3	10.9	47.6	–	–	–	–	–	–	Shoot	Sunflower	Najafi et al. (2012)
30 g kg^{-1} soil	Pot and greenhouse	Tabriz, Iran	–	77.8	−20.0	−12.2	−9.5	81.0	–	–	–	–	–	–	Shoot	Sunflower	Najafi et al. (2012)
15 g kg^{-1} soil	Pot and greenhouse	Tabriz, Iran	–	79.2	−19.2	−2.3	−5.2	8.6	–	–	–	–	–	–	Root	Sunflower	Najafi et al. (2012)

(continued)

Table 4 (continued)

Application of sewage sludge at the rates (ton ha^{-1})	Condition	Region	Increase of nutrients concentrations over untreated control (%)												Section of plant	Crop type	References
			N	P	K	Ca	Mg	Na	Fe	Zn	Cu	Mn	Pb	Cd			
30 g kg^{-1} soil	Pot and greenhouse	Tabriz, Iran	–	141.7	−3.3	7.7	−5.2	1.2	–	–	–	–	–	–	Root	Sunflower	Najafi et al. (2012)
15 g kg^{-1} soil	Pot and greenhouse	Gorgan, Iran	76.3	66.7	3.2	–	–	61.2	–	–	–	–	–	–	Shoot	Maize	Saadat et al. (2012)
20 g kg^{-1} soil	Pot and greenhouse	Gorgan, Iran	96.6	63.5	4.5	–	–	71.4	–	–	–	–	–	–	Shoot	Maize	Saadat et al. (2012)
25 g kg^{-1} soil	Pot and greenhouse	Gorgan, Iran	108.5	63.5	4.0	–	–	61.2	–	–	–	–	–	–	Shoot	Maize	Saadat et al. (2012)
25	Pot and greenhouse	Lorestan, Iran	–	–	–	–	–	–	7.0	–	−5.9	–	33.3	300	Shoot	Lettuce	Sohrabi et al. (2017)
50	Pot and greenhouse	Lorestan, Iran	–	–	–	–	–	–	5.0	–	29.4	–	83.3	500	Shoot	Lettuce	Sohrabi et al. (2017)
75	Pot and greenhouse	Lorestan, Iran	–	–	–	–	–	–	43.3	–	158.8	–	216.7	1100	Shoot	Lettuce	Sohrabi et al. (2017)
100	Pot and greenhouse	Lorestan, Iran	–	–	–	–	–	–	106.7	–	182.4	–	633.3	1800	Shoot	Lettuce	Sohrabi et al. (2017)
180	Greenhouse	Ardabil, Iran	36.0	112.3	86.6	–	–	–	–	–	–	–	–	–	Grain	Wheat	Fatholomi et al. (2015)
50	Greenhouse	Mashhad, Iran	–	–	–	–	–	–	–	–	–	–	46.7	80.0	Plant (shoot and Leaf)	Barley	Ghamari and Danesh (2007)
100	Greenhouse	Mashhad, Iran	–	–	–	–	–	–	–	–	–	–	53.3	215.0	Plant (shoot and Leaf)	Barley	Ghamari and Danesh (2007)
200	Greenhouse	Mashhad, Iran	–	–	–	–	–	–	–	–	–	–	66.7	215.0	Plant (shoot and Leaf)	Barley	Ghamari and Danesh (2007)
400	Greenhouse	Mashhad, Iran	–	–	–	–	–	–	–	–	–	–	166.7	400.0	Plant (shoot and Leaf)	Barley	Ghamari and Danesh (2007)

N nitrogen, *P* phosphorus, *K* potassium, *Ca* calcium, *Mg* magnesium, *Na* sodium, *Fe* Iron, *Zn* zinc, *Cu* copper, *Mn* manganese, *Pb* lead, *Cd* cadmium

maximum uptake of Na was revealed in 30 g kg^{-1} SS irradiated with 10 kGy absorbed dose. With the use of irradiated SS relative to non-irradiated in each level, the absorption of all studied nutrients increased in the root and shoot of basil.

A study found that the amounts of macro and microelements were increased in harvested corns by application of municipal compost and SS (organic fertilizers). It was clear that the those elements concentrations have increased in corn and soil after 5 years compared with 1 year of organic fertilizer consumption (Marjovvi and Mashayekhi 2018). Hosseinpour and Ghajar Sepanlou (2013) reported that the effects of the 3 years use of mineral and organic fertilizers were better than 1 and 2 years use of these fertilizers on increasing the micronutrients concentrations in soil and their uptake by plant. Increase in micronutrients amounts such as Zn, Cu, Fe, and Mn in lettuce (Hosseinpour et al. 2016) has been reported with increasing consumption of SS and more times. In this case, Boostani and Ronaghi (2011) reported that the addition of all SS (10, 20, 40, and 80 g kg^{-1} soil) levels caused a significant enhancement of concentrations of N, P, Zn, Cu, Fe, and Mn in spinach shoots. It was while with SS using, none of the nutrients concentrations in spinach reached the toxic level. The amounts of Cd and Pb were not detectable in shoots. Considering the Fe and Zn deficiency in calcareous soils, the application of SS can be effective for combating this deficiency. Bozkurt et al. (2006) stated that the 9.5 and 19 ton ha^{-1} SS applications did not considerably affect the heavy metals contents of grain and leaf of maize. But, 38.1 ton ha^{-1} SS applications increased Pb and Zn in maize leaf. The use of SS reduces pH, which increases the solubility of trace elements (such as Zn and Fe) and these elements are more available to the plant. So when using SS, more trace elements in the product are not unexpected. So that, Ahmad Abadi et al. (2012) reported that the fertilizer treatments (20 and 40 ton ha^{-1} SS, mix SS + CF, and only CF) on the amount of absorbent microelements such as Fe, Zn, Mn, and Cu in the soil, leaves, and petal of plant borage (*Borago officinalis*) were significant. They stated that the years of fertilizer consumption on all of the cases except the rate of Mn and Zn absorbent in leaves of the plant had a significant impact. The interaction between fertilizer treatments and years of consumption of fertilizers on the all of microelements concentrations except Mn in the soil was significant, in the leaves showed a significant effect on the rate of Cu and Fe absorbent only and in the petal had a significant impact on the amount of Cu and Mn absorbent.

Despite the positive impacts of SS on the quality of the product (both macro and microelements), in the following, the effect of SS on heavy metals in soil and plants is investigated.

4 Heavy Metals

4.1 Soil

The use of SS as a cheap and nutrients-rich fertilizer has been common in some regions of the world. But, the presence of potentially toxic heavy metals in some sludge's can restrict its application in agricultural lands (Ghamari and Danesh 2007). Also, using SS in large quantities can lead to heavy metals accumulation such as mercury (Hg), Pb, nickel (Ni), Cr, and Cd in the plant. It is, therefore, necessary to use the quantities of the elements introduced in soil and absorbed by plants in order to determine the toxicity level for each metal taking into account factors such as soil types, plant, and environmental conditions. This information can then be used to determine SS use quantities in each case.

The high heavy metals concentrations like Hg in SS can cause pollution of soil, crop, and chain of human food. Karimpour et al. (2010) reported that the SS application significantly increased total Hg concentration in 0–20 and 20–40 cm depths of soil. Total concentration of Hg in soil ranged from 20 to 1200μg kg^{-1} in control plots and plots with 500 Mg ha^{-1} SS use, respectively. According to the standards of USEPA, the maximum allowable amount of Cd and Pb through SS is 39 and 300 kg ha^{-1} and the maximum annual entry into the soil is 1.9 and 15 kg ha^{-1}, respectively (Anon 1995). Recent evidence demonstrates that using wastewater and SS led to increased heavy metals in soil, so, combination treatment of raw wastewater +100 ton ha^{-1} SS were increasing Ni, Pb, and Cd in soil by 304, 375, and 208%, respectively compared to control (Shakarami and Maroufi 2019). Also, Rahimi Alashty et al. (2011) indicated that the maximum amounts of Cd and Pb (available and total) in soil were accumulated in 40 ton ha^{-1} treatment, but it's in 1 and 2 years application treatments were significant compared to control treatments. Researchers observed an increase in the amount of absorbable Cd only in acidic soils under their experiments. In return, Fathololomi et al. (2015) reported that the effect of SS application rates (30, 60, 120, and 180 ton ha^{-1}) on the concentrations of Pb, Cr, Ni, and Cd was not significant in soil. Also, Marjovvi and Mashayekhi (2019) reported that the use of organic treatments of municipal compost fertilizer and SS had no impact on the enhancement of Pb and Cd in the soil after 1 year; however, the consumption of 5 consecutive years of municipal compost (50 ton ha^{-1}) increased the available Pb in the soil. The extent of changes of heavy metals in the soil due to different amounts of SS is shown in Table 2.

The interaction impact of heavy metals and some metals such as Fe can affect availability of heavy metals in soil. Baghaie (2018) reported that the increasing the loading amount of Arak municipal SS compost from 0 to 15 and 30 ton ha^{-1} in a Cd polluted soil (10 mg Cd kg^{-1} soil) caused a decrease in availability of Cd by 15 and 35%, respectively, but the availability of Fe increased by 5.6 and 8.4 times, respectively. Similar to this result, Cd concentration in root and shoot of pot marigold (*Calendula officinalis*) were decreased by 24 and 18%, respectively.

The rate of heavy metals transferred in the soil environment is a function of pH, CEC, clay content, OM, and etc. So that, the mobility of heavy metals decreases with increasing pH, carbonate, and OM in soil (Singh and Agrawal 2007). Sewage sludge has a dual effect that (1) increases the amount of transported heavy metals by increasing the acidity and increasing the CEC, but (2) reduces the amount of transported heavy metals by increasing the OM. This impact of SS depends on the amount of SS consumed and the amount of heavy metals in it. Zafarzadeh et al. (2015) reported that if the concentrations of heavy metals are less than the maximum allowable level in the SS. Therefore, wastewater and SS could be used to irrigate agricultural lands and green spaces in shortage of water; however, environmental regulations should be taken into account.

4.2 *Plant*

Due to the fertility attributes of SS and its application in agriculture, silviculture, and land reclamation and also the existence of some heavy metals in SS and their adverse impacts on soil and plant, assessment of the effects of the heavy metals on plants and agricultural products is needed.

Cadmium (Cd) is considered nonessential for living organisms and is one of the most toxic heavy metals. This metal is easily absorbed by plants and transported to the animal and chain of human food. Organic residuals such as SS used as fertilizers in agriculture are the important source of Cd in soil (Sharifi et al. 2013). Sharifi et al. (2013) reported that the source of Cd and species of plant are main factors in the assessment of uptake and translocation of Cd to the plants. In this case, they stated that the greatest Cd uptake by shoot of alfalfa was obtained in the compost treatment, and treatments of SS and cow manure were in the next ranks. The amount of Cd translocated to the alfalfa shoots in compost treatment was greater compared to the other treatments. The use of cow manure considerably decreased uptake and translocation of Cd to shoot. On the other hand, Chorom and Aghaei Foroushani (2007) reported that the SS increased uptake of heavy metals by barley plants but still below standard levels. So that, Hoshyar and Baghaie (2017) reported that applying 1.5 mmol DTPA chelate in a soil treated with 15 and 30 ton ha^{-1} polluted SS (15 mg Cd kg^{-1} soil) increased the corn plant Cd availability by 8.2 and 15%, respectively, and decreased P availability by 4 and 14%, respectively. Almost 79% of Cd uptake changes were estimated by the shoot Cd concentration of corn. Similarly, the contents of Cd, Cr, Mn, Cu, and Zn in eggplant were increased from 5 to 100% treatments of SS (Kumar and Chopra 2016). Recent evidence demonstrates that the addition of SS increased the contents of heavy metals in soil and crop. The Cd content in grain of rice was above the Indian safe limit at 20 ton ha^{-1} or higher rates of SS use (Latare et al. 2014). Rahimi Alashty et al. (2011) indicated that the application of SS was a significant effect on Cd and Pb content in root and shoot of lettuce and radish. The extent of changes of heavy metals in the various parts of the plant such as root, shoot, and leaf due to different amounts of SS is shown in Table 4.

On the other hand, research has shown that increasing Cd by 100 and 200 g Cd kg^{-1} soil reduces the absorption of Fe and Zn in the plant (Malekzadeh et al. 2012). Tabarteh Farahani and Baghaie (2017) reported in a study that with increasing Cd concentration in soil, the absorption of total Fe, Zn, and Mn by the plant and their accumulation in the shoots of corn plant is greatly reduced. Therefore, Cd as one of the heavy metals in addition to causing toxicity in the plant also interferes with the absorption of micronutrients such as Zn and Fe. Due to the geochemical similarity of

Cd to Fe (Malakootian and Khazaei 2014), the use of Fe compounds such as iron oxide can play an main role in changing the availability of heavy metals such as Cd. Enrichment of organic fertilizers alike SS with this element (Fe) is probably a good way to reduce the availability of Cd (Melali and Shariatmadari 2008; Baghaie 2018).

Karimpour et al. (2010) reported that the application of SS considerably increased Hg uptake in different plant parts such as roots, stems, and grains of corn. They stated that the average concentration of Hg in root, stem, and grain at the end of the fifth year were 91, 9, and 8μg kg^{-1}, respectively. They stated that the Hg concentration in the roots was higher than other parts of the plant (stems and grains). This is consistent with previous findings that more Hg accumulates in the roots (Eisler 2000), and the roots to some extent prevent the transfer of Hg to other parts of the plant (Patra and Sharma 2000).

Lead (Pb) has less mobility in soil and plants than other heavy metals. So despite the very high toxicity of Pb, its pollution potential is much lower than other heavy metals. Afyuni et al. (1998) reported that the concentrations of total metals showed an increasing trend with the adding SS. Concentrations of Zn, Cu, and Pb EDTA-extractable in soil and these elements concentrations in the plants increased considerably with SS level (from 22.5 to 45 ton ha^{-1}). They stated that the time of SS addition did not have any significant impact on EDTA-extractable and uptake of metals in plant. In another study, Saadat et al. (2012) stated that the application of SS significantly increased Pb and Cd concentrations in the roots of corn. Plant tissue analysis showed that SS application by 100 ton ha^{-1} significantly increased concentrations of Cu, Fe, Cd, and Pb in lettuce shoot, root, and stalk (Sohrabi et al. 2017). Shakarami and Maroufi (2019) indicate that using the wastewater (raw wastewater and treated wastewater) and SS (50 and 100 ton ha^{-1}), compared to the control, increase heavy metals (Ni, Pb, and Cd) of mint. The concentrations of heavy metals (mg kg^{-1} dry weight) in mint ranged from 0.01 to 0.57 for Pb, 0.02 to 0.71 for Ni, and 0.01 to 0.3 for Cd. Rahimi Alashty et al. (2011) reported that the amount of Pb and Cd accumulated in root and shoot of lettuce and radish were increased in 40 ton ha^{-1} SS treatment for 3 years, and also this increase was observed in treatments that were received SS for 1 and or 2 years. Moreover, the accumulation of Cd and Pb in radish root was more than lettuce root. Therefore, long-term use of SS due to accumulated heavy metals in the soil and uptake by plants must be carefully controlled. On the other hand, Karimpour and Afyuni (2007) stated that as the volume of SS used increases, the negative effects of this substance (such as the presence of heavy metals and disturbance of the C:N balance of the soil) gradually increases. As a result, when the volume of SS exceeds a certain amount, the negative effects of sludge override its positive effects and as a result, there is no significant difference among the yield of the plant grown in this treatment and the control. The concentrations of these elements (trace elements and heavy metals) due to the addition of SS increases significantly in wheat, barley, and corn (Nazari et al. 2006), broccoli (Perez-Murcia et al. 2006), lettuce (Rahimi Alashty et al. 2011), borage (Ahmad Abadi et al. 2012), radish (Hosseinpour and Ghajar Sepanlou 2013), rice-wheat system (Latare et al. 2014), etc.

Many SS and wastewaters alone cannot meet the plant's nutritional needs for high-consumption elements, especially N and K. Therefore, their application with appropriate amounts of CF to meet the nutritional needs of the plant should be considered (Sommers et al. 1976). In the following, is discussed the effect of simultaneous application of SS and CF on soil and plants.

5 Concomitant Use of Sewage Sludge with Chemical Fertilizers (CF)

Some studies have shown that the effect of SS on soil K is less than that of P and N (Zare et al. 2015). This can be due to the small amount of K in the SS. Because potassium salts have high solubility and during the process of separating sludge from wastewater (in wastewater treatment), K mostly remains soluble in the effluent and the sludge portion is poor in K (Tester et al. 1979). Potassium added to the soil through SS is often not enough for the plant and chemical fertilizers (CF, containing P) along with SS should be used to solve this problem. In this regard, Ahmadpoor et al. (2011) reported that the maximum amount of available K, P, and total N, activities of urease and alkaline phosphates were observed in 40 ton SS ha^{-1} + 1/2 chemical fertilizers (potassium sulfate, triple superphosphate (TSP), and urea to the 150, 150, and 200 kg ha^{-1}, respectively) treatment. A significant correlation was observed among enzyme activity, organic soil matter, total N, K, and P. Therefore addition of SS, by increasing the amount of OC, total N, available P, and K, caused to enhance the enzyme activity in the soil.

On the other hand, various results have been obtained with the combined application of SS and CF on crop production. Kazemalilou et al. (2018) reported that the application of TSP and SS increased all studied characteristics such as leaf chlorophyll index (LCI), LAI, biological yield (BY), plant height, and stem diameter of sunflower significantly compared to the control. They stated that in order to reduce the consumption of CF and achieving the optimum growth of sunflower plant, addition of 200 kg of TSP + 56.7 ton of SS ha^{-1} under optimum irrigation, and 100 kg of TSP + 56.7 ton of SS ha^{-1} under limited irrigation is recommended at similar conditions.

Various researches show the positive impact of the combined use of SS and CF on the soil fertility and quality of the product. So that, Hosseinpour and Ghajar Sepanlou (2013) reported that the addition of different fertilizer treatments had a significant impact on available and total micronutrient elements in soil and accumulation of these nutrients in radish root and shoot. By increasing the use of SS and CF, concentrations of micronutrients in soil and their uptake by plants were increased. Also, Hosseinpour et al. (2016) stated that the CF and SS treatments considerably affected the available of Cu and Zn in soil and concentrations of Cu, Zn, Fe, and Mn in root and Cu and Zn in shoots of lettuce. The highest enhancement in the amount of micronutrients in soil and lettuce belonged to the addition of 40 ton ha^{-1} SS individually and enriched with 50% CF. Hosseinpour and Ghajar Sepanlou (2013) showed that the highest increment Zn available in the soil belonged to the continuous 3 years use of 40 ton ha^{-1} SS. The highest uptake rate of Mn and Fe by root of radish belonged to the continuous 3 years adding 40 ton ha^{-1} SS + 50% NPK fertilizer. Also, they indicated that the highest concentration of Cu in radish shoots belonged to the continuous 3 years addition of 20 ton ha^{-1} SS + 50% NPK fertilizer. Finally, they stated that SS enriched with CF and individually, may be applied, but care must be taken to avoid pollution of soils by heavy metals and entry into the chain of human food.

6 Solutions to Eliminate the Disadvantages or Harms of Sewage Sludge

Sludge pretreatment is necessary prior to its use in order to eliminate the diverse pathogenic microorganisms, presence of pathogens, degradable organic contaminants, and heavy metals. Sludge disinfection can be done by long-term storage (Carl et al. 2002), liming, aerobic stabilization under thermophilic conditions, pasteurization, and irradiation (Limam et al. 2018; Asgari Lajayer et al. 2020). Of course, doing these methods is usually associated with a relatively high cost.

6.1 Composting of Sewage Sludge

Co-composting of SS and other waste (such as biological waste) is a method for healthy and safe disposal to reuse these (Carl et al. 2002). Recently, use of municipal solid waste compost and SS on the agriculture lands had received considerable attention. These materials provide a valuable source of OM and enhance yield of crop and soil fertility by improving soil biological and physico-chemical properties.

Regarding about composting of SS, Malakootian et al. (2014) stated that during the composting process from SS and pistachio hull waste, pH increased but the amounts of EC, total N, OC, P, K, Pb, and Cd decreased. They stated that the specifications of the produced compost comprising pH, EC, temperature, moisture ratio, C/N ratio, K, P, and heavy metals were in category one and two of the standard compost of Iran. The raw sludge in compost with organic waste and protein, fat, and cellulose (biochemical solids) are pure in terms of process evolution, destruction of biochemical compounds in composting materials, the potential for inactivation of pathogen, improved activity of biological, and nutrient retention (Zazouli and Ala 2019). Also, research has shown that adding SS compost enriched with 5% Fe pure can perhaps affect soil physico-chemical attributes that increasing Fe availability on soil and plant as a result led to decreasing availability of Cd (Baghaie 2018).

6.2 *Ionizing Radiation for the Disinfection of Sewage Sludge*

In Asian countries, due to the low cost of SS, it was used in the cultivation of plants, particularly fresh vegetables. The existence of various organic and inorganic pollutants in SS used in the cultivation of vegetables increased the risk of various diseases in consumers due to their direct consumption. Therefore, the use of risk assessment models has been suggested to predict the risks of crop consumption by various consumers and vegetable production with organic fertilizers containing pollutants such as SS. Lack of study about comparing plants grown in irradiated and non-irradiated sludge with risk assessment equations reveals the necessity of these equations and their mechanisms. Generally, Asgari Lajayer et al. (2015) reported that the selection of plant type for cultivation under SS using depends on the metabolism, nutritional needs, and stress tolerance and the change in biological, physical, and chemical properties products with radiation and type consumption of plants.

As mentioned, SS is an organic fertilizer and in addition to pathogens, containing macronutrients (including N, P, and K), micronutrients (for example Fe, Zn, Cu, Mn, and molybdenum (Mo)), and heavy metals such as arsenic (As), Cd, Cr, Pb, Hg, Ni, selenium (Se), etc. Using ionizing radiation is the best way for the disinfection of SS and absorption of radiation energy by water molecules of SS causes the ionization of their and formation of oxidizing and reducing free radicals. A comprehensive report does not exist on the effects of ionizing radiation such as gamma rays on the changes in chemical properties of SS; however, irradiation of SS could remove dangerous pathogenic organisms and plant growth inhibitors and could increase plant yield (Asgari Lajayer et al. 2015, 2020; Limam et al. 2018). Also, Limam et al. (2018) stated that 4.5 kGy is the optimal dose, and γ-irradiation pretreatment of anaerobic sludge added to the soil has considerably improved growth of broad bean (*Vicia faba*). Moreover, by degrading chemical pollutants and eliminating pathogens, γ-irradiation of anaerobic sludge provides promising for reusing SS as a safe fertilizer.

7 Conclusion

Generally, the application of sewage sludge (SS) with optimal dose led to improves the physico- chemical and biological characteristics of the soil. Improvement and fertility of soil due to use of SS leads to suitable conditions for growth of plant. Therefore, the plants grown in these soils are superior in terms of quantity and quality. In between, the effects of the several years application of SS are better than 1 and 2 years addition of this fertilizer on increasing the micronutrients concentrations in soil and their uptake by organs of plant. However, because of the concentrations of heavy metals in SS, its long-term application poses risk of heavy metals contamination in the soil and, consequently, the health risk to the human and animal food chain. Therefore, it is recommended that as much as possible SS should not be used for growing edible food crops. Indeed, before using SS in different farms, it is necessary to be examined for microbial load and heavy metals, and suitable recommendations should be made based on local conditions, soil properties, and vegetation type.

References

Afyuni M, Rezainejad Y, Khayambashi B (1998) Effect of sewage sludge on yield and heavy metal uptake of lettuce and spinach. J Water Soil Sci 2(1):19–30

Ahmad Abadi Z, Ghajar Sepanlou M, Bahmanyar MA (2012) The effect of sewage sludge application on the amount of microelements in soil and absorption in medicinal plant borage (*Borago officinalis* L.). J Water Wastewater 1:101–110

Ahmadpoor SR, Bahmanyar MA, Gilani SS, Forghani A (2011) Chemical characteristic, enzyme activity of urease and alkaline phosphatase in a soil amended with sewage sludge and sewage sludge enrichment with chemical fertilizer. J Water Soil Conserv 18(1):103–122

Akbarnejad F, Astaraei A, Fotovat A, Nasiri Mahalati M (2013) Effect of municipal solid waste compost and sewage sludge on chemical properties soil. J Water Soil 26(6):1329–1338

Al Zoubi MM, Arsalan A, Abdegawad GN, Pejon N, Tabbaa M, Jouzdan O (2008) The effect of sewage sludge on productivity of a crop rotation of wheat, maize and vetch and heavy metals accumulation in soil and plant Aleppo governorate. Am Eurasian J Agric Environ Sci 3 (4):618–625

Alcantara S, Perez DV, Almeida MRA, Silva GM, Polidoro JC, Bettiol W (2009) Chemical changes and heavy metal partitioning in an oxisol cultivated with maize (*Zea mays*, L.) after 5 years disposal of a domestic and an industrial sewage sludge. Water Air Soil Pollut 203:3–16. https://doi.org/10.1007/s11270-009-9986-y

Angin I, Yaganoglu V (2011) Effects of sewage sludge application on some physical and chemical properties of a soil affected by wind erosion. J Agric Sci Technol 13:757–768

Anon (1995) Standard for use or disposal of sewage sludge. USEPA. Federal Register. 25. October 1995 60(206):54763–54770

Asgari Lajayer H, Najafi N, Moghiseh E (2015) Study of the possibility of using gamma-irradiated sewage sludge in different plants production. J Radiat Nucl Technol 2(3):27–37

Asgari Lajayer B, Najafi N, Moghiseh E, Mosaferi M, Hadian J (2020) Effects of sewage sludge irradiation on uptake of macronutrients and sodium in basil plant under greenhouse conditions. J Nucl Sci Technol 41(91):151–166. https://doi.org/10.24200/nst.2020.1105

Baath E (1989) Effects of heavy metals in soil on microbial process and population: a review. Water Air Soil Pollut 47:335–379. https://doi.org/10.1007/BF00279331

Baghaie A (2018) Effect of iron slag and municipal sewage sludge compost addition to soil on decreasing cadmium availability in pot marigold. J Sabzevar Univ Med Sci 25(5):659–668

Bahremand MR, Afyuni M, Hajabbassi MA, Rezaeinejad Y (2003) Effect of sewage sludge on soil physical properties. J Water Soil Sci 6(4):1–9

Baker EG, Mathews PJ (1983) Metals in sewage sludge and their potential effects in agriculture. Water Sci Technol 15(1):209–225. https://doi.org/10.2166/wst.1983.0018

Balkcom KS, Adams JF, Hartzog DL, Woodm CW (2001) Mineralization of composted municipal sludge under field condition. Commun Soil Sci Plant Anal 32(9–10):1589–1605. https://doi.org/10.1081/CSS-100104215

Barati Rashvanlou R, Farzadkia M, Marousi M (2018) Selection of suitable guidelines for the application of sewage sludge in agricultural using hierarchical analysis system (AHP). J Environ Health Eng 5(3):299–308. https://doi.org/10.29252/jehe.5.3.299

Boostani HR, Ronaghi A (2011) Comparison of sewage sludge and chemical fertilizer application on yield and concentration of some nutrients in spinach (*Spinosa olerace* L.) in three textural classes of a calcareous soil. J Soil Plant Interact 2(2):65–74
Bozkurt MA, Hakki Akdeniz H, Keskin B, Yilmaz I (2006) Possibilities of using sewage sludge as nitrogen fertilizer for maize. Plant Soil Sci 56(2):143–149. https://doi.org/10.1080/09064710510031113
Carl A, Trine E, Jane H, Henrik L (2002) Degradation of organic contaminants in sewage sludge during windrow composting, No. 3036, Drammen, Norway
Casado-Vela J, Selles S, Navarro J, Bustamante MA (2006) Evaluation of composted sewage sludge as nutritional source for horticultural soils. Waste Manag 26(9):946–952. https://doi.org/10.1016/j.wasman.2005.07.016
Cheng H, Xu W, Liu J, Zhao Q, He Y, Chen G (2007) Application of composted sewage sludge (CSS) as a soil amendment for turfgrass growth. Ecol Eng 29(1):96–104. https://doi.org/10.1016/j.ecoleng.2006.08.005
Chiba MK, Mattiazzo ME, Oliveira FC (2008) Sugarcane cultivation in a sewage-sludge treated ultisol: II - soil fertility and plant nutrition. Rev Bras Ciênc Solo 32(2):653–662. https://doi.org/10.1590/S0100-06832008000200020
Chorom M, Aghaei Foroushani M (2007) Effects of amended sewage sludge application on yield and heavy metal uptake of barley: a case study of Ahvaz sewage treatment plant. J Water Wastewater 18(2):53–63
Dehghan Manshadi H, Bahmanyar MA, Lakzian A, Salek Gilani S (2012) Effect application of sewage sludge and sewage sludge enriched with chemical fertilizer on the rate of organic carbon, respiration and enzyme activity of soil under basil cultivation. J Water Soil 26 (3):554–562
Eisler R (2000) Handbook of chemical risk assessment: health hazards to humans, plants, and animals, vol 1-Metals. Lewis, Boca Raton, FL
Fathololomi S, Asghari SH (2015) Effects of Ardabil municipal sewage sludge on some physical and hydraulic properties of a coarse-textured soil under wheat cultivation. Water Soil Sci 24 (4):169–183
Fathololomi S, Asghari S, Goli Kalanpal E (2015) Effects of municipal sewage sludge on the concentration of macronutrients in soil and plant and some agronomic traits of wheat. J Soil Manag Sustain Prod 5(2):49–70
Fernandes SAP, Bettiol W, Cerri CC (2005) Effect of sewage sludge on microbial biomass, basal respiration, metabolic quotient and soil enzymatic activity. Appl Soil Ecol 30(1):65–77. https://doi.org/10.1016/j.apsoil.2004.03.008
Franco A, Abreu Junior CH, Perecin D, Oliveira FC, Granja ACR, Braga VS (2010) Sewage sludge as nitrogen and phosphorus source for cane plant and first ratoon crops. Rev Bras Ciênc Solo 34 (2):553–561. https://doi.org/10.1590/S0100-06832010000200029
Gardio S, Martin del Champio G, Esteller MV, Vaca R, Lugedo J (2005) Heavy metals in soil treated with sewage sludge composting, their effect on yield and uptake of broad bean seeds (*Vicia faba* L.). Water Air Soil Pollut 166:303–319. https://doi.org/10.1007/s11270-005-5269-4
Gascó G, Lobo MC (2007) Composition of a Spanish sewage sludge and effects on treated soil and olive trees. Waste Manag 27(11):1494–1500. https://doi.org/10.1016/j.wasman.2006.08.007
Ghamari N, Danesh S (2007) Effects of sewage sludge application and leaching on soil chemical properties, yield and quality of barley. Irrig Drain Struct Eng Res 8(3):65–80
Ghanavati N, Nadian H, Moezy A, Rejali F (2013) Effect of arbuscular-mycorrhizal fungi on growth in berseem clover plant under different levels of sewage sludge. Crop Physiol J 5 (17):17–30
Ghorbani S, Karimian N, Raeisi F (2017) The effect of urban sewage sludge on chemical properties, soil basal respiration and microbial biomass carbon in a calcareous silty clay loam soil. J Hydrol Soil Sci 21(1):255–264
Hadas A, Kautsky L, Mustafa G, Kara EE (2004) Rates of decomposition of plant residues and available nitrogen in soil, related to residue composition through simulation of carbon and

nitrogen turnover. Soil Biol Biochem 36(2):255–266. https://doi.org/10.1016/j.soilbio.2003.09.012

Hargreaves JC, Adl MS, Warman PR (2008) A review of the use of composted municipal solid waste in agriculture. Agric Ecosyst Environ 123(1–3):1–14. https://doi.org/10.1016/j.agee.2007.07.004

Hemmat A, Aghilinategh N, Rezainejad Y, Sadeghi M (2010) Long-term impacts of municipal solid waste compost, sewage sludge and farmyard manure application on organic carbon, bulk density and consistency limits of a calcareous soil in center Iran. Soil Till Res 108(1–2):43–50. https://doi.org/10.1016/j.still.2010.03.007

Herman DJ, Johnson KK, Jaeger CH, Schwartz E, Firestone MK (2006) Root influence on nitrogen mineralization and nitrification in avena barbata rhizosphere soil. Soil Sci Soc Am J 70 (5):1504–1511. https://doi.org/10.2136/sssaj2005.0113

Hernandez T, Moral R, Prez-Espinosa A, Moreno-Caselles A, Perez-Murica MD, Garcia C (2002) Nitrogen mineralization potential in a calcareous soil amended with sewage sludge. Bioresour Technol 83(3):213–219. https://doi.org/10.1016/S0960-8524(01)00224-3

Hojati S, Nourbakhsh F, Khavazi K (2006) Microbial biomass index, enzyme activites and corn yield in a soil amended with sewage sludge. Iran J Soil Water Sci 20(1):84–93

Hoshyar P, Baghaie AH (2017) Effect of Arak municipal sewage sludge application on corn Cd uptake in a loamy soil. Iran J Soil Res 31(2):303–314

Hossain MK, Strezov V, Nelson PF (2015) Comparative assessment of the effect of wastewater sludge biochar on growth, yield and metal bioaccumulation of cherry tomato. Pedosphere 25 (5):680–685. https://doi.org/10.1016/S1002-0160(15)30048-5

Hosseinpour R, Ghajar Sepanlou M (2013) Effects of integrated municipal sewage sludge with chemical fertilizer on micronutrient concentration in soil and radish (*Raphanus sativus*). Iran J Soil Res 27(3):373–385

Hosseinpour R, Ghajar Sepanlou M, Salek Gilani S (2016) The effects of sewage sludge and chemical fertilizers on concentration of some microelements in soil and lettuce (*Lactuca sativa* L.). J Agric Sci Sustain Prod 26(2):31–43

Karimpour M, Afyuni M (2007) Use of sewage sludge as fertilizer in agriculture and its effect on plant yield. 2nd National Congress of Ecological Agriculture, 17 October 2007, Gorgan University of Agricultural Sciences and Natural Resources, Gorgan, Iran

Karimpour M, Afyuni M, Esmaili Sari A (2010) Effect of sewage sludge on mercury concentration in soil and corn. J Water Soil Sci 14(52):115–123

Karimzadeh HR, Rezae R, Mehrabi S (2016) Investigating the effects of soil and sewage sludge on some vegetative traits of Garadagh plant (*Nitraria schoberi*). J Soil Water Res Conserv 5 (4):81–95

Kassray R, Saedi S (2010) Effects of Tabriz petrochemical sewage sludge on tomato growth. J Water Soil 24(1):10–20

Kassray R, Saedi S, Aliasgharzadeh N (2008) The study of biochemical effects of Tabriz petrochemical factory biological sludge application on a sample of Ahar region soils. J Agric Sci Nat Res 15(2):67–75

Kazemalilou S, Najafi N, Reyhanitabar A, Ghaffari M (2018) Effects of integrated application of phosphorus fertilizer and sewage sludge on leaf chlorophyll index and some growth characteristics of sunflower under water deficit conditions. J Soil Manag Sustain Prod 7(4):1–18

Kumar V, Chopra AK (2016) Agronomical performance of high yielding cultivar of eggplant (*Solanum melongena* L.) grown in sewage sludge amended soil. Res Agric 1(1):1–24. https://doi.org/10.22158/ra.v1n1p1

Latare AM, Kumar O, Singh SK, Gupta A (2014) Direct and residual effect of sewage sludge on yield, heavy metals content and soil fertility under rice-wheat system. Ecol Eng 69:17–24. https://doi.org/10.1016/j.ecoleng.2014.03.066

Limam RD, Limam I, Clérandeau C, Khouatmia M, Djebali W, Cachot J, Chouari R (2018) Assessment of the toxicity and the fertilizing power from application of gamma irradiated anaerobic sludge as fertilizer: effect on *Vicia faba* growth. Radiat Phys Chem 150:163–168. https://doi.org/10.1016/j.radphyschem.2018.05.004

Malakootian M, Khazaei A (2014) Comparison of nano zerovalent iron particles and manganese compounds efficiency in cadmium ion removal from aqueous environments. J Ilam Univ Med Sci 22(2):93–103

Malakootian M, Mobini M, Nekoonam GA (2014) Evaluation of the compost produced from mixed sludge of municipal wastewater treatment plant and pistachio hull waste. J Mazandaran Univ Med Sci 24(116):172–183

Malekzadeh E, Alikhani HA, Savaghebi-Firoozabadi GR, Zarei M (2012) Bioremediation of cadmium-contaminated soil through cultivation of maize inoculated with plant growth-promoting rhizobacteria. Biorem J 16(4):204–211. https://doi.org/10.1080/10889868.2012.703258

Marjovvi AR, Mashayekhi P (2018) Effects of repeated applications of sewage sludge and MSW compost on soil chemical characteristics and yield of *Zea mays* L. in Rudasht-Esfahan. Iran J Soil Water Res 49(3):653–664. https://doi.org/10.22059/ijswr.2017.238780.667727

Marjovvi A, Mashayekhi P (2019) Effects of sewage sludge and municipal compost application on bio availability of soil nutrients in onion (*Allium cepa* L.) cultivation. Environ Sci 17 (3):189–208

Melali AR, Shariatmadari H (2008) Application of steel making slag and converter sludge in farm manure enrichment for corn nutrition in greenhouse conditions. J Water Soil Sci 11 (42):505–513

Mirhosseini G, Alavimoghaddam MR, Maknon R (2007) Investigation of application of Tehran municipal WWTPs' dried sludge in agriculture. Environ Sci 4(4):47–56

Mokhtari M, Salehi Vaziri A, Zareyi T, Jalili M (2017) An investigation on feasibility of composting of Yazd WWT dewatered sludge with windrow method using different treatments. Iran J Health Environ 9(4):493–502

Najafi NA, Mardomi S (2013) Effects of sunflower cultivation, manure and sewage sludge on availability of elements, pH and EC of an alkaline soil. Appl Soil Res 1(1):1–16

Najafi N, Mohammadnejad A (2016) Differential concentrations of some nutrient element in forage of corn (*Zea mays* L.) as affected by organic fertilizers and soil compaction. J Crop Ecophysiol 9 (4):561–582

Najafi N, Mardomi S, Oustan S (2012) The effect of waterlogging, sewage sludge and manure on selected macronutrients and sodium uptake in sunflower plant in a loamy sand soil. J Water Soil 26(3):619–636. https://doi.org/10.22067/jsw.v0i0.14927

Nazari MA, Shariatmadari H, Afyuni M, Mobli M, Rahili S (2006) Effect of industrial sewage-sludge and effluents application on concentration of some elements and dry matter yield of wheat, barley and corn. J Water Soil Sci 10(3):97–111

Patra M, Sharma A (2000) Mercury toxicity in plants. Bot Rev 66:379–409. https://doi.org/10.1007/BF02868923

Perez-Murcia MD, Moral R, Moreno-Caselles J, Perez-Espinosa A, Paredes C (2006) Use of composted sewage sludge in growth media for broccoli. Bioresour Technol 97(1):123–130. https://doi.org/10.1016/j.biortech.2005.02.005

Rahimi Alashty S, Bahmanyar MA, Ghajar Sepanlou M (2011) The effects of sewage sludge application on pH, EC, O.C, Pb and Cd in soil and lettuce and radish plants. J Water Soil Conserv 18(3):133–147

Rahmani HR, Moayyeri M, Mazaheri Kouhanestani Z, Khodabakhsh N, Sharifi H (2014) Studying some of the qualitative properties and concentration of heavy metals in dried sewage sludge from Shahin-Shahr WWTP in Isfahan. J Environ Sci Technol 16(2):55–66

Rasouli Sadaghiani MH, Sepehr E (2011) The effect of sewage sludge and manures application on nitrogen mineralization and rhizosphere characteristics in corn and sunflower plants. J Water Soil 25(2):327–337. https://doi.org/10.22067/jsw.v0i0.9386

Reeves DW (1997) The role of soil organic matter in maintaining soil quality in continuous cropping systems. Soil Till Res 2(1–2):113–123. https://doi.org/10.1016/S0167-1987(97)00038-X

Rezaii R, Karimzadeh HR, Mehrabi S, Heydarnezhad S (2016) Impact of soil, wastewater and sewage sludge on growth characteristics of *Haloxylon pp* and *Nitraria schoberi* plant. Desert Ecosyst Eng J 5(11):35–46

Ros M, Pascual JA, Garcia C, Hernandez MT, Insam H (2006) Hydrolase activities, microbial biomass and bacterial community in a soil after long-term amendment with different composts. Soil Biol Biochem 38(12):3443–3452. https://doi.org/10.1016/j.soilbio.2006.05.017

Saadat K, Barani Motlagh M, Dordipour E, Ghasemnezhad A (2012) Influence of sewage sludge on some soil properties, yield and concentration of lead and cadmium in roots and shoots of maize. J Soil Manag Sustain Prod 2(2):27–48

Salehi A, Mirghaffari N, Koohnejad H, Mahmoudi F (2013) Evaluating the role of industrial sludge as a source of heavy metal pollution. J Feyz 16(7):719–720

Shafieepour S, Ayati B, Ganjidoust H (2011) Reuse of sewage sludge for agricultural soil improvement (Case study: Kish island). J Water Wastewater 22(2):85–93

Shahbazi F, Ghasemi S, Sodaeizadeh H, Pourdara H, Ayaseh K (2018) The effects of sewage sludge on some physical and chemical properties of soil and wheat yield. Water Soil Sci 28(2):193–206

Shakarami M, Maroufi S (2019) Wastewater and sewage sludge effects on the absorption of some heavy metals in soil and mint. J Environ Stud 45(1):1–15

Sharifi M, Afyuni M, Khoshgoftarmanesh AH (2013) Uptake and root-to-shoot transfer of cadmium in alfalfa from a soil treated with sewage sludge, municipal waste compost and cow manure. J Feyz 16(7):691–692

Shirani H, Haj Abbasi MA, Afyouni M, Dashti H (2010) Cumulative effects of sewage sludge on soil physical and chemical characteristics. J Water Wastewater 3:8–36

Singh RP, Agrawal M (2007) Effect of sewage sludge amendment on heavy metal consequent responses of *Beta vulgaris* plants. Chemosphere 67(11):2229–2240. https://doi.org/10.1016/j.chemosphere.2006.12.019

Sohrabi N, Alinejadian Bidabadi A, Feizian M, Maleki A (2017) A comparative study of the effects of sewage sludge on heavy metals concentrations and some morphological characteristics of lettuce. Iran J Soil Res 31(2):291–302

Sommers LE, Nelson DW, Yost KJ (1976) Variable nature of chemical composition of sewage sludge. J Environ Qual 5(3):303–306. https://doi.org/10.2134/jeq1976.00472425000500030017x

Tabarteh Farahani N, Baghaie AH (2017) Effect of enriched cow manure with converter sludge on Fe bio-availability in a lead polluted soil. J Water Soil Conserv 24:205–220

Tester CF, Sikora LJ, Taylor JM, Parr JF (1979) Decomposition of sewage sludge compost in soil: III. Carbon, nitrogen, and phosphorus transformations in different sized fractions. J Environ Qual 8(1):79–82. https://doi.org/10.2134/jeq1979.00472425000800010017x

Torabian A, Momeni Farahani L (2002) Urban waste sludge management plan (Case study of Ghods town wastewater treatment plant). J Environ Stud 28(30):29–78

Vaseghi S, Afyuni M, Shariatmadari H, Mobli M (2005) Effect of sewage sludge on some macronutrients concentration and soil chemical properties. J Water Wastewater 16(1):15–22

Veeresh H, Tripathy S, Chaudhuri D, Ghosh BC, Hart BR, Powell MA (2003) Changes in physical and chemical properties of three soil types in India as a result of amendment with fly ash and sewage sludge. Environ Geol 43:513–520. https://doi.org/10.1007/s00254-002-0656-2

Warrick AW (2002) Soil physics companion. CRC Press, New York

Whalen JK, Chang C (2002) Phosphorus sorption capacities of calcareous soil receiving cattle manure application for 25 years. Commun Soil Sci Plant Anal 33(7–8):1011–1026. https://doi.org/10.1081/CSS-120003870

Xie Q, Zhang X, Wang D, Li J, Qin Y, Chen Y (2001) Research on the effect of sludge fertilizer on farmland and the safety of heavy metals in a karst area. Environ Geol 41:352–357. https://doi.org/10.1007/s002540100393

Zafarzadeh A, Mehdinejad MH, Amanidaz N (2015) Accumulation of heavy metals in agricultural soil irrigated by sewage sludge and industrial effluent (Case study: Agh ghallah industrial estate). J Mazandaran Univ Med Sci 25(121):217–226

Zare M, Chorom M, Moallemi N (2015) Effect of treated urban sewage sludge on soil essential nutrients, soil chemical properties and physiological properties of olive tree. J Agric Eng (Soil Sci Agric Mech) 37(2):1–15

Zazouli M, Ala A (2019) Review of co-composting municipal sewage sludge with a variety of biological waste. J Health 10(1):19–33

Municipal Waste Management: Current Research and Future Challenges

Arash Hemati, Khatereh Nobaharan, Arian Amirifar, Ebrahim Moghiseh, and Behnam Asgari Lajayer

1 Introduction

Annually, millions of tons of municipal and agricultural organic wastes are produced due to human activities and the lack of principled management of these materials causes severe environmental damage such as greenhouse gas production, accumulation of municipal waste and pollution of water and soil (Srivastava et al. 2015). Proper control of wastes is one of the determinant factors in cleanliness of the city and maintaining people's health (Collivignarelli et al. 2004). Waste materials are all wastes resulting from human and animal activities that are usually solid and useless. During the last few decades, planning in the field of waste management has attracted a lot of attention. Firstly, most of the efforts in the field of waste management planning have been due to the amount of production and less attention has been paid to waste compounds. However, in the last few years, with the knowledge of these materials management has also changed (Department for Environment, Food and Rural Affairs 2007). Large amounts of municipal solid waste are produced in modern society and its disposal has serious environmental, social and economic

A. Hemati
Department of Soil Science, Faculty of Agriculture, University of Tabriz, Tabriz, Iran

K. Nobaharan
12/55-57 Doncaster east Road, VIC, Mitcham, Melbourne, Australia

A. Amirifar
Department of Biology, Islamic Azad University of Urmia, Urmia, Iran

E. Moghiseh
Nuclear Agriculture Research School, Nuclear Science and Technology Research Institute, Karaj, Iran

B. Asgari Lajayer (✉)
Health and Environment Research Center, Tabriz University of Medical Sciences, Tabriz, Iran

V. D. Rajput et al. (eds.), *Sustainable Management and Utilization of Sewage Sludge*, https://doi.org/10.1007/978-3-030-85226-9_16

problems. The rapid expansion of the cities, the widespread migration of population from rural areas to urban centers has caused a significant increase in per capita production of municipal solid waste (Srivastava et al. 2015). Increasing solid waste production, limited landfill space and stricter environmental regulations for landfill and incineration sites have increased waste disposal costs, especially in developing countries. Therefore, municipalities and local governments are under heavy pressure to find sustainable and cost effective municipal solid waste management practices (Saha et al. 2010; Sharholy et al. 2008). Composting of these materials is one of the proper solutions (Tosun et al. 2008).

2 Solid Waste Production in Different Societies

The amount and composition of municipal waste varies significantly from city to city, which depends on people's incomes and their habits. Countries with high levels of income and wealth and their people are affluent, produce more waste than the people of poor countries. The amount of waste production in poor countries is about 0.4 to 0.6 kg per person. However, this rate for developed countries is about 0.7 to 1.8 kg per person. In poor and low-income countries, the percentage of degradable waste production is higher than in developed countries, so that this rate is between 40% and 85% in poor countries and between 20% and 50% in developed countries. Most of the percentage of waste in developed countries include paper and plastic (Giusti 2009; Abdel-Shafy and Mansour 2018).

3 Production Resources and Solid Waste

In general, solid waste sources depend on land use and zoning. Although many categories can be provided, the following categories have been found to be more useful (Giusti 2009; Srivastava et al. 2015):

1. Household
2. Commercial
3. Urban
4. Industrial
5. Open zones
6. Treatment plant
7. Agriculture

3.1 Types of Solid Waste

The term solid waste is general and includes all sources and types of categories and compounds with different properties. Some waste that is discarded, may be precious to others or in another situation. These materials include residues of animal waste, fruits and vegetables that are resulted from displacement, preparation, cooking and eating. These waste materials are called garbage. The most important characteristic of these substances is that they are highly degradable. Their degradability is often associated with a nasty odor. In most regions, degradability of these materials is quite effective in designing and functioning waste collection systems. In addition to the extras of foodstuffs in residential areas, a considerable amount of this kind of waste is also produced in restaurants, deli, and large institutions such as prisons, hospitals, fruit and vegetable fields and grocery stores (Hoornweg and Laura 1999).

3.1.1 Garbage

These materials are usually a result of domestic activity, commercial institutions and etc. are divided in two parts: flammable and non-flammable. Flammable materials include paper, cardboard, plastic, textiles, rubber, leather, wood and gardening additions, non-flammable materials such as glass, aluminum cans, tin cans, ferrous and non-ferrous metals and dirt (Suchitra 2007).

3.1.2 Ashes and Residues

The remaining materials from burning wood, coal, charcoal and other waste materials from burning in houses, shops, institutions, industries and urban facilities are ashes or residual materials. The residual materials of power plants usually include powdery soft materials, ash, weirs and small amounts of semi burned materials, glass, porcelain and metals (Assi et al. 2020).

3.1.3 Construction Waste

These materials include materials from destructed buildings, repairing them or residues from construction. The amount of these materials can hardly be estimated, but includes materials such as soil, stone, concrete, brick, wood, iron, pipes and other construction equipment (Abdel-Shafy and Mansour 2018).

3.1.4 Special Wastes

These materials include waste from sweeping streets, garbage, dead animals and scrap from vehicles. Because it is not possible to predict where the dead animal or the car will be found, these materials will come from non-specific and extensive sources (Abdel-Shafy and Mansour 2018).

3.1.5 Waste Treatment Plants

Solid and semi-solid wastes obtained from water and wastewater treatment plants and industries are in this classification. The properties of these materials depend entirely on the treatment processes (Abdel-Shafy and Mansour 2018).

3.1.6 Agricultural Waste

Waste from agricultural activities, milk production, barns, slaughterhouses, pastures and etc. are generally agricultural wastes (Hargreaves et al. 2008).

3.1.7 Hazardous Waste

Chemical, biological, flammable, explosive and radioactive wastes that can immediately or overtime pose risks to human, animal and plant life. Usually, these materials are liquid, but they are also found in the form of gas, solids and sludge (Johnson 1999).

4 Factors Affecting Waste Recognition

The identification of waste is done by determining the following characteristics (Obe et al. 2017; Abdel-Shafy and Mansour 2018):

4.1 Garbage Compounds

For all disposal methods such as burial, incineration or compost, understanding waste compounds is essential. So that every once in a while, the waste should be physically and chemically decomposed and informed of changes in their content. Even researching on different types of waste and their compounds can obtain useful information about the amount and pattern of people's consumption, which is one of the most important foundations in the economy. Therefore, collecting waste analysis statistics has a great value (Abdel-Shafy and Mansour 2018).

4.2 Garbage Grading

Grading means determining the magnitude and small size of waste grains that are normally carried out in most parts of the world. Therefore, they pass the waste through sieves with diameters of 8, 40 and 120 mm. Waste that passes through 8 mm holes is called fine waste, which mostly includes shards of glass, ash, sand, breadcrumbs, foodstuffs and such materials and due to the lack of air in the holes and pores, burning and composing them causes problems. But composing them is better than burning because lots of energy is going to be needed for burning them. Waste that passes through holes 40 mm in diameter is called medium waste, which includes kitchen residues, foods, vegetables, paper and etc. that are suitable for burning and composing. This type of waste has a balanced state compared to the population according to the statistics of the world's major cities and almost the amount remains the same during the year with a little difference. The part of the waste that passes through holes with a diameter of 120 mm and the part that cannot be scaled are plastic, cardboard, paper, wood, stone and metals. Bunch of these materials can be both burned and composted such as wood, thick cardboards, paper and some materials such as rubber and plastic materials are also suitable for burning. The other category is neither compostable nor burnable, including glass, stone, tiles, clay fragments, metals etc. Iron and ferrous metals can be separated by magnets, but some non-ferrous and diamagnetic metals cannot be easily separated (International Energy Agency 2014; Abdel-Shafy and Mansour 2018).

4.3 Water Content of Garbage

The amount of waste water is obtained from compostable organic materials. Thus, they heat a certain amount of waste at 100–150 °C resulting waste water to vaporize. The difference between initial weight and dry weight indicates the amount of waste water is generally higher in summer due to the consumption of vegetables and fruits than in winter (Mor et al. 2006).

4.4 Organic and Mineral Waste

Ordinary waste is generally made up of water, organic and mineral materials. To obtain the adequate amount of minerals and organic compounds, it is enough to burn a certain amount of waste at high temperatures so that it is weighed before and after burning. The initial and final weight difference will determine the amount of water and organic materials, and the number obtained from weighing the remaining materials or ashes almost indicates the amount of minerals in the waste materials. In the case of more accurate results, it is necessary to take glass shards, clay and stone out of the sample. Also, for determining how much organic matter is burned and annihilated, it should be noted that the organic matter in the waste may remain in the ash and exit at higher temperatures (about 1000 °C). On the other hand, some minerals such as carbonates and bicarbonates escape in high temperatures. The most suitable temperatures for incineration of waste is to determine the amount of organic and minerals materials between 600 and 750 °C. The ratio of these three compositions is called water, organic materials and minerals as the basis of chemical waste tests, also called waste base numbers. The base numbers of waste could be measured monthly and the curve of their changes should be determined and recorded in different months of the year (Sen and Chandra 2007; Assi et al. 2020).

4.5 *Waste Latent Heat*

The latent heat of waste depends on the amount of water, minerals and organics in the waste. The less waste water content, the more effective the heat is, so that in winter the waste water is less and the more effective the heat is. Also, the less residual garbage ash, the higher is the latent heat of waste. The amount of effective heat of dry organic materials is very different. Materials such as paper, wood, cloth etc. have heat capacity of 4500–5000 kCal and materials like plastic have heat capacity of 10,000 kCal/kg. To calculate heat capacity of waste, it is enough to calculate the heat from fuel (HO) with the help of a calorimeter and reduce the amount of heat used to vaporize the water in the waste (HU) (Xu et al. 2019).

5 Percentage of Elements in the Composition of Waste Materials

Understanding the elements and properties of solid wastes brings extensive information about proper management of solid wastes. Table 1 contains comprehensive information about the percentage of elements in different solid waste compounds (Abdel-Shafy and Mansour 2018).

Table 1 Percentage of elements in solid waste compounds

	Sulfur	Nitrogen	Oxygen	Hydrogen	Carbon
Food waste	0.4	2.6	37.6	6.4	48
Fruit waste	0.2	1.4	39.5	6.2	48.5
Meat waste	0.2	1.2	24.7	9.4	59.6
Paper	0.2	0.3	44.3	5.8	43.4
Garden waste	0.1	0.1	42.3	6.4	50.1

6 Classifications in the Field of Waste

Old information is not suitable guide for new planning for waste management because of people's lifestyle and industrialization, as well as various laws and regulations, including environmental laws, are constantly changing. Therefore, the first action is classification of different wastes in terms of their physical and mechanical properties. The classifications that exist today are either based on the production flow of waste materials such as paper, plastic, metals etc. or based on organic and non-organic compounds, in this case we are facing two smaller classifications, corrupting and non-corrupting materials. The specificity of this division is that it demonstrates the decomposition of materials or its compatibility potential, but does not provide comprehensive information about the shape of materials or the mechanical properties of materials. Another classification is expressed by Kolsch, which is defined based on the size and dimensions of waste particles. The characteristics of this division are that it expresses more comprehensive information about the physical properties of materials and the weakness of this classification is that firstly, its use needs a lot of awareness and secondly, this classification does not contain comprehensive information about degradability of materials (Sen and Chandra 2007; Abdel-Shafy and Mansour 2018; Assi et al. 2020).

The following information about the classification of waste materials is required (Sen and Chandra 2007; Abdel-Shafy and Mansour 2018; Assi et al. 2020):

1. Ratio of size and weight of different components of each group of waste materials
2. Mechanical properties of materials such as compatibility and tensile and shear stresses
3. Compatibility and finally the potential of deformation of materials, especially at the time of burial
4. Measurement of material degradation potential for organic and non-organic waste materials (Fig. 1)

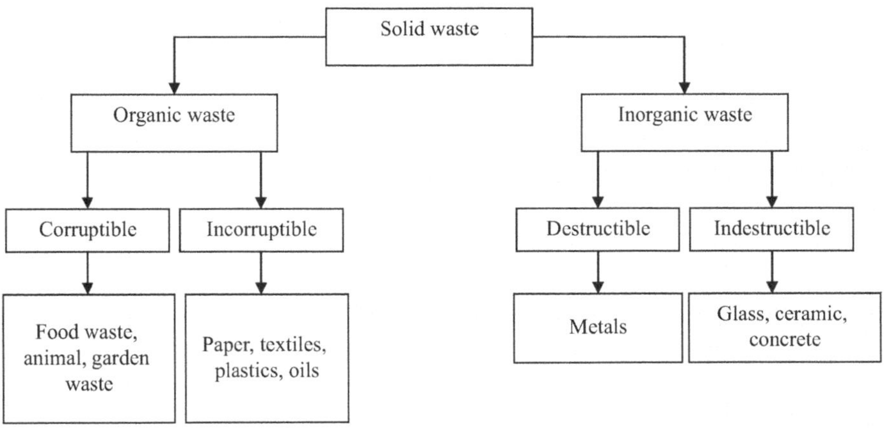

Fig. 1 Classification of wastes in organic and inorganic terms

7 Solid Waste Disposal

Municipal solid waste management systems include production elements, on-site production and storage of elements, collection, transportation, processing recycling and disposal of materials. Ways of behaving with production, storage, collection, transportation, processing and recycling elements depend on the type of disposal method. Waste disposal means cleaning waste from human environment or converting it into substances that no longer have the specificity of waste. This stage of management is very important environmentally, because using appropriate methods for waste disposal prevents numerous problems and prevents all kind of pollution for a long time. Disposal is seventh element in Municipal solid waste management system, common methods of disposal include: 1—hygienic landfill 2—incineration 3—composting (Datta 1997; Ghose et al. 2006).

Also there are other methods such as land rehabilitation, grinding food wastes and feeding livestock for municipal waste disposal, which have not yet been public due to their limited use under certain conditions. Compost and incineration, in addition to Municipal waste disposal methods, are also mentioned as methods of producing materials and energy (Ghose et al. 2006).

8 Principles of Choosing the Method of Disposal

The disposal methods used in a city depend on the construction of the urban system, because it affects the amount and type of waste produced. The quantity and quality of waste produced and disposed varies according to the industrial, civil or commercial community. Geographical location, climate, living standards, population density, main sources of income, people's habits, different seasons of the year and the number of collecting times, all affect the amount and type of waste produced. Basically, in a society, daily, weekly, monthly and yearly changes in the amount and type of waste produced are observed, depending on economic, cultural and other changes (Datta 1997; Ghose et al. 2006).

In general, three main factors are considered in the selection of disposal method:

1. Minimizing environmental hazards
2. Compatibility of the selected method with local conditions
3. System flexibility

9 Comparison and Investigation of Effective Factors in Recycling and Disposal of Domestic Solid Wastes

Domestic waste has more moisture and organic matter compared to industrial wastes in industrial countries. Moisture content and organic materials in wastes of industrial countries are about 30% and domestic waste is about 70%. Energy production from municipal waste is not recommended for cities of the country due to low thermal value of materials and high moisture content. In addition, the lack of imported design of incinerators based on the quality of municipal solid wastes in the country and the high costs of purchasing and maintaining urban incinerators are other factors that make the use of incinerators to produce energy from municipal wastes of Iran with numerous technical, navigational, economic and environmental problems. In addition to the production of these gasses, rain or groundwater that are also in the landfill sites when decomposing and crushing waste materials, a variety of gasses (such as carbon dioxide, nitrogen, methane, hydrogen and hydrogen sulfide) are removed from wastes derived from houses. The concentration of these gasses will depend on factors such as the type of waste, moisture content and activity of bacteria or other microorganisms that degrade the waste. Intrusion of effluents [produced by chemical and biological activities]. These effluents can be contaminated sources of rivers and other water resources due to their high organic materials and toxic metals. Physical degradation of solid wastes in cities of Iran shows that organic fertilizer production and plastic and paper separation can be considered as solid waste management programs. In other words, in terms of percentage of waste constituents, organic fertilizer production is at the top of programs (Manser and Keeling 1996; Mehta 2013; Suthar 2010).

10 Recycling of Waste Materials

During the 1970s, many countries used sophisticated processing equipment to separate special parts of the waste flow, which required a large investment in the construction of such equipment. In fact, in the design of these equipment and machinery, other industries such as the mining industry have been modeled, these machines have been gradually modified and used for mechanical separation of heterogeneous compounds of waste materials. However, an overview of these primary technologies will lead to a better understand of the processing equipment of the waste materials. Studies of preliminary models show that logical and technical processing theories can only be used according to economic issues. The aim of each recycling process is to reuse the materials in the waste to reduce the amount of waste produced. In practice, recyclable materials must always be separated in the source of production, because not only there are no equipment for the complete processing of waste materials, but separation at the source will increase the efficiency of the material processing equipment. In the following, some methods and equipment for processing municipal solid wastes are discussed (Manser and Keeling 1996; Mehta 2013).

11 Processing Principles

Processing involves physical treatment of materials in which physical techniques are used to change the composition and specificity of materials. Two common processing technologies in solid waste management are mechanical and thermal methods. Mechanical treatments are described using mechanical rules and thermal treatments of materials using thermodynamic rules. Conversion and processing of raw materials and waste is carried out in separate units. Waste processing equipment includes units that perform consecutive conversions to the waste, during which the properties and compositions of the wastes are changed. Each of the equipment is made of different models that represent the basic structure of a processing unit (Manser and Keeling 1996). Recycling process follows the production process and its product is considered as a secondary product, resulting in reusing waste materials. Usually, the recycling process requires one or more in-processes and a variety of products are produced. The design and performance of processing equipment requires a complete understand of the principles of recycling material separation from the flow of waste materials. Depending on the collection system or separation in place, different streams of waste materials will be produced, which are the raw materials of the recycling process. The type or in other words, the components of the material flows determine the general aspects of the production equipment, which are the material processing equipment. There is usually an entrance flow and several exit flows during the separation of the waste materials. For example, separation of ferrous metals using magnetic separator creates two outflows while the material's sorting, using several sieve plates will cause several outflow currents (Ghose et al. 2006).

Nowadays, different methods are used for the sorting of waste, which are the use of sieves, mills, ballistic separators, disc sieves, zig zag air sorting, rotary air sorting, suction cap, cross current air sorting and airbed flooring. These methods are essential for separation (sorting). The advantage of sorting is that separation of materials based on their unique physical properties. Different types of separation methods include density separation, material floating, and optical separation, magnetic and electrical separation. These methods are used to separate glass, aluminum, copper and plastic. By the way price of these equipment is very high (Ghose et al. 2006; Mehta 2013).

12 Compost Production Process

The process of producing compost is the degradation of organic wastes in the presence of oxygen, which produces products including water, ammonia carbon dioxide and heat. The process of compost production is an aerobic process (Zazouli et al. 2009). On the other hand, anaerobic processes of fixation of wastes are carried out in an oxygen free environment. The final product of which includes methane gas, carbon dioxide, ammonia and other gasses (Kutzner 2000). Also both in aerobic and anaerobic processes, light weighted molecules of organic acids are also produced. In the process of compost production due to the production of high thermal energy, the decomposition rate of materials in this process is high. The aerobic process in the production of compost is a slow process and leads to the production of lots of unpleasant odor. The final produced compounds consist of three main ingredients: Humus, microbial mass and ash (Pietro and Paola 2004; Raynal et al. 1998). Humus contains lignin compounds, which are lignin compounds, containing organic compounds that are resistant to degradation, plus humic acid that is a result of degradation is one of the constituents of Humus. Microbial mass also includes dead and living microorganism inside the mass (Fels et al. 2014). Ash also contains minerals inside the compost mass that have a stable form. But the materials used to produce compost must have the following four characteristics (Vargas-García et al. 2010; Pietro and Paola 2004):

1. Energy
2. Food
3. Water
4. Suitable structure

The first three parameters are important for the activity and growth of microorganisms and the last parameter will be useful for suitable aeration and oxygen flow among microorganisms in the mass. The term aerobic and anaerobic are the dominant conditions in the production stages of compost. In certain circumstances, the mass may also have an anaerobic state. Some of the production processes of the compost have a combined state so that in early stages of compost production, the process is aerobic, but in the later stages, especially in the period of completion and maturation of mass, process is anaerobic. The compost batch production process is continuous or semi-continuous. When the temperature is the basis of process, the compost production process can be divided into thermophilic and mesophilic conditions. In mesophilic conditions, the temperature is between 40 and 25 °C but in thermophilic conditions the temperature is between 50 and 65 °C (Raynal et al. 1998; Pietro and Paola 2004; Vargas-García et al. 2010).

12.1 Properties of Raw Materials in the Production of Compost

The quality of raw materials needed in recycling processes is very important in technical, health and environmental aspects. The quality of raw material directly affects recycled products so that products should be prepared according to market standards and customer's opinions, on the other hand the economic issues. The amount of waste composition varies depending on seasons and locations. In farms and rural areas, the type of compostable materials depends on the population, type of agriculture and livestock, diet, social practices, economic and climatic conditions of the region. But in cities, In addition to the cases mentioned above, the quantity and quality of compost materials are related to hospital waste, the existence of incinerators, waste separation and the number of collection times (Vargas-García et al. 2010; Fels et al. 2014).

Precise study of the area with using the information obtained from the study of similar areas to the desired location helps to determine the quantity and quality of waste. On the other hand, in the process of preparing organic fertilizers, depending on the use of degradation in the process of compost or biomass, the quality of raw materials and products is different (Campbell et al. 1997). According to the definition of the compost, in which solid waste enters the factory without separation and in the factory, the process of separating the organic materials and non-degradable materials is done, the quality of the raw materials is not much different and it is only tried to transfer these wastes from places to factory that has more corruptible materials. In addition, the entry of industrial and hospital wastes is prevented as much as possible. However, in the processes of biocompost preparation, dry and wet waste separation operations are carried out at the source so that with minimal separation operations and using less manpower, non-degradable materials such as polyethylene garbage bags. Almost all wastes composed of plant residues, animals and microorganisms are suitable for the compost process provided that its C/N ratio and humidity are desirable. These materials include household waste (after separating metal, glass, and plastic), waste from fields and forests, leaves and grass cut from parks and gardens. Paper and sawdust are nitrogen-free, so they can only be used as additives that have a lower C/N ratio than desired. The presence of organic materials or in disposable ones causes their chemical involvement in biological operations, thus endangering the quality of final product. It is better to separate such materials from other wastes before starting the compost process. In general, suitable primary sources for the production of compost are: (1) human waste, (2) animal waste, (3) slaughterhouse waste, (4) sewage sludge, (5) food industry wastes such as vegetables, fruit peel etc., (6) household, restaurants and grocery stores (Epstein 1997).

12.2 Advantages and Limitations of Compost Production

The most important goals and advantages of compost production are classified as follows:

12.2.1 Waste Stabilization

Biological reactions stabilize the wastes during compost production. Essentially, mineralization of waste causes less contamination at the time of discharge or its use in the environment (Bernal et al. 1998).

12.2.2 Deactivation of Pathogens

Heat production, which is biologically produced during the compost production process, can bring the temperature to more than 60 °C, which is enough to deactivate pathogens over a period of 1 day. By creating such conditions, compost could be used in agricultural purposes with ease. The higher the temperature, the less time it takes to eliminate pathogens (Lopez and Foster 1985).

12.2.3 Application as a Nutrient and Soil Improvement

The main nutrients in the form of nitrogen, phosphorus and potassium are often present in wastes in complex compounds, which are difficult to consume by plants. In the process of producing compost, these nutrients are converted mineral form (PO_4^- and NO_3^-) that could be easily uptake-able for plants. The use of compost as a nutrient in soil has the advantage that food losses from the compost are very low and insignificant because the minerals in the compost are insoluble and less prone to get washed. In addition, its use improves the physical properties of the soil, resulting in the easy growth of the plant roots in the soil, resulting in better uptake of the nutrients. The use of compost can also solve the problem of land fertility to a larger extent (Ahmad et al. 2008).

12.2.4 Drying the Waste Water

Animal, human and sewage wastes have about 80 to 95% water, on the other hand, these materials cost a lot of money during collection and transport. Drying waste water at the time of compost production is an alternative method in which the heat produced by the waste itself leads to evaporation of water in waste water. But the main concerns during the production process of the compost are that the situation of produced compost in terms of pathogens or nutrients in it. Because the quality of the produced compost is significantly dependent on temperature, climate and the type of system used for the production of compost. The presence of heterogeneous wastes used in compost production causes the temperature to not to be distributed uniformly, resulting in the growth of pathogens. Another limitation of this system is economic factors. The majority of farmers still prefer chemical fertilizer consumption because in addition to being cheap, it has faster results in yield growth (Epstein 1997).

12.3 Reasons for the Acceptability of Compost Production

Solid waste disposal methods used in a city depends on how the management system works. Because this depends on the amount and the type of waste produced. In general, three main factors are considered in disposal method (Hargreaves et al. 2008):

1. Minimizing environmental hazards
2. Adaptation of selected methods with local conditions
3. System flexibility

There are three ways to minimize environmental risks (Cointreau 2006; Srivastava et al. 2015; Kaur et al. 2018):

1. Reusing and recycling as much as possible
2. Reducing the volume as much as possible
3. Using methods in which the permissible limits and standards are observed.

As mentioned above, currently the most important methods of waste disposal in the world are incineration, sanitary landfill and waste recycling. In some industrial countries, due to the lack of land and consequently its value and increase in the value and the global price of energy generating materials, supplying energy from waste is considered as a suitable solution (Cointreau 2006; Srivastava et al. 2015). In addition, if the incinerator furnace is well designed and properly used, the combustible waste disposal problem will be solved. Bacteria and insects have also been destroyed during process, and the remaining ashes and metals are less important in terms of hygienic disposal. But the waste disposal through burning is not only a costly and very expensive method, also its environmental impacts are significant despite the use of various pollution control technologies. Incinerating substances containing chlorine during the burning process causes the formation of dioxins and furans, some of them are among the most toxic materials. In our country, especially in the desert regions, due to the abundance of arid area, sufficiency of workers, fuels and desperate need of the soil for organic materials, incineration is not necessary. Burning a part of hospital waste that it is in contact with patients is inevitable. On the other hand, in most countries, sanitary landfill is considered as a suitable and inexpensive method. In this method, all waste forming materials are transported to a special place and buried in a way that does not pose any danger to the environment (Suchitra 2007). The advantages of using sanitary burial methods are (Kaur et al. 2018):

1. In areas where the land is accessible and suitable, sanitary burial is the most economic method.
2. The initial investment of this method is much less than other known methods.
3. In sanitary landfill method except toxic and dangerous waste, all types of waste can be disposed.
4. In this method, more waste can be disposed of with minimum personnel and equipment

Disadvantages of using this method are (Kaur et al. 2018):

1. Destruction of natural landscapes
2. Soil and water pollution

3. Creating dissatisfaction in residential and public areas adjacent to burial site
4. Limited useful and unused sanitary burial sites, at least for a certain period of time.
5. Dangerousness of burial sites in terms of the ability to detonate methane gas and create fire hazards.

In recycling method, some solids in municipal waste can be retrieved and used. By recovering waste, not only can resources be protected, but also the volume of waste that should be buried will be reduced. Usually, materials are separated from the composition of municipal waste that is valuable and found in waste abundantly (Giusti 2009). Materials such as plastics, glass, ferrous and non-ferrous metals can be recycled. On the other hand, since about 75% of waste in the country are organic materials, therefore, the recovery of energy from waste has high potential (Hargreaves et al. 2008; Srivastava et al. 2015).

13 Conclusion

Old information is no longer a suitable guide for new planning of waste management, because the lifestyle of people and industrialization, as well as various laws and regulations, including environmental laws, are constantly changing. Therefore, the first action is classification of different types of waste in terms of their physical and mechanical properties. Physical degradation of solid wastes show that organic fertilizer production and separation of plastic and paper can be considered as solid waste management programs. The aim of each recycling process is to reuse the disposal materials to reduce the amount of waste produced. Two common processing technologies in solid waste management are mechanical and thermal methods. Waste that must be discarded may be valuable to individuals in other situations, but pointless to the owner and must be discarded. Recycling process follows the production process and the product is considered as a secondary product, resulting in reused waste materials. Compost production and energy production from incinerators is one of the most important recycling discussions in modern society. The most important goals and advantages of compost production are classified as: fixation of wastes, inactivation of pathogens, application as foodstuff and soil improvement and sewage drying.

References

Abdel-Shafy HI, Mansour MS (2018) Solid waste issue: sources, composition, disposal, recycling, and valorization. Egypt J Pet 27(4):1275–1290

Ahmad R, Khalid A, Arshad M, Zahir ZA, Mahmood T (2008) Effect of compost enriched with N and L-tryptophan on soil and maize. Agron Sustain Dev 28(2):299–305

Assi A, Bilo F, Zanoletti A, Ponti J, Valsesia A (2020) Rita La Spina 2, Laura E. Depero 1 and Elza Bontempi. Review of the reuse possibilities concerning ash residues from thermal process in a medium-sized urban system in northern Italy. Sustainability 12:4193

Bernal MP, Paredes C, Monedero MA, Cegarra J (1998) Maturity and stabili parameters of composts prepared with a wide range of organic waste. Bioresour Technol 63:91–99

Campbell AG, Flok RL, Tripcpi R (1997) Wood ash as an amendment in municipalsludge and yard waste composting process. Compost Sci Util 5(1):62–73

Cointreau S (2006) Occupational and environmental health issues of solid waste management: special emphasis on middle and lower-income countries. The World Bank, Urban Solid Waste Management, Washington, DC. http://www.worldbank.org/urban/uswm/pubs.html

Collivignarelli C, Sorlini S, Vaccari M (2004) Solid wastes management in developing countries. CD-ROM of ISWA World Congress, Rome

Datta M (1997) Waste disposal in engineered landfills. Narosa Publishing House

Department for Environment, Food and Rural Affairs (2007) Incineration of municipal solid waste, waste management technology brief, the new technologies work steam of the defra Waste Implementation Programme

Epstein E (1997) The science of composting. Technomic Publishing Company, Lancaster

Fels LE, Lemee L, Ambles A, Hafidi M (2014) Identification and biotransformation of lignin compounds during cocomposting of sewage sludge-palm tree waste using pyrolysis-GC/MS. Int Biodeterior Biodegr 92:26–35

Ghose MK, Dikshit AK, Sharma SK (2006) A GIS based transportation model for solid waste disposal—a case study on Asansol municipality. Waste Manag 26(11):1287–1293

Giusti L (2009) A review of waste management practices and their impact on human health. Waste Manag 29(8):2227–2239

Hargreaves JC, Adl MS, Warman PR (2008) A review of the use of composted municipal soild waste in agriculture. Agric Ecosyst Environ 123:1–14

Hoornweg D, Laura T (1999) What a waste: solid waste management in Asia. Working Paper Series Nr. 1

Johnson BL (1999) A review of the effects of hazardous waste on reproductive health. Am J Obstet Gynecol 181(1):S12–S16

Kaur R, Kaur K, Rajinder Kaur R (2018) Menstrual hygiene, management, and waste disposal: practices and challenges faced by girls/women of developing countries. J Environ Public Health 1730964. doi:https://doi.org/10.1155/2018/1730964

Kutzner HJ (2000) Microbiology of composting. In: Klein J, Winter J (eds) Biotechnology, environmental processes III e solid waste and waste gas treatment, preparation of drinking water, vol 11c, 2nd edn. Wiley-VCH, Weinheim

Lopez-Real J, Foster M (1985) Plant pathogen survival during the composting of agricultural wastes. Composting of agriculture and other wastes. Elsevier Applied Science Publishers, London

Manser AGR, Keeling AA (1996) Processing and recycling municipal waste. CRC, Boca Raton, FL

Mehta AM (2013) Waste management: threats and opportunities. In: Proceedings technical seminar on "plastic recycling & waste management" Varanasi, 22 Feb 2013

Mor S, Ravindra K, Dahiya RP, Chandra A (2006) Leachate characterization and assessment of groundwater pollution near municipal solid waste landfill site. Environ Monit Assess 118 (1–3):435–456

Obe RKD, De Brito J, Lynn CJ, Silva RV (2017) Sustainable construction materials: municipal incinerated bottom ash (1st edn). Woodhead Publishing, Sawston. ISBN 9780081009970

Pietro M, Paola C (2004) Thermal analysis for the evaluation of the organic matter evolution during municipal solid waste aerobic composting process. Thermochim Acta 413:209–214

Raynal J, Delgenès JP, Moletta R (1998) Two-phase anaerobic digestion of solid wastes by a multiple liquefaction reactors process. Bioresour Technol 65:97–103

Saha JK, Panwar N, Singh MV (2010) An assessment of municipal solid waste compost quality produced in different cities of India in the perspective of developing quality control indices. Waste Manag 30:192–201

Sen B, Chandra TS (2007) Chemolytic and solid-state spectroscopic evaluation of organic matter transformation during vermicomposting of sugar industry wastes. Bioresour Technol 98 (8):1680–1683

Sharholy M, Ahmad K, Mahmood G, Trivedi RC (2008) Municipal solid waste management in Indian cities-a review. Waste Manag 28:459–467

Srivastava V, Ismail SA, Singh P (2015) Urban solid waste management in the developing world with emphasis on India: challenges and opportunities. Rev Environ Sci Biotechnol 14:317–337

Suchitra M (2007) Outside: burnt or buried, garbage needs land. Down Earth 15:22–24

Suthar S (2010) Recycling of agro-industrial sludge through vermitechnology. Ecol Eng 36:703–712

Tosun I, Gonullu MT, Arslankaya E, Gunay A (2008) Co-composting kinetics of rose processing waste with OFMSW. Bioresour Technol 99:6143–6149

Urban Development Se International Energy Agency, IEA BIOENERGY TASK 37. 2014. Biowaste and sewage sludge recovery: separate digestion, common gas upgrading and heat supply, Case Study. Available from: http://www.iea-biogas.net/country-reports.html

Vargas-García MC, Suárez-Estrella F, López MJ, Moreno J (2010) Microbial population dynamics and enzyme activities in composting processes with different starting materials. Waste Manag 30:771–778

Xu H, Lin WY, Dal Magro F, Li T, Py X, Romagnoli A (2019) Towards higher energy efficiency in future waste-to-energy plants with novel latent heat storage-based thermal buffer system. Renew Sust Energ Rev 112:324–337

Zazouli M, Bagheri Ardebilian M, Ghahramani E, Ghorbanian AM (2009) Principles of compost production technology, vol 270. Khaniran, Tehran

Management of Sewage Sludge for Environmental Sustainability

Deeksha Krishna, Hirdesh Kumar Sachan, and Hanuman Singh Jatav

1 Introduction

Environmental Pollution is a world-wide problem. The world's attention is currently focused mostly on worries about harmful changes to the physical, biochemical, and microbiological qualities of air, water, and soils, which influence human, animal, and plant life (Wong 2012; Ukaogo et al. 2020). Sludge from wastewater treatment plants has been described as a potential sustainable energy source and material recovery (Tyagi and Lo 2013; Bora et al. 2020). Sewage sludge management is important for avoiding emissions and mitigating negative effects on the atmosphere, human health, and long-term development. Cities generate hundreds of tons of waste annually, the bulk of which is sewage sludge, as the global population expands, and urbanization occurs. In lower-middle-income nations and low-income developing nations, respectively, the proportion of the population living in urban areas is expected to expand to 59% and 50% by 2050 (UN 2019). Rapid industrialization and economic expansion over the previous three decades have created extreme environmental scenarios such as deteriorating air quality, river water contamination, illegal dumping, and unhygienic landfilling of industrial and municipal solid wastes (Ferronato and Torretta 2019).

Sludge from sewage treatment plants is a blend of solid and water that is an unavoidable part of the process. These wastes can be either be left untreated or can be collected, treated, and finally used directly, indirectly, or not at all, depending on the local context. Around the world, over 80% of wastewater is released into the environment untreated (Opec 2018). Not only does failing to manage effluents damage the ecosystem, weather, atmosphere, and human health, but it also

D. Krishna (✉) · H. K. Sachan
Fiji National University, Suva, Fiji

H. S. Jatav
Department of Soil Science and Agriculture Chemistry, Sri Karan Narendra Agriculture University, Jobner, Rajasthan, India

V. D. Rajput et al. (eds.), *Sustainable Management and Utilization of Sewage Sludge*, https://doi.org/10.1007/978-3-030-85226-9_17

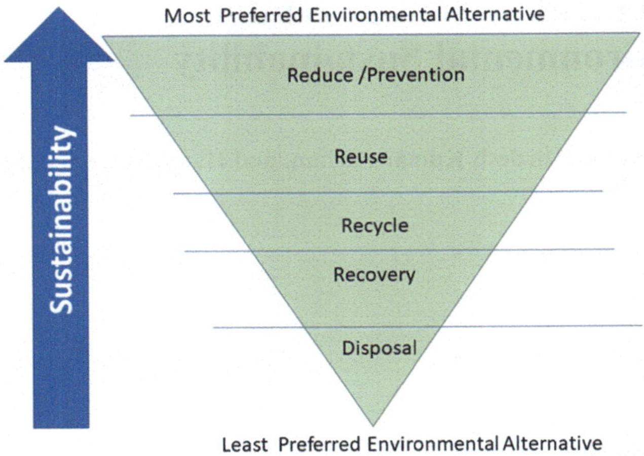

Fig. 1 The waste management hierarchy

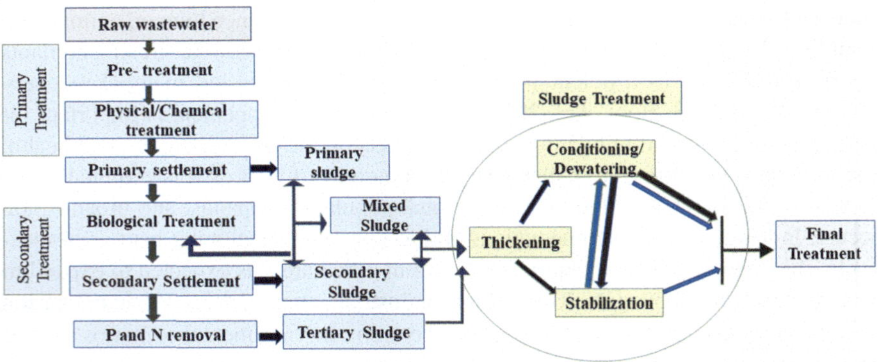

Fig. 2 Schematic flowchart of a typical wastewater treatment and types of sludge

constitutes a costly waste. Waste management hierarchy indicates diagram in the form of pyramid to reduce and manage waste (UNEP 2013) (Fig. 1). The hierarchy depicts the movement of a product through distinct steps of wastewater treatment and represents the final step of each product's life cycle waste (UNEP 2013). Waste hierarchy is used to prevent emissions GHG, reduces pollutants, save energy, conserves resources.

Figure 2 depicts a simplified schematic representation of the major processes of sewage treatment. Prescreening to remove grit, plastic, and other debris; primary sedimentation of settleable solids; and secondary aerobic biological treatment of settled sewage, or primary effluent, utilizing aerated reactors (activated sludge process) or biological filters (percolating filter). Co-settled with the primary sludge is biological sludge from these aerobic stages, as well as tertiary treatment sludge that may be required for additional effluent cleansing, such as nitrogen and/or phosphorus removal operations. It has therefore not been classified a waste but has become an alternative for the recycling of energy, matter and two into the environment.

Sewage sludge from treatment plants can be reused in a variety of ways since it contains nutrients and organic materials. As a result, all conceivable solutions must meet the criteria for eco-innovation. As a result, they would conserve natural resources and minimize the release of harmful chemicals, resulting in significant reductions in adverse environmental effects (Rorat and Kacprzak 2017; Rorat et al. 2019). When sewage sludge is recycled on land, a number of concerns develop, including nutrient leakage, loss of soil biodiversity, and greenhouse gas emissions. After sludge and other bio-waste are recycled onto agricultural fields, both methane and nitrous oxide are released as active greenhouse gases. It is critical to establish procedures and methods that will enable them to reduce their unregulated production and emissions during treatment and recycling. When comparing the global warming potentials (GWPs) of various treatment, recycling, and disposal paths, it is often established that successful treatment and recycling to agricultural land have a lower GWP than other techniques. In some localized circumstances, such as land use or sludge composition, the cumulative environmental consequences, be it in terms of emissions or in conjunction with other ecological issues, show that non-agricultural pathways may be more favourable (Pradel and Reverdy 2012).

Sewage sludge contains a variety of organic compounds, macro- and micronutrients, non-essential trace metals, organic micro-contaminants, and micro-organisms (Ozcan et al. 2013). It is suitable for use as a fertilizer in agriculture because it has high organic (OM) and plant (N, P) nutrients. With the appropriate uses, the beneficial OM, N, and P can be recycled (Haynes et al. 2009). It decomposes anaerobically if not properly treated, resulting in air pollution. Zn, Cu, Ni, Pb, Cd, Cr, and Hg are some of the heavy metals present in sewage sludge. Persistent organic contaminants (POPs) include polycyclic aromatic hydrocarbons (PAHs) and polychlorinated biphenyls (PCBs) (Zennegg et al. 2013). Industrial wastewater and precipitation runoff add these pollutants into the combined drainage system. They are the most important barrier to the natural usage of water sludge. A variety of methods have been used to convert these hazardous compounds into non-toxic forms or to decrease their potential for release into the environment (Rizzardini and Goi 2014; Fernández et al. 2009). On a global scale, the proper disposal of sewage sludge seems to have become a serious environmental concern (Leng et al. 2014; Li et al. 2014). Effluent consumers are primarily concerned with the visual benefits of effluent, which include increased agricultural production, low-cost water supplies, effluent disposal that is effective, and a source of nutrients and organic matter. They are, however, unaware of the adverse consequences, which include soil and agricultural contamination, as well as health and safety risks. According to studies, prolonged irrigation with this sewage effluent contaminates the soil and crops to the point where they become harmful to plants and deteriorate the soil (Khan et al. 2008). Heavy metals accumulate in the crop's many food components, causing a variety of clinical problems in both animals and humans worldwide. Heavy metals have a wide range of harmful effects, causing diseases in many important parts in the body (Engwa et al. 2019). Due to the serious threat that heavy metals in the nutrient cycle have presented to health and environment integrity, contamination problem must be addressed immediately and decisively. Nonetheless, the relatively low lignin and cellulose content of sewage sludge facilitates decomposition of the organic matter. As a result, sewage sludge readily degrades and can cause a sudden increase in soil concentrations of nitrate and pollutants, if not pretreated. By contrast, the contaminants detected in wastewater sludge are classified as organic (polycyclic aromatic hydrocarbons, PCBs, adsorbable organo halogenes, pesticides (AOX), surfactants, hormonal, and medications); inorganic (metals and nanoparticles); and harmful living organisms, including microorganisms, protozoa, viruses, and other pathogens (Fijalkowski et al. 2017). The health and environmental threats posed by the existence of such sludge pollutants and treatment options are discussed to suggest suitable sewage sludge reuse. This chapter covers these topics in detail, highlighting the threats to humanity and the ecology posed by the occurrence of such sludge pollutants and treatment methods, which will be analyzed to recommend an appropriate sewage sludge reuse method.

2 Sewage Sludge Production Worldwide

The world's population is growing and concentrated in cities. This is particularly evident in developing countries, where an estimated 2.1 billion people will live in urban areas by 2030 (UN 2012; Javier Mateo-Sagasta et al. 2015; Kumar et al. 2017). While removal of suspended solids from waste water and conversion of organic soluble compound to bacterial biomass, sewage treating plants produce a semi-solid substance known as sewage sludge (Javier Mateo-Sagasta et al. 2015; Ning-Yi Wang et al. 2013; Kumar et al. 2017). The highest greenhouse gas emissions are produced by land filling (296.9 kg CO_2/t sludge), followed by mono-incineration (232.2 kg CO_2/t sludge) and carbonization (146.1 kg CO_2/t sludge) (Kumar et al. 2017). It has been demonstrated that municipal solid waste co-incineration reduces greenhouse gas emissions (about 15.4 kg CO_2 equivalent/t sludge reduction). As a result, with enhanced energy generation, distribution, and incorporation processes, carbonization may be an extremely beneficial method of treating sewage sludge. The EU-27's annual sludge production expected to increase about 1.5 million tons of DS from 2010 to 2020 (Kumar et al. 2017). Production and disposal of sewage sludge for selected countries has been depicted in Tables 1 and 2. Although the legislation of Europe and US needs nations to reform management policies for wastewater sludge, there is still a research on the treatment options of many poor and developing countries. In 2012, for example, nearly 30% of produced loam was dumped directly into the Pacific Ocean by the Federal States of Micronesia (Rouse 2013). This year, the world's population has exceeded 7.5 billion by the year 2050. The population of urban areas will almost double over the next 10 years, from

Table 1 Wastewater and sludge production in selected countries worldwide

Country	Municipal wastewater (10^9 m^3/year)	Sludge production Sewage sludge - total (thousand Mg DM/year)	Year
Austria	NA	237.94	2016
Finland	0.40	115.70	2017
France	NA	1174.00	2017
Germany	NA	1794.44	2016
Greece	NA	119.77	2016
Hungary	NA	264.71	2017
Netherlands	1.93	351.00	2010
Poland	2.17	540.30	2013
Portugal	NA	119.17	2016
Romania	0.98	283.34	2017
Sweden	1.00	204.30	2016
China	48.51	6250	2013
Japan	NA	2260	2015
USA	60.41[a]	13,840[b]	–

Municipal wastewater and sewage sludge production (FAO-AQUASTAT 2020; EUROSTAT 2020; Rorat et al. 2019; Seiple et al. 2017; Takaoka et al. 2018; Yang et al. 2015). *NA* data not available

[a]Data for 2008

[b]Data for 2017

Table 2 Sewage sludge disposal (thousand tonnes)in selected countries worldwide

Country	Year	Total disposal	Agricultural use	Composting	Landfill
Belgium	2017	153.09	24.93	n.a.	0
Bulgaria	2017	45.3	22.5	3.8	6.8
Czechia	2018	228.22	108.31	78.01	19.56
Ireland	2017	58.773	46.487	10.065	0.087
France	2017	809	299	318	13
Croatia	2018	3.954	1.548	0.153	0.776
Latvia	2018	24.128	4.288	8.842	0.071
Lithuania	2018	38.684	17.506	15.892	3.402
Luxembourg	2017	8.618	1.138	4.557	n.a.
Hungary	2018	231.349	34.08	166.948	1.513
Poland	2018	583.07	118.33	25.196	10.638
Romania	2018	247.76	46.39	4.15	128.31
Slovenia	2018	38	0	0.6	0.3
Slovakia	2018	55.93	0	25.45	11.27
Norway	2018	111.7	65.4	26.7	8.7

Sewage sludge disposal from urban WWTPs (EUROSTAT 2020)

3.4 to 6.4 billion in 2012, specifically in developing nations, doubling the slum population rapidly, from 1 billion to 1.4 billion in less than 10 years (Matiasi 2012).

2.1 Characters of Sewage Sludge

The source and volume of flushing water (public and private toilets), storage technique (on-site and off-site), and additional treatment phase, such as digestion, all have an impact on the sludge's characteristics (Tables 3 and 4). Fresh sludge contains a high concentration of microorganisms, a high water content, a high biochemical oxygen demand (BOD), and is typically rotten and odorous if not treated. Sludge, on the other hand, contains important nutrients to plants (such as phosphorus and nitrogen) and can be a highly effective fertilizer (Kumar et al. 2017). After stabilization, the organic matter in the sludge can be used as a soil amendment to improve soil quality for plant roots, or it can be converted to energy through bio-digestion or cremation. Given the risk of dangerous chemicals (e.g., heavy metal ions, medications) gaining and accumulating in sewage sludge from manufacturers and other sources, sludge collected from on-site systems is normally regarded safe for reusing, unless consumers utilize their bathrooms for routine disposal sites (Arthurson 2008; Kumar et al. 2017).

Sludge processing and management is while the technology for creating sewage sludge vary greatly depending on the kind of sewage, the overall goal is to lower the volume of sediments as well as moisture content and, the amount of tools in future operations (Turlej and Banaś 2018). There are three stages of treating sewage sludge (Fig. 2) including primary treatment, secondary treatment, and final treatment (Wójtowicz 2013). Primary treatment: Wastewater treatment consists of a number of techniques dependent on available resources which vary in effectiveness and complexity. This stage can also be subdivided into a series of phases for the processing of sewage. The primary treatment (BODs); secondary treatment (biological that capture organic dissolved compounds missed during primary treatment; and treatment at the third stage is usual for the collection of suspended solid waste and for the reduction of the biochemical need for oxygen) (sophisticated technology to further remove contaminants or specific pollutants). The initial process phase is meant to drain as much sludge as possible, which reduces the volume of sewage sludge (Turlej and Banaś 2018).

Thickening and dewatering is by increasing the solids content of sewage sludge by thickening and dewatering operations, the material's treatability and transportability are improved. Gravity thickening, gravity belt thickeners along with dissolved air flotation, are all technologies that are frequently employed to achieve a 3–6% solids content. Centrifuges, belt filter presses, as well as sludge drying beds are used to dewater the sludge cake, resulting in a 10–30% solids content (Tchobanoglous et al. (2003). It is well established that the composition of sludge changes as a result of thickening and dewatering procedures. In wastewater, there are dissolved and particulate elements. For instance, digested sludge has a large amount

Table 3 Characteristics of municipal sewage sludge (Source: Kacprzak et al. 2017 and Kwarciak-Kozłowska 2019 modified)

Item (% dry weight)	TS (Total dry solids %)	Volatile solids (%Total dry solids)	N (% of Total dry solids)	P_2O_5 (% Total dry solids)	K_2O (% of Total dry solids)	pH	Cellulose (% Total dry solids)	Iron (not as sulfide)	Silica (SiO_2, % Total dry solids)
Untreated primary sludge (range)	2–8	60–80	1.5–4.0	0.8–2.8	0–1	5–8	8.0–15.0	2.0–4.0	15.0–20.0
Digested primary sludge (range)	6–12	30–60	1.6–6.0	1.5–4.0	0–3	6.5–7.5	8.0–15.0	3.0–8.0	10.0–20.0
Secondary sludge	0.5–2%	50–70%	2.4–5	2.8–11	0.5–0.7	6.5–8	7–9.7	nr	nr

nr not reported

Table 4 Ultimate analysis of primary and secondary sludge (Manara and Zabaniotou 2012, Liew et al. 2021 modified)

	C	H	N	O	S	Protein	Fat
Item (% dry weight)	%VM					%DM	
Primary sludge	52	7	4.5	35.5	1.5	24	18
Secondary sludge	53	6.7	6.3	33	1	34	10
Mixed sludge	51	7.4	7.1	33	1.5	30	14

of dissolved ammonium or nitrate nitrogen along with particles linked heavy metals and phosphorus (Hjorth et al. 2009). Particulates are selectively collected by thickening and dewatering the sludge, which results in an increase in the concentration of particulate-bound components. Houillon and Jolliet (2005), as well as Lederer and Rechberger (2010), considered the constituents removal by assigning each constituent in the sludge a transfer coefficient. Along with the changing sludge composition, Soda et al. (2010) incorporated fugitive emissions of CH_4 and N_2O from thickening with dewatering operations. Stabilization: Sludge stabilization occurs after thickening. It is important for future applications, with the primary objective of reducing pathogens in organic matter and thus minimizing potential risks. There are two distinct methods of stabilizing liquid sewage sludge (1) Chemical stabilization by increasing the pH to greater than 11; this eliminates the risk of microorganisms. (2) Stabilization of the biological system by either aerobic or anaerobic digestion.

Aerobic digestion: It is used for the treatment of secondary sludge from the processing of biological waste water as sludge activated or trickled filters. Anaerobic digestion can be performed in solid anaerobic digestion systems which are low (10%), medium (15–20%) or high (22–40%). Dewatering is the filter presses or centrifuges are often used for the dewatering of stabilized sludge. This is generally followed by a conditioning step to ensure appropriate dewatering, which is critical. Conditioners, synthetic organic polymers, or metal ions coagulate colloids in to sludge. This accelerates removal of water from sludge (Novak 2006). Only high efficiency of water removal enables final sewage sludge treatment and disposal technologies to be applied effectively. The wastewater sludge composition varies significantly during the treatment phase and significantly amongst wastewater systems. Raw (untreated) wastewater loads usually contain from 2.0 and 8.0% dry solids, 60 to 80% volatile solids, fatty solid (VS), protein, nitrogen, phosphorus, potassium, iron, silica, cellulose and organic acids (Tables 4 and 5) (Metcalf & Eddy in 1991). Since pathogens and other pollutants are present, sewage-to-matter disposal solutions include significant hazards through untreated sewage sludge (i.e. sludge that has not been stabilised but is still mechanically treated). In consequence, such stabilization protocols will be implemented in WWTPs. The raw sludge's properties have a significant impact on the processing technology. For instance, the pH, organic acids and alkalinities all inhibit the digestion of anaerobic substances in the body (Metcalf et al. 2013; Suleiman et al. 2017).

Table 5 Physical-chemical properties of sewage sludge in selected countries worldwide

Sampling site	Iran	Portugal	Spain	Brazil	Egypt	Japan	Turkey
pH	7.8	7.1	8.73	7.8	5.2	6.43	8.22
TN (%)	2.4	6.2	4.5	3.5	nr	nr	1.75
TP (%)	nr	5.9	1.72	1.1	nr	1.715	0.1148
TK (%)	nr	5.9	0.275	0.1	nr	nr	0.21
OM (%)	43.5	67.5	57.9	nr	nr	46.09	21.4
Description	Anaerobic digested sewage sludge	Dehydrated sludge	Anaerobic sewage sludge	Dehydrated sludge	Original sewage sludge	Dewatered anaerobically digested sewage sludge	Obtained from sewage sludge treatment facility
References	Nafez et al. (2015)	Alvarenga et al. (2015)	Walter et al. (2006)	Moretti et al. (2016)	Ashmawy et al. (2012)	Shi et al. (2013)	Tuefenkci et al. (2006)

nr not reported

2.2 *Sewage Sludge Final Treatment and Disposal*

There are critical guidelines for a long-term control and management of sewage sludge such as (a) matter recovery for agriculture use (as fertilizer) and reclamation of degraded or contaminated areas along with recovery of incineration energy and other heat involving processes like pyrolysis, quasi-pyrolysis, gasification, or co-incineration (such as cement plants). (b) Chemical energy contained in wastewater must be converted to usable energy in order to satisfy a portion of the world's demand for sources of renewable energy (Puyol et al. 2017). Numerous methods exist for the conversion of surplus sewage sludge to electricity. This technique has recently garnered a great deal of attention. It primarily enables the utilization of sewage sludge's ability without introducing contaminants into the soil through land applications. (c) Others, like dumping in landfills or at sea, are prohibited in the majority of countries but mostly developing nations continue to practice in some regions.

2.2.1 Anaerobic Digestion

Initially, anaerobic digestion was discovered as a simple pathogen-removing procedure. Over time, however, it has proven to be advanced and efficient at biogas production (Cao and Pawłowski 2012). This biogas can be utilized for a WWTP to provide 0.78 kWh of electricity per m^3 wastewater after treatment (Cano et al. 2015). Anaerobic digestion has an exceptionally high ability to meet this high demand. The so-called co digest method was proposed to integrate the additional ingredients to increase biogas production. Grosser (2018), for example, cosubstrating sewage sludge with an organic portion of municipality waste material including grease trap sludge, increased the efficiency of anaerobic digestion. Also, previously wastewater derived from discarded food, cheese whey, and olive mill, were successfully studied (Rorat et al. 2019; Maragkaki et al. 2018). Sewage sludge co-digestion, therefore, be perceived as a technique for the treatment of a number of agricultural wastes. However, it cannot be considered a final sludge disposal mechanism because it produces digested sludge (digestate), which often includes nutrients in excesses and contamination and must be treated. This has been demonstrated that digestates with a heavy concentration of plant-available nutrients (ammo) can be used in place of or in addition to inorganic fertilizers in agronomic plant cultivation (Sogn et al. 2018). However, if the product meets applicable standards, in-land use as a biofertilizer is permitted, which are typically governed by soil conservation, fertilizer, or waste legislation. In the absence of that, alternate solutions would be deliberated. Peng et al. (2018) recently proposed that older landfill leachate be extracted of nitrogen using digestate in landfill bioreactors. Digestate has also been successfully used in vermicomposting with other organic wastes such as urban wastes, sawdust, and renewable wastes (Rorat et al. 2017). Since digestate has an identical chemical composition to that of substrates used, long-term soil incorporation effects must be studied to determine their influence on the soil function (microbial cycles and biodiversity of soil). The risk of air pollution (emission of ammonium and nitrous oxide), pollution (excess nitrogen, phosphorus), and the contaminated soil (chemical/biological contamination) is the most common risk of the application of digestate in soil Nkoa (2014).

2.2.2 Composting

Composting, described by way of the biological breakdown of biowaste in the presence of oxygen, is critical for recycling and conserving a variety of major and minor nutrients of sewage sludge. In contrast, vermicomposting is an advanced, low-cost, and environmentally sustainable biotechnology that utilizes earthworms as natural bioreactors to break down organic material (Suleiman et al. 2017). Their metabolism in conjunction with microrganism reduced the volume of the products by 40%–60%, the nutrient bioavailability of plants increased, the C:N ratio decreased, and dangerous metal pollutants were decreased (Rorat et al. 2015). Even though composting may be seen as a highly advantageous and low-cost method of converting sewage to matter allowing the ecosystem's organic mineral nutrients to regenerate, it does raise some significant environmental concerns. Significant nitrogen losses and greenhouse gas (GHG) emissions have been observed as a result of nitrogenous organic matter rapidly degrading (Sánchez-Monedero et al. 2010). The adding of bulky materials such as agricultural waste and alkaline alterations like lime, zeolite, and bentonite can partially lower these effects. Recently, biochar has been recognized as a key technique for mitigating greenhouse gas, ammonia, and emissions of extractable ammonia (Malińska et al. 2014; Awasthi et al. 2016).

While some repercussions are controllable, researchers concerns are about the entry of metal ions with potential danger to enter into the soil horizon and their long-term accumulation (Fang et al. 2017). Similarly, when additional chemical substances and medications are present, as well as when certain organisms survive the process, the same type of risk exists.

2.2.3 Thermal Treatment

Both thermal procedures are energy-saving systems because they result in complete oxidation of volatile substances and residue formation (ash). The most well-known technologies include incinerating, gasifying, pyrolysing, and plasma gasification. The most suitable waste sludge disposal methods are considered in Europe and some other countries are combustion and/or incineration (EC 2008). The result is a risk to the degradation of landfills and agricultural policies. This also reduces sewage sludge considerably, mitigating microbiological hazards, reducing odors and simultaneously recovering renewable energy. This approach is implemented in the following three different ways: in-house incineration, urban solid waste co-incineration, and concrete kilns incineration. These systems are mainly due to their high energy use and output of hazardous gas (dioxins and furans) their environmental expense (Garrido-Baserba et al. 2015). Additionally, the ashes produced have ability to accumulate chemical pollutants, and be considered a concentrated pollutant. A cement substitute is an intriguing option for ash disposal. After combustion, sewage sludge retains a high concentration of calcium oxide, iron oxide, alumina and silica making it suitable for use for manufacture of materials for construction. Additionally, metals are stabilized and solidified in this state, lowering the risk (Samolada and Zabaniotou 2014). As a low-cost and highly efficient technology, Pyrolysis gains acceptance.

The method essentially converts organic material into bioenergy (oil/gas), developing by-products of so-called biochar. The pyrolyse therefore allows a significant proportion of bio-oil to be generated which can be used as a fuel or chemical product source. In general, pyrolysis and related processes of sewage sludge combustion are considered endothermic. Nonetheless, it was recently demonstrated that the need for pretreating this substrate (dehydration) categorizes it toward exothermic processes, despite the fact that this procedure remains the most costly in the scenario. Thus, optimizing the steps involved in removing water is critical for the sustainability of the process. Several studies conducted independently have revealed several benefits of using biochar in managing environment. It can be used in a variety of ways, for instance to improve soils, to increase resource production, remedy and/or protect lands and to reduce greenhouse gas potential emissions (Lehmann and Joseph 2015). Thus, biochar is a solid carbon-containing material derived by zero- or low-oxygen pyrolysis of a variety of feedstocks derived from carbon and applied to soils in order to maintain carbon levels in a sustainable manner, hence enhancing soil quality over time (Verheijen et al. 2010).

While the combustion process mitigates the microbiological risk associated with sewage sludge land application, concerns regarding chemical pollutants persist. These must be assessed on an individual basis, taking into account not only the biochar itself, not only the soil, but also the environment (Verheijen et al. 2010). However, the adding biochar to the sewage sludge has recently been recognizable as immobilizing bioavailable, PAHs and metals (Kończak and Oleszczuk 2018; Rorat et al. 2019). Frišták and Soja (2015) and Rorat et al. (2019) further acknowledged that the adding to the sewage sludge of biochar made of wood chips and garden residues, and its application as modifier of the soil, improved the amount of available phosphorus. During the sewage sludge vermicomposting, the benefit effects of adding sewage sludge biochar also were observed in significantly reducing the bio-available Cadmium as well as Zinc for earthworms (Malińska et al. 2017; Rorat et al. 2019).

3 Sewage Sludge as Pollutant and Its Pathways

Numerous countries have identified sewage sludge as a suitable substrate material for agricultural fertilization and/or zone remediation, owing to the abundance of rich resources contained inside (organic matter, available plant culture nutrients).

However, the usage of sewage sludge on farms might have a dispersal into soils that may be used to manufacture food of many undesirable constituents. These unwanted pollutants (e.g. metals, trace organic (TrOC) and pathogens) may present a risk to the environment and health (Andreoli et al. 2017). If not adequately treated, toxic chemicals in sewage sludge can also exacerbate pre-existing environmental problems and result in secondary environmental pollution and poisonings.

Our wastewater and sewage sludge reflect the industrial culture we live in based on chemicals. These contain the chemicals we use, produce, release and discharge. The composition of the Sludge, the agricultural value and contamination level can vary greatly. These criteria are different by source of wastes (farming, industrial or urban), by domestic and consumer activity, by wastewater collection (splitting and runoff or not), by regional legislation, by season and, of course, by the size and method of the deliberated WWTP.

3.1 Chemical Contamination in Sludge

Sludge contaminants posing environmental risk and their concentrations in soil following farm application are governed by their initial levels, application rate management strategies, and losses. As a result, volatile and rapidly degradable pollutants can pose a threat to the environment even at initially higher levels with regular uses (Harrison et al. 2006). Concentrations at the beginning (in soils and sludge), rates of application (accumulation effects), management measures, and losses all contribute to a determination of the environmental risks posed by sludge contaminants and their soil levels following land application. Thus, at high initial levels and regular applications, volatile and quickly degradable contaminants can constitute an environmental concern (Harrison et al. 2006). The use or disposal of biosolids from wastewater treatment plants raises issues about the environment and human health, such as the presence of potentially hazardous elements and organic contaminants (OCs), the majority of which are persistent chemical products. Concerns include the risk of trophic transfer and potential pollution of groundwater (via cultivated plants). PTE stands for 'metal trace elements,' which are sometimes referred to as 'heavy metals'. These occur naturally in soils and are currently a source of the usage of fertilizers (sludge, slurries, manure, pesticides, and inadequate waste management, which results in human-caused soil contamination. Arsenic, cadmium, chromium, copper, lead, mercury, molybdenum, nickel zinc was traditionally metals and metalloids (Zn). Environmental hazards associated with these PTEs, as well as environmental considerations and regulations for sludge land application have been widely examined. “They have been apprehended. Numerous OCs are regulated similarly to polycyclic aromatic hydrocarbons (POPs like PCBs, dioxins, and associated compositions (PCDD/F)). Additionally, there are several persistent organic contaminants. In Europe, the list and limits for metal element and organic compound concentrations that should restrict sewage sludge usage in agriculture were recently checked (EC 2000)”. “They propose to regulate LAS, di-phtalate, di(2-ethylhexyl) phthalate (DEHP), nonylphenol and nonylphenol ethoxylates (NP (E)), organic halogenation compounds (e.g. organic adsorbable halides (AOX)), HAP, PCB, and polychlorinated dibenzo-p-d. Additionally, they are proposed for use in regulating LAS”. In the 1970s, research on Sludge emissions began. The most chemicals that we utilize on a daily basis and in industry are likely to end up as biosolids or byproducts of human activity. Only highly volatile and rapidly degradable products may we exclude. Along with the OCs in urban waste, the surface runoff in artificial (concreted and paved) areas by atmospherically accumulated environmental pollutants leads accumulation of lipophilic compounds in sewage sludge. The natural tendency of lipophilic compounds to adsorb solids is strong. The sediment situation in the marine environment is close to this. The non-polar and some persistent compounds with sludge recycling can pose an environmental risk. A most specific example is that of surfactant fluorooctanosulfonic acid (PFOS) and perfluorooctanoic acid (PFNA) (which is not especially relevant when considering current uses but is concerned with real risk assessments (Schowanek et al. 2007) and more (Smith 2009).

Clarke and Smith (2011) claimed in their introduction to their new OC analysis in sludge that out of one lakh chemicals identified in the EU, The majority are almost found in wastewater treatment plant sludge. Wilson et al. (1996) suggested a series consisting three hundred sludge experiments and surveys in 1996. Eriksson et al. (2008) recently discovered that 541 OCs present in sewage sludge could be present as a result of its use in manufacturing products of personal care and pharmaceuticals (Kumar et al. 2017).

In recent times, following a review, Eriksson et al. (2008) determined that, due to their use in various construction goods, pharmaceuticals as well as personnel items, 541 OCs may be found in sewage sludge. However, in matrices like sludge and biosolids some OC concentrations are not properly described, as their analyses are contained in priority lists of regulated or identified contaminants which account for a small proportion of current OCs (Harrison et al. 2006).

Certain concentrations of OC, on the other hand, are not adequately detected in matrices such as sludge and biosolids because their studies include only controlled or contaminants listed on priority lists, which constitute a small portion of all OCs (Harrison et al. 2006). However, certain OC levels in matrices such as biosolid and sludge are not adequately characterized, as their studies are based on lists of priority controlled or known pollutants that account for small amount of current OCs (Harrison et al. 2006). Reports emphasize paucity of information on OCs, like nitrosamins, in comparison to other families of chemicals such as pesticides, polychlorinated biphenyls, and polycyclic aromatic hydrocarbons that pose significant environmental threats. Therefore, for pollutants such as metals, PCBs, and

PAH, a global perspective is therefore likely for contaminants in sludge. In addition, sludge compounds begin to decrease in concentrations. Table 6 depicts contaminations of sludge from around the world. Global inventories of PCBs and PCDD/Fs are also feasible, but not for the majority of other TrOCs. Metal concentrations in sludge is that these elements enjoy genuine scientific consensus (Table 2). In terms of concentrations, zinc remains the most abundant metal. PTE is one of the few pollutant families for which the critical issue of bioavailability can be tackled using sequential or selective extraction approaches in contaminant risk assessment. Additionally, the source of PTE in waste water is well documented and recognized, and several technical options have been developed and proposed for sludge metal removal (Babel and Del Mundo Dacera 2006).

Due to its possible toxicity to biota and ability to cause cancer in humans, PAHs (Table 7) are considered priority pollutants in the ecosystem. These are excessively lipophilic and biodegradable, which results in their accumulation in sludge, sedimentation, and soils. Industrial waste, domestic sewage, ambient rainfall, contaminant deposition in the air, and surface of the road and abrasion of tires are the principal sources of sludge polycyclic aromatic hydrocarbons (PAHs) (Bomboi and Hernandez 1991). Stefaniuk et al. (2018) suggests measuring the concentrations of freely dissolved PAHs rather than levels of total PAH in order to obtain a more reliable estimation of their potential environmental availability. At the turn of the decade, experimental works on so-called "emerging" contaminants have emerged, and a global structure has begun to evolve. These contaminants are considered emergent because sophisticated analytical methods now enable their detection in sludge or because applications in modern industry and domestic lives for certain products enhance their levels in matrices of the environment. This category of substances contains well-known pharmaceutical products, personal care products, indocrine disruptors, nanoparticles, and microplastics. Clarke and Smith (2011) identified compounds in decreasing order of significance: PFOS, poly(compounds), polyhalogens, polycyclicotin, polyethers, and antibiotics, as well as polyphenolic antimycotics synthesized from bisphenols. Sludge's fate and actions are poorly understood in the literature, and case-by-case tests of ecotoxicity should be used.

3.2 Pathogens

Sewage sludge comprises biological agents that may pose a risk to living species because of pathogenic potential of some or the potential for disruption of natural ecosystem. Sewage sludge typically includes four pathogens types viz. bacteria, fungi, viruses and parasites. Because of highly organic nature, sewage sludge may contain a wide variety of bacteria and fungi (Fijalkowski et al. 2017). Additionally, sewage sludge often contains additional species such as viruses and parasites (Frąc et al. 2014). Pathogen type and concentration are determined according to waste water treatment plant type (WWTP), waste water source including a few

Table 6 Concentrations of heavy metals in sewage sludge in a few countries throughout the world

	Total element concentrations (mg/kg of dry sludge solids; except for a few values that have units)							
Countries	Cd	Cr	Cu	Fe	Ni	Pb	Zn	References (mean values)
USA Hawai	4.4	67.6	443	n/a	32	41	1360	Van Wesenbeeck et al. (2014)
Poland-Malopolska	2.4	495	324	n/a	100	61	1478	Gondek et al. (2014)
Brazil	0.75	143.72	255.39	n/a	41.99	80.37	688.83	LeBlanc et al. (2008), Tyagi et al. (1988), Benmoussa et al. (1997), Filali-Meknassi et al. 2000). In Pathak et al. (2009)
Canada	2.3–10	66–2021	180–2300	n/a	37–179	26–465	354–640	
	1	50	460	n/a	16	51	593	LeBlanc et al. (2008)
	2.3	50.7	888	n/a	26.4	56	588	
	0.5	n/a	137	n/a	9	27	223	
UE countries	0.3–5.1	10.8–1542.2	27.3–578.1%	n/a	8.6–310	4.0–429.8	0–0.01(%)	Fijalkowski et al. (2017)
Sweden	2.10	n/a	323	n/a	17.3	45	720	Östman et al. (2017)
Iran	6.1–15.3	2782–8071	57.5–163	n/a	17.9–59.3	n/a	260–2077	Feizi et al. (2018)
Ireland	12	35	520	n/a	18	252	886	Healy et al. (2016)
India	41–54	102–8810	280–543	n/a	192–293	91–129	870–1510	Pathak et al. (2008) In Pathak et al. (2009)
Hongkong	663	112–255	n/a	n/a	44.5–622	52.5–57	1009–2823	Xiang et al. (2000), Wong and Selvam (2006). In Pathak et al. (2009)
South Africa	0.13 ± 0.003	n/a	0.74 ± 0.214	39 ± 4.125	n/a	0.099 ± 0.037	n/a	Agoro et al. (2020)
	0.1 ± 0.002	n/a	0.44 ± 0.25	40 ± 9.48	n/a	n/a	n/a	
	0.5 ± 0.001	n/a	0.53 ± 0.101	45 ± 2.335	n/a	0.065 ± 0.02	n/a	
Australia-Sydney	2.07	81	810	n/a	70	86.5	1350	Hossain et al. (2011)

Malaysia Domestic	1.57	n/a	127	n/a	19	36	841	Abu Bakar et al. (2008)
Venezuela	6.8	72.8	226.01	n/a	76.46	304.29	1474.79	García et al. (2006), Sánchez et al. (2017)
Thailand Bangkok	2.5	385	4673	n/a	156	139	2387	Babel et al. (2009)
Italy	2.70	107	628	n/a	49	129	1250	Seggiani et al. (2012)
China Shandong	4.58	150.65	181.70	n/a	140.79	150.72	1126.28	Liu et al. (2018)

n/a not available

Table 7 PAHs concentrations in sewage sludge from selected countries worldwide percentage to total PAHs in parentheses (Adapted from Chen et al. 2019)

Countries	ΣPAHs (mg/kg)	ΣCPAHs (mg/kg)	References
United Kingdom	18–94 44.80	4.5–27.60 13.20	Stevens et al. (2003), Chen et al. (2019)
Poland	2.04–36.44 11.61 ± 8.72	4.30	Baran and Oleszczuk (2003), Chen et al. (2019)
India (Delhi)	14.9–24.20 20.67 ± 4.14	9.81 ± 2.35	Khillare et al. (2018), Chen et al. (2019)
Italy (Venice)	1.26–1.44 1.35 ± 0.13	0.57–0.73 0.65 ± 0.11	Busetti et al. (2006), Chen et al. (2019)
Korea	1.30–44.90 10.4 ± 17.0	0.23–25.60 4.8 ± 10.20	Ju et al. (2009), Chen et al. (2019)
Tunisia	0.096–7.72 1.25 ± 2.45	0.005–1.37 0.21 ± 0.44	Khadhar (2010), Chen et al. (2019)
Turkey (Bursa)	1.78–19.90 6.08 ± 4.69	1.31–11.57 4.18 ± 2.77	Salihoglu et al. (2010), Chen et al. (2019)
Taiwan	0.53–1.07 0.75 ± 0.26	0.021–0.037 0.028 ± 0.007	Chen et al. (2019)

environmental variables (Romdhana et al. 2009). On the other hand, the many of these harmful microbes originated in waste of animals or human (Bloem et al. 2017).

Owing to sewage sludge with high organic matter content, the microbial flora is extremely diverse and abundant. Most of these bacteria are saprophytes and non-pathogenic, making flocs and degrading such contaminants an important contribution to the waste water treatment method (Tozzoli et al. 2017; Rorat et al. 2019). However, several of these bacteria are pathogenic. "Huang et al. (2018) identified two hundred fourty three sludge-pathogenic bacteria, *including Bacillus anthracis, Clostridium perfringens*, *Enterococcus faecalis*, and *Escherichia coli* (*Pseudomonas*, *Pseudomonas aeruginosa* and *Vibrio cholera*)". "Further, *Salmonella*, *Shigella*, *Klebsiella*, *Serratia*, *Enterobacter* and *Proteus* pathogens have been reported (Korzeniewska 2011)". "The majority of these bacteria are capable of causing a variety of diseases, including urinary tract infections (*E. coli*), pneumonia (Klebsiella and Enterobacteriaceae), blood infections (Enterobacteriaceae), and gastrointestinal infections (Klebsiella and Enterobacteriaceae) (*E. coli*, *Salmonella*). These conditions can be caused by gastrointestinal, respiratory, urinary, or bile duct contamination (Korzeniewska 2011)".

Salmonella is also one of the bacteria that has been examined the most intensively in sludge of WWTP (Krzyzanowski Jr et al. 2016). When released into environment, these bacteria can flourish, in part due to the spread of sludge on cultivated land (Krzyzanowski Jr et al. 2016; Bloem et al. 2017; Ellis et al. 2018; Rorat et al. 2019). As a result, consuming food from these lands will contaminate it. Certain plants, like tomatoes (Manios et al. 2013), (Asplund and Nurmi 1991) and lettuce (Manios et al. 2013) have been shown to contain certain bacteria in their tissue despite low salmonella concentrations in sludge (Krzyzanowski Jr et al. 2016). Antibiotics have been shown to amplify the risk of pathogenic bacteria in wastewater. As a result, the number of bacteria resistant to antibiotics grows. Due to the high concentration of bacteria in WWTP reactors, genetic material is more likely to be passed among bacteria (Turolla et al. 2018). "Galler et al. (2018) isolated three enterobacteria with multiple resistance mechanisms (ESBLs) from Austrian activated sludge: Gram-negative bacilli, methicillin-resistant Staphylococcus aureus (MRSA), and vancomycin-resistant enterococci (VRE)." Antibiotic-resistant bacteria can spread through the food chain and the environment, causing serious health dangers (Fijalkowski et al. 2017; Reinthaler et al. 2013; Tozzoli et al. 2017). Additionally, fungi are prevalent in the sewage sludge microflora (Frąc et al. 2014; Rorat et al. 2019) (Table 8).

Fungi assist breakdown a wide variety of contaminants,therefore, are crucial in wastewater treatment (Tozzoli et al. 2017). However, many are plant pathogenic. For instance, *M. circinelloides* and *G. citri-aurantii* are both ordinary plant pathogens regularly encountered and cause illnesses in fruits and egetables to have a negative impact on crop yield. Apart from this ecological and environmental/agronomic hazard, fungus may be pathogenic to humans and animals (Frąc et al. 2014).

In addition, the most frequent pathogenic viruses in some US samples of sludge have been herpes viruses, Bibby and Peccia (2013). 90% of the samples tested positive for DNA viruses (adenovirus, herpes virus, HPV, and bocavirus), while 80% of the sludge samples tested positive for RNA viruses (coronavirus, klasevirus or rotavirus). These viruses can lead to serious human and animal diseases of the respiratory and gastrointestinal system. When introduced into an area, viruses like bacteria can survive. The survival of enteric viruses on soil has been demonstrated for roughly 100 days by Bloem et al. (2017). Furthermore, investigations with sewage sludge have identified parasites like nematodes and cestodes, human and livestock pathogens causing a number of diseases (Chaoua et al. 2018). Sludge has frequently been detected in helminths (Ascaris, Trichuris, Toxocara) and is one of the most resistant to sludge treatment (Da Rocha et al. 2016). Their longevity was seen several years after biosolids were applied to the soil (Bloem et al. 2017). There are also additional protozoan parasites. In more than half of the samples, 100% of the protozoan cysts and oocysts were recorded, Correa Medeiros and Antonio Daniel (2018). There were no effects of sludge treatment on their concentration or viability. Many families containing animal and human pathogenics, including Cryptosporidium, Entamoeba, and Giardia have been described (Khouja et al. 2010; Rorat et al. 2019, Sabbahi et al. 2018) (Table 9).

4 Conclusions

Sustainable development in sewage sludge management should begin with recovery and then move on to disposal. Recent progress in water and sewage sludge technologies has led to a major increase in sewage management. This is especially true when

Table 8 Pathogens in sludge

	Pathogen										Parasites					
	Virus				Bacteria						Protozoa		Nematode	Fungi		
Name of species	Enteric viruses	Enteroviruses	Rotavirus	Adenovirus	Enteric bacteria	Salmonella	*Escherichia coli*	Shigella	Clostridium	Mycobacterium	Cryptosporidium	*Giardia Lamblia*	Eggs	Filamentous	Yeast-like	Aspergillus
Concentration	103–104 MPN/100 ml	4.8 × 102–2.1 × 103 cfu/10 g dry matter	30–26 × 104 cfu/l	5800–32,500 cfu/l	5.6 × 106 ml − 1	103 gene/ml	107–108 gene/ml	1–103 MPN/100 ml	102–104 gene/ml	107 cfu/100 ml	10–103 MPN/100 ml	103–104 MPN/100 ml	2–53 eggs/10 g dry matter)	1.2 × 103–9 × 103 colonies/ml	104–106 colonies/ml	102–103 cfu/g dry weight
References	Tchobanoglous et al. (2003)	Guzmán et al. (2007)	Bosch et al. (1986)	Williams and Hurst (1988)	Cooke (1970)	Wéry et al. (2008)	Wéry et al. (2008)	Tchobanoglous et al. (2003)	Wéry et al. (2008)	EPA (1979)	Tchobanoglous et al. (2003)	Tchobanoglous et al. (2003)	Gantzer et al. (2001)	Cooke (1970)	Cooke (1970)	Millner et al. (1977)

Table 9 Commonly detected organic compounds in sewage sludge.(modified Dubey et al. 2021)

Polycyclic aromatic hydrocarbons (PAH)	Phthalates	Polychlorinated biphenyls (PCBs)	Polychloinated dibenzo-p-dioxins and -furans (PCDD/Fs)
Concentration (no. of compounds, no. of samples) 67–370 mg/kg dw (24 compounds, $n = 14$); 19–219μg/L (13 compounds, $n = 5$); 5.2–11.57 mg/kg dw (13 compounds, n = 5)	Concentration (no. of compounds, no. of samples) DEHP-130 mg/kg dw, DBP-1094 mg/kg dw; 23.9–506.3 mg/kg dw (5 compounds, n = 3)	Concentration (no. of compounds, no. of samples) 0–5100μg/L (35 compounds, n = 9); 0.11–0.44 mg/kg dw (37 compounds, n = 14)	Concentration (no. of compounds, no. of samples) Toxic equivalency 0.73 and 7348.40 pg/g dw (compounds, n = 14), 104.0–1661 pg/g dw (17 compounds, n = 24)
Stevens et al. (2003), Barret et al. (2010), Benabdallah El-Hadj et al. (2006), Dubey et al. (2021)	Salaudeen et al. (2018), Dubey et al. (2021)	Stevens et al. (2003), Dubey et al. (2021)	Balasubramani and Rifai (2015), Lu et al. (2012), Dubey et al. (2021)

it comes to the usage of sewage sludge processing byproducts such organic material, biomass, phosphorus, nitrogen, or volatile acids. It is indeed a good product for agriculture and natural usage. Composting made from sewage sludge treatment can be used to repair and protect polluted areas and soil. Materials recovered during sewage sludge treatment are significant resources, such as recycled phosphorous, which is a scarce resource. Additionally, waste sludge should be used to cultivate energy crops to help mitigate emitting greenhouse gasses and fossil fuels dependence. If sewage sludge does not meet a country's legal criteria, thermal treatment is the most cost-effective method of disposal (incineration, co-incineration). The optimal solution for municipal sewage sludge treatment would balance environmental and economic concerns, as well as an overview of the demand for finished products, market size, and time required to implement a given solution. Every country has been obliged to comply with the same hierarchy of waste management by legislative pressure, with prevention, reusability, recycling and recovery rigorously barred by preferable approaches and disposal and trash disposal.

References

Abu Bakar R, Ramahsamy K, Ishak C (2008) Characterization of Malaysian sewage sludge and nitrogen mineralization in three soils treated with sewage sludge. Malaysian J Soil Sci 12:103–112

Agoro MA, Adeniji AO, Adefisoye MA, Okoh OO (2020) Heavy metals in wastewater and sewage sludge from selected municipal treatment plants in eastern cape province, south africa. Water 12(10):2746. https://doi.org/10.3390/w12102746

Alvarenga P, Mourinha C, Farto M, Santos T, Palma P, Sengo J et al (2015) Sewage sludge, compost and other representative organic wastes as agricultural soil amendments: benefits versus limiting factors. Waste Manag 40:44–52. https://doi.org/10.1016/j.wasman.2015.01.027

Andreoli CV, Von Sperling M, Fernandes F, Ronteltap M (2017) Sludge treatment and disposal. IWA Publishing

Arthurson V (2008) Proper sanitization of sewage sludge: a critical issue for a sustainable society. Appl Environ Microbiol 74(17):5267–5275. https://doi.org/10.1128/AEM.00438-08

Ashmawy AM, Ibrahim HS, Moniem SMA, Saleh TS (2012) Immobilization of some metals in contaminated sludge by zeolite prepared from local materials. Toxicol Environ Chem 94(9):1657–1669. https://doi.org/10.1080/02772248.2012.727819

Asplund K, Nurmi E (1991) The growth of salmonellae in tomatoes. Int J Food Microbiol 13:177–181

Awasthi MK, Wang Q, Ren X, Zhao J, Huang H, Awasthi SK, Lahori AH, Li R, Zhou L, Zhang Z (2016) Role of biochar amendment in mitigation of nitrogen loss and greenhouse gas emission during sewage sludge composting. Bioresour Technol 219:270–280

Babel S, Del Mundo Dacera D (2006) Heavy metal removal from contaminated sludge for land application: a review. Waste Manag 26:988–1004

Babel S, Sae-Tang J, Pecharaply A (2009) Anaerobic co-digestion of sewage and brewery sludge for biogas production and land application. Int J Environ Sci Technol 6(1):131–140. https://doi.org/10.1007/BF03326067

Balasubramani A, Rifai HS (2015) Polychlorinated dibenzo-p-dioxins and polychlorinated dibenzofurans in sewage sludge: congener patterns, toxic equivalency and comparison with effluent concentrations and toxic equivalency. Water Resour Manag 8:399

Baran S, Oleszczuk P (2003) The concentration of polycyclic aromatic hydrocarbons in sewage sludge in relation to the amount and origin of purified sewage. Pol J Environ Stud 12(5):523–530

Barret M, Carrère H, Delgadillo L, Patureau D (2010) PAH fate during the anaerobic digestion of contaminated sludge: do bioavailability and/or cometabolism limit their biodegradation? Water Res 44(13):3797–3806. https://doi.org/10.1016/j.watres.2010.04.031

Benabdallah El-Hadj T, Dosta J, Mata-Alvarez J (2006) Biodegradation of PAH and DEHP micropollutants in mesophilic and thermophilic anaerobic sewage sludge digestion. Water Sci Technol 53(8):99–107. https://doi.org/10.2166/wst.2006.240

Benmoussa H, Tyagi RD, Campbell PGC (1997) Simultaneous sewage sludge digestion and metal leaching using an internal loop reactor. Water Res 31(10):2638–2654. https://doi.org/10.1016/S0043-1354(97)00112-7

Bibby K, Peccia J (2013) Identification of viral pathogen diversity in sewage sludge by metagenome analysis. Environ Sci Technol 47:1945–1951

Bloem E, Albihn A, Elving J, Hermann L, Lehmann L, Sarvi M, Schaaf T, Schick J, Turtola E, Ylivainio K (2017) Contamination of organic nutrient sources with potentially toxic elements, antibiotics and pathogen microorganisms in relation to P fertilizer potential and treatment options for the production of sustainable fertilizers: a review. Sci Total Environ 607:225–242

Bomboi M, Hernandez A (1991) Hydrocarbons in urban runoff: their contribution to the wastewaters. Water Res 25:557–565

Bora AP, Gupta DP, Durbha KS (2020) Sewage sludge to biofuel: a review on the sustainable approach of transforming sewage waste to alternative fuel. Fuel 259:116262. https://doi.org/10.1016/jfuel.2019.116262

Bosch A, Lucena F, Jofre J (1986) Fate of Human Enteric Viruses (Rotaviruses and Entero Viruses) in Sewage after Primary Sedimentation. Water Sci Technol 18(10):47–52. https://doi.org/10.2166/wst.1986.0110

Busetti F, Heitz A, Cuomo M, Badoer S, Traverso P (2006) Determination of sixteen polycyclic aromatic hydrocarbons in aqueous and solid samples from an Italian wastewater treatment plant. J Chromatogr A 1102(1–2):104–115. https://doi.org/10.1016/j.chroma.2005.10.013

Cano R, Pérez-Elvira SI, Fdz-Polanco F (2015) Energy feasibility study of sludge pretreatments: a review. Appl Energy 149:176–185

Cao Y, Pawłowski A (2012) Sewage sludge-to-energy approaches based on anaerobic digestion and pyrolysis: brief overview and energy efficiency assessment. Renew Sust EnergRev 16:1657–1665

Chaoua S, Boussaa S, Khadra A, Boumezzough A (2018) Efficiency of two sewage treatment systems (activated sludge and natural lagoons) for helminth egg removal in Morocco. J Infect Public Health 11(2):197–202

Chen C-F, Ju Y-R, Lim YC, Hsieh S-L, Tsai M-L, Sun P-P, Katiyar R, Chen C-W, Dong C-D (2019) Determination of polycyclic aromatic hydrocarbons in sludge from water and wastewater treatment plants by GC-MS. Int J Environ Res Public Health 16(14):2604. https://doi.org/10.3390/ijerph16142604

Clarke BO, Smith SR (2011) Review of "emerging" organic contaminants in biosolids and assessment of international research priorities for the agricultural use of biosolids. Environ Int 37:226–247

Cooke WB (1970) Fungi associated with the activated-sludge process of sewage treatment at the Lebanon, Ohio, sewage-treatment plant. http://hdl.handle.net/1811/5536

Corrêa Medeiros R, Antonio Daniel L (2018) Quantification and analysis of the viability of (oo) cysts of pathogenic protozoa in sewage sludge. Acta Scientiar Technol 40

Da Rocha MCV, BarÉS ME, Braga MCB (2016) Quantification of viable helminth eggs in samples of sewage sludge. Water Res 103:245–255

Dubey M, Mohapatra S, Tyagi VK, Suthar S, Kazmi AA (2021) Occurrence, fate, and persistence of emerging micropollutants in sewage sludge treatment. Environ Pollut 273:116515. https://doi.org/10.1016/j.envpol.2021.116515

EC (2000) European Commission working document on sludge. In: Third Draft, Brussels

EC (2008) Environmental, economic and social impacts of the use of sewage sludge on land. Final report. Part I: overview report. In: Milieu Ltd, W. A. R. F. T. E. C, editor. DG Environment Under Study Contract DG ENV.G.4/ETU/2008/0076R (ed.)

Ellis S, Tyrrel S, O'leary E, Richards K, Griffiths B, Ritz K (2018) Proportion of sewage sludge to soil influences the durvival of Salmonella Dublin and Escherichia coli. Clean Soil Air Water 46: 1800042

Engwa GA, Ferdinand PU, Nwalo FN, Unachukwu MN (2019) Mechanism and health effects of heavy metal toxicity in humans. In: Poisoning in the modern world-new tricks for an old dog, vol 10. IntechOpen, London. https://doi.org/10.5772/intechopen.82511

Environmental Protection Agency- USA (1979) Design of wastewater treatment facilities major systems. Environmental Protection Agency, Office of Water Program Operations

Eriksson E, Christensen N, Schmidt JE, Ledin A (2008) Potential priority pollutants in sewage sludge. Desalination 226:371–388

EUROSTAT (2020) Regional statistics: reference guide. Luxembourg. https://appsso.eurostat.ec.europa.eu/nui/show.do?dataset=env_ww_spd&lang=en#. Database accessed 24 February 2020

Fang W, Delapp RC, Kosson DS, Van Der Sloot HA, Liu J (2017) Release of heavy metals during long-term land application of sewage sludge compost: percolation leaching tests with repeated additions of compost. Chemosphere 169:271–280

FAO (2020) AQUASTAT Core Database. Food and Agriculture Organization of the United Nations. http://www.fao.org/nr/water/aquastat/wastewater/index.stm. Database accessed 24 February 2020

Feizi M, Jalali M, Renella G (2018) Nanoparticles and modified clays influenced distribution of heavy metals fractions in a light-textured soil amended with sewage sludges. J Hazard Mater 343:208–219. https://doi.org/10.1016/j.jhazmat.2017.09.027

Fernández J, Palza C, García-Gil JC, Polo A (2009) Biochemical properties and barley yield in a semiarid Mediterranean soil amended with two kinds of sewage sludge. Appl Soil Ecol 42:18–24

Ferronato N, Torretta V (2019) Waste mismanagement in developing countries: a review of global issues. Int J Environ Res Public Health 16(6):1060. https://doi.org/10.3390/ijerph16061060
Fijalkowski K, Rorat A, Grobelak A, Kacprzak MJ (2017) The presence of contaminations in sewage sludge—the current situation. J Environ Manag 203:1126–1136
Filali-Meknassi Y, Tyagi RD, Narasiah KS (2000) Simultaneous sewage sludge digestion and metal leaching: effect of aeration. Process Biochem 36(3):263–273. https://doi.org/10.1016/S0032-9592(00)00213-2
Frąc M, Oszust K, Lipiec J, Jezierska-Tys S, Nwaichi EO (2014) Soil microbial functional and fungal diversity as influenced by municipal sewage sludge accumulation. Int J Environ Res Public Health 11:8891–8908
Frišták V, Soja G (2015) Effect of wood-based biochar and sewage sludge amendments for soil phosphorus availability. Nova Biotechnologica et Chimica 14:104
Galler H, Feierl G, Petternel C, Reinthaler FF, Haas D, Habib J, Kittinger C, Luxner J, Zarfel G (2018) Multiresistant bacteria isolated from activated sludge in Austria. Int J Environ Res Public Health 15:479
Gantzer C, Gaspard P, Galvez L, Huyard A, Dumouthier N, Schwartzbrod J (2001) Monitoring of bacterial and parasitological contamination during various treatment of sludge. Water Res 35(16):3763–3770. https://doi.org/10.1016/S0043-1354(01)00105-1
García H, El Zauahre M, Morán H, Acosta Y, Senior A, Fernández N (2006) Comparative analysis of two digestion techniques for the determination of heavy metals in sewage sludge. Multiciencias 6:234–243
Garrido-Baserba MMSPG, Molinos-Senante M, Abelleira-Pereira JM, Fdez-Güelfo LA, Poch M, Hernández-Sancho F (2015) Selecting sewage sludge treatment alternatives in modern wastewater treatment plants using environmental decision support systems. J Clean Prod 107:410–419. https://doi.org/10.1016/j.jclepro.2014.11.021
Gondek K, Baran A, Kopeć M (2014) The effect of low-temperature transformation of mixtures of sewage sludge and plant materials on content, leachability and toxicity of heavy metals. Chemosphere 117:33–39. https://doi.org/10.1016/j.chemosphere.2014.05.032
Grosser A (2018) Determination of methane potential of mixtures composed of sewage sludge, organic fraction of municipal waste and grease trap sludge using biochemical methane potential assays. A comparison of BMP tests and semi-continuous trial results. Energy 143:488–499
Guzmán C, Jofre J, Blanch AR, Lucena F (2007) Development of a feasible method to extract somatic coliphages from sludge, soil and treated biowaste. J Virol Methods 144(1–2):41–48. https://doi.org/10.1016/j.jviromet.2007.03.017
Harrison EZ, Oakes SR, Hysell M, Hay A (2006) Organic chemicals in sewage sludges. Sci Total Environ 367:481–497
Haynes RJ, Murtaza G, Naidu R (2009) Inorganic and organic constituents and contaminants of biosolids: implications for land application. Adv Agron 104:165–267
Healy MG, Fenton O, Forrestal PJ, Danaher M, Brennan RB, Morrison L (2016) Metal concentrations in lime stabilised, thermally dried and anaerobically digested sewage sludges. Waste Manag 48:404–408. https://doi.org/10.1016/j.wasman.2015.11.028
Hjorth M, Nielsen AM, Nyord T, Hansen MN, Nissen P, Sommer SG (2009) Nutrient value, odour emission and energy production of manure as influenced by anaerobic digestion and separation. Agron Sustain Dev 29(2):329–338
Hossain MK, Strezov V, Chan KY, Ziolkowski A, Nelson PF (2011) Influence of pyrolysis temperature on production and nutrient properties of wastewater sludge biochar. J Environ Manag 92(1):223–228. https://doi.org/10.1016/j.jenvman.2010.09.008
Houillon G, Jolliet O (2005) Life cycle assessment of processes for the treatment of wastewater urban sludge: energy and global warming analysis. J Clean Prod 13:287–299
Huang K, Mao Y, Zhao F, Zhang X-X, Ju F, Ye L, Wang Y, Li B, Ren H, Zhang T (2018) Free-living bacteria and potential bacterial pathogens in sewage treatment plants. Appl Microbiol Biotechnol 102:2455–2464

Ju JH, Lee IS, Sim WJ, Eun H, Oh JE (2009) Analysis and evaluation of chlorinated persistent organic compounds and PAHs in sludge in Korea. Chemosphere 74(3):441–447. https://doi.org/10.1016/j.chemosphere.2008.09.059

Kacprzak M, Neczaj E, Fijałkowski K, Grobelak A, Grosser A, Worwag M, Rorat A, Brattebo H, Almås Å, Singh BR (2017) Sewage sludge disposal strategies for sustainable development. Environ Res 156:39–46. https://doi.org/10.1016/j.envres.2017.03.010

Khadhar 1S, Higashi T, Hamdi H, Matsuyama S, Charef A (2010) Distribution of 16 EPA-priority polycyclic aromatic hydrocarbons (PAHs) in sludges collected from nine Tunisian wastewater treatment plants. J Hazard Mater 183(1–3):98–102. https://doi.org/10.1016/j.jhazmat.2010.06.112

Khan S, Cao Q, Zheng YM, Huang YZ, Zhu YG (2008) Health risks of heavy metals in contaminated soils and food crops irrigated with wastewater in Beijing, China. Environ Pollut 152(3):686–692

Khillare PS, Sattawan VK, Jyethi DS (2018) Profile of polycyclic aromatic hydrocarbons in digested sewage sludge. Environ Technol. https://doi.org/10.1080/09593330.2018.1512654

Khouja LBA, Cama V, Xiao L (2010) Parasitic contamination in wastewater and sludge samples in Tunisia using three different detection techniques. Parasitol Res 107:109–116

Kończak M, Oleszczuk P (2018) Application of biochar to sewage sludge reduces toxicity and improve organisms growth in sewage sludge-amended soil in long term field experiment. Sci Total Environ 625:8–15

Korzeniewska E (2011) Emission of bacteria and fungi in the air from wastewater treatment plants—a review. Front Biosci 3:393–407

Krzyzanowski F Jr, De Souza Lauretto M, Nardocci AC, Sato MIZ, Razzolini MTP (2016) Assessing the probability of infection by Salmonella due to sewage sludge use in agriculture under several exposure scenarios for crops and soil ingestion. Sci Total Environ 568:66–74

Kumar V, Chopra AK, Kumar A (2017) A review on sewage sludge (Biosolids) a resource for sustainable agriculture. Arch Agric Environ Sci 2(4):340–347. https://doi.org/10.26832/24566632.2017.020417

Kwarciak-Kozłowska A (2019) 15 - Co-composting of sewage sludge and wetland plant material from a constructed wetland treating domestic wastewater. In: Prasad MNV, de Campos Favas PJ, Vithanage M, Mohan SV (eds) Industrial and municipal sludge. Butterworth-Heinemann, pp 337–360. https://doi.org/10.1016/B978-0-12-815907-1.00015-5

LeBlanc RJ, Matthews P, Richard RP (2008) Global atlas of excreta. In: Wastewater sludge, and biosolids management: moving forward the sustainable and welcome uses of a global resource. United Nations Humans Settlements Program, Nairobi

Lederer J, Rechberger H (2010) Comparative goal-oriented assessment of conventional and alternative sewage sludge treatment options. Waste Manag 30:1043–1056. https://doi.org/10.1016/j.wasman.2010.02.025

Lehmann J, Joseph S (2015) Biochar for environmental management: an introduction. Routledge, pp 33–46. https://doi.org/10.4324/9780203762264

Leng L, Yuan X, Huang H, Jiang H, Chen X, Zeng G (2014) The migration and transformation behavior of heavy metals during the liquefaction process of sewage sludge. Bioresour Technol 167:144–150

Li J, Luo G, Gao J, Yuan S, Du J, Wang Z (2014) Quantitative evaluation of potential ecological risk of heavy metals in sewage sludge from three wastewater treatment plants in main urban area of Wuxi, China. Chem Ecol:1–28

Liew C-S, Kiatkittipong W, Lim J-W, Lam M-K, Ho Y-C, Ho C-D, Ntwampe SKO, Mohamad M, Usman A (2021) Stabilization of heavy metals loaded sewage sludge: reviewing conventional to state-of-the-art thermal treatments in achieving energy sustainability. Chemosphere 277:130310. https://doi.org/10.1016/j.chemosphere.2021.130310

Liu T, Liu Z, Zheng Q, Lang Q, Xia Y, Peng N et al (2018) Effect of hydrothermal carbonization on migration and environmental risk of heavy metals in sewage sludge during pyrolysis. Bioresour Technol 247:282–290

Lu H, Zhang W, Yang Y, Huang X, Wang S, Qiu R (2012) Relative distribution of Pb2+ sorption mechanisms by sludge-derived biochar. Water Res 46(3):854–862. https://doi.org/10.1016/j.watres.2011.11.058

Malińska K, Zabochnicka-Świątek M, Dach J (2014) Effects of biochar amendment on ammonia emission during composting of sewage sludge. Ecol Eng 71:474–478

Malińska K, Golańska M, Caceres R, Rorat A, Weisser P, Ślęzak E (2017) Biochar amendment for integrated composting and vermicomposting of sewage sludge–the effect of biochar on the activity of Eisenia fetida and the obtained vermicompost. Bioresour Technol 225:206–214. https://doi.org/10.1016/j.biortech.2016.11.049

Manara P, Zabaniotou A (2012) Towards sewage sludge based biofuels via thermochemical conversion - a review. Renew Sustain Energy Rev 16:2566–2582. https://doi.org/10.1016/j.rser.2012.01.074

Manios SG, Konstantinidis N, Gounadaki AS, Skandamis PN (2013) Dynamics of low (1–4 cells) vs high populations of Listeria monocytogenes and Salmonella typhimurium in fresh-cut salads and their sterile liquid or solidified extracts. Food Control 29:318–327

Maragkaki A, Vasileiadis I, Fountoulakis M, Kyriakou A, Lasaridi K, Manios T (2018) Improving biogas production from anaerobic co-digestion of sewage sludge with a thermal dried mixture of food waste, cheese whey and olive mill wastewater. Waste Manag 71:644–651

Mateo-Sagasta J, Raschid-Sally L, Thebo A (2015) Global wastewater and sludge production, treatment, and use. In: Drechsel P, Qadir M, Wichelns D (eds) Wastewater: economic asset in an urbanizing world. Springer, Dordrecht, pp 15–38

Matiasi TM (2012) Wastewater Management and reuse for sustainable development. Scholar J Agric Sci 2(11):269–276

Metcalf E, Eddy P (1991) Wastewater engineering: treatment, disposal, reuse, 3rd edn. McGraw Hill, Singapore

Metcalf, Inc, Tchobanoglous G, Burton FL, Stensel HD (2013) In: Metcalf E (ed) Wastewater engineering: treatment and resource recovery. McGraw-Hill Education

Millner PD, Marsh PB, Snowden RB, Parr JF (1977) Occurrence of Aspergillus fumigatus during composting of sewage sludge. Appl Environ Microbiol 34(6):765–772. https://doi.org/10.1128/aem.34.6.765-772.1977

Moretti SML, Bertoncini EI, Vitti AC, Alleoni LRF, Abreu-Junior CH (2016) Concentration of Cu, Zn, Cr, Ni, Cd, and Pb in soil, sugarcane leaf and juice: residual effect of sewage sludge and organic compost application. Environ Monit Assess 188(3):163. https://doi.org/10.1007/s10661-016-5170-1

Nafez AH, Nikaeen M, Kadkhodaie S et al (2015) Sewage sludge composting: quality assessment for agricultural application. Environ Monit Assess 187:709. https://doi.org/10.1007/s10661-015-4940-5

Nkoa R (2014) Agricultural benefits and environmental risks of soil fertilization with anaerobic digestates: a review. Agron Sustain Dev 34:473–492

Novak JT (2006) Dewatering of sewage sludge. Dry Technol 24:1257–1262

Opec FFID (2018) Wastewater Report 2018: the reuse opportunity. IWA, International Water Association

Östman M, Lindberg RH, Fick J, Björn E, Tysklind M (2017) Screening of biocides, metals and antibiotics in Swedish sewage sludge and wastewater. Water Res 115:318–328. https://doi.org/10.1016/j.watres.2017.03.011

Ozcan S, Tor A, Aydin ME (2013) Investigation on the levels of heavy metals, polycyclic aromatic hydrocarbons, and polychlorinated biphenyls in sewage sludge samples and ecotoxicological testing. Clean Soil Air Water 41:411–418

Pathak A, Dastidar MG, Sreekrishnan TR (2008) Bioleaching of heavy metals from anaerobically digested sewage sludge. J Environ Sci Health A 43(4):402–411. https://doi.org/10.1080/10934520701795624

Pathak A, Dastidar MG, Sreekrishnan TR (2009) Bioleaching of heavy metals from sewage sludge: a review. J Environ Manag 90(8):2343–2353. https://doi.org/10.1016/j.jenvman.2008.11.005

Peng W, Pivato A, Lavagnolo MC, Raga R (2018) Digestate application in landfill bioreactors to remove nitrogen of old landfill leachate. Waste Manag 74:335–346

Pradel M, Reverdy AL. Assessing GHG emissions from sludge treatment and disposal routes: the method behind GESTABoues tool. ORBIT2012, Global assessment for organic resources and waste management, Jun 2012, Rennes, France. 9 p. ffhal-00781673f

Puyol D, Batstone DJ, Hülsen T, Astals S, Peces M, Krömer JO (2017) Resource recovery from wastewater by biological technologies: opportunities, challenges, and prospects. Front Microbiol 7:2106. https://doi.org/10.3389/fmicb.2016.02106

Reinthaler FF, Galler H, Feierl G, Haas D, Leitner E, Mascher F, Melkes A, Posch J, Pertschy B, Winter I (2013) Resistance patterns of Escherichia coli isolated from sewage sludge in comparison with those isolated from human patients in 2000 and 2009. J Water Health 11:13–20

Rizzardini CB, Goi D (2014) Sustainability of domestic sewage sludge disposal. Sustainability 6(5):2424–2434. https://doi.org/10.3390/su6052424

Romdhana MH, Lecomte D, Ladevie B, Sablayrolles C (2009) Monitoring of pathogenic microorganisms contamination during heat drying process of sewage sludge. Process Saf Environ Prot 87:377–386

Rorat A, Kacprzak M (2017) Eco-innovations in sustainable waste management strategies for smart cities. In: Brdulak A, Brdulak H (eds) Happy city—how to plan and create the best livable area for the people. Springer International Publishing, Cham

Rorat A, Suleiman H, Grobelak A, Grosser A, Kacprzak M, Płytycz B, Vandenbulcke F (2015) Interactions between sewage sludge-amended soil and earthworms—comparison between Eisenia fetida and Eisenia andrei composting species. Environ Sci Pollut Res 23:3026–3035

Rorat A, Wloka D, Grobelak A, Grosser A, Sosnecka A, Milczarek M, Jelonek P, Vandenbulcke F, Kacprzak M (2017) Vermiremediation of polycyclic aromatic hydrocarbons and heavy metals in sewage sludge composting process. J Environ Manag 187:347–353. https://doi.org/10.1016/j.jenvman.2016.10.062

Rorat A, Courtois P, Vandenbulcke F, Lemiere S (2019) Sanitary and environmental aspects of sewage sludge management. Emerging concerns and scope for resource recovery industrial and municipal sludge, pp 155–180

Rouse JD (2013) Sustainability of wastewater treatment and excess sludge handling practices in the federated states of Micronesia. Sustainability 5:4183–4194

Sabbahi S, Trad M, Ayed LB, Marzougui N (2018) Occurrence of intestinal parasites in sewage samples and efficiency of wastewater treatment systems in Tunisia. Water Qual Res J 53(2):86–101

Salaudeen T, Okoh O, Agunbiade F, Okoh A (2018) Fate and impact of phthalates in activated sludge treated municipal wastewater on the water bodies in the Eastern Cape, South Africa. Chemosphere 203:336–344. https://doi.org/10.1016/j.chemosphere.2018.03.176

Salihoglu NK, Salihoglu G, Tasdemir Y, Cindoruk SS, Yolsal D, Ogulmus R, Karaca G (2010) Comparison of polycyclic aromatic hydrocarbons levels in sludges from municipal and industrial wastewater treatment plants. Arch Environ Contam Toxicol 58(3):523–534. https://doi.org/10.1007/s00244-009-9389-5

Samolada M, Zabaniotou A (2014) Comparative assessment of municipal sewage sludge incineration, gasification and pyrolysis for a sustainable sludge-to-energy management in Greece. Waste Manag 34(2):411–420

Sánchez-Monedero MA, Serramiá N, Civantos CG-O, Fernández-Hernández A, Roig A (2010) Greenhouse gas emissions during composting of two-phase olive mill wastes with different agroindustrial by-products. Chemosphere 81:18–25

Sánchez CH, Gutiérrez Á, Galindo JM, González-Weller D, Rubio C, Revert C et al (2017) Contenido de metales pesados en lodos de depuradora: estrategia de gestión para una isla oceánica. Revista de Salud Ambiental 17(1):3–9

Schowanek D, David H, Francaviglia R, Hall J, Kirchmann H, Krogh PH, Schraepen N, Smith S, Wildemann T (2007) Probabilistic risk assessment for linear alkylbenzene sulfonate (LAS) in sewage sludge used on agricultural soil. Regul Toxicol Pharmacol 49:245–259

Seggiani M, Puccini M, Raggio G, Vitolo S (2012) Effect of sewage sludge content on gas quality and solid residues produced by cogasification in an updraft gasifier. Waste Manag 32:1826–1834. https://doi.org/10.1016/j.wasman.2012.04.018

Seiple TE, Coleman AM, Skaggs RL (2017) Municipal wastewater sludge as a sustainable bioresource in the United States. J Environ Manag 197:673–680. https://doi.org/10.1016/j.jenvman.2017.04.032

Shi W, Liu C, Ding D, Lei Z, Yang Y, Feng C, Zhang Z (2013) Immobilization of heavy metals in sewage sludge by using subcritical water technology. Bioresour Technol 137:18–24. https://doi.org/10.1016/j.biortech.2013.03.106

Smith S (2009) Organic contaminants in sewage sludge (biosolids) and their significance for agricultural recycling. Philos Trans A Math Phys Eng Sci 367:4005–4041

Soda S, Iwai Y, Sei K, Shimoda Y, Ike M (2010) Model analysis of energy consumption and green house gas emission of sewage sludge treatment systems with sifferent processes and scales. Water Sci Technol 61(2):365–373

Sogn TA, Dragicevic I, Linjordet R, Krogstad T, Eijsink VGH, Eich-Greatorex S (2018) Recycling of biogas digestates in plant production: NPK fertilizer value and risk of leaching. Int J Recycl Org Waste Agric 7:49–58

Stefaniuk M, Tsang DC, Ok YS, Oleszczuk P (2018) A field study of bioavailable polycyclic aromatic hydrocarbons (PAHs) in sewage sludge and biochar amended soils. J Hazard Mater 349:27–34

Stevens JL, Northcott GL, Stern GA, Tomy GT, Jones KC (2003) PAHs, PCBs, PCNs, organochlorine pesticides, synthetic musks, and polychlorinated n-alkanes in UK sewage sludge: survey results and implications. Environ Sci Technol 37(3):462–467. https://doi.org/10.1021/es020161y

Suleiman H, Rorat A, Grobelak A, Grosser A, Milczarek M, Płytycz B et al (2017) Determination of the performance of vermicomposting process applied to sewage sludge by monitoring of the compost quality and immune responses in three earthworm species: Eisenia fetida, Eisenia andrei and Dendrobaena veneta. Bioresour Technol 241:103–112. https://doi.org/10.1016/j.biortech.2017.05.104

Takaoka M, Oshita K, Okada M, Watanabe T, Tanida K (2018) Mercury behaviour in flue gas from sewage sludge incinerators and melting furnace. Water Sci Technol 2017(3):782–790. https://doi.org/10.2166/wst.2018.268

Tchobanoglous G, Burton FL, Stensel HD (2003) Wastewater engineering treatment and reuse. Metcalf and Eddy, McGraw Hill, NY. Technol 241:103–112

Tozzoli R, Di Bartolo I, Gigliucci F, Brambilla G, Monini M, Vignolo E, Caprioli A, Morabito S (2017) Pathogenic Escherichia coli and enteric viruses in biosolids and related top soil improvers in Italy. J Appl Microbiol 122:239–247

Tuefenkci S, Türkmen Ö, Sönmez F, Erdinc C, Sensoy S (2006) Effects of humic acid doses and aplication times on the plant growth, nutrient and heavy metal contents of lettuce grown on sewage sludge-applied soils. Fresenius Environ Bull 15(4):295–300

Turlej T, Banaś M (2018) Sustainable management of sewage sludge. E3S Web Conf 49:00120

Turolla A, Cattaneo M, Marazzi F, Mezzanotte V, Antonelli M (2018) Antibiotic resistant bacteria in urban sewage: role of full-scale wastewater treatment plants on environmental spreading. Chemosphere 191:761–769

Tyagi VK, Lo SL (2013) Sludge: a waste or renewable source for energy and resources recovery? Renew Sust Energ Rev 25:708–728. https://doi.org/10.1016/j.rser.2013.05.029

Tyagi RD, Couillard D, Tran F (1988) Heavy metals removal from anaerobically digested sludge by chemical and microbiological methods. Environ Pollut 50(4):295–316. https://doi.org/10.1016/0269-7491(88)90194-7

Ukaogo PO, Ewuzie U, Onwuka CV (2020) 21 - Environmental pollution: causes, effects, and the remedies. In: Chowdhary P, Raj A, Verma D, Akhter Y (eds) Microorganisms for sustainable environment and health. Elsevier, pp 419–429

United Nations (2012). World population prospects: the 2012 revision. United Nations, Department of Economic and Social Affairs, Population Division, NY

United Nations, Department of Economic and Social Affairs, Population Division (2019) World urbanization prospects: the 2018 revision (ST/ESA/SER.A/420). United Nations, New York

United Nations Environmental Program (2013) Guidelines for national waste management strategies moving from challenges to opportunities (PDF). ISBN 978-92-807-3333-4

Van Wesenbeeck S, Prins W, Ronsse F, Antal MJ Jr (2014) Sewage sludge carbonization for biochar applications. Fate of heavy metals. Energy Fuel 28(8):5318–5326. https://doi.org/10.1021/ef500875c

Verheijen F, Jeffery S, Bastos A, Van Der Velde M, Diafas I (2010) Biochar application to soils. A critical scientific review of effects on soil properties, processes, and functions. European Commision, p. 162. 24099

Wang N-Y, Shih C-H, Chiueh P-T, Huang Y-F (2013) Environmental effects of sewage sludge carbonization and other treatment alternatives. Energies 6:871–883

Walter I, Martinez F, Cala V (2006) Heavy metal speciation and phytotoxic effects of three representative sewage sludges for agricultural uses. Environ Pollut 139(3):507–514. https://doi.org/10.1016/j.envpol.2005.05.020

Wéry N, Lhoutellier C, Ducray F, Delgenès JP, Godon JJ (2008) Behaviour of pathogenic and indicator bacteria during urban wastewater treatment and sludge composting, as revealed by quantitative PCR. Water Res 42(1–2):53–62. https://doi.org/10.1016/j.watres.2007.06.048

Williams FP Jr, Hurst CJ (1988) Detection of environmental viruses in sludge: Enhancement of enterovirus plaque assay titers with 5-iodo-2′-deoxyuridine and comparison to adenovirus and coliphage titers. Water Res 22(7):847–851. https://doi.org/10.1016/0043-1354(88)90022-X

Wilson S, Duarte-Davidson RZ, Jones K (1996) Screening the environmental fate of organic contaminants in sewage sludges applied to agricultural soils: 1. The potential for downward movement to groundwaters. Sci Total Environ 185:45–57

Wójtowicz A (2013) Modelowe rozwiązania w gospodarce osadowej, Bydgoski Dom Wydawniczy "Margrafsen"

Wong MH (2012) Environmental contamination: health risks and ecological restoration. Taylor & Francis Group

Wong JWC, Selvam A (2006) Speciation of heavy metals during co-composting of sewage sludge with lime. Chemosphere 63(6):980–986. https://doi.org/10.1016/j.chemosphere.2005.08.045

Xiang L, Chan LC, Wong JWC (2000) Removal of heavy metals from anaerobically digested sewage sludge by isolated indigenous iron-oxidizing bacteria. Chemosphere 41(1–2):283–287. https://doi.org/10.1016/S0045-6535(99)00422-1

Yang G, Zhang G, Wang H (2015) Current state of sludge production, management, treatment and disposal in China. Water Res 78:60–73. https://doi.org/10.1016/j.watres.2015.04.002

Zennegg M, Munoz M, Schmid P, Gerecke AC (2013) Temporal trends of persistent organic pollutants in digested sewage sludge (1993–2012). Environ Int 60:202–208

Global Scenario of Sewage- Sludge Management

Prerna Negi, Himanshu Verma, S. P. Singh, B. S. Mahapatra, and Hanuman Singh Jatav

1 Introduction

Most residents drained their polluted water or sewage into lakes and waterways without any care until 1950. As urban populations increased, the natural capacity of streams and rivers to handle sewage sludge was exceeded, causing water quality to deteriorate in many areas. During the 1950s and 1960s, sewage sludge treatment facilities were built in thousands of communities throughout the United States.

This increased the consistency of stream and river water greatly, but it also produced a new substance to work with: waste sludge. About 99% of wastewater entering a treatment facility is discharged. The balance is a dilute suspension of solids that has been captured by the treatment procedure. Sewage sludge is the common name for these solids used in sewage sludge treatment.

Sewers are now built to convey trash to a wastewater treatment facility instead of a water supply. Present waste treatment methods may include the separate handling of surface drainage from sewage, the separate handling of grey water from black water (flush toilets), and the improved handling of irregular incidents.

P. Negi · H. Verma (✉)
Department of Agronomy, SGRR University, Dehradun, Uttarakhand, India

S. P. Singh
Department of Agronomy, College of Agriculture, GBPUAT, Pantnagar, Uttarakhand, India

B. S. Mahapatra
Bidhan Chandra Krishi Viswavidyalaya (BCKV), Mohanpur, West Bengal, India

H. S. Jatav
Department of Soil Science and Agriculture Chemistry, Sri Karan Narendra Agriculture University, Jobner, Rajasthan, India

V. D. Rajput et al. (eds.), *Sustainable Management and Utilization of Sewage Sludge*, https://doi.org/10.1007/978-3-030-85226-9_18

2 Sewage Sludge: Concept

The concept of sewage sludge (also referred to as bio solids) is the semi-solid, residual material formed as a by-product of industrial or municipal wastewater treatment during sewage treatment (Kumar and Chopra 2016). More precisely, wastewater sludge is produced as a by-product of the various treatment stages of domestic wastewater which often also contains agricultural and commercial wastewater (Williams 2005). The pH is generally ranging from 6.5–7.0 as reported by Xu 2014.

Sewage sludge is adequately handled and processed to create nutrient-rich organic materials from wastewater treatment plant bio solids (Kumar et al. 2016; Kumar and Chopra 2013). These activities improve the usability of sewage sludge or bio solids and reduce hazardous substances in sewage sludge to prevent discharge into marine resources such as rivers, lakes, streams, and other bodies of water (Spinosa 2008; Rogers 2012).

The properties of waste sludge or bio solids typically differ and include organic and inorganic materials, radioactive metals and pathogens. When approaching the sewer treatment system for treatment, the sewage sludge, which has approximately 1% of waste water, is anaerobically digested and the wastewater is separated from the sludge. Sewage sludge, which is present at the output vent after mechanical drying, has nearly 80% moisture and 20% dry matter. Sludge usage in agriculture is prohibited by these factors because its deposition is not good for ecology (Kumar and Chopra 2012, 2016).

Effluents from commercial, urban or rural sources are collected. The sludge from the sewage is derived from processes of major, secondary and tertiary treatment. Sludge can be used as fertilizer, but it is difficult to use sewage.

3 Constituents of Sewage- Sludge

Water is the most important part of sludge (Table 1). The type of sludge (primary, secondary or tertiary) and the stabilisation process have an impact on its consistency (aerobic, anaerobic). The water content of raw sludge ranges from 93 to 99%. For further usage, dewatering (up to approximately 35% dry substance content) or drying (up to over 85% dry substance content) may be needed.

Table 1 C, H, O, N, P and K in the Sewage- Sludge

SI No.	Particular	Percent
1	C	50–70
2	H	6.5–7.3
3	O	21–24
4	N	15–18
5	P	1.0–1.5
6	S	0–2.4

Table 2 Expression of major elements as the median and 95th percentiles values in Pennsylvania sewage sludge

SI No.	Major elements	Median	95th Percentile
1	Nitrogen (N)	4.8	7.7
2	Phosphorus (P)	2.2	3.9
3	Potassium (K)	0.22	0.7
4	Calcium (Ca)	3.1	18.0
5	Magnesium (Mg)	0.4	0.8

Table 3 Expression of trace elements as the median and 95th percentiles values in Pennsylvania sewage sludge

SI No.	Trace elements	Median (%)	95th Percentile (%)
1	As	0.00036	0.0018
2	Cd	0.00023	0.00074
3	Cr	0.0035	0.0314
4	Cu	0.0511	0.1382
5	Hg	0.00015	0.0006
6	Mo	0.00082	0.0036
7	Ni	0.0022	0.0085
8	Pb	0.0065	0.0202
9	Se	0.00043	0.00085
10	Zn	0.0702	0.1985

The dried product, which is made up of organic and inorganic compounds, is the second major component. Sludge contains a broad range of trace components that have been extracted from waste water, in addition to the main ingredient (Table 2). Organic and inorganic trace elements contained in waste water are enriched in the sludge.

The composition of sewage sludge varies greatly depending on the consistency of the waste water and the treatment methods used. Plant nutrient levels at the median and 95th percentiles, as well as some of the some of the trace elements found are provided in Table 3.

Any material detected in small or minute quantities on the earth's surface is referred to as a "trace element." It refers to any of the possible inorganic pollutants discussed in this fact sheet. Since inorganic pollutants like arsenic and selenium are not metals, the term "heavy metal" is used instead of "trace element" or "trace metal."

4 Sewage

Sewage, like residential and municipal waste water, is a type of waste water created by a group of people. It is described using terms like flow rate or volume, physical status, chemical and poisonous components, and bacteriological status. Grey water (from sinks, bathtubs, showers, dishwashers, and clothes washers), and black water (from toilets, mixed with human waste).

4.1 Composition of Sewage

The composition of sewage is largely determined by per capita water intake, which varies by location and season.

4.1.1 Chemical Composition

In its suspended and soluble forms, sewage is 99% water and 1% inorganic and organic materials. Ligno-cellulose, proteins, lipids, cellulose and other inorganic particulate matter are soluble sources of fatty acids, sugars, alcohols, amino acids. On average, however, city water contains about 350 parts per million of biodegradable organic matter, 52 parts per million of nitrogen, 45 parts per million of potassium, and 16 parts per million of phosphorus. High metal salts, such as Zn, Cr, Ni, Pb, and other heavy metals, are also present in excess.

4.1.2 Microorganisms

The number of microbes per millilitre might range from a few thousand to a few million lacs. Microorganisms such as fungus, protozoa and bacteria together known as 'sewage fungus' have been found to thrive in sewage. Viruses have also been found in sewage and a variety of micro-algal genera. *Micrococci, Proteus, Pseudomonas, Coliforms, streptococci, Clostridia* and *lactobacilli* are common types of sewage bacteria that live in the intestine and soil.

4.2 Contaminants of Sewage

4.2.1 Pollutants and Nutrients from Organic Sources

Sewage is a complex chemical mixture including a variety of distinct chemical components. High amounts of nitrogen, ammonium, nitrate and phosphorus, as well as higher magnitude of conductivity and alkalinity are among them. The quantity of organic matter contained in sewage is measured using the biological oxygen demand (BOD) or chemical oxygen demand (COD).

4.2.2 Pathogens

It contains four types of pathogens mentioned in Table 4.

4.2.3 Micro-contaminants

The setting's recurrent pharmaceutical pollutants are also found in sewage. As a result of prior disinfection, trihalomethanes may be present. The relative rates of prescription and illicit drug usage by municipal inhabitants were also studied using sewage.

4.3 Sewage Characteristics

- In sewage, the levels of biological oxygen demand and consumption of oxygen are extraordinarily more.
- Anaerobic or partial breakdown of sewage organic matter leads in the production of toxic gases such as Co, H_2S and CH_3 under anaerobic conditions. These gases are poisonous and generate acids when they combine with water.
- Higher rate of acid production increases the acidity, rendering it unfit for life-sustaining activities.
- Heavy metals can also be found in sewage in different concentrations (Table 5).

Table 4 Details of microbial colonies in Sewage waste

SI No.	Particular	Example	Source
1	Bacteria	*Salmonella* *Campylobacter* *Vibrio cholerae* *Shigella*	–
2	Viruses	Hepatitis A Corona Virus Rotavirus	Naddeo and Liu (2020)
3	Protozoa	*Cryptosporidium parvum* *Entamoebahistolytica* *Giardia lamblia*	–
4	Parasite	Ascaris (Roundworm) Ancylostoma (Hookworm) Trichuris (Whipworm)	–

Table 5 Sewage- sludge: a source of toxic chemicals and heavy metals in the soil

Heavy metal/ toxic matter	Particular	References
Cd, Zn, DTPA and Mn	Ultisol and vertisol	Ramachandran and D'Souza (1998)
Cr, Cd	–	Shrivastava and Banerjee (2003)
Ni, Mo, Mn	*Trifolium pratense*	McBride et al. (2004)
Cd, Cr, Pb, Zn	Soil treated with sewage- sludge compost	Selivanovskaya et al. (2006)
Cd	Seeds of leaf beet	Datta and Young (2005)
K	South Florida	Sigua et al. (2005)
Zn, Fe, Pb, Cd	*Foeniculum vulgare*	El-Motaium and Abo El-Seoud (2007)
Cd, Cr, Cu	*Sorghum*	Jamali et al. (2008)
Cd	Indian mustard, Cabbage and Cauliflower	Sikka et al. (2009)
Ni, Cu, Zn Fe, Cr, Cd, Pb, Cr	Potato	Pakhnenkoa et al. (2009)
Cd, Pb, Mg, Na, K, P, Fe, Cr, Cu, Mn, Ni, Zn, N	*Pinusradiate*	Rodríguez et al. (2010)
Cu, Zn, Ni, Cd, Cr, Pb, Cd, Cr, Ni	*Brassica juncea,* Radish, Turnip, Carrot, Potato, Tomato, Bean, Cauliflower, Brinjal, Cabbage, Spinach, Coriander	Dede et al. (2012), Ghosh et al. (2012)
Mn, Zn	–	Nogueirol et al. (2013)
Cu, Pb, Cr	French Bean	–
Zn, Ni, Cd	–	Ullah and Khan (2015)
Zn	Sunflower	
Zn, Cd	Willow	
Zn, Fe	Paddy	Meena et al. (2016)
Cu, Zn, Cr	*Saccharum*	LeiteMoretti et al. (2016)
Cd, Ni, Pb, Co, Cr, Ca, Mg, Fe, Mn, Cu, Zn	Tomato	Alghobar and Suresha (2017)
Zn, Cu	Clay loam soil	Tziachris et al. (2017)
Pb, Cr, Fe, Mn	*Brassica*	Ullah et al. (2017)
Pb, Fe, Mn, Cu, Zn, Cd,	Wheat	Shahbazi et al. (2017)
Cr, Cu, Ni, Co, Cd, Fe, Pb, Zn, Mn,	Cucumbers	Eid et al. (2017)

5 Sludge

Sludge is a solid or semi-solid waste product produced by wastewater treatment. It can be primary or secondary in development, suggesting that both primary and secondary sludge are present. It formed by a variety of manufacturing processes, such as sewage treatment, wastewater treatment, or on-site drainage. Sludge's effectiveness is challenged by high-end fertilisers due to its low performance as a fertiliser. However, since it produces fewer toxins and is more accessible, farmers are more likely to use it. Sludge from water tanks, for example, is completely useless. It has no agricultural value and cannot be used to generate electricity, heat, or cooking purposes. Sewage sludge, on the other hand, may be processed anaerobically to create energy and can also be utilised in agriculture. It is critical to understand the source of any sludge, the pathogens present, and how it is treated before utilising it.

Primary Sludge Processes including sedimentation, chemical precipitation, and other primary processes produce this type of sludge. It is generated during the mechanical wastewater treatment process. It occurs after the grit screen and the grit chamber and is characterised by the contamination of unsolved waste water. The sludge collecting at the bottom of the main sedimentation basin is often referred to as primary sludge. Composition is determined by the characteristics of the catchment area. The consistency is a dense liquid with a water content ranging from 93 to 97%.

5.1 Characteristics of Sludge

Sewage sludge is a form of waste generated by municipal wastewater treatment plants. These ones handle large quantities of water every day to kill bacteria, viruses, and toxins from animals. The primary outputs of these plants are treated water and waste sludge. Sludge is digested anaerobically and dehydrated, accounting for 1% of the waste water entering the plant. After mechanical drying, approximately 80% of sewage sludge is moisture, with the remaining 20% being dry matter (Table 6).

Table 6 Country wise overview of sewage and sludge

Properties	India (Delhi)	China	Brazil	Spain
pH	6.41	7.45	6.60	7.90
EC (dSm^{-1})	1.39	–	–	1.20
Organic Carbon (%)	7.64	42.55	24.80	18.30
N (%)	1.48	6.48	2.10	2.22
P (%)	1.61	1.65	1.0	1.66
K (%)	0.20	0.49	–	0.47

5.1.1 Physical Properties

Sludge is essentially very dense particle dispersion with a broad range of particle sizes. The interactions of these particles, both with each other and with soluble constituents, are crucial in deciding the properties of the sludge.

6 Main Contaminants of Sewage- Sludge

6.1 Organic Content

It is one of the most crucial factors to consider while designing and operating sewage treatment plants. BOD levels in industrial sewage can be several times higher than in residential sewage. The BOD of storm sewage is especially important when it is mixed with residential sewage in combined sewerage systems.

6.2 Suspended Solids

Suspended particles are another significant feature of sewage. The total suspended particles in the sewage are proportional to the quantity of sludge created in the treatment facility. Suspended solids concentrations in industrial and storm sewage can be greater than in household sewage. The quantity of suspended particles and BOD collected by the treatment plant determines the efficacy of a treatment technique.

6.3 Nutrients from Plants

Both nitrogen and phosphorus molecules, which are important plant nutrients, may be found in sewage. Algae may grow fast in lakes with high levels of nitrates and phosphates. Algal blooms, which are frequently generated by sewage discharges, promote eutrophication, which is the natural ageing of lakes.

6.4 Microorganisms

Per gallon, domestic sewage includes millions of bacteria. The bulk of the bacteria are coliform bacteria from the human gut, although sewage can also include other microorganisms. Coliforms are employed to identify contamination in water. A high coliform level typically means that sewage has recently been polluted.

7 Sewage- Sludge Treatments

Sludge treatments attempt to minimise sludge weight and thickness in order to lower disposal costs and the health concerns associated with other disposal techniques. Pathogens can be killed by heating during thermophilic digestion, composting, or burning, but the most frequent way to lose weight and volume is by water absorption. When choosing a sludge treatment system, the quantity of sludge generated and the treatment expenses necessary for the various disposal options must be taken into account.

7.1 Treatment Processes

7.1.1 Thickening

It is frequently the initial stage in the treatment procedure. Sludge can be separated from primary or secondary clarifiers to generate bigger aggregates that settle more easily. Clarifiers are sometimes mistaken for thickeners with a stirring motor (Steel and McGhee 1979). Sludge thickened to fewer than 10% solids undergo further sludge treatment, while the liquid thickener output is returned to the sewage treatment cycle.

7.1.2 Dewatering

To minimise transportation costs and increase composting compatibility, sludge's water content can be lowered by centrifugation, filtering, and/or evaporation. Centrifugation can be used as a preliminary step before filtering or evaporation to reduce sludge volume.

7.1.3 Digestion

The biological process of turning organic materials into stable chemicals is known as sludge digestion. Digestion lowers the total bulk of solids, kills germs, and smooths out the dewatering and drying of sludge. Digested sludge, which resembles a rich potting soil in look and properties, is completely safe.

7.1.3.1 Anaerobic Digestion

In the absence of oxygen, anaerobic digestion is a bacterial process. The procedure can be either thermophilic digestion, which ferments sludge in tanks at 55 °C, or mesophilic digestion, which ferments sludge in tanks at 36 °C.

Mesophilic anaerobic digestion is widely used to remediate sludge from water treatment facilities (MAD). To allow the digestion process to complete all four phases, the sludge is put into huge tanks and kept there for at least 12 days. Processes such as methanogenesis, hydrolysis, acidogenesis, and acetogenesis all occur.

7.1.3.2 Aerobic Digestion

Aerobic and traditional anaerobic digestion converts the vast majority of biological sludge solids to liquids and gases. Thermal hydrolysis and anaerobic digestion will convert 60–70% of solid materials to liquids and gases. Not only are the particles released less than in traditional digestion, but enhanced biogas processing might make certain wastewater treatment plants energy self-sufficient (Fig. 1).

7.1.4 Composting

It is an aerobic process that involves mixing sewage sludge with carbon-rich materials such as sawdust, compost, or wood chips. Bacteria that digest sewage

Fig. 1 Thermal hydrolysis system

sludge and plant debris produce heat in the presence of oxygen, which kills disease-causing germs and parasites (Primer for Municipal Wastewater Treatment Systems Report 2004).

Bulking agents that allow air to pass through the fine sludge particles are used to maintain aerobic conditions of 10–15% oxygen. Maize cobs, nut shells, shredded tree-pruning waste, or paper mill bark are stiffer than lighter leaves and grass clippings in detecting ventilation sludge (Reed et al. 1988; Use of Composting for Bio-solids Management: Report 2002).

In a composting mixture, the initial carbon-to-nitrogen ratio should be between 26 and 30:1. The amount necessary to dilute hazardous chemical concentrations in sludge to levels suitable for compost consumption will, however, be utilised to determine the composting ratio of agricultural by products (Reed et al. 1988). Although most agricultural by-products are low in toxicity, residual herbicide levels in suburban grass clippings may become toxic for agricultural uses, and prevent seedling emergence (Aslam et al. 2008).

7.1.5 Incineration

Sludge incineration is becoming less common owing to concerns about air pollution and the additional fuel needed to burn the low calorific content of sludge while also evaporating remaining water.

Stepped multiple hearth incinerators with extended residence times and fluidized bed incinerator is a common gadget for combusting wastewater sludge. Co-firing is an alternative as it is less expensive if solid waste facilities are already in place and no auxiliary fuel is required (Primer for Municipal Waste water Treatment Systems (Report) 2004).

7.1.6 Drying Beds

Simple drying beds are used in many countries, particularly in impoverished countries, since they are a cheap and simple method of drying sewage sludge. Drainage water must be collected, and drying beds are sometimes shaded but generally open. Devices that flip the sludge over in the early stages of the drying process are now available in the market.

These are typically constructed with four layers of gravel and sand. The coarse gravel layer, which is 150–200 mm deep, is the initial layer. There is also fine grit that is 10 cm thick. Sand, which can range in size from 100 to 150 mm. Sludge dries out and moisture seeps through to the first sheet, accumulating beneath all drains.

8 Wastewater Treatments

It is the process by which organic matter and other contaminants are separated from wastewater. These are designed to keep wastewater clean and safe for release into the environment without harming the local ecology or inhabitants.

A wastewater treatment system could be used by a city to disinfect sewage and flood water. An industrial manufacturing plant may have an on-site wastewater treatment system or partner with nearby wastewater treatment facilities to decontaminate the chemical-filled process water.

Despite the fact that environmental variables have an impact on water quality, pollution implies that contamination is caused by human action. As a result, the major cause of water pollution is the discharge of untreated wastewater (Fig. 2).

8.1 Types of Treatment Systems for Waste Water

8.1.1 Sewage Treatment Plants (STPs)

A sewage treatment plant (STP) is a type of treatment facility that may be found in a large American metropolis. This facility receives an excess of household waste water and gathers untreated trash from houses and businesses. Rainwater is collected, as well as debris from flood drains.

A STP like this keeps people safe and healthy by cleaning their waste water before releasing it into the environment, using a combination of physical, mechanical, and biological treatment. When wastewater reaches a sewage treatment facility, it must first pass through a simple filtering process.

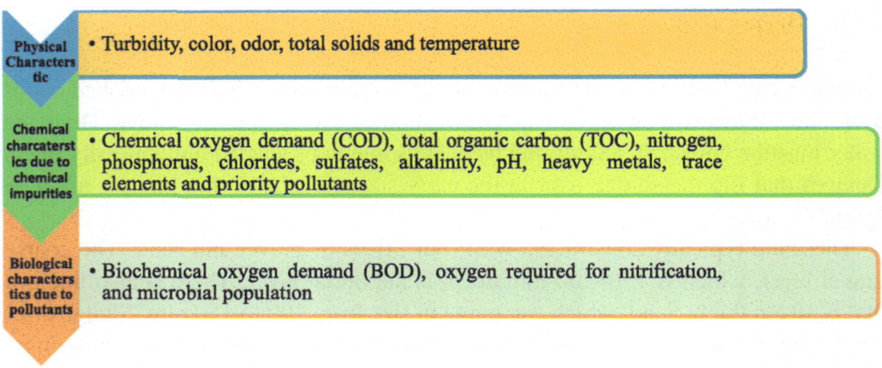

Fig. 2 Physical, chemical and biological characteristics of waste water

8.1.1.1 Primary Treatment

Initial clarifiers filter the wastewater during primary treatment. Wastewater flows slowly in these tanks, as it does in the settlement basins. The tanks' design allows organic materials to settle at the bottom and lighter substances to ascend to the surface for removal, facilitating settling.

A main sludge blanket is made up of organic materials that fall to the ground. This primary sludge is transported to aeration basins for the activated sludge process, which is the backbone of most STPs, after a few hours in the clarifying tanks.

8.1.1.2 Secondary Treatment

It is a wastewater (or sewage) treatment system that uses a physical phase separation sewage treatment plant to separate settleable solids from dissolved and suspended organic compounds and a biological process for removing suspended organic materials. Waste water is referred to be secondary-treated waste water after this type of application.

Secondary treatment is a step in the wastewater treatment process that eliminates dissolved and colloidal chemicals and determines how much biochemical oxygen is required. Temperature has an effect on biological oxidation cycles, and between 0 and 40 °C, the pace of biological responses rises. Temperatures in aerated surface vessels typically vary from 4 to 32 °C.

8.1.1.3 Tertiary Treatment

Tertiary treatment is intended to offer a last degree of treatment before the effluent is released to the receiving zone. The removal of biological nutrients, disinfection, and the elimination of micro-pollutants such as persistent pharmaceutical pollutants from the atmosphere are all examples of tertiary treatment.

8.1.2 Effluent Treatment Plants (ETPs)

Leading pharmaceutical and medicine companies employ effluent treatment facilities (ETPs) to eliminate hazardous chemicals. All companies utilise these plants to preserve the environment. An effluent treatment plant (ETP) is a facility that treats wastewater and waste water.

A number of effluents and pollutants are produced during the manufacturing of medicines. Organic compounds, dirt, dust, grit, trash, dangerous and non-toxic chemicals, polymers, and other pollutants are removed from medicines and other medicinal goods using effluent treatment plants. Evaporation and drying, as well as centrifugation, filtration, chemical process incineration, and effluent disposal, among other auxiliary techniques (Fig. 3).

Fig. 3 Sludge drying bed

Proper management of wastewater is essential for preventing water contamination. Biodegradable organics, if left unresolved, will kill the bacterial treatment beds, resulting in contamination of controlled waters during the purifying process.

8.1.3 Activated Sludge Plants (ASPs)

Activated sludge is made up of flocks (incorrectly referred to as flocks) produced by active bacteria cells suspended and diluted in waste water; bacteria extract biodegradable organic matter and nutrients from civilian black and grey domestic or comparable waste water.

8.1.4 CETPs

CETPs are treatment systems for treating small-scale industrial wastewater generated by a group of industries. Individual effluent treatment plants (ETPs) encounter problems owing to a lack of capacity, manpower, capital expenses, and a skilled operation and maintenance workforce in general, which is compounded for small-scale industrial facilities.

These problems are alleviated by treating wastewater from a wide range of small-scale facilities at a single location, where the effluent is treated in the same manner as that which would be treated individually. This allows for more treatment to be done at a single location, as well as easier management and maintenance and cheaper land protection expenses.

9 Wastewater Treatment Process

To work, sewage must be delivered to a treatment facility via appropriate pipelines and equipment, and the process must be monitored and regulated. Any waste water must be treated using specialised waste water treatment equipment. The initial stage in sewage disposal and the treatment of most waste water is the separation of particles from liquids, which is usually accomplished by sedimentation.

9.1 Phase Separation

In this, impurities are transferred to a non-aqueous level. This can occur at intermediate stages in a treatment series to remove particles generated during oxidation or polishing. Grease may either be recycled as fuel or saponified. Sludge dewatering is commonly necessary in a wastewater treatment facility due to particles.

9.2 Sedimentation

It is a physical water treatment method that removes suspended particles from water using gravity. Due to sedimentation, strong particles entrained by turbulence in moving water may naturally dissolve in the calm water of lakes and seas. Settling basins are sediment-filled reservoirs that may be used to remove entrained materials from the environment.

9.3 Oxidation

It lowers the toxicity of contaminants while reducing the demand for oxidative oxygen in waste water. Chemical compounds are converted to CO_2, H_2O and biosolids during secondary treatment.

9.4 Polishing

These treatments can also be used on their own to treat some municipal waste water. After chemical oxidation, chemical reduction or pH adjustment reduces the chemical reactivity of waste water. Chemical absorption removes any residual toxins and impurities from activated carbon during carbon filtering.

10 Advantages of Waste- Water Treatment

This not only provides clean, reused water, but it also having the potential to deliver a number of additional benefits. It may reduce trash creation in a given region, generate electricity from methane extraction, and produce natural fertiliser from waste generated during the process (Walker 1994; Walker and Bernal 2008)

10.1 Reducing Waste

Trash water disposal decreases the quantity of waste that is now released into the ecosystem, therefore enhancing environmental health. In exchange, the government avoids the health hazards connected with air pollution and reduces waste-related water loss.

10.2 Making Fertilizer

In "drying lagoons," whatever biodegradable material that remains is dried and turned into natural fertiliser. This cuts down on the use of chemical fertilisers, which pollute the area's marine and surface ecosystems (Table 7).

Dubey et al. (2006) discovered higher levels of arsenic (8–23 mg kg^{-1}), Cd (2–9 mg kg^{-1}), Cr (66–1098 mg kg^{-1}), Hg (7–32 mg kg^{-1}), Ni (12–596 mg kg^{-1}) and Pb (12–596 mg kg^{-1}) (26–154 mg kg^{-1}). Kumar and Chopra (2013) conceptualize that handsome amounts of nutrients in urban sewage.

Table 7 Sewage-sludge and growth and yield

Crop	Effect on crop growth and yield	Source
Rice	Different rates @ 3.0, 4.5, 6.0, 9.0 and 12.0 kg m^{-2} increase the grain yield by 60, 111, 125, 134 and 137% respectively	Singh and Agrawal (2010a)
Wheat	–	NEPA (1992)
Sunflower	Application of sewage sludge @ 0.7 & 1.4 kg m^{-2} increase the yield of sunflower by 30% and 31% respectively	Lavado (2006)
Maize	Application of sewage sludge @ 0, 22.7, 45.5 and 91.0 g pot^{-1} increase yield of the maize	Chitdeshwari et al. (2002)
Mung Bean	–	Singh and Agrawal (2010b)
Broad Bean	3 times higher yield of broad bean grown in sludge treated soil @ 4.5 kg m^{-2}	Garrido et al. (2005)

11 Conclusion

Sewage sludge management's primary aim is to protect the environment while still taking into account environmental health and socioeconomic issues. Sewage and sludge may be treated in a variety of ways before being disposed of, including primary, secondary and tertiary treatments. Furthermore, sewage sludge can be used for a variety of uses, including composting, energy production, pesticide production, environmental amendment, and more.

Acknowledgement First author would like to thank all the co-authors especially Dr. Himanshu Verma for his valuable suggestions, cooperation and guidance during chapter writing.

References

Alghobar MA, Suresha A (2017) Evaluation of metal accumulation in soil and tomatoes irrigated with sewage water from Mysore city, Karnataka, India. J Saudi Soc Agric Sci 16:49–59

Aslam DN, Vandergeynst JS, Rumsey TR (2008) Development of models for predicting carbon mineralization and associated phytotoxicity in compost-amended soil. Bioresour Technol 99 (18):8735–8741

Chitdeshwari T, Savithri P, Mahimairaja S (2002) Effect of sewage biosolid composts on the yield of crops. Indian J Environ Protect 21(10):911–912

Datta SP, Young SD (2005) Predicting metal uptake and risk to the human food chain from leaf vegetables grown on soils amended by long-term application of sewage sludge. Water Air Soil Pollut 163:119–136

Dede G, Ozdemir S, Dede OH (2012) Effect of soil amendments on phytoextraction potential of Brassica juncea growing on sewage sludge. Int J Environ Sci Technol 9:559–564

Dubey SK, Yadav RK, Chaturvedi PK, Goyel, B, Yadav R, Minhas PS (2006) Agricultural uses of sewage sludge and water and their impact on soil water and environmental health in Haryana, India. Abstract of 18th world congress of soil science, Philadelphia, 9–15 July

Eid EM, Alrumman SA, El-Bebany AF, Hesham AEL, Taher MA, Fawy KF (2017) The effects of different sewage sludge amendment rates on the heavy metal bioaccumulation, growth and biomass of cucumbers (*Cucumis sativus* L.). Environ Sci Pollut Res 24:16371–16382

El-Motaium RA, Abo El-Seoud MA (2007) Irradiated sewage sludge for production of fennel plants in sandy soil. Nutr Cycl Agroecosyst 78:133–142

Garrido S, Del Campo GM, Esteller MV, Vaca R, Lugo J (2005) Heavy metals in soiltreated with sewage sludge composting, their effect on yield and uptake of broad bean seeds (*Vicia faba* L.). Water Air Soil Pollut 166(1):303–319

Ghosh AK, Bhatt MA, Agrawal HP (2012) Effect of long-term application of treated sewage water on heavy metal accumulation in vegetables grown in Northern India. Environ Monit Assess 184:1025–1036

Jamali MK, Kazi TG, Arain MB, Afridi HI, Jalbani N, Baig JA, Niaz A (2008) Effect of liming on the distribution of heavy metals in untreated industrial sewage sludge produced in Pakistan for the cultivation of Sorghum bicolor (L.). Environmentalist 28:366–375

Kumar V, Chopra AK (2012) Translocation of micronutrients in French bean (*Phaseolus vulgaris* L.) grown on soil amended with paper mill sludge. J Chem Pharmaceut Res 4(11):4822–4829

Kumar V, Chopra AK (2013) Accumulation and translocation of metals in soil and different parts of French bean (*Phaseolus vulgaris* L.) amended with sewage sludge. Bull Environ Contam Toxicol 92(1):103–108. https://doi.org/10.1007/s00128-013-1142-0

Kumar V, Chopra AK (2016) Effects of sugarcane press mud on agronomical characteristics of hybrid cultivar of eggplant (*Solanum melongena* L.) under field conditions. Int J Recycl Organ Waste Agric 5:149–162. https://doi.org/10.1007/s40093-016-0125-7

Kumar V, Chopra AK, Srivastava S (2016) Assessment of heavy metals in spinach (Spinaciaoleracea L.) grown in sewage sludge amended soil. Commun Soil Sci Plant Anal 47 (2):221–236. https://doi.org/10.1080/00103624.2015.1122799

Lavado RS (2006) Effects of sewage sludge application on soils and sunflower yield: quality and toxic element accumulation. J Plant Nutr 29(6):975–984

LeiteMoretti SM, Bertoncini EI, Vitti AC, Alleoni LRF, Junior CHA (2016) Concentration of Cu, Zn, Cr, Ni, Cd, and Pb in soil, sugarcane leaf and juice: residual effect of sewage sludge and organic compost application. Environ Monit Assess 188(3):163. https://doi.org/10.1007/s10661-016-5170-1

McBride MB, Richards BK, Steenhuis T (2004) Bioavailability and crop uptake of trace elements in soil columns amended with sewage sludge products. Plant Soil 262:71–84

Meena R, Datta SP, Golui D, Dwivedi BS, Meena MC (2016) Long-term impact of sewage irrigation on soil properties and assessing risk in relation to transfer of metals to human food chain. Environ Sci Pollut Res 23:14269–14283

Naddeo V, Liu H (2020) Editorial perspectives: 2019 novel coronavirus (SARS-CoV-2): what is its fate in urban water cycle and how can the water research community respond. Environ Sci Water Res Technol 6(5):1213–1216. https://doi.org/10.1039/D0EW90015J

National Environmental Protection Agency (NEPA) (1992) Study on municipal wastewater reuse. Science Press, Beijing

Nogueirol RC, De Melo WJ, Bertoncini EI, Alleoni LRF (2013) Concentrations of Cu, Fe, Mn, and Zn in tropical soils amended with sewage sludge and composted sewage sludge. Environ Monit Assess 185:2929–2938

Pakhnenkoa EP, Ermakova AV, Ubugunovb LL (2009) Influence of sewage sludge from sludge beds of Ulan Ude on the soil properties and the yield and quality of potatoes. Moscow Univ Soil Sci Bull 64(4):175–181

Primer for Municipal Waste water Treatment Systems (Report) (2004). Washington, DC: US Environmental Protection Agency (EPA). EPA 832-R-04-001

Ramachandran V, D'Souza TJ (1998) Plant uptake of cadmium, zinc, and manganese in soils amended with sewage sludge and city compost. Bull Environ Contam Toxicol 61:347–354

Reed SC, Middle Brooks EJ, Crites RW (1988). Natural systems for waste management and treatment. McGraw-Hill, New York, pp 268–290. ISBN 0070515212. OCLC 16087827

Rodríguez AR, Castro S, Losada MRM (2010) Effects of dose and period of sewage sludge application on soil tree and pasture components in a Pinusradiata D. Don silvopastoral system. Agrofor Syst 79:237–247. https://doi.org/10.1007/s10457-010-9281-3

Rogers H (2012) Perspective for biogas in Europe. Oxford Institute for Energy Studies

Selivanovskaya SY, Latypova VZ, Gubaeva LA (2006) Microbiological processes in gray forest soil treated with sewage sludge compost. Eurasian Soil Sci 39(4):443–449

Shahbazi F, Ghasemi S, Sodaiezadeh H, Ayaseh K, Ahmadmahmoodi RZ (2017) The effect of sewage sludge on heavy metal concentrations in wheat plant (Triticumaestivum L.). Environ Sci Pollut Res 24:15634–15644

Shrivastava SK, Banerjee DK (2003) Speciation of metals in sewage sludge and sludge amended soils. Water Air Soil Pollut 152:219–232

Sigua GC, Adjei MB, Rechcigl JE (2005) Cumulative and residual effects of repeated sewage sludge applications: forage productivity and soil quality implications in South Florida, USA. Environ Sci Pollut Res 12(2):80–88

Sikka R, Nayyar V, Sidhu SS (2009) Monitoring of Cd pollution in soils and plants irrigated with untreated sewage water in some industrialized cities of Punjab, India. Environ Monit Assess 154:53–64

Singh RP, Agrawal M (2010a) Variations in heavy metal accumulation, growth and yield ofrice plants grown at different sewage sludge amendment rates. Ecotoxicol Environ Saf 73 (4):632–641

Singh RP, Agrawal M (2010b) Effect of different sewage sludge applications on growth and yield of Vignaradiata L. field crop: metal uptake by plant. Ecol Eng 36(7):969–972

Spinosa L (2008) Status and perspectives of sudge management. CNR

Steel EW, McGhee, Terence J (1979) Water supply and sewerage (5th edn). McGraw-Hill, New York, pp 535–545. ISBN 0070609292. OCLC 3771026

Tziachris P, Lekakis P, Zambetoglou K, Metaxa I, Papadopoulos F (2017) A case study of the effects of sewage sludge application on soil properties and heavy metal availability in the Thessaloniki Plain (Greece). Waste Biomass Valori 8:1803–1811

Ullah H, Khan I (2015) Effects of sewage water irrigation of cabbage to soil geochemical properties and products safety in periurban Peshawar, Pakistan. Environ Monit Assess 187:126. https://doi.org/10.1007/s10661-015-4344-6

Ullah H, Khan NU, Ali F, Shah ZA, Ullah Q (2017) Health risk of heavy metals from vegetables irrigated with sewage water in peri-urban of Dera Ismail Khan, Pakistan. Int J Environ Sci Technol. https://doi.org/10.1007/s13762-017-1384-1

Use of Composting for Bio-solids Management (Report). Bio-solids Technology Fact Sheet. EPA. September 2002. EPA 832-F-02-024

Walker JM (1994) Production, use, and creative design of sewage sludge biosolids. In: Clapp CE, Larson WE, Dowdy RH (eds) Sewage Sludge: land utilization and the environment. American Society of Agronomy, Inc., Crop Science Society of America, Inc., Soil Science Society of America, Inc., Madison, WI, pp 67–74

Walker DJ, Bernal MP (2008) The effects of olive mill waste compost and poultry manure on the availability and plant uptake of nutrients in a highly saline soil. Bioresour Technol 99:396–403

Williams PT (2005) Waste treatment and disposal. Wiley

Biological and Thermo-chemical Treatment Technologies for Sustainable Sludge Management

Sobia Riaz, Zia Ur Rahman Farooqi, Predrag Ilic, Muhmmad Saad Bin Zafar, Hanan Ahmad, Moaz Khursheed, Noman Gulzar, Mohsin Raza, and Rizwan Mehmood

1 Introduction

Natural resources are under severe pressure due to exponential growth of population, improved quality of lifestyle and an increasing need of neat, pure and worth living environment. Wastewater in the form of effluents from different industrial and domestic activities contains higher amount of organic, inorganic and heavy metal contaminants which needs to be removed physically, chemically or biologically prior its disposal into natural water resources of dumping in open surface, however wastewater is rich source of energy (Gude 2015). Most of the wastewater methods in practice these days are energy consuming and are not economical such as aerated systems. Sewage sludge (SS) is the end-product of wastewater treatment that is produced in bulk quantities and is used as chief source of energy feedstock (Rulkens 2008). Wastewater treatment includes primary, secondary and tertiary treatment. Primary treatment includes physical and chemical separation through sedimentation pH adjustment and aeration, being followed by secondary treatment (biological and chemical treatment by organic substrate degradation with the help of microbes along with biochemical procedures) and subsequently other treatment approaches such as disinfection, filtration and aeration, called tertiary treatment (Costa et al. 2019).

Such treatment is used as standard method of wastewater treatment plants globally and usually named as biological or secondary treatment. The process of

S. Riaz · Z. U. R. Farooqi (✉) · M. S. B. Zafar · H. Ahmad · M. Khursheed · N. Gulzar · M. Raza
Institute of Soil and Environmental Sciences, University of Agriculture, Faisalabad, Pakistan

P. Ilic
PSRI Institute for Protection and Ecology of the Republic of Srpska, Banja Luka, Republic of Srpska, Bosnia and Herzegovina

R. Mehmood
Department of Entomology, University of Agriculture, Faisalabad, Pakistan

V. D. Rajput et al. (eds.), *Sustainable Management and Utilization of Sewage Sludge*, https://doi.org/10.1007/978-3-030-85226-9_19

wastewater treatment might be energy intensive but there are several approaches that can be adopted to make it energy-efficient and energy producing (Longo et al. 2016).

Sewage sludge may contain sludge from both sources i.e., from industrial and municipal but authorities have more concern with municipal sewage sludge because it may contain a huge content of human excreta as it comes from municipal wastewater treatment plants (WWTPs). Usually, there is 1% wastewater inflow in municipal WWTPs before the processes of dewatering and thickening. Sewage sludge contains a reasonable amount of fecal components, nutrients, and organic components hence it has diverse application as soil conditioner and medium-level fertilizer. Application of SS as soil conditioner or fertilizer may have better results compared to artificial fertilizer because it holds more nutrients and make those nutrients available for plants for longer time (Wiśniowska et al. 2019). However, there are many health risks associated with the soil application of SS are because of higher content of pathogens including bacteria, virus, protozoa or helminths and other organic or inorganic micro-pollutants (Wiśniowska et al. 2019). In general, there is about 3% of total nitrogen (N), 30–55% of organic matter (OM) content, 0.7% total potassium (K), 0.7–1.5% phosphorus (P), 10–20% carbon to nitrogen (C/N) ratio and different concentration of heavy metals content in stabilized SS, while its pH ranges from 6.5 to 7.5. Total energy yield from dry SS can be 12–15 kJ kg^{-1}. SS may contain a variety of plant essential nutrients and other components including sulfur (S), potassium (K), and magnesium (Mg) (Kijo-Kleczkowska et al. 2012).

Upon comparison, it is found that fecal sludge has more nutrient contents compared to sludge produced from WWTPs as it has higher amount of suspended solids (SS) 3% versus 1%, greater chemical oxygen demand (COD) (can be higher than 10,00 mg/L) (Niwagaba et al. 2014). Sewage sludge may have higher content of fecal coliform bacteria and helminths eggs i.e., 1×10^5 CFU/100 mL and up to 16,000 number/L, respectively. We can treat such SS in same way. Sludge produced from industrial activities can be very toxic as it may possess a varied physical and chemical properties and may contain a variety of hazardous compounds in higher quantity (Wiśniowska et al. 2019). When SS is treated in plants through the processes of mechanical dewatering, anaerobic or aerobic digestion and incineration, it normally costs about 50% of total cost in the whole facility. Treatment of SS in plants has significant importance because it reduces the volume and hence decreasing the disposal cost. Consequently, while planning SS management strategies, it is vital to take in consideration all the possible alternative technologies for SS management and removal (Zhang et al. 2019). As a result of industrialization and improvement in lifestyle, increased sewerage, building of new and up-gradation of existing infrastructure has increased the rate of sludge production hence the modern society is facing the perilous problem of SS management in a way that is sustainable in both ways economically and environmentally. Along with this, there are more problems regarding finding the disposal facilities as well as fulfilling the legislative requirements concerning environmental quality. Additionally, from the last few decades there is an increasing trend of energy recovery approaches and re-use of waste materials as well as making a global strategy to follow regarding prioritization of the different waste (Zangmo 2017).

2 Composition of Sewage Sludge

Generally, many factors directly affect the characteristics and composition of SS such as the coagulants in use, source of wastewater stream, treatment approach used for wastewater, as well as time and prevailing conditions during storage. SS consists-of microbes, micronutrients and macronutrients, non-essential trace elements and organic compounds and micro-pollutants (Sun et al. 2019).

2.1 Strategies in Sludge Management

The end-product of wastewater treatment processes is a semi-solid filtrate called SS. It may consist of a large variety of organic or inorganic compounds, biodegradable compounds, along with a reasonable amount of heavy metals and other pathogenic content. Sludge is also seen as a potential source of energy and nutrients which can be restored by the application of economically practicable techniques. The total energy efficiency can be increased, and carbon footprints can be decreased by reusing and recovering the energy content of by-products (Zhang et al. 2014).

The management of SS is very of prime importance because of its concerned environmental contamination and health hazards. The optimization of handling techniques is a serious concern and hence an extremely detailed rheological compositional properties of SS has been given in order to aid in SS management, treatment and disposal services. While concerning with the disposal activities, selection of both procedure and equipment in order to transport the SS or for the direct land application is stalwartly inter-linked with consistency of sludge. However, instead of solid contents, the shear strength of sludge directly affects the formation of SS pile for composting or landfilling (Spinosa and Lotito 1999).

From past few decades, researchers have conducted detailed research work on the handling, treatment and disposal techniques of the SS and meaningful innovations in both technical and administrative context have been accomplished. But on the other hand, convincing the community about the use, nutritient value and importance of sludge is another challenge. It is also stated that a preliminary decrease in the total quantity or improving the quality of SS may have a direct or indirect positive impact on the further processes such as energy recovery or extraction of useful component (Spinosa 2004a). After a series of research work and innovative technologies we are standing at the point stating that SS is a renewable source of nutrients, organic matter, water and energy; hence not a waste. Co-digestion of SS along with organic waste such as food materials in order to enhance biogas productivity, phosphorus recovery and for other agricultural applications, is one of the central aim of energy value harvesting technologies mainly energy recovery by thermal process and anaerobic digestion on which sludge treatment and management is based. Application of SS in agriculture is routinely practiced but it will be more difficult because there is a general shift in trend of SS quality for land use hence limiting concentrations of certain pathogens, contaminants, and heavy metals (Silva et al. 2018).

Another expected alteration is the change in wastewater treatment processes to decrease the SS production in future either by the conversion of aerobic to anaerobic process or by the use of preliminary treatment technologies. Moreover, innovation is also required in the processes of both thickening and dewatering technologies. Ultimately, SS consists of water up to 95–99% and tis water content is determinant of further treatment processes to be selected as well as the viability of future land use, landfilling and incineration (IWA 2019).

2.2 Sludge Treatment and Disposal

Basically, SS is the by-product of WWTPs, and it may be solid, semi-solid or any residue in the form of slurry that can be categorized into primary and secondary SS. Primary processes such as sedimentation, chemical precipitation and others produce primary SS while secondary processes such as biological treatment forms secondary SS. Some on-site wastewater treatment system loads SS plants with septic tanks solids. Meanwhile, sometimes SS from both primary and secondary sources are mixed prior additional disposal or treatment (Aradelli and Cantù 2016). Wastewater treatment plants are mainly designed on the basis of treatment and disposal approach to be used for SS. Most of the times, SS is only treated before disposal in order to decrease its final volume as well as for the stabilization of organic materials. After stabilization, SS becomes less odorous and its handling is easy in context of health risks. Eventually, less volume lessens the pumping, transportation, and storage costs (Britannica 2020).

2.3 Treatment Methods

Sewage sludge management methods include treatment, recovery, recycling or disposal.

These terms were defined by Directive 2008/98/EC of European parliament and Council (Directive 2008/98/EC, 2008) as:

- *Recovery* refers to any process that results in the useful function of waste or discarded materials by the direct use in replacement of any raw materials to be used or being processed in plat to meet a specific purpose.
- *Recycling* is any further reprocessing or waster or discarded materials which results in new products, substances or materials that can be re-used for specific

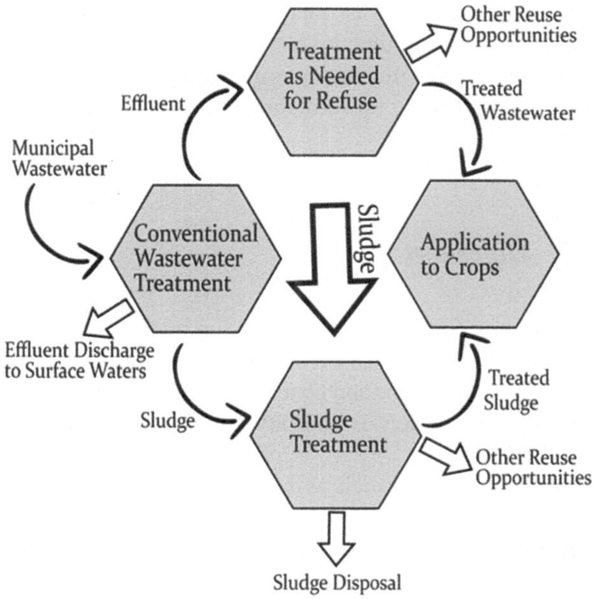

Fig. 1 Fate of sewage sludge

purpose. In this technique organic materials are reprocessed for further use as fuels and other needs such as backfilling activities with an exemption of energy recovery.

- Disposal refers to such processes that are not listed under recovery even in secondary value the operation is not done for the energy or material recovery.
- *Treatment* includes processes done before the final disposal or recovery hence it refers to the disposal or recovery operations.

Primary and secondary sludge have total energy content of 15–15.9 MJkg^{-1} and 12.4–17.3 MJkg^{-1} respectively. Hence, if the sludge is subject to co-digestion after primary and secondary treatment in combination with other organic waste (including fats, oils and grease) from different industrial processing units may increase the energy recovery rendering it even more energy positive as shown in Fig. 1. When co-digestion is unviable, primary sludge recovery can be done using many processes including amendments. Sewage sludge treatment involves a variety of thermal, chemical, and biological processes along with thickening, dewatering and digestion processes.

2.4 Thickening

When SS treatment starts, it is necessary to do thickening because handling of raw SS having solid suspending in water in the form of a thin slurry is unpractical and this whole procedure is done in a tank named gravity thickener. Thickening can decrease the SS volume by 50% of its original volume. Another procedure that can be done instead of thickening is dissolved-air flotation including air bubbles carrying suspended solids to upward surface forming a layer of thickened sludge (Van Lier et al. 2008).

2.5 Digestion

Digestion of SS is a process involving microorganisms decomposing organic solids into stable compounds. As a result, a sludge is produced with less total mass of solids, less pathogens and hence dewatering or drying of such sludge is more feasible as well as less odorous and physically resembling the fertile potting soil (Elalami et al. 2019).

Microorganisms such as bacteria are used for the anaerobic metabolization of organic substances in most of the large SS treatment units involving two-stage digestion system. First of all, the SS is thickened and reduced to 5% and then kept in a closed tank with heating and mixing for many days. During this process, microbiological activity occurs, and acid-forming bacteria are involved in this action by hydrolyzing and breaking larger molecules including lipids and proteins into simpler ones eventually forming different fatty acids by the fermentation of these simple molecules. This thickened and microbiologically decomposed sludge then moves to another tank, second phase, in which conversion of dissolved matter into biogas; mixture of methane and carbon dioxide is done by some other bacteria. Methane is inflammable and hence it can be used in primary digestion tank as a fuel and it can generate electricity for other processing units as well (Meng et al. 2017) (Fig. 2).

Anaerobic can be influenced by many factors including acidity, temperature, pH and many more hence it need careful handling and control. In order to enhance the activity of bacteria involved in the process of digestion, sometimes SS is injected with some additional hydrolytic enzymes at the initial stage. This enzymatic inoculation is useful in destroying the harmful pathogens present in SS as well as it also generates more quantity of methane and carbon dioxide eventually biogas during the process of digestion. This conventional two-stage digestion process can be enhanced in another way and that is thermal hydrolysis or the use of heat to breakdown the complex molecules into simpler ones and it is done prior digestion at a separate stage (Lin et al. 2018). Usually, thermal hydrolysis of such SS is done that is dewatered and having a solid part up to 15%. First of all, steam is combined and homogeneously mixed with steam in a tank called pulper that is further supplied to another reactor having temperature about 165 °C and kept under pressure for almost 30 minutes over there. Subsequently, produced steam is drained off to the pulper and as this hydrolytic breakdown completes some of the sludge under high pressure is curtly released in another flash tank. This abrupt change in pressure causes disruption in the cell walls of solid material. The resulting hydrolyzed sludge is

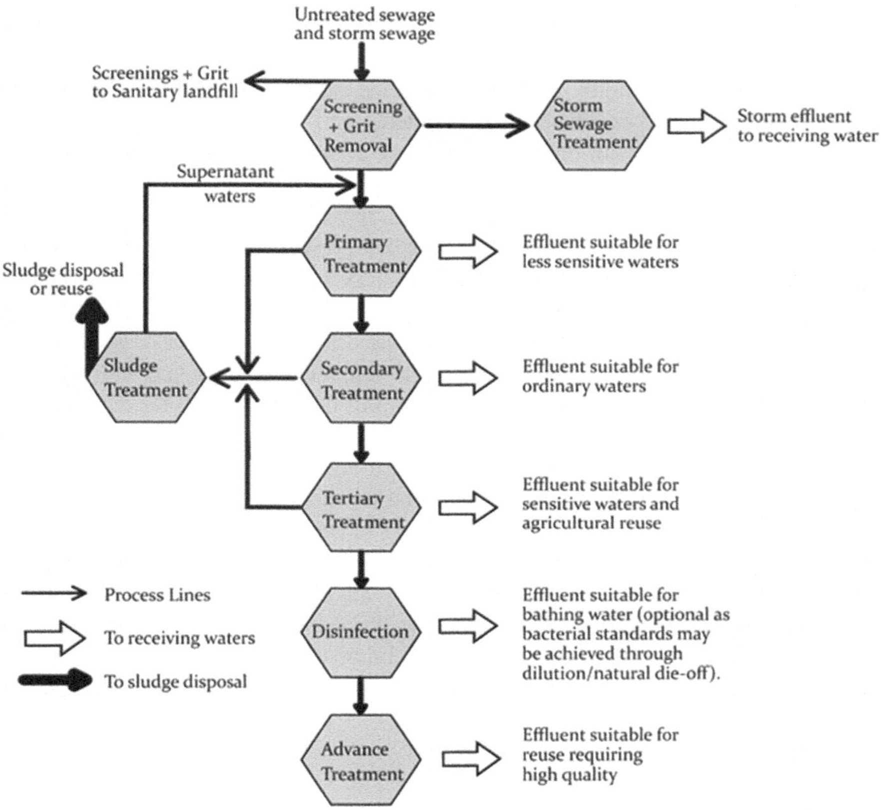

Fig. 2 Sewage sludge treatment methods

cooled and diluted using water and subjected to second stage of anaerobic digestion that may occur aerobically as well. Aeration of SS is done in open tank for the duration of 20 days, but this process does not produce methane (Kumar and Samadder 2020).

Generally, aerobic systems are easier to handle and operate as compared to anaerobic but anaerobic systems are economical when it comes to cost because aerobic systems need more power supply for continuous aeration. Anaerobic digestion is sometimes coupled with stabilization systems or some extended aeration. When we treat SS using aerobic or typical anaerobic digestion processes then most of its solid organic contents is converted into gases or liquids. Anaerobic digestion after the thermal hydrolysis of SS converts 60–70% of solid content into liquid or gas. Hence, the quantity of solid end-products is lessened compared to conventional digestion as well as the biogas production can be used to fuel the wastewater treatment plants self-sufficient in energy (Ward et al. 2014).

2.6 Dewatering

Sewage sludge is commonly dewatered or dried prior to its disposal. In spite of dewatering, sludge still contain moisture content up to 70% but meanwhile it cannot be seen as liquid even with this much moisture content and hence can treated and managed as solid material. Dewatering can be done with minimal cost by the use of sludge drying beds in which digested sludge slurry is openly placed over the sand and permitted to remain over there till it dries. Evaporation and gravity drainage can be the major factors in drying (Zhen et al. 2017). A wide range of piping network is installed under sand for the collection of water that is directly sent towards the head of wastewater treatment plant. SS after being dried for almost 6 weeks is converted into SS cake having solid content of almost 40% that can be detached with the help of front-end loader. During wet or cold weather this drying time can be decreased by building a glass or plastic tunnel over sand beds. This process of drying or dewatering is usually done in suburban or rural areas instead of urban or densely populated areas as a large land area is required for this purpose.

Sewage sludge drying beds include a variety of processes such as centrifuge, the belt filter papers and the rotary drum filters. Such mechanical systems are cost-effective and occupy less space as well as they have better operational control. As compared to SS-drying beds. Though these procedures are led by another step named sludge conditioning that involves addition of chemicals in slurry for the coagulation of solid contents and subsequently enhance the drainage (McGonigle et al. 2012).

2.7 Disposal

The ultimate destination of treated and improved SS is land disposal. Dewatered and dried sludge is directly buried into any dump site as well as it can be used as soil conditioner and fertilizer for agricultural lands. But this agricultural use of SS has some limitations as it may contain some hazardous toxicants for human health so we cannot spread it over that land that is intended for future staple crop production. Incineration is an alternative technique for SS disposal in an area where a suitable dumping site is inviable such as in densely populated urban areas. It absolutely removes the moisture through evaporation and transforms the organic content (solids) into stolid ash hence decreases the volume and cost of SS that can be disposed-off later on in a cost-effective way. Air pollution is the major concern when incineration of sewage sludge is done for this reason use of proper air pollution control devices such as wet and dry scrubbers is mandatory. Direct dumping of sewage sludge in oceans and sea was once considered as most economical and cheapest way by residents in coastal areas but it is not an appropriate method anymore. As United States has put ban on many coastal communities for waste disposal in ocean (Britannica 2020).

2.8 Biological Method

When the biological treatment of SS is done, some carbon containing solid materials are converted and degraded into another forms such as such as carbon dioxide gas emission causing more environmental pollution as well as a potential source of biofuel or energy is also lost. An increase in algae and biomass production, nutrient removal and carbon uptake can be done by the use of algae-bacteria mix consortium. Use of algae for sequestration of carbon dioxide and inclusion of sunlight results in more SS that can be used for biogas production. This procedure is totally mixotrophic as a pure medium of bacteria and algae is not required and it may work with a mixture medium of photoautotrophic and heterotrophic algal species. This in-situ algal reaction produces more oxygen that will be enough to meet the energy demands by making it available for both organic oxidation processes and microbial respiration hence this process is economical. Biological method of SS treatment can produce energy and co-digestion using algal SS is considered a favorable technique for alternative energy production (Dogaris et al. 2020).

As it is stated that treatment units using algae are cost effective as they need less energy for its oxygen demand but meanwhile it produces more quantity of SS for co-digestion with the help of microbial biomass. Such as, wastewater treatment plants involving the algae integration is more validated than other methods because it lessens the energy and aeration cost of wastewater treatment. Biofuel generation can be enhanced by the use of algae biomass in secondary effluents containing higher amount of nutrients or dissolved carbon dioxide. Exposure of algae biomass to a series of thermochemical processes may generate higher amount of biofuel and the energy recovery can be enhanced by exposing their residues to anaerobic digestion (Dogaris et al. 2020). Hydrothermal or thermochemical processes are more pledging as it generate more quantity of biofuel by catching all of the carbon matter of biomass. Integrative studies about the biogas and biofuel are not reported as much. Bio-electrochemical processes such as microbial fuel or microbial electrolysis cells can be indirectly practiced, decreasing the sludge volume in which sludge growth is dripped down to almost 50–70% as compared other conventional activated sludge procedures. Subsequently, such treatment processes directly result into valuable bioproducts as well as clean energy from organic biomass. But the problem is that current operating systems are not efficient with minimal output and less economics to support the large-scale applications. Extensive research is needed to be done in this area for the development of sustainable energy-positive bio-electrochemical systems (Britannica 2020).

3 Preventing Actions

Most of the SS contents are worthy to be recycled but they come up along with a variety of hazardous compounds in from of contaminants such as heavy metals, pesticides, polycyclic aromatic hydrocarbons, and pathogens etc., hence limiting the recycling options. The basic issue is that all of these hazardous compounds are found in a slurry, whereas sustainable treatment processes include the re-use and recovery of important products and degradation of toxic hazardous compounds. Hence, it is obvious that energy or valuable product recovery from sludge can be positively affected by decreasing total volume or amount of and/or by improving the quality of sludge. Such biological treatment of wastewater has more advantages over disposal techniques such as landfilling that directly results in increasing air pollution by producing on of the major greenhouse gas, methane. Hence, biological treatment converts organic waste into a safer end-product and reduces the environmental impacts of waste materials (Marmo 2002).

This treatment can be done both in aerobic (in the presence of oxygen) and anaerobic (in the absence of oxygen) conditions. Digestion that is done in the presence of oxygen generally results in the generation of gaseous emissions such as methane, carbon dioxide and water droplets that can subsequently be used to meet the need of fuel consumption or green fuel but meanwhile this whole approach is not economical or cost effective and subtle towards some of the environmental factors as compared to composting. Hence, there is more focus on decreasing the volume of sludge prior to its treatment or management by digestion ultimately cutting the sludge volume that is to be dumped later on. Specially, the microbial cell degeneration eradicate the need of hydrolysis stage that would otherwise be the limiting step of digestion. Thus, we can make the process of digestion more efficient and the rate of step also increases. Breakdown or disintegration of microbial cell walls can be done in many ways including biologically, thermos-chemically, and mechanically (Weemaes and Verstraete 2001; Müller 2001). There is a list of additional preventing actions that can be done to enhance the efficiency of systems and they include the utilization of complex organisms like metazoan and protozoa, other techniques such as vermicomposting, anaerobic and aerobic composting, advanced dewatering techniques involving electro-osmotic dewatering, advanced drying processes, Carver Greenfield evaporation, sludge conditioning through freeze-drying method as well as the to prohibit the direct release of toxic micro-pollutants into sewerage system, pretreatment of slurry including elimination of suspended and colloidal solids, and to remove the heavy metals from sludge by different means such as using organic and inorganic leachates, complex chelating agents or by microbiological leaching (Alavi et al. 2019).

3.1 Material Recovery

Material recovery ranges from the manufacturing or production of carbon source, organic matter, coagulants, pumice, bricks, artificial lightweight aggregate, slag, and Portland cement (Mañosa et al. 2021).

3.2 Organic Matter

The major fractions of SS being organic and inorganic content, and water out of which organic part depicts the energy source as well as beneficial to enhance the soil fertility specially with low humic constituents. Though the nutrients content of organic sludge is lower than the conventional sludge but meanwhile it is also carrying lesser heavy metals (Hansen 2001). Organic content of SS can also be used as raw material for activated carbon. Studies have revealed that specific surface of SS is 30–40% lower compared to commercial activated carbon but can be used as conventional activated carbon (Hagström et al. 2007).

3.3 Nutrients

The major nutrients of interest in SS are nitrogen (N) and phosphorus (P) (Hansen 2001). Nitrogen is generally available in the form of ammonium and organic nitrogen. Disintegration of nitrogen from SS can be done in many diverse methods but in handling of sludge the nitrogen mainstream is present in discarded liquid during dewatering, it is well treated. Ammonium separation can be done by stripping and/or struvite. Stripping is commonly practiced method for nitrogen recovery that results in formation of ammonium sulphate or ammonium nitrate, both of these products have significant application in agriculture. Precipitation of discarded water from dewatering can formulate magnesium ammonium phosphate or struvite, whereas ammonium separation can also be done by the means of ion exchange or adsorption methods such as by zeolites (Caraguay 2018).

Phosphorus is considered to be the most valued content in sludge; additionally, it is the not a renewable resource and within almost 150 years of time span the current phosphorus apatite mines are expected to be exhausted and there will be no potential source of phosphorus left for us hence the phosphorus content of sludge being the mainstream source of society is highly inviable to recover. Regrettably, enforcement of new rules and regulations and strict ultimatum regarding the quality of sludge can be the major factors limiting the sludge applications in agriculture sector. Subsequently, the phosphorus removal or recovery from sludge should be in a cleaner and less polluted manner. Phosphorus can be recovered from SS in many possible ways in today's advanced era and most convenient way of phosphorus recover is using biological methods. In anaerobic digestion, phosphorus separateor eliminates the phosphorus fraction that stays in the discarded water of the SS treatment stage called dewatering (Arun et al. 2020). SS is considered to be the purest source of phosphorus because all the heavy metals content of SS are lingering in dewatered or dried sludge and the phosphorus recovery output using this approach is about 50%. We can increase the yield proportion by using further physical or chemical processes. Keeping in mind the sludge quality, the nutrient recovery from SS specifically of phosphorus can reach up to 90% even only the acidification of sludge can yield more than 70% of P content. Unfortunately, acidification of sludge or treatment of sludge with acid dissolves not only phosphorus but the precipitates of other available heavy metals or chemicals may also be formed. So, we need to focus on the process that yields phosphorus content only so firstly the other metals have to be separated prior to phosphorus removal or phosphorus recovery should be done in an atmosphere that is not favorable for other metals to form precipitates. So, the end product is the purest and can be used as commercial fertilizer with zero or minimum pollution potential. Phosphorus recovery can also be done from the ash of incineration processes by mixing the ash with acid and phosphorus can be separated by dissolution, by formation of precipitates or by ion-exchange method (Gunarathne et al. 2020).

3.4 Carbon Source

Sewage sludge can yield a valuable "carbon source" upon hydrolyzation that can be used for biogas generation as well as to enhance the efficiency of denitrification (nitrogen removal), additionally other chemical, biological, and mechanical methods can also do the same (Kristensen and Jørgensen 1992). Primary SS is usually treated by the process of anaerobic fermentation within a closed reactor for short period of time and temperature lesser than digester ultimately forming the biological carbon source. Methane is not generated through this process, but SS can be incompletely degraded. Volatile fatty acids are the end-products of fermentation as in digester and hence the efficiency of bio-P process or denitrification improvement can be done using this process. Carbon source can be produced using other treatment processes such as chemical, enzymatic and mechanical sludge processes with lower or higher pH. Dissolved organic compounds within the cell are liberated upon the mechanical disintegration of SS because it directly damages the cell membrane. Sludge volume reduction and enhancement of biogas generation can also be done using mechanical disintegration (Hansen 2001).

3.5 Coagulants

Enhancement in efficiency of wastewater treatment plants and phosphorus recovery can be done using coagulants or precipitating chemicals and it is a rectifiable portion.

Coagulant can be dissolved by the process of acidification and it is recycled in wastewater treatment. Dissolution of phosphorus and other heavy metals occurs at comparatively lower pH and these compounds are then subsequently propelled back towards influent along with coagulants. After the recovery of coagulant, there are variety of processes available to separate the phosphorus and other heavy metals at lower concentration such as ion-exchange technique (Hansen 2001).

3.6 Bricks

The very first fully equipped brick plant was manufactured and operated successfully in Tokyo in 1991 having a total capacity of almost 5500 bricks per day from 15,000 kg of sludge incinerated ash. Notably, there was no leaching of heavy metals or other toxic compounds were reported from refined bricks even in extreme environmental conditions such as lowest pH (Okuno 2001). Using 100% ash to make bricks without using additives is mainly influenced by the process called molding, so it should be done cautiously as well as temperature should be continuously examined. Black core is the phenomenon take places upon the poor oxidization of organic substance hence to avoid this portent the temperature of brick kiln is once stopped when it touches 900 °C and then it is steadily enhanced and maintained for about 20 min at almost 1030 °C for the final modulation and heating of ash. This process is followed by a steady cooling stage of about 4 h to prevent the thermal strain breaking and in order to minimize the air temperature. SS bricks are comparatively more efficient and superior as compared to other conventional bricks in many ways as when considering their water absorption rate, 5 bending strength, compression strength and abrasion strength. Keeping in mind these properties, SS bricks are well welcomed and widely used in public corridors and walkways however some other issues can be faced such as growth of moss, ice and whitening of bricks with the passage of time (Dharma and Boora 2019).

3.7 Pumice

Pumice is made in same way as the SS bricks are manufactured but along with some additional processes such as crushing and sieving with an immense focus on reuse of end-product in athletic fields. Usually, needs of athletic fields are met with natural raw materials such as volcanic gravel because it has properties of draining extra water meanwhile holding enough moisture content hence maintains the athletic fields condition when it rains (Zeyad et al. 2019). Conversely, volcanic gravel is not sufficiently available, so sewage pumice is the best alternative (Wang et al. 2018).

3.8 Slag

Slag is the promising solution when the basic aim of sewage sludge treatment is the reduction of volume and heavy metals immobilization as it drops the waste volume to 4% of its original mass. This process is energy efficient with less fuel demands and the rich fatty greasy content of raw material serves as heat for furnace but this whole process needs to be skillfully operated and efficient drying system at the end (Gao et al. 2020). If the maximum temperature of incinerator is kept at or below 800 °C, it results in the persistence of almost 80% of metal contents in ash available in raw SS as it is examined during operational date. Slag can be formed in two different ways; water cooled and air-dried slag. Both of these slags are translucent and can be used as raw crushed material for concrete but on the other hand the compression strength is not good as of natural gravel. Air-cooled slag can be a promising substitute for natural coarse aggregate such as back-filling material, concrete aggregate, roadbed raw materials, interlocking tiles, permeable pavement, and many other concrete products (Cong et al. 2020).

3.9 Artificial Lightweight Aggregate (ALWA)

The first ever treatment plant working on "artificial lightweight aggregate (ALWA) became functional at the Nambu plant in Tokyo in 1996 having production capacity of 500 kg of sludge/h (Spinosa 2004b). Ash after being incinerated is vigorously mixed with water content with a ratio of 23% w/w and along with a little quantity of binding agent such as alcohol distillation waste. This mixture is then forwarded to another processing unit called centrifugal pelletizer where these pallets are kept for maximum 10 min for drying purpose at a temperature of 270 °C and then moved to fluidized-bed kiln for quick heating at 1050 °C (Ramanathan 2015).

When heating is done, pallets are subjected to air for drying that makes a thin film and inside surface stays porous subsequently forming an end-product having specific gravity of 1.5 and spherical shape. This ALWA is having properties of higher spherical shape, low specific gravity and lesser compression strength as compared to other commercially lightweight aggregates. This artificial lightweight aggregate can be used as fillers for removal between kerosene storage tanks and room walls, flower vase additives, thermal insulator panel, planter soils, water-infiltrating pavements and as an alternative of anthracite material in rapid sand filters. In a survey, pedestrians appreciated the walkways that were paved by ALWA because they have more pleasant appearance, more elastic, and have less penetration or standing of rainwater (Okuno et al. 2004).

3.10 Portland Cement

Sludge can be used as an alternative raw material for "Portland cement" instead of some chief components like silicon dioxide (SiO_2), iron oxide (F_2O_3) and calcium oxide (CaO) that conventionally used as natural source of limestone and clay (Okuno 2001). The concentration of phosphorus pentoxide (P_2O_5) is the most critical factor that determines the use of SS as Portland cement as manufacturers take SS in any form be it dewatered sludge cake, dried sludge or incinerated ash. The maximum permissible limit for P_2O_5 is 0.4% and there is no standard value given by WHO or any other global institution. Incinerated ash comprises up to 15% of P_2O_5 content and this incinerated ash can be mixed about 2% of concrete raw material. Other relevant use is lime blending which involves blending of dewatered SS cake and lime in equal quantity. After vigorous mixing, the water contents are separated from SS cake using chemical reactions and some extra heating is done that subsequently forms a dry powder. This dried cake is an efficient raw material as well as fuel in Portland cement processing (Boniardi 2020).

3.11 Thermal Conversion

Thermal conversion is based on three chief conversion processes and they are; thermo-chemical conversion, thermo-chemical liquefaction and conversion or combustion. The process of thermal conversion of dewatered or dried SS into oil or low to medium temperature conversions resembles natural processes to those producing liquid hydrocarbons from organic raw materials, and it includes conversion by the means of thermal cracking and catalytic conversions being the chief processes making it more complex than an ordinary pyrolysis process. This conversion is carried out anaerobically (in the absence of oxygen) at temperature of about 400–500 °C and at atmospheric pressure. Upon the conversion of sewage sludge from an industrial source it produces 30–70% char, 15–40% oil, 10–15% reaction water, and 7–10% gaseous content. The oil generated through this process have properties similar as an ordinary fuel and have applications in electricity generation sector. As compared to biological conversion processes, thermo-chemical processes are less likely to be affected by organic or inorganic impurities persisting in the SS (Veluchamy 2018). This technology has many other pros such as availability of instantly usable and storable liquid fuels, production of greenhouse credits, total energy outputs facility, control over the heavy metal contents, destruction of hazardous compounds such as organochlorides, pathogens and viruses, more control over the gaseous emissions and odor, less footprint of processing unit, less volume to deal while disposal and dumping, as well as cost effective or economic. Major drawbacks of this process are the requirement of full-scale processing unit and complex methodology (Ronda et al. 2019).

The direct conversion of wet sludge into oil was firstly done at Batelle-Northwest Laboratories in the United States of America (USA) during 1980s at experimental scale using thermo-chemical liquefaction. In this process, wet sludge containing 20% total solids is directly subjected to heat at about 300 °C and 10 Mpa pressure for almost 90 minutes that starts the liquefaction process ultimately yielding char, heavy oil, gas and reaction water. This technology was later named as Sludge-to-Oil Reaction System (STORS) that usually produces up to 10–20% of oil and 5–30% char with respect to weight. Oil produced in this way has greater viscosity and behaves as solid at room temperature. Gas produces an average 14% of carbon dioxide (CO_2) and the reseals in this process are wastewater (reaction water). Wastewater has greatest biological oxygen demand being as high as 30,000 mg/L. This approach has many advantages over others such as production of reusable liquid fuel, less footprints for processing units, cost effective, more economic, probable degradation of viruses and pathogens, and minimization of waste needing transportation to dispose off-site.

On the other hand, this technology has many cons such as complex operation, complicated equipments, exertion in handling and separation of products, and additionally this process is only tested or operated at an experimental stage yet (Pawlak-Kruczek et al. 2019). During the combustion or conversion process, the combustion of gaseous products occurs rather than condensation eventually forming the liquid fuel. Most of the experiments have been carried out using organic waste with an addition or in absence of minor concentration of SS. First of all, SS is tattered and then mixed thoroughly with dewatered or dried SS and then subjected to high temperature of 450 °C anaerobically, subsequently it forms carbonized solid content and gases. Then, product segregation is done where liquid, solid, glass, metals and ceramics are removed from carbonized material and then incorporated into gaseous phases. In this phase, combustion of all products is done with a high temperature as 1300 °C that yields granulated slag and steam. This steam has two fates; i.e., can be converted into an electricity source using steam turbine or can act as heat source. Combusted-off gases are then subjected to further cleansing mechanism before being released to avoid any kind of atmospheric pollution by the use of wet scrubbers, dry scrubbers, fabric filters and electro-static precipitators (Chen et al. 2019).

It is reported in previous studies that the fate of all products is well-managed such as air emissions standards are followed, solid content is recycled but the gas cleansing residues are identified as hazardous waste. According to available data, it is stated that use of this technology can yield up to 1050 kWh/t of waste treated (Gao et al. 2020).

3.12 Deep-Shaft Wet Air Oxidation

The "deep shaft wet air oxidation" involves the use of air or oxygen (as an oxidant) to oxidize the organic waste content of SS in an aqueous phase that occurs at a standard temperature and pressure of 260 °C and 150 Mpa respectively. A distinctive characteristic of this approach is that the pressure is maintained or achieved using the suspension of reactor in a 1500 m deep shaft. The resultant aqueous stream consists up to 30% of SS organic contents and enough nitrogen gas (N) hence it is preferred to return it to wastewater treatment plant or disposed-off after basic treatment (Debellefontaine and Foussard 2000). Further, solid remains are passive substances with no potential hazards such as silicates, phosphates, carbonates and un-leachable heavy metal contents and additionally, resulting hot water can act as an energy source. This technique comes with multiple ecological benefits such as devastation of many viruses and pathogens, comparatively simpler operating method, lesser footprint, more control over heavy metal contents and odor, and less volume of waste to be dealt for final disposal and transportation. Disadvantages varies form manual shut-down of plant is needed to de-scale the reactor because of inorganic components and wide-spread treatment is required for large volume of wastewater (Tungler et al. 2015).

3.13 Gasification

The modified form of ravenous air combustion at minimum temperature of 900 °C is called "Gasification". Sludge is incorporated with sub-stoichiometric amount of oxygen to initiate the combustion of carbon content into carbon dioxide (CO_2) that further reacts with solid carbon content eventually producing carbon monoxide (CO). The chief components of gases produced by gasification process of SS are H_2, CO, CO_2, N_2, and hydrogen sulfide (H_2S). Typical process of SS gasification produces solid products (especially char) containing a certain amount of volatile compounds. It is not previously reported in data about what happens to heavy metals in gasification (Lin et al. 2021). An early experimental stage, advanced pressurized entrained flow gasifier is established in Germany. Pure oxygen is required as an oxidant as well as high temperature and pressure ranging from 1400–1700 °C and 0.6–2.6 Mpa respectively for the gasification process. At such a high temperature, molten slag from SS is formed that is sent to gasifier ultimately converted into granulated slag particle. High quality syn-gas is produced as a result of raw gas cleansing by the removal of ammonia, hydrogen sulfide, and carbon to nitrogen ratio. This vitrified gas is pure inert material and has application as a constituent of concrete mixtures. Wide-spread examination of this procedure has exposed that most of the heavy metals and organochlorides are totally under control. According to experimental stage testing, it is stated that gasification can result in high control over heavy metals such as mercury, degradation of organochlorides, control of odor, energy recovery, complete devastation of viruses and pathogens, reduction in probable greenhouse credits and decrease in volume of material for final disposal and off-site transportation. On the other hand, there are many cons of this technology such as unknown costs, comparatively complicated system to operate, and additionally it is not proven at full equipped stage (Parés Viader et al. 2017; Lin et al. 2019).

4 Conventional Management and Recycling Options

Conventional management and recycling options involves incineration, land application and land filling because of its proficiency to recover energy by biogas production.

4.1 Land Application

Sewage sludge has diverse application in agricultural sector and after biological treatment it can be directly applied on soil. This results into many advantages such as nutrients recycling and organic matter that indirectly benefits the soil structure by improving the stability of aggregate, water retention capacity, pH, porosity, and cation exchange capacity. Although, the land application results in many issues such as application of SS containing hazardous organic and inorganic phytotoxic contaminants, and pathogen. Though there are a number of technologies and processes claiming to be completely disinfecting, but there is no such steady approach that ensure the removal or neutralization of heavy metals hence authorities have set the permissible limit of heavy metals content in SS and soil (Usman et al. 2012). The annual demand of a specific nutrient varies from crop to crop. It is seen in many cases that SS being the source of primary plant nutrients such as nitrogen and phosphorus but if the quantity exceeds the required amount it may be the potential source of ground or surface water pollution. So it is essential to calculate and consider the potential nutrients amount that are bioavailable and will persist for the very next year. SS that is derived from small and medium wastewater works with a potential of land disposal and distributing in residential areas are suitable for land applications. On the other hand, direct application of sludge in field or land depends on many long-term factors such as crop type, and weather circumstances whilst the production of sludge is going on. Thus, sludge composting can be suitable option over sludge land filling because it generates the material with more simple storage, transportation needed from the production area to application site. Composting results in safer, environmentally friendly, and hygienic material (Yoshida et al. 2018).

4.2 Incineration

Incineration of sewage sludge needs proper evaluation of its costs, but it is an economical option to deal with waste in densely populated large urban zones where the vicinity between land application and disposal of waste is too far away for an economical disposal. Thermal treatment could be the possible solution for the waste that have minimum chances of yielding beneficial products upon recovery or recycling as well as it offers a permanent solution with confined space for storage in bad weather conditions (Werther and Ogada 1999).

The probable advantages of this high temperature treatment process are:

1. Significantly reduces the weight and volume of waste to be handled.
2. Degradation of hazardous organic substances.
3. Energy recovery

Incineration results in absolute oxidization of organic materials involving volatile matter and ultimately produces ash. Before the sludge incineration, an efficient dewatering or drying is needed because even if the enough water has been removed the organic material will still continue to combust. The major issue with incineration is the emission of potential toxic gases but meanwhile there are a large number of devices available that can significantly cut the emissions. As compared to dewatered sludge, the reduction in volume of SS incinerated is up to 90%. The ash produced by this process is pesticides, viruses and pathogens free plus the metal content is less soluble and oxidized form and hence less bioavailable. Efficiency of incineration plant is directly related to availability of proper supplementary machinery and equipments, such as receiving and storage system, preliminary treatment devices, heat recovery, feeding devices, flue gas cleaning, ash management, wastewater dumping and process examination (Donatello and Cheeseman 2013).

4.3 Landfilling

Landfilling is the viable solution to the waste in an area where there is sufficient room nearby available with cost effective manner and it is basic need for all other systems for disposing or dumping the waste with no more potential recycling or recovery options. It is casually thought to be a disposal solution because most of the organic matter content of SS is rationally lost but a specific amount of recovery might be another option to adopt by yielding the biogas. If this methane in the form of biogas is not captured it can potentially be a greenhouse effect source and methane is almost 20 times more hazardous as compared to carbon dioxide. Landfilling only supports SS that is well-dewatered; around 20–25% solid concentrations are typically required but it can exceed up to 30–35% to ensure an effective physical constancy to strengthen the cover material. Efficient biological management is also mandatory to avoid unpleasant odors and other emissions. Another issue with landfilling is the leaching that can be easily controlled naturally by the use of amendments as liners/covers, imported soils, membrane liners and collection and management. Use of imported clays or synthetic compounds such as high-density polyethylene (HDPE) can be effective to control or lower the natural permeability of soil. Right after few months of deposition, the biogas production typically starts and touches the peak points after 5–7 years and carry on for a period of many years at a declining rate (Spinosa 2007; Spinosa et al. 2007).

4.4 Conclusion and Future Perspectives

In the previous sections, it has been stated the capacity and potential of sewage sludge, even if containing some contaminants, can be act as a chief source of many products and substances as well as energy source, so as an alternative of waste, these chances offered by sludge should me acknowledged and more research should be done regarding its diverse applications and possibilities. Choosing an effective system for sludge management should have on priority basis i.e., (1) maximization of material and energy content recovery and, (2) minimization of total energy input and processing cost. Till date, there have been a variety of treatment options recommended, suggested and available in the market but selecting the appropriate system is directly affected by many other vital factors such as regional geography and local economy, weather and climatic conditions, regulatory restraints and many practices need public recognition as well. It is obvious that proper management and handling of sewage sludge demands extensive development of "multiple and diversified options" strategies, that is a joint challenge mutually for city administration, industries and citizens. It is recommended that these parties should focus on production of less sewage sludge for disposal on one hand and production of high-quality sludge on another hand. Hence, while deciding the best option for optimum sludge management, the local and site-specific considerations should also be considered as it is important for the optimization of the whole system that upon choosing an optimal disposal or reuse approach. It is also stated that sludge management varies from country to country according to many factors such as land area, population density, cost and public acceptance are the key factors. Developed nations have widespread legislative systems regarding sludge handling and management and in such countries major focus is on waste minimization and then recycling, meanwhile landfilling being the most undesirable solution. In less developed nations, more waste or sludge is disposed-off or landfilled directly without any primary treatment. Application of sludge in agricultural sector could be the potential future solution of waste as well as circular economy approach including recovery of phosphorus. Among all options, the most provocative approach is thermal treatment of sludge because of its high cost. The sewage sludge can be taken as impartial, when it comes to CO_2 emissions, a source of energy that makes it more valuable to some industrial processing units such as cement factories.

References

Alavi N, Sarmadi K, Goudarzi G, Babaei AA, Bakhshoodeh R, Paydary P (2019) Attenuation of tetracyclines during chicken manure and bagasse co-composting: degradation, kinetics, and artificial neural network modeling. J Environ Manag 231:1203–1210

Aradelli P, Cantù G (2016) Sewage sludge disposal routes: thermal treatments and energy recovery

Arun J, Gopinath KP, Vigneshwar SS, Swetha A (2020) Sustainable and eco-friendly approach for phosphorus recovery from wastewater by hydrothermally carbonized microalgae: study on spent bio-char as fertilizer. J Water Process Engg 38:101567

Boniardi G (2020) Phosphorous recovery from sewage sludge ashes via wet chemical leaching

Britannica (2020). https://www.britannica.com/technology/wastewater-treatment/Sludge-treatment-and-disposal. Accessed on 22 Dec 2020

Caraguay DEG (2018) Evaluation of phosphate and ammonium removal and valorization from urban waste waters by impregnated metal hydrated oxides inorganic natural zeolites. Universitat Politècnica de, Catalunya

Chen H, Rao Y, Cao Y, Shi Y, Hao S, Luo G, Zhang S (2019) Hydrothermal conversion of sewage sludge: focusing on the characterization of liquid products and their methane yields. Chem Engg J 357:367–375

Cong X, Lu S, Gao Y, Yao Y, Elchalakani M, Shi X (2020) Effects of microwave, thermomechanical and chemical treatments of sewage sludge ash on its early-age behavior as supplementary cementitious material. J Clean Prod 258:120647

Costa F, Lago A, Rocha V, Barros OS, Costa L, Vipotnik Z, Silva B, Tavares T (2019) A review on biological processes for pharmaceuticals wastes abatement-a growing threat to modern society. Environ Sci Technol 53:7185–7202

Debellefontaine H, Foussard JN (2000) Wet air oxidation for the treatment of industrial wastes. Chemical aspects, reactor design and industrial applications in europe. Waste Manag 20:15–25

Dharma A, Boora A (2019) Utilization of waste material in concrete for the construction of rigid pavement

Dogaris I, Ammar E, Philippidis GP (2020) Prospects of integrating algae technologies into landfill leachate treatment. World J Microbiol Biotech 36:1–25

Donatello S, Cheeseman CR (2013) Recycling and recovery routes for incinerated sewage sludge ash (issa): a review. Waste Manag 33:2328–2340

Elalami D, Carrere H, Monlau F, Abdelouahdi K, Oukarroum A, Barakat A (2019) Pretreatment and co-digestion of wastewater sludge for biogas production: recent research advances and trends. Renew Sust Energ Rev 114:109287

Gao N, Kamran K, Quan C, Williams PT (2020) Thermochemical conversion of sewage sludge: a critical review. Prog Energy Combust Sci 79:100843

Gude VG (2015) Energy and water autarky of wastewater treatment and power generation systems. Renew Sust Energ Rev 45:52–68

Gunarathne V, Rajapaksha AU, Vithanage M, Alessi DS, Selvasembian R, Naushad M, You S, Oleszczuk P, Ok YS (2020) Hydrometallurgical processes for heavy metals recovery from industrial sludges. Crit Rev Environ Sci Technol:1–41

Hagström JA, Granéli E, Maneiro I, Barreiro A, Petermann A, Svensen C (2007) Release and degradation of amnesic shellfish poison from decaying Pseudo-nitzschia multiseries in presence of bacteria and organic matter. Harmful Algae 6(2):175–188

Hansen B (2001) Other energy and resource recovery. In: Spinosa L, Vesilind PA (eds) Sludge into biosolids. IWA Publishing, London, pp 163–171

International Water Association (IWA) (2019) Data can be accessed at: https://iwa-network.org/groups/sludge-management/

Kijo-Kleczkowska A, Otwinowski H, Środa K (2012) Properties and production of sewage sludge in Poland with reference to the methods of neutralizing. ArchiwumGospodarki OdpadamiiOchronyŚrodowiska 14(4)

Kristensen GG, Jørgensen PE (1992) Pre-precipitation followed by biological denitrification supported by addition of biological or thermal/chemical hydrolysis products. In: Klute R, Hahn H (eds) Chemical water and wastewater treatment II. Springer, Hidelberg, pp 313–328

Kumar A, Samadder S (2020) Performance evaluation of anaerobic digestion technology for energy recovery from organic fraction of municipal solid waste: a review. Energy 197:117253

Lin L, Xu F, Ge X, Li Y (2018) Improving the sustainability of organic waste management practices in the food-energy-water nexus: a comparative review of anaerobic digestion and composting. Renew Sust Energ Rev 89:151–167

Lin J, Sun S, Cui C, Ma R, Fang L, Zhang P, Quan Z, Song X, Yan J, Luo J (2019) Hydrogen-rich bio-gas generation and optimization in relation to heavy metals immobilization during pd-catalyzed supercritical water gasification of sludge. Energy 189:116296

Lin J, Liao Q, Hu Y, Ma R, Cui C, Sun S, Liu X (2021) Effects of process parameters on sulfur migration and h2s generation during supercritical water gasification of sludge. J Hazard Mater 403:123678

Longo S, d'Antoni BM, Bongards M, Chaparro A, Cronrath A, Fatone F, Lema JM, Mauricio-Iglesias M, Soares A, Hospido A (2016) Monitoring and diagnosis of energy consumption in wastewater treatment plants. A state of the art and proposals for improvement. Appl Energy 179:1251–1268

Mañosa J, Formosa J, Giro-Paloma J, Maldonado-Alameda A, Quina M, Chimenos J (2021) Valorisation of water treatment sludge for lightweight aggregate production. Constr Build Mater 269:121335

Marmo L (2002) Management of biodegradable waste - Current practice and future perspectives in Europe. Waste Manag World Rev 3:75–79

McGonigle T, Beauchamp E, Evans L, Wells C (2012) Mineralization and crop uptake of nitrogen from textile manufacturing wastewater sludge cake. Appl Environ Soil Sci

Meng X, Yuan X, Ren J, Wang X, Zhu W, Cui Z (2017) Methane production and characteristics of the microbial community in a two-stage fixed-bed anaerobic reactor using molasses. Bioresour Technol 241:1050–1059

Müller JA (2001) Prospects and problems of sludge pre-treatment processes. Water Sci Technol 44 (10):121–128

Niwagaba CB, Mbéguéré M, Strande L (2014) Faecal sludge quantification, characterization and treatment objectives. In: Faecal sludge management: systems approach for implementation and operation. IWA Publishing, London, pp 19–44

Okuno N (2001) Production of usable materials. In: Spinosa L, Vesilind PA (eds) Sludge into biosolids. IWA Publishing, London, pp 147–162

Okuno N, Ishikawa Y, Shimizu A, Yoshida M (2004) Utilization of sludge in building material. Water Sci Technol 49:225–232

Parés Viader R, Jensen PE, Ottosen LM, Thomsen TP, Ahrenfeldt J, Hauggaard-Nielsen H (2017) Comparison of phosphorus recovery from incineration and gasification sewage sludge ash. Water Sci Technol 75:1251–1260

Pawlak-Kruczek H, Wnukowski H, Krochmalny K, Kowal M, Baranowski M, Zgóra J, Czerep M, Ostrycharczyk M, Niedzwiecki L (2019) The staged thermal conversion of sewage sludge in the presence of oxygen. J Energy Res Technol 141

Ramanathan T (2015) Alkaline bioleaching of municipal solid waste incineration fly ash with indigenous extremophiles, National University of Singapore (Singapore)

Ronda A, Gómez-Barea A, Haro P, de Almeida V, Salinero J (2019) Elements partitioning during thermal conversion of sewage sludge. Fuel Process Technol 186:156–166

Rulkens W (2008) Sewage sludge as a biomass resource for the production of energy: overview and assessment of the various options. Energy and Fuels 22(1):9–15

Silva FM, Mahler CF, Oliveira LB, Bassin JP (2018) Hydrogen and methane production in a two-stage anaerobic digestion system by co-digestion of food waste, sewage sludge and glycerol. Waste Manag 76:339–349

Spinosa L (2004a) EU strategy and practice for sludge utilization in agriculture, disposal and landfilling. Residuals Sci Technol 1(1):7–14

Spinosa L (2004b) From sludge to resources through biosolids. Water Sci Technol 50:1–9

Spinosa L (2007) Sewage sludge: from waste to resource. Paper read at V: Strokovni seminar, Simpozij vodni dnevi

Spinosa L, Lotito V (1999) Rheological measurements and their application in biosolids management. Water 21

Spinosa L, Leschber R, Wichmann K (2007) Sewage sludge co-management: developments in EU regulations and characterization procedures. Water Pract Technol 2(1):1–8

Sun Y, Wang D, Tsang DC, Wang L, Ok YS, Feng Y (2019) A critical review of risks, characteristics, and treatment strategies for potentially toxic elements in wastewater from shale gas extraction. Environ Int 125:452–469

Tungler A, Szabados E, Hosseini EM (2015) Wet air oxidation of aqueous wastes. Wastewater Treat Engg 153

Usman K, Khan S, Ghulam S, Khan MU, Khan N, Khan MA, Khalil SK (2012) Sewage sludge: an important biological resource for sustainable agriculture and its environmental implications

Van Lier JB, Mahmoud N, Zeeman G (2008) Anaerobic wastewater treatment. Biological wastewater treatment: principles, modelling and design, pp 415–456

Veluchamy C (2018) Anaerobic digestion of pulp and paper mill sludge. Indian Institute of Technology, Guwahati

Wang K, Mao H, Li X (2018) Functional characteristics and influence factors of microbial community in sewage sludge composting with inorganic bulking agent. Bioresour Technol 249:527–535

Ward A, Lewis D, Green F (2014) Anaerobic digestion of algae biomass: a review. Algal Res 5:204–214

Weemaes M, Verstraete W (2001) Other treatment techniques. In: Sludge into biosolids. IWA Publishing, London

Werther J, Ogada T (1999) Sewage sludge combustion. Prog Energy Combust Sci 25:55–116

Wiśniowska E, Grobelak A, Kokot P, Kacprzak M (2019) Sludge legislation-comparison between different countries. In: Prasad MNV, de Campos Favas PJ, Vithanage M, Mohan SV (eds) *Industrial and municipal sludge*. Butterworth-Heinemann, pp 201–224

Yoshida H, ten Hoeve M, Christensen TH, Bruun S, Jensen LS, Scheutz C (2018) Life cycle assessment of sewage sludge management options including long-term impacts after land application. J Clean Prod 174:538–547

Zangmo C (2017) Life cycle assessment of solid waste management in paro district, bhutan. Doctoral dissertation, M.Sc. Thesis, Thammasat University, Thailand, p 111

Zeyad AM, Tayeh BA, Yusuf MO (2019) Strength and transport characteristics of volcanic pumice powder based high strength concrete. Constr Build Mater 216:314–324

Zhang L, Xu C, Champagne P, Mabee W (2014) Overview of current biological and thermo-chemical treatment technologies for sustainable sludge management. Waste Manag Res 32 (7):586–600

Zhang H, Rigamonti L, Visigalli S, Turolla A, Gronchi P, Canziani R (2019) Environmental and economic assessment of electro-dewatering application to sewage sludge: a case study of an italian wastewater treatment plant. J Clean Prod 210:1180–1192

Zhen G, Lu X, Kato H, Zhao Y, Li YY (2017) Overview of pretreatment strategies for enhancing sewage sludge disintegration and subsequent anaerobic digestion: current advances, full-scale application and future perspectives. Renew Sust Energ Rev 69:559–577

The manufacturer's authorised representative in the EU is Springer Nature Customer Service Centre GmbH, Europaplatz 3, 69115 Heidelberg, Germany. If you have any concerns regarding our products, please contact ProductSafety@springernature.com

Printed and bound by CPI Group (UK) Ltd, Croydon, CR0 4YY
30/09/2026
02225687-0009